ELECTRONIC COMMERCE

Sixth Annual Edition

ELECTRONIC COMMERCE

Sixth Annual Edition

Gary P. Schneider, Ph.D., CPA
University of San Diego

THOMSON

COURSE TECHNOLOGY

Australia • Canada • Mexico • Singapore • Spain • United Kingdom • United States

THOMSON
COURSE TECHNOLOGY™

Electronic Commerce: Sixth Annual Edition
by Gary P. Schneider, Ph.D, CPA

Executive Editor:
Mac Mendelsohn

Senior Acquisitions Editor:
Maureen Martin

Product Manager:
Beth Paquin

Development Editor:
Amanda Brodkin

Production Editor:
Danielle Slade

Marketing Manager:
Karen Seitz

Associate Product Managers:
Mirella Misiaszek and Jennifer Smith

Editorial Assistant:
Jennifer Smith

Text Designer:
Anne Small-Wills

Cover Designer:
Laura Rickenbach

Proofreader:
Harry Johnson

Indexer:
Rich Carlson

Manufacturing Manager:
Laura Burns

This publication is designed to provide accurate and authoritative information regarding the subject matter covered. It is sold with the understanding that neither the publisher nor the author are engaged in rendering legal, accounting, or other professional services. If legal, accounting, or other expert advice is required, the services of a competent professional person should be sought.

Cases in this book that mention company or organization names were written using publicly available information to provide a setting for student learning. They are not intended to provide commentary on or evaluation of any party's handling of the situation described.

ISBN 0-619-21704-9

BRIEF CONTENTS

TABLE OF CONTENTS

Electronic Commerce, Sixth Annual Edition provides complete coverage of the key business and technology elements of electronic commerce. The book does not assume that readers have any previous electronic commerce knowledge or experience.

In 1998, having spent several years doing electronic commerce research, consulting, and corporate training, I began developing both an undergraduate business school course and an MBA-level course in electronic commerce. Although I had used a variety of books and other materials in my corporate training work, I was concerned that those materials would not work well in university courses because they were written at widely varying levels and did not have the pedagogic organization and features, such as review questions, that are so important to students.

After searching for a textbook that offered balanced coverage of both the business and technology elements of electronic commerce, I concluded that no such book existed. The first edition of *Electronic Commerce* was written to fill that void. In the subsequent editions, I have worked to improve the book and keep it current with the rapid changes in this dynamic field. The sixth edition includes many updates to the content that reflect the rapid changes that are occurring in electronic commerce today.

ORGANIZATION AND COVERAGE

Electronic Commerce: Sixth Annual Edition introduces readers to both the theory and practice of conducting business over the Internet and World Wide Web. The book is organized into four sections: an introduction, business strategies, technologies, and integration.

New to this Edition

The most important change in this edition is the set of 12 new cases that have been added throughout the book. Each chapter now concludes with two cases. One of the cases uses a fictional setting to illustrate specific points raised in the chapter. The other case challenges students to apply what they have learned to an actual situation that has occurred in a real company or organization. This edition includes the usual updates to keep the content current with the rapidly occurring changes in electronic commerce and includes new material on the following topics:

- A detailed comparison of the first wave of electronic commerce to the second wave.

- Advances in wireless networking technologies and their applications in metropolitan area networks, integration of cellular and wireless networks, and the increased availability of wireless "hot spots"

- The renewed profitability of online advertising, including the emergence of new revenue models based on keyword and paid placement advertising that are supplanting banner advertising approaches

- Increasing use of targeting and specialized Web sites in online marketing campaigns

- New developments in contextual and localized online advertising

- New revenue models being used by distributors of downloadable movies and by online disk storage providers

- Updated research on the effects of intrusive online advertising formats such as pop-up ads

- Materials tracking technologies now used in managing supply chains, especially radio-frequency identification (RFID) devices

- Increased competition between eBay Stores and Amazon.com's zShops

- Second wave virtual Web communities for social and business networking such as craigslist and Friendster

- Expanded coverage of unsolicited commercial e-mail (UCE or spam), including technical solutions such as filtering techniques and legal solutions such as the U.S. CAN-SPAM law

- New developments in customer relationship (CRM) management software and strategies

- New enterprise application integration (EAI) strategies

- Expanded coverage of cookies, Web bugs, and certification authorities

- Threats posed to online financial services providers by phishing attacks and countermeasures that show promise for reducing those threats

- Internet technologies in bank check processing, including the U.S. Check 21 initiative

Introduction

The book's first section includes two chapters. Chapter 1, "Introduction to Electronic Commerce," defines electronic commerce and describes how companies use it to create new products and services, reduce the cost of existing business processes, and improve the efficiency and effectiveness of their operations. The concept of the second wave of electronic commerce is presented and developed in this chapter. Chapter 1 also describes the history of the Internet and the Web, explains the international environment in which electronic commerce exists, provides an overview of the economic structures in which businesses operate, and describes how electronic commerce fits into those structures. Two themes are introduced in this chapter that recur throughout later chapters: examining a firm's value chain can suggest opportunities for electronic commerce initiatives and reductions in transaction costs are important elements of many electronic commerce initiatives.

Chapter 2, "Technology Infrastructure: The Internet and the World Wide Web," introduces the technologies used to conduct business online, including topics such as Internet infrastructure, protocols, and packet-switched networks. Chapter 2 also describes the markup languages used on the Web (HTML and XML) and discusses Internet connection options and tradeoffs, including wireless technologies.

Business Strategies for Electronic Commerce

The second section of the book includes five chapters that describe the business strategies that companies and other organizations are using to do business online. Chapter 3, "Selling on the Web: Revenue Models and Building a Web Presence," describes revenue models that companies are using on the Web and explains how some companies have changed their revenue models as the Web has matured. The chapter also describes how firms that understand the nature of communication on the Web can identify and reach the largest possible number of qualified customers.

Chapter 4, "Marketing on the Web," provides an introduction to Internet marketing and online advertising. It includes coverage of market segmentation, technology-enabled customer relationship management, rational branding, contextual advertising, localized advertising, viral marketing, and permission marketing. The chapter also explains how online businesses can share and transfer brand benefits through affiliate marketing and cooperative efforts among brand owners.

Chapter 5, "Business-to-Business Strategies: From Electronic Data Interchange to Electronic Commerce," explores the variety of methods that companies are using to improve their purchasing and logistics primary activities with Internet and Web technologies. Chapter 5 also provides an overview of EDI and explores how the Internet now provides an inexpensive EDI communications channel that allows smaller businesses to reap EDI's benefits. Chapter 5 also explains how the Internet and the Web have become an important force driving the adoption of technologies such as e-procurement, radio-frequency identification, and reverse auctions in the practice of supply chain management.

Chapter 6, "Online Auctions, Virtual Communities, and Web Portals," outlines how companies now use the Web to do things that they have never done before, such as operating auction sites, creating virtual communities, and serving as Web portals. The chapter describes how firms are using Web auction sites to sell goods to their customers and generate advertising revenue. The chapter explains how businesses are creating virtual communities that facilitate social and business networking.

Chapter 7, "The Environment of Electronic Commerce: Legal, Ethical, and Tax Issues," discusses the legal and ethical aspects of intellectual property usage and the privacy rights of customers. Online crime, terrorism, and warfare are covered as well. The chapter also explains that the large number of government units that have jurisdiction and power to tax makes it essential that companies doing business on the Web understand the potential liabilities of doing business with customers in those jurisdictions.

Technologies for Electronic Commerce

The third section of the book includes four chapters that describe the technologies of electronic commerce and explains how they work. Chapter 8, "Web Server Hardware and Software," describes the computers, operating systems, e-mail systems, utility programs, and Web server software that organizations use in the operation of their electronic commerce Web sites. Web site hosting options are also discussed in this chapter. The chapter also describes the problem of unsolicited commercial e-mail (UCE, or spam) and outlines both technical and legal solutions to the problem.

Chapter 9, "Electronic Commerce Software," describes the basic functions that all electronic commerce Web sites must accomplish and explains the various software options available to companies of various sizes. This chapter includes a completely new overview of Web services, a set of technologies that could become an important electronic commerce infrastructure element in the near future.

Chapter 10, "Electronic Commerce Security," discusses security threats and countermeasures that organizations can use to ensure the security of client computers, communications channels, and Web servers. The role of industry organizations in promoting computer, network, and Internet security is also outlined. The chapter emphasizes the importance of a written security policy and explains how encryption and digital certificates work.

Chapter 11, "Payment Systems for Electronic Commerce," presents a discussion of electronic payment systems, including electronic cash, electronic wallets, stored-value cards, credit cards, debit cards, and charge cards. The chapter describes how payment systems operate, including approval of transactions and disbursements to merchants, and describes new developments in how banks are using Internet technologies to improve check clearing and payment processing operations.

Integration

The fourth section of the book includes one chapter that integrates the business and technology strategies used in electronic commerce. Chapter 12, "Planning for Electronic Commerce," presents an overview of key elements that are typically included in business plans for electronic commerce implementations. These elements include the setting of objectives and estimated costs and benefits of the project. The chapter describes outsourcing strategies used in electronic commerce and covers the use of project management as a formal way to plan and control specific tasks and resources used in electronic commerce projects. This chapter concludes with discussions of change management and staffing strategies.

FEATURES

The sixth annual edition of *Electronic Commerce* includes a number of features and offers additional resources designed to help readers understand electronic commerce. These features and resources include:

- **Business Case Approach** The introduction to each chapter includes a real business case that provides a unifying theme for the chapter. The case provides a backdrop for the material described in the chapter. Each case illustrates an important topic from the chapter and demonstrates its relevance to the current practice of electronic commerce.

- **Learning From Failures** Not all electronic commerce initiatives have been successful. Each chapter in the book includes a short summary of an electronic commerce failure related to the content of that chapter. We all learn from our mistakes—this feature is designed to help readers understand the missteps of electronic commerce pioneers who learned their lessons the hard way.

- **Summaries** Each chapter concludes with a Summary that concisely recaps the most important concepts in the chapter.

- **Online Companion** The Online Companion is a set of Web pages maintained by the publisher for readers of this book. The Online Companion complements the book and contains links to Web sites referred to in the book and to other online resources that further illustrate the concepts presented. The Web is constantly changing and the Online Companion is continually monitored and updated for those changes so that its links continue to lead to useful Web resources for each chapter. You can find the Online Companion for this book by visiting Course Technology's Web site at *www.course.com/* and searching on Electronic Commerce.

- **Online Companion References in Text** Throughout each chapter, there are Online Companion References that indicate the name of a link included in the Online Companion. Text set in bold, sans-serif letters (**"Metabot Pro"**) indicates a like-named link in the Online Companion. The links in the Online Companion are organized under chapter and subchapter headings that correspond to those in the book. The Online Companion also contains many supplemental links to help students explore beyond the book's content.

- **Review Questions and Exercises** Every chapter concludes with meaningful review materials including both conceptual discussion questions and hands-on exercises. The review questions are ideal for use as the basis for class discussions or as written homework assignments. The exercises give students hands-on experiences that yield a computer output or a written report.

- **Cases** Each chapter concludes with two comprehensive cases. One case uses a ficticious setting to illustrate key learning objectives from that chapter. The other case gives students an opportunity to apply what they have learned from the chapter to an actual situation that a real company or organization has faced. The cases offer students a rich environment in which they can apply what they have learned and provide a motivation for doing further research on the topics.

- **For Further Study and Research** Each chapter concludes with a comprehensive list of the resources that were consulted during the writing of the chapter. These references to publications in academic journals, books, and the IT industry and business press provide a sound starting point for readers who want to learn more about the topics contained in the chapter.

- **Key Terms and Glossary** Terms within each chapter that may be new to the student or have specific subject-related meaning are highlighted by boldface type. The end of each chapter includes a list of the chapter's key terms. All of the book's key terms are compiled, along with definitions, in a Glossary at the end of the book.

TEACHING TOOLS

When this book is used in an academic setting, instructors may obtain the following teaching tools from Course Technology:

- **Instructor's Manual** The Instructor's Manual has been carefully prepared and tested to ensure its accuracy and dependability. The Instructor's Manual is available through the Course Technology Faculty Online Companion on the World Wide Web (call your customer service representative for the exact URL and to obtain your username and password).

- **ExamView©** This textbook is accompanied by ExamView, a powerful testing software package that allows instructors to create and administer printed, computer (LAN-based), and Internet exams. ExamView includes hundreds of questions that correspond to the topics covered in this text, enabling students to generate detailed study guides that include page references for further review. The computer-based and Internet testing components allow students to take exams at their computers, and also save the instructor time by grading each exam automatically.

- **PowerPoint Presentations** Microsoft PowerPoint slides are included for each chapter as a teaching aid for classroom presentations, to make available to students on a network for chapter review, or to be printed for classroom distribution. Instructors can add their own slides for additional topics they introduce to the class. The presentations are included on the Instructor's CD.

- **Distance Learning** Course Technology is proud to present online content in WebCT and Blackboard to provide the most complete and dynamic learning experience possible. For more information on how to bring distance learning to your course, contact your local Course Technology sales representative.

ACKNOWLEDGMENTS

I owe a great debt of gratitude to my good friends at Course Technology who made this book possible. Course Technology remains the best publisher with which I have ever worked. Everyone at Course Technology put forth tremendous effort to publish this edition on a very tight schedule. My heartfelt thanks go to Kristen Duerr, Executive Vice President, Publisher; Mac Mendelsohn, Executive Editor; Maureen Martin, Acquisitions Editor, Beth Paquin, Product Manager; Danielle Slade, Production Editor; and Mirella Misiaszek and Jennifer Smith, Associate Product Managers, for their tireless work and dedication. I am deeply indebted to Amanda Brodkin, Development Editor extraordinaire, for her outstanding contributions to all six editions of this book. Amanda performed the magic of turning my manuscript drafts into a high-quality textbook and was always ready with encouragement and fresh ideas when I was running low on them. Many of the best elements of this book resulted from Amanda's ideas and inspirations. In particular, I want to thank Amanda for contributing the Dutch auction example in Chapter 6 and the ideas for the cases in Chapters 7 and 8.

I want to thank the following reviewers for their insightful comments and suggestions on this and previous editions: Paul Ambrose, University of Wisconsin, Milwaukee; Tina Ashford, Macon State College; Robert Chi, California State University-Long Beach; Chet Cunningham, Madisonville Community College; Roland Eichelberger, Baylor University; Mary Garrett, Michigan Virtual High School; Barbara Grabowski, Benedictine University; Milena Head, McMaster University; Perry M. Hidalgo, Gwinnett Technical Institute; Brent Hussin, University of Wisconsin, Green Bay; Cheri L. Kase, Legg Mason Corporate Technology; Rick Lindgren, Graceland University; William Lisenby, Alamo Community College; Diane Lockwood, Albers School of Business and Economics, Seattle University; Jane Mackay, Texas Christian University; Michael P. Martel, Culverhouse School of Accountancy, University of Alabama; William E. McTammany, Florida Community College at Jacksonville; Leslie Moore, Jackson State Community College; Martha Myers, Kennesaw State University; Pete Partin, Forethought Financial Services; Andy Pickering, University of Maryland University College; David Reavis, Texas A&M University; and Barbara Warner, University of South Florida. Special thanks go to reviewer A. Lee Gilbert of Nanyang Technological University in Singapore, who provided extremely detailed comments and many useful suggestions for improving Chapter 12. My thanks also go to the many professors who have used the previous editions in their classes and who have sent me suggestions for improving the text. In particular, I want to acknowledge the detailed recommendations made by David Bell of Pacific Union College regarding the coverage of IP addresses in Chapter 2.

I appreciate the role that the University of San Diego had in making this book possible. The University provided research funding that allowed me to work on the first edition of this book and gave me fellow faculty members who were always happy to discuss and critically evaluate ideas for the book. Of these faculty members, my thanks go first to Jim Perry for his contributions as co-author on the first two editions of this book. Tom Buckles, now a professor of marketing at Biola University, provided many useful suggestions, pointed out a number of valuable research resources, and was willing to sit and discuss ideas for this book long after everyone else had left the building. Rahul Singh, now teaching at the University of North Carolina-Greensboro, provided suggestions regarding the book's coverage of electronic commerce infrastructure. Carl Rebman made recommendations on a number of networking, telecommunications, and security topics. The University of San Diego School of Business Administration also provided the research assistance of many graduate students. Among those students were Sebastian Ailioaie, a Fulbright Fellow who did substantial work on the Online Companion, Anthony Coury, who applied his considerable legal knowledge to reviewing Chapter 7 and suggesting many improvements, and Dima Ghawi, who shared her significant background research on reverse auctions and helped me develop many of the ideas presented in Chapters 5 and 6. I am grateful to Robin Lloyd for her help with the Lonely Planet case (in Chapter 3) and to Zu-yo Wang for his help with the Alibaba.com case (Chapter 6). Other students who provided valuable assistance and suggestions include Maximiliano Altieri, Adrian Boyce, Karl Flaig, Kathy Glaser, Emilie Johnson Hersh, Chad McManamy, Dan Mulligan, Firat Ozkan, Suzanne Phillips, Susan Soelaiman, Carolyn Sturz, and Leila Worthy.

Finally, I want to express my deep appreciation for the support and encouragement of my wife, Cathy Cosby, and our children, Ben, Annie, and Maggie. Without their support and patience, writing this book would not have been possible.

DEDICATION

To Trinity, Ella, and Justin

ABOUT THE AUTHOR

Gary Schneider is a Professor of Accounting and Information Systems at the University of San Diego, where he teaches courses in electronic commerce, database design, supply chain management, and management accounting. He has won several teaching awards and has served as academic director of the school's graduate programs in electronic commerce and information systems. Gary has published more than 45 books and 70 research papers on a variety of accounting, information systems, and management topics. His books have been translated into Chinese, French, Italian, Korean, and Spanish. Gary's research has been funded by the Irvine Foundation and the U.S. Office of Naval Research. His work has appeared in the *Journal of Information Systems, Interfaces,* and the *Information Systems Audit & Control Journal.* He has served as editor of the *Accounting Systems and Technology Reporter,* as associate editor of the *Journal of Global Information Management,* and on the editorial boards of the *Journal of Information Systems,* the *Journal of Electronic Commerce in Organizations,* the *Journal of Database Management,* and the *Information Systems Audit & Control Journal.* Gary has lectured on electronic commerce topics at universities and businesses in the United States, Europe, South America, and Asia. He has provided consulting and training services to a number of major clients, including Gartner, Gateway, Honeywell, the National Science Foundation, Qualcomm, and the U.S. Department of Commerce. In 1999, he was named a Fellow of the Gartner Institute. In 2003, he was awarded the Clarence L. Steber professorship by the University of San Diego. Gary is a licensed CPA in Ohio, where he practiced public accounting for 14 years. He holds a Ph.D. in accounting information systems from the University of Tennessee, an M.B.A. in accounting from Xavier University, and a B.A. in economics from the University of Cincinnati.

PART 1

INTRODUCTION

CHAPTER **1**

INTRODUCTION TO ELECTRONIC COMMERCE

LEARNING OBJECTIVES

In this chapter, you will learn about:

- What electronic commerce is and how it is poised for a second wave of growth and a new focus on profitability
- Why business models have given way to revenue models and the analysis of business processes as key elements of electronic commerce initiatives
- How economic forces have created a business environment that is fostering a rebirth of electronic commerce
- How businesses use value chains and SWOT analysis to identify electronic commerce opportunities
- Why electronic commerce is international by its very nature and what challenges arise in doing global electronic commerce

INTRODUCTION

Very few people in the United States truly enjoy their hunt for a new or used car. Although many auto

dealers have worked to improve their customers' experiences by introducing fixed pricing and "no-haggle"

policies, a number of auto dealers continue to use aggressive sales approaches that can leave buyers

exhausted, confused, and even worried that they might have been cheated in the transaction. In 1995,

Autobytel (Note: This typeface indicates a corresponding link to a related Web page in the book's Online

Companion; Autobytel's URL is http://www.autobytel.com) launched an online car-buying service that

promised purchasers a haggle-free experience and offered car dealers a way to increase new vehicle sales volumes and reduce selling costs. Autobytel also acquired the operations of several competitors, and continues to operate their Web sites, including **Autoweb.com**, **AutoSite.com**, and **CarSmart**, as part of its business.

Buying a car with the assistance of Autobytel requires that the buyer register with an Autobytel Web site and specify the desired auto in detail, usually after researching the vehicle's options and features on the Internet or by visiting local dealers. More than 90 percent of car buyers today do research on the Internet before buying their cars. Autobytel provides the buyer with a firm price quote for the selected car, then forwards the buyer's contact information to a local participating dealer. Dealers pay Autobytel a subscription fee to receive exclusive rights to referrals from a particular geographic area for the brands of vehicles that they sell. The dealer contacts the buyer, who then completes the purchase transaction at the dealer's location.

The buyer benefits from a speedy, straightforward, and predictable buying process. The dealer benefits by selling more automobiles and not paying a commission to a salesperson. Autobytel receives the monthly subscription fee from each dealer that it has under contract and sells advertising to insurance and finance companies on its Web site. Autobytel currently has contracts with more than 29,000 auto dealers. Autobytel's revenue from fees paid by auto dealers on these transactions is more than $100 million per year. Internet sales referrals to dealers from Autobytel and companies like it are expected to account for approximately 10 percent of all U.S. new vehicle sales by 2007.

Autobytel established a business by replacing the salesperson in consumer car-buying transactions. Although auto manufacturers are moving to build effective online selling opportunities on their Web sites, Autobytel believes it can continue to offer value to both buyers and dealers that the manufacturers' sites cannot. Several research studies have concluded that the average Internet car buyer pays between 2 and 4 percent less for a car than other buyers. These studies attribute the savings to a combination of negotiating power and transaction efficiency, which means that the dealer can make more profit, despite giving the buyer a better price and paying a subscription fee to Autobytel. Auto dealers spend an average of $420 in marketing each new vehicle they sell. If they use Autobytel, those costs are reduced to about $150 per car, including Autobytel's referral fee.

Autobytel experienced rapid growth in sales from its inception in 1995 through 2002, when sales growth flattened. Like many other companies launched during the early boom years of electronic commerce, Autobytel had to change its focus. Instead of pursuing a strategy of revenue growth at all costs, it began to examine its costs carefully. The company also took steps to improve the quality of its service by ending relationships with a number of dealers who were generating significant numbers of customer complaints.

After a year of cost cutting and finding other ways to generate revenue, such as selling advertising space on its Web site, Autobytel began growing again. Autobytel has been earning a profit since 2003. Thus, Autobytel emerged from the difficult years of 2001 through 2003 as a growing and profitable participant in the second wave of electronic commerce that you will learn about in this chapter.

ELECTRONIC COMMERCE: THE SECOND WAVE

The business phenomenon that we now call electronic commerce has had an interesting history. From humble beginnings in the mid-1990s, electronic commerce grew rapidly until 2000, when a major downturn occurred. Many people have seen news stories about the "dot-com boom" followed by the "dot-com bust" or the "dot-bomb." In the 2000 to 2003 period, many industry observers were writing obituaries for electronic commerce. Just as the unreasonable expectations for immediate success fueled the high expectations during the boom years, overly gloomy news reports colored perceptions during this time. Beginning in 2003, with the general economy still in the doldrums, electronic commerce began to show signs of new life. Companies that had survived the downturn were not only seeing growth in sales again, but many of them were showing profits. Although the rapid expansion and high levels of investment of the boom years are not likely to be repeated, the second wave of electronic commerce is well underway. This section defines electronic commerce and describes how it is growing once again in its second wave.

Electronic Commerce and Electronic Business

To many people, the term "electronic commerce" means shopping on the part of the Internet called the World Wide Web (the Web). However, **electronic commerce** (or **e-commerce**) also includes many other activities, such as businesses trading with other businesses and internal processes that companies use to support their buying, selling, hiring, planning, and other activities. Some people use the term **electronic business** (or **e-business**) when they are talking about electronic commerce in this broader sense. For example, IBM defines electronic business as "the transformation of key business processes through the use of Internet technologies." Most people use the terms "electronic commerce" and "electronic business" interchangeably. In this book, the term electronic commerce (or e-commerce) is used in its broadest sense and includes all business activities conducted using electronic data transmission technologies. The most common technologies used are the Internet and the World Wide Web, but other technologies, such as wireless transmissions on mobile telephone and personal digital assistant (PDA) devices, are also included.

Categories of Electronic Commerce

Some people find it useful to categorize electronic commerce by the types of entities participating in the transactions or business processes. The five general electronic commerce categories are business-to-consumer, business-to-business, business processes, consumer-to-consumer, and business-to-government. The three categories that are most commonly used are:

- Consumer shopping on the Web, often called **business-to-consumer** (or **B2C**)
- Transactions conducted between businesses on the Web, often called **business-to-business** (or **B2B**)
- Transactions and business processes that companies, governments, and other organizations undertake on the Internet to support selling and purchasing activities

To understand these categories better, consider a company that manufactures stereo speakers. The company might sell its finished product to consumers on the Web, which would be B2C electronic commerce. It might also purchase the materials it uses to make the speakers from other companies on the Web, which would be B2B electronic commerce. Businesses often have entire departments devoted to negotiating purchase transactions with their suppliers. These departments are usually named **supply management** or **procurement**. Thus, B2B electronic commerce is sometimes called **e-procurement**.

In addition to buying materials and selling speakers, the company must also undertake many other activities to convert the purchased materials into speakers. These activities might include hiring and managing the people who make the speakers, renting or buying the facilities in which the speakers are made and stored, shipping the speakers, maintaining accounting records, purchasing insurance, developing advertising campaigns, and designing new versions of the speakers. An increasing number of these transactions and business processes can be done on the Web. Manufacturing processes (such as the fabrication of the speakers) can be controlled using Internet technologies within the business. All of these communication, control, and transaction-related activities have become an important part of electronic commerce. Some people include these activities in the B2B category; others refer to them as underlying or supporting business processes.

Figure 1-1 shows the three main elements of electronic commerce. The figure presents a rough approximation of the relative sizes of these elements. In terms of dollar volume and number of transactions, B2B electronic commerce is much greater than B2C electronic commerce. However, the number of supporting business processes is greater than the number of all B2C and B2B transactions combined.

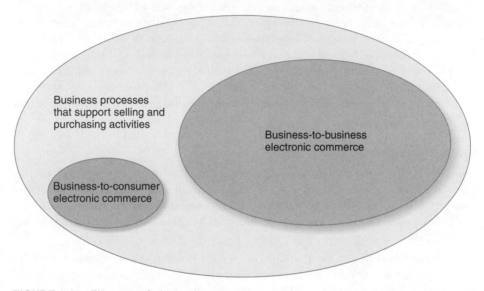

FIGURE 1-1   Elements of electronic commerce

The large oval in Figure 1-1 that represents the business processes that support selling and purchasing activities is the largest element of electronic commerce. This section provides some background and explains how business processes are built from their component parts, activities, and transactions.

For more than 70 years, business researchers have been studying the ways people behave in businesses. This research has helped managers better understand how workers do their jobs. The research results have also helped managers, and, increasingly, the workers themselves, improve job performance. By changing the nature of jobs, managers and workers can, as the saying goes, "work smarter, not harder." An important part of doing these job studies is to learn what activities each worker performs. In this setting, an **activity** is a task performed by a worker in the course of doing his or her job. For a much longer time—centuries, in fact—business owners have kept records of how well their businesses are performing. The formal practice of accounting, or recording transactions, dates back to the 1400s. A **transaction** is an exchange of value, such as a purchase, a sale, or the conversion of raw materials into a finished product. By recording transactions, accountants help business owners keep score and measure how well they are doing. All transactions involve at least one activity, and some transactions involve many activities. Not all activities result in measurable (and therefore recordable) transactions. Thus, a transaction always has one or more activities associated with it, but an activity might not be related to a transaction.

The group of logical, related, and sequential activities and transactions in which businesses engage are often collectively referred to as **business processes**. Transferring funds, placing orders, sending invoices, and shipping goods to customers are all types of activities or transactions. For example, the business process of shipping goods to customers might include a number of activities (or tasks, or transactions), such as inspecting the goods, packing the goods, negotiating with a freight company to deliver the goods, creating and printing the shipping documents, loading the goods onto the truck, and sending a check to the freight company. One important way that the Web is helping people work more effectively is by enabling employees of many different kinds of companies to work at home. In this arrangement, called **telecommuting** or **telework**, the employee logs in to the company computer through the Internet instead of traveling to an office.

Some researchers define a fourth category of electronic commerce, called **consumer-to-consumer** (or **C2C**), which includes individuals who buy and sell items among themselves. For example, C2C electronic commerce occurs when a person sells an item through a Web auction site to another person. In this book, C2C sales are included in the B2C category because the person selling the item acts much as a business would for purposes of the transaction.

Finally, some researchers also define a category of electronic commerce called **business-to-government** (or **B2G**); this category includes business transactions with government agencies, such as paying taxes and filing required reports. An increasing number of states have Web sites that help companies do business with state government agencies. For example, the **CAL-Buy** site makes it easy for businesses to conduct online transactions with the State of California. In this book, B2G transactions are included in our discussions of B2B electronic commerce. Figure 1-2 summarizes these five categories of electronic commerce.

Category	Description	Example
Business-to-consumer (B2C)	Businesses sell products or services to individual consumers.	Walmart.com sells merchandise to consumers through its Web site.
Business-to-business (B2B)	Businesses sell products or services to other businesses.	Grainger.com sells industrial supplies to large and small businesses through its Web site.
Business processes that support buying and selling activities	Businesses and other organizations maintain and use information to identify and evaluate customers, suppliers, and employees. Increasingly, businesses share this information in carefully managed ways with their customers, suppliers, employees, and business partners.	Dell Computer uses secure Internet connections to share current sales and sales forecast information with suppliers. The suppliers can use this information to plan their own production and deliver component parts to Dell in the right quantities at the right time.

FIGURE 1-2   Electronic commerce categories

Category	Description	Example
Consumer-to-consumer (C2C)	Participants in an online marketplace can buy and sell goods to each other. Because one party is selling, and thus acting as a business, this book treats C2C transactions as part of B2C electronic commerce.	Consumers and businesses trade with each other in the eBay.com online marketplace.
Business-to-government (B2G)	Businesses sell goods or services to governments and government agencies. This book treats B2G transactions as part of B2C electronic commerce.	CAL-Buy portal allows businesses to sell online to the State of California.

FIGURE 1-2 Continued   Electronic commerce categories

The Development and Growth of Electronic Commerce

Over the thousands of years that people have engaged in commerce with one another, they have adopted the tools and technologies that became available. For example, the advent of sailing ships in ancient times opened new avenues of trade to buyers and sellers. Later innovations, such as the printing press, steam engine, and telephone, have each changed the way in which people conduct commerce activities. The Internet has changed the way people buy, sell, hire, and organize business activities in more ways and more rapidly than any other technology has in the history of business.

Although the Web has made online shopping possible for many businesses and individuals, in a broader sense, electronic commerce has existed for many years. For more than 30 years, banks have been using **electronic funds transfers** (**EFTs**, also called **wire transfers**), which are electronic transmissions of account exchange information over private communications networks.

Businesses also have been engaging in a type of electronic commerce, known as electronic data interchange, for many years. **Electronic data interchange** (**EDI**) occurs when one business transmits computer-readable data in a standard format to another business. In the 1960s, businesses realized that many of the documents they exchanged were related to the shipping of goods, for example, invoices, purchase orders, and bills of lading. These documents included the same set of information for almost every transaction. Businesses also realized that they were spending a good deal of time and money entering this data into their computers, printing paper forms, and then reentering the data on the other side of the transaction. Although the purchase order, invoice, and bill of lading for each transaction contained much of the same information—such as item numbers, descriptions, prices, and quantities—each paper form usually had its own unique format for presenting that information. By creating a set of standard formats for transmitting that information electronically, businesses were able to reduce errors, avoid printing and mailing costs, and eliminate the need to reenter the data.

Businesses that engage in EDI with each other are called **trading partners**. The standard formats used in EDI contain the same information that businesses have always included in their standard paper invoices, purchase orders, and shipping documents.

Firms such as General Electric, Sears, and Wal-Mart have been pioneers in using EDI to improve their purchasing processes and their relationships with suppliers.

The U.S. government, which is one of the largest EDI trading partners in the world, also was instrumental in bringing businesses into EDI. For nine years, ending in 2001, the Defense Logistics Agency operated a number of Electronic Commerce Resource Centers (ECRCs) throughout the country. The ECRCs provided free assistance to many businesses, especially smaller businesses, so they could do EDI with the U.S. Defense Department and other federal agencies. The Georgia Institute of Technology continues to operate one of these centers as the **Georgia Tech Electronic Commerce Resource Center**, which serves businesses in Alabama, Georgia, and Tennessee.

One serious problem that potential adopters of EDI faced was the high cost of implementation. Until the late 1990s, doing EDI meant buying expensive computer hardware and software and then either establishing direct network connections (using leased telephone lines) to all trading partners or subscribing to a value-added network. A **value-added network** (**VAN**) is an independent firm that offers connection and transaction-forwarding services to buyers and sellers engaged in EDI. Before the Internet came into existence as we know it today, VANs provided the connections between most trading partners and were responsible for ensuring the security of the data transmitted. VANs usually charged a fixed monthly fee plus a per-transaction charge, adding to the already significant expense of implementing EDI. Many smaller firms were unable to afford to participate in EDI and lost important customers, who went elsewhere to buy. The companies that operated VANs have gradually moved EDI traffic to the Internet, but many other companies have developed other ways to do EDI types of transactions on the Internet. You will learn more about EDI, VANs, and new B2B transaction technologies in Chapter 5.

The Dot-Com Boom, Bust, and Rebirth

Between 1997 and 2000 more than 12,000 Internet-related businesses were started with more than $100 billion of investors' money. In an extended burst of optimism and what many came to describe as irrational exuberance, investors feared that they might miss the money-making opportunity of a lifetime. As more investors competed for a fixed number of good ideas, the price of those ideas increased. Worse, a number of bad ideas were proposed and funded. More than 5000 of these companies went out of business or were acquired in the downturn that began in 2000. The media coverage of the "dot-com bust" was extensive. However, between 2000 and 2003, more than $200 billion was invested in purchasing electronic commerce businesses that were in trouble and starting new online ventures, according to industry research firm WebMergers. This second wave of financial investment has not been reported extensively in either the general or business media, but it is fueling a rebirth of growth in online business activity.

After seeing so many news stories during the period from 2000 through 2002 proclaiming the death of electronic commerce, many people are surprised to find that the growth in online B2C sales had continued through that period, although at a slower pace than during the boom years of the late 1990s. Thus, the "bust" that was so widely reported in the media was really more of a slowdown than a true collapse. After four years of doubling or tripling every year, growth in online sales slowed to an annual rate of 20 to 30 percent starting in 2001. Most experts expect this growth rate to continue over the next several years.

One force driving the growth in online sales to consumers is the ever increasing number of people who have access to the Internet. The **Pew Internet & American Life Project** (funded by the Pew Charitable Trusts) began conducting several long-term research projects in 2000 to study the growth of the Internet and its effects on society. You can consult its Web site for the latest reports on these projects. In 2004, a Pew research project found that two-thirds of Internet users have purchased at least one item online.

In addition to the renewed growth in the B2C sector, B2B sales online have been increasing steadily also. B2B online sales have been more impressive than B2C sales because EDI was already well established in 1995, with more than $400 billion per year in transactions, so B2B has been growing from a larger base. In this book, we include business processes in the B2B category, so companies' transactions with other businesses, with their employees, and with governmental agencies (for example, when they pay their taxes) are all candidates for the application of Internet technologies. The dollar amount of these transactions, even for individual businesses, is substantial. Intel is a good example of a company that sells its products to other businesses rather than to consumers. Intel accepts more than 95 percent of its orders (more than $30 billion per year) through the Internet. Intel also purchases billions of dollars' worth of supplies and raw materials on the Web each year. The total volume of all worldwide business activities on the Web is expected to exceed $6 trillion by 2007. Figure 1-3 summarizes the growth of actual and estimated online sales for the B2C and B2B categories.

Year	B2C Sales: Actual and Estimated $ Billions	B2B Sales (including EDI): Actual and Estimated $ Billions
2007	240	6800
2006	190	5300
2005	150	4100
2004	130	2800
2003	100	1600
2002	80	900
2001	70	730
2000	50	600
1999	25	550
1998	10	520
1997	5	490
1996	Less than 1	460

Adapted from reports by CyberAtlas (http://cyberatlas.internet.com/); eMarketer (http://www.emarketer.com/); Forrester Research (http://www.forrester.com); and the *Statistical Abstract of the United States*, 2004, Washington: U.S. Census Bureau.

FIGURE 1-3   Actual and estimated online sales in B2C and B2B categories

The Second Wave of Electronic Commerce

Economists Chris Freeman and Francisco Louçã describe four waves that occurred in the Industrial Revolution in their book *As Time Goes By* (see the For Further Study and Research section at the end of this chapter). Many researchers predict that electronic commerce and the information revolution brought about by the Internet will go through similar waves. Those researchers agree that the second wave of electronic commerce has begun. This section outlines the defining characteristics of the first wave of electronic commerce and describes how the second wave is different.

The first wave of electronic commerce was primarily a U.S. phenomenon. Web pages were primarily in English, particularly on commerce sites. As the second wave begins, it is clear that the future of electronic commerce will be international in scope and will allow sellers to do business in many countries and in many languages. The problems of language translation and handling currency conversion will need to be solved to allow efficient conduct of business in the second wave. You will learn more about the issues that arise in global electronic commerce later in this chapter, in Chapter 7, and also in Chapter 11, which concerns online payment systems.

In the first wave, easy access to startup capital led to an over-emphasis on creating new large enterprises to exploit electronic commerce opportunities. Investors were excited about electronic commerce and wanted to participate, no matter how much it cost or how bad the underlying ideas were. In the second wave, established companies are using their own internal funds to finance gradual expansion of electronic commerce opportunities. These measured and carefully considered investments are helping electronic commerce grow more steadily, although more slowly.

The Internet technologies used in the first wave, especially in B2C commerce, were slow and inexpensive. Most consumers connected to the Internet using dial-up modems. The increase in broadband connections in homes is a key element in the second wave. Although these connections are more expensive, they are more than 10 times faster. This increased speed not only makes Internet use more efficient, it can alter the way people use the Web. You will learn more about types of connections in Chapter 2 and how connection speed can affect consumers' online shopping experiences in Chapters 3 and 4.

In the first wave, Internet technologies were integrated into B2B transactions and internal business processes by using bar codes and scanners to track parts, assemblies, inventories, and production status. These tracking technologies were not well integrated. Also, companies sent transaction information to each other using a patchwork of communication methods including fax, e-mail, and EDI. In the second wave, radio-frequency identification (RFID) devices and smart cards are being combined with biometric technologies such as fingerprint readers and retina scanners to control more items and people in a wider variety of situations. These technologies are increasingly integrated with each other and with communication systems that allow companies to communicate with each other and share transaction, inventory level, and customer demand information effectively. You will learn more about how these technologies are integrated with B2B electronic commerce in Chapter 5.

The use of electronic mail (or e-mail) in the first wave was as a tool for relatively unstructured communication. In the second wave, sellers are using e-mail as an integral part of their marketing and customer contact strategies. You will learn about e-mail marketing in Chapter 4, and about e-mail and more advanced Internet communications technologies in Chapter 8.

Online advertising was the main revenue source of many failed dot-com businesses in the first wave. After a two-year dip in online advertising activity and revenues, companies are beginning the second wave with a renewed interest in making the Internet work as an effective advertising medium. Some categories of online advertising, such as employment services (job wanted ads) are growing rapidly and are replacing traditional advertising outlets. You will learn about second wave advertising strategies in Chapter 4.

The sale of digital products was fraught with difficulties during the first wave of electronic commerce. The music recording industry was unable (or, some would say, unwilling) to devise a way to distribute digital music on the Web. This created an environment in which digital piracy—the theft of musical artists' intellectual property—became rampant. The promise of electronic books was also unfulfilled. The second wave offers the promise of legal distribution of music, video, and other digital products on the Web. Apple Computer's **iTunes** site is one of the early second wave attempts at digital product distribution. You will learn more about digital product distribution strategies in Chapter 3 and about the related legal issues in Chapter 7.

Figure 1-4 shows a summary of some key characteristics of the first wave and the second wave of electronic commerce. This list is not complete because every day brings new technologies and combinations of existing technologies that make additional second wave opportunities possible.

Large businesses, both existing and new businesses that had obtained large amounts of capital early on, dominated the first wave. As the second wave begins, more than 60 percent of U.S. businesses with fewer than 200 employees do not have Web sites. Thus, the second wave of electronic commerce will include a larger proportion of smaller businesses. Enabling those businesses to use electronic commerce will also be a substantial business.

Not all of the future of electronic commerce is based in its second wave. Some of the first wave companies were successful, such as Amazon.com, eBay, and Yahoo!. The second wave of electronic commerce will provide new opportunities for these businesses, too.

Electronic Commerce Characteristic	First Wave	Second Wave
International character of electronic commerce	Dominated by U.S. companies	Global enterprises in many countries participating in electronic commerce
Languages	Most electronic commerce Web sites in English	Many electronic commerce Web sites available in multiple languages
Funding	Many new companies started with outside investor money	Established companies funding electronic commerce initiatives with their own capital
Connection technologies	Many electronic commerce participants using slow Internet connections	Rapidly increasing use of broadband technologies for Internet connections
B2B technologies	B2B electronic commerce relied on a patchwork of disparate communication and inventory management technologies	B2B electronic commerce increasingly integrated with radio-frequency identification and biometric devices to manage information and product flows effectively
E-mail contact with customers	Unstructured e-mail communication with customers	Customized e-mail strategies now integral to customer contact
Advertising and electronic commerce integration	Over-reliance on simple forms of online advertising as main revenue source	Use of multiple sophisticated advertising approaches and better integration of electronic commerce with existing business processes and strategies
Distribution of digital products	Widespread piracy due to ineffective distribution of digital products	New approaches to the sale and distribution of digital products

FIGURE 1-4   Key characteristics of the first two waves of electronic commerce

BUSINESS MODELS, REVENUE MODELS, AND BUSINESS PROCESSES

A **business model** is a set of processes that combine to yield a profit. In the first wave of electronic commerce, many investors sought out start-up companies with appealing business models. A good business model was expected to lead to rapid sales growth and market dominance. The idea that the key to success was simply to copy the business model of a successful dot-com business led the way to many business failures, some of them quite dramatic.

In the wake of the dot-com debacle that ended the first wave of electronic commerce, many business researchers analyzed the efficacy of the business model approach and began to question the advisability of focusing great attention on a company's business model. One of the main critics, Harvard Business School professor Michael Porter, argued that business models not only did not matter, they probably did not exist (you can read more about Porter's criticisms of the business model approach in the articles cited in the For Further Study and Research section at the end of this chapter).

It has become clear to many companies that copying or adapting someone else's business model is neither an easy nor wise road map to success. Instead, companies should examine the elements of their business; that is, they should identify business processes that they can streamline, enhance, or replace with processes driven by Internet technologies.

Companies and investors do still use the idea of a **revenue model**, which is a specific collection of business processes used to identify customers, market to those customers, and generate sales to those customers. The revenue model idea is helpful for classifying revenue-generating activities for communication and analysis purposes. The details of revenue models that are used on the Web are presented in Chapter 3.

Focus on Specific Business Processes

In addition to the revenue model grouping of business processes, companies think of the rest of their operations as specific business processes. Those processes include purchasing raw materials or goods for resale, converting materials and labor into finished goods, managing transportation and logistics, hiring and training employees, managing the finances of the business, and many other activities.

An important function of this book is to help you learn how to identify those business processes that firms can accomplish more effectively by using electronic commerce technologies. In some cases, business processes use traditional commerce activities very effectively, and technology cannot improve upon them. Products that buyers prefer to touch, smell, or examine closely can be difficult to sell using electronic commerce. For example, customers might be reluctant to buy items such as high-fashion clothing or antique jewelry if they cannot closely examine the products before agreeing to purchase them.

This book will help you learn how to use Internet technologies to improve existing business processes and identify new business opportunities. An important aspect of electronic commerce is that firms can use it to help them adapt to change. The business world is changing more rapidly now than ever before. Although much of this book is devoted to explaining technologies, the book's focus is on the business of electronic commerce; the technologies only enable the business processes.

Role of Merchandising

Retail merchants have years of traditional commerce experience in creating store environments that help convince customers to buy. This combination of store design, layout, and product display knowledge is called **merchandising**. In addition, many salespeople have developed skills that allow them to identify customer needs and find products or services that meet those needs.

The skills of merchandising and personal selling can be difficult to practice remotely. However, companies must be able to transfer their merchandising skills to the Web for their Web sites to be successful. Some products are easier to sell on the Internet than others, because the merchandising skills related to those products are easier to transfer to the Web.

Product/Process Suitability to Electronic Commerce

Some products, such as books or CDs, are good candidates for electronic commerce because customers do not need to experience the physical characteristics of the particular item before they buy it. Because one copy of a new book is identical to other copies, and because the customer is not concerned about fit, freshness, or other such qualities, customers are usually willing to order a title without examining the specific copy they will receive. The advantages of electronic commerce, including the ability of one site to offer a wider selection of titles than even the largest physical bookstore, can outweigh the advantages of a traditional bookstore—for example, the customer's ability to browse the pages of the books. In later chapters, you will learn how to evaluate the advantages and disadvantages of using electronic commerce for specific business processes. Figure 1-5 lists examples of business processes categorized as to how well suited they are to electronic commerce and traditional commerce.

Well Suited to Electronic Commerce	Suited to a Combination of Electronic and Traditional Commerce Strategies	Well Suited to Traditional Commerce
Sale/purchase of books and CDs	Sale/purchase of automobiles	Sale/purchase of impulse items for immediate use
Online delivery of software	Online banking	Small-denomination purchases and sales
Sale/purchase of travel services	Roommate-matching services	
Online shipment tracking	Sale/purchase of residential real estate	
Sale/purchase of investment and insurance products	Sale/purchase of high-value jewelry and antiques	

FIGURE 1-5   Business process suitability to type of commerce

The classifications shown in the figure depend on the current state of available technologies, and thus will change as new tools emerge for implementing electronic commerce. For example, low-denomination transactions are not well suited to electronic commerce because no standard method for transferring small amounts of money on the Web has become generally accepted (although such standards are taking shape; Chapter 11 contains a more detailed discussion of this issue). If a company or group of companies could create a standard that gains general acceptance among buyers and sellers, low-denomination transactions could move from the traditional commerce column to the electronic commerce column.

One business process that is especially well suited to electronic commerce is the selling of commodity items. A **commodity item** is a product or service that is hard to distinguish from the same products or services provided by other sellers; its features have become standardized and well known. Gasoline, office supplies, soap, computers, and airline transportation are all examples of commodity products or services, as are the books and CDs sold by Amazon.com.

Another key factor that can make an item well suited to electronic commerce is the product's shipping profile. A product's **shipping profile** is the collection of attributes that affect how easily that product can be packaged and delivered. A high value-to-weight ratio can help by making the overall shipping cost a small fraction of the selling price. An airline ticket is an excellent example of an item that has a high value-to-weight ratio. Products that are consistent in size, shape, and weight can make warehousing and shipping much simpler and less costly. The shipping profile is only one factor, however. Expensive jewelry has a high value-to-weight ratio, but many people are reluctant to buy it without examining it in person unless the jewelry is sold under a well-known brand name and with a generous return policy.

A product that has a strong brand identity—such as a Sony CD player—is easier to sell on the Web than an unbranded item, because the brand's reputation reduces the buyer's concerns about quality when buying that item sight unseen. Other items that are well suited to electronic commerce are those that appeal to small, but geographically dispersed, groups of customers. Collectible comic books are an example of this type of product.

When personal selling skills are a factor, as in commercial real estate sales, or when the condition of the products is difficult to determine without making a personal inspection, as in purchases of high-fashion clothing, antiques, or perishable food products, traditional commerce or a combination of traditional commerce (for the inspection) and electronic commerce can be a better way to sell the items or services.

A combination of electronic and traditional commerce strategies works best when the business process includes both commodity and personal inspection elements. For example, many people are finding information on the Web about new and used automobiles. As you learned in the beginning of this chapter, Autobytel has had much success handling new car transactions. Most consumers who use the Autobytel service have already visited auto dealers and test-driven the cars in which they are interested. They are willing to take delivery of a particular make and model of a new vehicle even if they did not test-drive the specific car they are purchasing through Autobytel. In contrast, fewer people are willing to buy a used car without driving that specific car and personally inspecting it. In the case of used cars, electronic commerce provides a good way for buyers to obtain information about available models, features, reliability, prices, and dealerships; but the variability in the condition of used cars makes the traditional commerce component of personal inspection a key part of the transaction negotiation. The next two sections summarize some advantages and disadvantages of electronic commerce.

Advantages of Electronic Commerce

Firms are interested in electronic commerce because, quite simply, it can help increase profits. All the advantages of electronic commerce for businesses can be summarized in one statement: Electronic commerce can increase sales and decrease costs. Advertising done well on the Web can get even a small firm's promotional message out to potential customers in every country in the world. A firm can use electronic commerce to reach small groups of customers that are geographically scattered. The Web is particularly useful in creating virtual communities that become ideal target markets for specific types of products or services. A **virtual community** is a gathering of people who share a common interest, but instead of this gathering occurring in the physical world, it takes place on the Internet. You will learn more about virtual communities in Chapter 6.

Just as electronic commerce increases sales opportunities for the seller, it increases purchasing opportunities for the buyer. Businesses can use electronic commerce to identify new suppliers and business partners. Negotiating price and delivery terms is easier in electronic commerce because the Internet can help companies efficiently obtain competitive bid information. Electronic commerce increases the speed and accuracy with which businesses can exchange information, which reduces costs on both sides of transactions. Many companies are reducing their costs of handling sales inquiries, providing price quotes, and determining product availability by using electronic commerce in their sales support and order-taking processes.

Cisco Systems, a leading manufacturer of computer networking equipment, currently sells almost all its products online. Because no customer service representatives are involved in making these sales, Cisco operates very efficiently. In 1998, the first year in which its online sales initiative was fully operational, Cisco made 72 percent of its sales on the Web. Cisco avoided handling 500,000 calls per month and saved $500 million in that year alone. Today, Cisco conducts more than 99 percent of its purchase and sales transactions online.

Electronic commerce provides buyers with a wider range of choices than traditional commerce because buyers can consider many different products and services from a wider variety of sellers. This wide variety is available for consumers to evaluate 24 hours a day, every day. Some buyers prefer a great deal of information in deciding on a purchase; others prefer less. Electronic commerce provides buyers with an easy way to customize the level of detail in the information they obtain about a prospective purchase. Instead of waiting days for the mail to bring a catalog or product specification sheet, or even minutes for a fax transmission, buyers can have instant access to detailed information on the Web.

Some digital products, such as software, music and video files, or images, can even be delivered through the Internet, which reduces the time buyers must wait to begin enjoying their purchases. The ability to deliver digital products online is not just a cost-reduction opportunity. It can increase sales, too. Intuit sells its TurboTax income tax preparation software online and lets customers download the software immediately if they wish. Intuit sells a considerable amount of TurboTax software late in the evening on April 14 each year (April 15 is the deadline for filing personal income tax returns in the United States).

The benefits of electronic commerce extend to the general welfare of society. Electronic payments of tax refunds, public retirement, and welfare support cost less to issue and arrive securely and quickly when transmitted over the Internet. Furthermore, electronic payments can be easier to audit and monitor than payments made by check, providing protection against fraud and theft losses. To the extent that electronic commerce enables people to telecommute, everyone benefits from the reduction in commuter-caused traffic and pollution. Electronic commerce can also make products and services available in remote areas. For example, distance education is making it possible for people to learn skills and earn degrees no matter where they live or which hours they have available for study.

Disadvantages of Electronic Commerce

Some business processes may never lend themselves to electronic commerce. For example, perishable foods and high-cost, unique items, such as custom-designed jewelry, might be impossible to inspect adequately from a remote location, regardless of any technologies that might be devised in the future. Most of the disadvantages of electronic commerce today, however, stem from the newness and rapidly developing pace of the underlying technologies. These disadvantages will disappear as electronic commerce matures and becomes more available to and accepted by the general population.

Many products and services require that a critical mass of potential buyers be equipped and willing to buy through the Internet. For example, online grocers such as **Peapod** offer their delivery services only in a few cities. As more of Peapod's potential customers become connected to the Internet and begin to feel comfortable with purchasing online, the business might be able to expand into more geographic areas. But even the expansion of online grocery shopping is subject to limits; most online grocers focus their sales efforts on packaged goods and branded items. Perishable grocery products, such as fruit and vegetables, are much harder to sell online because customers want to examine and select specific items that are still fresh and appealing.

Peapod is a good example of how challenging it can be to build a business in an industry that requires this kind of critical mass. Although it was one of the first online grocery stores, Peapod has had a difficult time staying in business, and was even offline for a few weeks in mid-2000. Peapod was then acquired by a European firm that was willing to invest additional cash to keep it in operation. Two of Peapod's major competitors, WebVan and HomeGrocer, were unable to stay in business long enough to attract a sufficient customer base. Three of the most successful online grocery efforts in the world are **Grocery Gateway** in Toronto, **Disco Virtual** in Buenos Aires, and **Tesco** in the United Kingdom. Grocery Gateway and Disco Virtual operate in densely populated urban environments that offer sufficiently large numbers of customers within relatively small geographic areas, which make their delivery routes profitable. Tesco started its operations in London, which offers a similar densely populated urban area. However, Tesco has also expanded its operations to selected rural areas that are near a Tesco supermarket.

Established traditional grocery chains in the United States such as **Albertsons** and **Safeway** also now offer online ordering and delivery services in a second wave of using Internet technologies in the grocery business. By using their existing infrastructure (including warehouses, purchasing systems, and physical stores in multiple locations), they are able to avoid having to make the large capital investment in facilities that led to the demise of dot-com grocers such as WebVan and HomeGrocer.

One online grocer that has successfully implemented an updated version of the WebVan and HomeGrocer operational approach is **FreshDirect**. By limiting its service area to the densely populated region in and around New York City, FreshDirect has found the right combination of operating scale and market. The company started in 2002 and achieved profitability in 2004 on sales of $90 million. This is a much smaller sales volume than either WebVan or HomeGrocer would have needed to be profitable.

Businesses often calculate return on investment numbers before committing to any new technology. This has been difficult to do for investments in electronic commerce, because the costs and benefits have been hard to quantify. Costs, which are a function of technology, can change dramatically even during short-lived electronic commerce implementation projects because the underlying technologies are changing so rapidly. Many firms have had trouble recruiting and retaining employees with the technological, design, and business process skills needed to create an effective electronic commerce presence. You will learn more about return-on-investment calculations and employee recruitment and retention issues in Chapter 12.

Another problem facing firms that want to do business on the Internet is the difficulty of integrating existing databases and transaction-processing software designed for traditional commerce into the software that enables electronic commerce. Although a number of companies offer software design and consulting services that promise to tie existing systems into new online business systems, these services can be expensive. You will learn more about how companies deal with these software issues in Chapter 9.

In addition to technology and software issues, many businesses face cultural and legal obstacles to conducting electronic commerce. Some consumers are still fearful of sending their credit card numbers over the Internet and having online merchants—merchants they have never met—know so much about them. You will learn more about electronic commerce security, privacy issues, and payment systems later in this book. Other consumers are simply resistant to change and are uncomfortable viewing merchandise on a computer screen rather than in person. The legal environment in which electronic commerce is conducted is full of unclear and conflicting laws. In many cases, government regulators have not kept up with technologies. As you will learn in Chapter 7, laws that govern commerce were written when signed documents were a reasonable expectation in any business transaction. However, as more businesses and individuals find the benefits of electronic commerce to be compelling, many of these technology and culture-related disadvantages will be resolved or seem less problematic.

LEARNING FROM FAILURES

PETS.COM

In February 1999, Pets.com launched its Web site with the hopes of making substantial sales to the 60 percent of U.S. households that own pets and spend more than $20 billion each year feeding, entertaining, and caring for them. More than 10,000 stores sold pet supplies. These stores included small retail outlets, grocery stores, discount retailers (such as Wal-Mart and Costco), and a new generation of pet superstores. Pets.com had acquired an excellent domain name and intended to exploit the opportunities presented by high levels of investor interest in funding electronic commerce companies. The plan for Pets.com was to spend heavily to develop a brand and a Web presence that would rapidly make the company the premier online source for pet-related products.

continued

After launching the site, Pets.com raised $110 million from private investors in 1999, and another $80 million in a public sale of stock in early 2000. Pets.com spent more than $100 million of the money on advertising during its short life. It also spent significant sums to create a Web store that offered more than 12,000 different products. In November 2000—less than two years after launching its Web site—Pets.com went out of business.

Pets.com had created an electronic commerce initiative in an industry in which online business offered few advantages over traditional commerce. The products had a very low value-to-weight ratio. The shipping costs for pet food, one of the company's best-selling product categories, caused it to lose money on every sale. Pet products come in all shapes, sizes, and weights, and are, therefore, difficult to pack and ship efficiently. Pets.com was also spending money rapidly at a time when investors were beginning to question the long-run viability of all electronic commerce businesses. The lesson here is that Pets.com could not develop any sustainable advantage over traditional pet stores. Without such an advantage, the business was doomed.

In the years following the Pets.com failure, a number of companies began selling pet food and related items online. These companies were more careful than Pets.com was about what they offered for sale. By selling only items that had an appropriate shipping profile, many of these companies have now become successful. For example, veterinarians who formulate foods that meet the needs of specific pet diets are finding they can charge enough for those products to make online sales profitable.

ECONOMIC FORCES AND ELECTRONIC COMMERCE

Economics is the study of how people allocate scarce resources. One important way that people allocate resources is through commerce (the other major way is through government actions, such as taxes or subsidies). Many economists are interested in how people organize their commerce activities. One way people do this is to participate in markets. Economists use a formal definition of **market** that includes two conditions: first, that the potential sellers of a good come into contact with potential buyers, and second, that a medium of exchange is available. This medium of exchange can be currency or barter. Most economists agree that markets are strong and effective mechanisms for allocating scarce resources. Thus, one would expect most business transactions to occur within markets. However, much business activity today occurs within large **hierarchical business organizations**, which economists generally refer to as **firms**, or **companies**.

Most hierarchical organizations are headed by a top-level president or chief operating officer. Reporting to the president are a number of executives who, in turn, have a larger number of middle managers who report to them, and so on. An organization can have a relatively flat hierarchy, in which there are only a few levels of management, or it can have many reporting levels. In either case, the bottom level includes the largest number of employees and is usually made up of production workers or service providers. Thus, the hierarchical organization always has a pyramid-shaped structure.

These large firms often conduct many different business activities entirely within the organizational structure of the firm and participate in markets only for purchasing raw materials and selling finished products. If markets are indeed highly effective mechanisms for allocating scarce resources, these large corporations should participate in markets at every stage of their production and value-generation processes. Nobel laureate Ronald Coase wrote an essay in 1937 in which he questioned why individuals who engaged in commerce often created firms to organize their activities. He was particularly interested in the hierarchical structure of these business organizations. Coase concluded that transaction costs were the main motivation for moving economic activity from markets to hierarchically structured firms.

Transaction Costs

Transaction costs are the total of all costs that a buyer and seller incur as they gather information and negotiate a purchase-sale transaction. Although brokerage fees and sales commissions can be a part of transaction costs, the cost of information search and acquisition is often far larger. Another significant component of transaction costs can be the investment a seller makes in equipment or in the hiring of skilled employees to supply the product or service to the buyer.

To understand better how transaction costs occur in markets, consider the following example: A sweater dealer could obtain sweaters by engaging in market transactions with a number of independent sweater knitters. Transaction costs incurred by the dealer would include the costs of identifying the independent knitters, visiting them to negotiate the purchase price, arranging for delivery of the sweaters, and inspecting the sweaters on arrival. The knitters would also incur costs, such as the purchase of knitting tools and yarn. Since individual knitters could not know whether any sweater dealer would ever buy sweaters from them, the investments they would need to make to enter the sweater-knitting business would have an uncertain yield. This risk is a significant transaction cost for the knitters.

After purchasing the sweaters, the dealer takes them to a different market in which sweater dealers meet and do business with the retail shops that sell sweaters to the consumer. The dealers can use these market negotiations to find out which sweater colors and patterns are in demand and can then use that information to negotiate price and other terms in the knitters' market. A diagram of this set of markets appears in Figure 1-6.

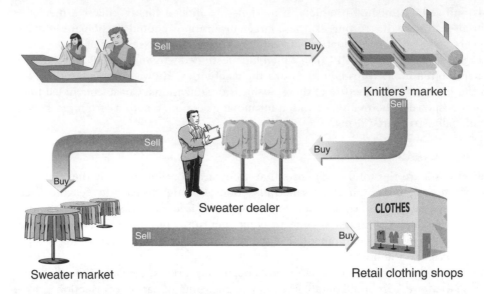

FIGURE 1-6   Market form of economic organization

Markets and Hierarchies

Coase reasoned that when transaction costs were high, businesspeople would form organizations to replace market-negotiated transactions. These organizations would be hierarchical and would include strong supervision and worker-monitoring elements. Instead of negotiating with individuals to purchase sweaters they had knit, a hierarchical organization would hire knitters, and then supervise and monitor their work activities. This supervision and monitoring system would include flows of monitoring information from the lower levels to the higher levels of the organization. It would also have control of information flowing from the upper levels of the organization to the lower levels. Although the costs of creating and maintaining a supervision and monitoring system are high, they can be lower than transaction costs in many instances.

In the sweater example, the sweater dealer would hire knitters, supply them with yarn and knitting tools, and supervise their knitting activities. This supervision could be done mainly by first-line supervisors, who might be drawn from the ranks of the more skilled knitters. The practice of an existing firm replacing one or more of its supplier markets with its own hierarchical structure for creating the supplied product is called **vertical integration**. Figure 1-7 shows how the wool sweater example would look after the knitters were vertically integrated into the hierarchical structure of the sweater dealer's organization.

Oliver Williamson, an economist who extended Coase's analysis, noted that industries with complex manufacturing and assembly operations tended to include many firms that used hierarchical structures and that were substantially vertically integrated. Many of the manufacturing and administrative innovations that occurred in businesses during the 20th century increased the efficiency and effectiveness of hierarchical monitoring

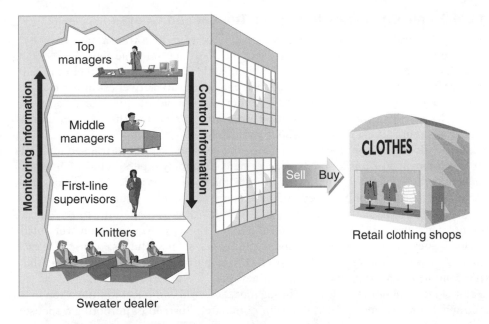

FIGURE 1-7 Hierarchical form of economic organization

activities. Assembly lines and other mass production technologies allowed work to be broken down into small, easily supervised procedures. The advent of computers brought tremendous increases in the ability of upper-level managers to monitor and control the detailed activities of their subordinates. Some of these direct measurement techniques are even more effective than the first-line supervisors on the shop floor.

During the years from the Industrial Revolution through the present, as improvements in monitoring became commonplace, the size and level of vertical integration of firms have increased. In some very large organizations, however, monitoring systems have not kept pace with the organization's increase in size. This has created problems because the economic viability of a firm depends on its ability to track operational activities effectively at the lowest levels of the firm. These firms have instituted decentralization programs that allow business units to function as separate organizations, negotiating transactions with other business units as if they were operating in a market rather than as part of the same firm. A **strategic business unit**, or simply **business unit**, is one particular combination of product, distribution channel, and customer type. These decentralization approaches are simply a return to the highly effective market mechanisms that worked so well before the firm vertically integrated itself.

Exceptions to the general trend toward hierarchies do exist. Many commodities, such as wheat, sugar, and crude oil, are still traded in markets. The commodity nature of the products traded in these markets significantly reduces transaction costs. There are a large number of potential buyers for an agricultural commodity such as wheat, and the farmer does not make any special investment in customizing or modifying the product for a particular customer. Thus, neither buyers nor sellers in commodity markets experience significant transaction costs.

Using Electronic Commerce to Reduce Transaction Costs

Businesses and individuals can use electronic commerce to reduce transaction costs by improving the flow of information and increasing the coordination of actions. By reducing the cost of searching for potential buyers and sellers and increasing the number of potential market participants, electronic commerce can change the attractiveness of vertical integration for many firms. It is not clear yet whether widespread adoption of electronic commerce will cause hierarchical organization structures to revert to their former market-based structures, but it certainly is a distinct possibility.

To see how electronic commerce can change the level and nature of transaction costs, consider an employment transaction. The agreement to employ a person has high transaction costs for the seller—the employee who sells his or her services. These transaction costs include a commitment to forego other employment and career development opportunities. Individuals make a high investment in learning and adapting to the culture of their employers. If accepting the job involves a move, the employee can incur very high costs, including actual costs of the move and related costs, such as the loss of a spouse's job. Much of the employee's investment is specific to a particular job and location; the employee cannot transfer the investment to a new job.

If a sufficient number of employees throughout the world can telecommute, then many of these transaction costs could be reduced or eliminated. Instead of uprooting a spouse and family to move, a worker could accept a new job by simply logging on to a different company server!

Network Economic Structures

Some researchers argue that many companies and strategic business units operate today in an economic structure that is neither a market nor a hierarchy. In this **network economic structure**, companies coordinate their strategies, resources, and skill sets by forming long-term, stable relationships with other companies and individuals based on shared purposes. These relationships are often called **strategic alliances** or **strategic partnerships**, and when they occur between or among companies operating on the Internet, these relationships are also called **virtual companies**.

In some cases, these entities, called **strategic partners**, come together as a team for a specific project or activity. The team dissolves when the project is complete; however, the partners maintain contact with each other through the ensuing period of inactivity. When the need for a similar project or activity arises, the same organizations and individuals build teams from their combined resources. In other cases, the strategic partners form many intercompany teams to undertake a variety of ongoing activities. Later in this book, you will see many examples of strategic partners creating alliances of this sort on the Web. In a hierarchically structured business environment, these types of strategic alliances would not last very long because the larger strategic partners would buy out the smaller partners and form a larger single company.

Network organizations are particularly well suited to technology industries that are information intensive. In our sweater example, the knitters might organize into networks of smaller organizations that specialize in certain styles or designs. Some of the particularly skilled knitters might leave the sweater dealer to form their own company to produce custom-knit sweaters. Some of the sweater dealer's marketing employees might form an independent firm that conducts market research on what the retail shops plan to buy in the upcoming months. This firm could sell its research reports to both the sweater dealer and the custom-knitting firm. As market conditions change, these smaller and more nimble organizations could continually reinvent themselves and take advantage of new opportunities that arise in the sweater markets. An illustration of such a network organization appears in Figure 1-8.

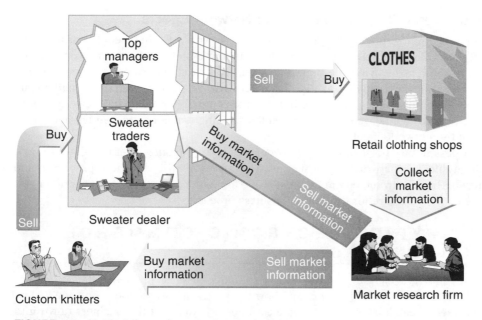

FIGURE 1-8 Network form of economic organization

Electronic commerce can make such networks, which rely extensively on information sharing, much easier to construct and maintain. Some researchers believe that these network forms of organizing commerce will become predominant in the near future. One of these researchers, Manuel Castells, even predicts that economic networks will become the organizing structure for all social interactions among people. Thomas Petzinger, a columnist for *The Wall Street Journal*, has written extensively about these new patterns of work and commerce in his newspaper columns and in his book, *The New Pioneers*.

Network Effects

Economists have found that most activities yield less value as the amount of consumption increases. For example, a person who consumes one hamburger obtains a certain amount of value from that consumption. As the person consumes more hamburgers, the value provided by each hamburger decreases. Few people find the fifth hamburger as enjoyable as the first. This characteristic of economic activity is called the **law of diminishing returns**. In networks, an interesting exception to the law of diminishing returns occurs. As more people or organizations participate in a network, the value of the network to each participant increases. This increase in value is called a **network effect**.

To understand how network effects work, consider an early user of a fax machine. When fax machines were first introduced, few companies had fax machines. The value of each fax machine increased as more companies purchased fax machines. As the network of fax machines grew, the capability of each individual fax machine increased because it could be used to communicate with more companies. The increase in the value of each fax machine is the result of a network effect.

Using Electronic Commerce to Create Network Effects

Your e-mail account, which gives you access to a network of other people with e-mail accounts, is another example of a network effect. If your e-mail account were part of a small network, it would be less valuable than it is. Most people today have e-mail accounts that are part of the Internet (which is a large global network). In the early days of e-mail, most e-mail accounts only connected people in the same company or organization to each other. Internet e-mail accounts are far more valuable than single-organization e-mail accounts because of the network effect.

Regardless of how businesses in a particular industry organize themselves—as markets, hierarchies, or networks—you will need a way to identify business processes and evaluate whether electronic commerce is suitable for each process. The next section presents one useful structure for examining business processes.

IDENTIFYING ELECTRONIC COMMERCE OPPORTUNITIES

Internet technologies can be used to improve so many business processes that it can be difficult for managers to decide where and how to use it. One way to focus on specific business processes as candidates for electronic commerce is to break the business down into a series of value-adding activities that combine to generate profits and meet other goals of the firm. In this section, you will learn one popular way to analyze business activities as a sequence of activities that create value for the firm.

Commerce is conducted by firms of all sizes. Smaller firms can focus on one product, distribution channel, or type of customer. Larger firms often sell many different products and services through a variety of distribution channels to several types of customers. In these larger firms, managers organize their work around the activities of strategic business units. Multiple business units owned by a common set of shareholders make up a firm, or company, and multiple firms that sell similar products to similar customers make up an **industry**.

Strategic Business Unit Value Chains

In his 1985 book, *Competitive Advantage*, Michael Porter introduced the idea of value chains. A **value chain** is a way of organizing the activities that each strategic business unit undertakes to design, produce, promote, market, deliver, and support the products or services it sells. In addition to these **primary activities**, Porter also includes **supporting activities**, such as human resource management and purchasing, in the value chain model. Figure 1-9 shows a value chain for a strategic business unit engaged in manufacturing a product, including both primary and supporting activities.

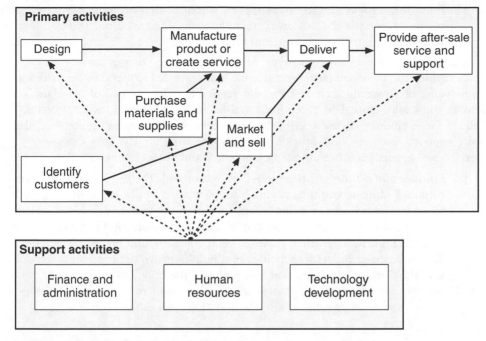

Primary activities

Design → Manufacture product or create service → Deliver → Provide after-sale service and support

Purchase materials and supplies

Market and sell

Identify customers

Support activities

Finance and administration | Human resources | Technology development

FIGURE 1-9 Value chain for a strategic business unit

The left-to-right flow in Figure 1-9 does not imply a strict time sequence for these processes. For example, a business unit may engage in marketing activities before purchasing materials and supplies. For each business unit, the primary activities are as follows:

- *Identify customers*: activities that help the firm find new customers and new ways to serve existing customers, including market research and customer satisfaction surveys
- *Design*: activities that take a product from concept to manufacturing, including concept research, engineering, and test marketing
- *Purchase materials and supplies*: procurement activities, including vendor selection, vendor qualification, negotiating long-term supply contracts, and monitoring quality and timeliness of delivery
- *Manufacture product or create service*: activities that transform materials and labor into finished products, including fabricating, assembling, finishing, testing, and packaging
- *Market and sell*: activities that give buyers a way to purchase and that provide inducements for them to do so, including advertising, promoting, managing salespersons, pricing, and identifying and monitoring sales and distribution channels
- *Deliver*: activities that store, distribute, and ship the final product, including warehousing, handling materials, consolidating freight, selecting shippers, and monitoring timeliness of delivery

- *Provide after-sale service and support*: activities that promote a continuing relationship with the customer, including installing, testing, maintaining, repairing, fulfilling warranties, and replacing parts

The importance of each primary activity depends on the product or service the business unit provides and to which customers it sells. If a strategic business unit provides a service, its value chain would include a Provide service activity instead of the Manufacture activity shown in Figure 1-9. The other activities in its value chain would be similar to those for a product manufacturing business unit. Each business unit must also undertake support activities that provide the infrastructure for the unit's primary activities. These support activities appear in Figure 1-9 and are as follows:

- *Finance and administration*: activities that provide the firm's basic infrastructure, including accounting, paying bills, borrowing funds, reporting to government regulators, and ensuring compliance with relevant laws
- *Human resources*: activities that coordinate the management of employees, including recruiting, hiring, training, compensation, and providing benefits
- *Technology development*: activities that help improve the product or service that the firm is selling and that help improve the business processes in every primary activity, including basic research, applied research and development, process improvement studies, and field tests of maintenance procedures

Industry Value Chains

Porter's book also identifies the importance of examining where the strategic business unit fits within its industry. Porter uses the term **value system** to describe the larger stream of activities into which a particular business unit's value chain is embedded. However, many subsequent researchers and business consultants have used the term **industry value chain** when referring to value systems. When a business unit delivers a product to its customer, that customer may, for example, use the product as purchased materials in its value chain. By becoming aware of how other business units in the industry value chain conduct their activities, managers can identify new opportunities for cost reduction, product improvement, or channel reconfiguration. An example of an industry value chain appears in Figure 1-10. This value chain is for a wooden chair and traces the life of the product from trees in a forest to its grave in a landfill or at a sawdust recycler.

Each business unit (logger, sawmill, lumberyard, chair factory, retailer, consumer, and recycler) shown in Figure 1-10 has its own value chain. For example, the sawmill purchases logs from the tree harvester and combines them in its manufacturing process with inputs, such as labor and saw blades, from other sources. Among the sawmill customers are the chair factory, shown in Figure 1-10, and other users of cut lumber. Examining this industry value chain could be useful for the sawmill that is considering entering the tree-harvesting business or the furniture retailer who is thinking about partnering with a trucking line. The industry value chain identifies opportunities up and down the product's life cycle for increasing the efficiency or quality of the product.

As they examine their industry value chains, many managers are finding that they can use electronic commerce and Internet technologies to reduce costs, improve product quality, reach new customers or suppliers, and create new ways of selling existing products.

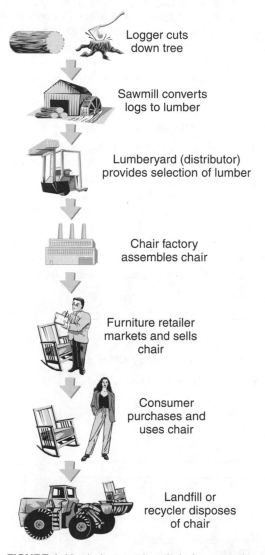

Logger cuts
down tree

Sawmill converts
logs to lumber

Lumberyard (distributor)
provides selection of lumber

Chair factory
assembles chair

Furniture retailer
markets and sells
chair

Consumer
purchases and
uses chair

Landfill or
recycler disposes
of chair

FIGURE 1-10 Industry value chain for a wooden chair

For example, a software developer who releases annual updates to programs might consider removing the software retailer from the distribution channel for software updates by offering to send the updates through the Internet directly to the consumer. This change would modify the software developer's industry value chain and would provide an opportunity for increasing sales revenue (the software developer could retain the margin a retailer would have added to the price of the update), but it would not appear as part of the software developer business unit value chain. By examining elements of the value chain outside the individual business unit, managers can identify many business opportunities, including those that can be exploited using electronic commerce.

The value chain concept is a useful way to think about business strategy in general. When firms are considering electronic commerce, the value chain can be an excellent way to organize the examination of business processes within their business units and in other parts of the product's life cycle. Using the value chain reinforces the idea that electronic commerce should be a business solution, not a technology implemented for its own sake.

SWOT Analysis: Evaluating Business Unit Opportunities

Now that you have learned how to identify industry value chains and break each value chain down into strategic business units, you can learn one popular technique for analyzing and evaluating business opportunities. Most electronic commerce initiatives add value by either reducing transaction costs, creating some type of network effect, or a combination of both. In **SWOT analysis** (the acronym is short for strengths, weaknesses, opportunities, and threats), the analyst first looks into the business unit to identify its strengths and weaknesses. The analyst then reviews the environment in which the business unit operates and identifies opportunities presented by that environment and the threats posed by that environment. Figure 1-11 shows questions that an analyst would ask in conducting a SWOT analysis.

Strengths
- What does the company do well?
- Is the company strong in its market?
- Does the company have a strong sense of purpose and the culture to support that purpose?

Weaknesses
- What does the company do poorly?
- What problems could be avoided?
- Does the company have serious financial liabilities?

Opportunities
- Are industry trends moving upward?
- Do new markets exist for the company's products/services?
- Are there new technologies that the company can exploit?

Threats
- What are competitors doing well?
- What obstacles does the company face?
- Are there troubling changes in the company's business environment (technologies, laws, and regulations)?

FIGURE 1-11   SWOT analysis questions

By considering all of the issues that it faces in a systematic way, a business unit can formulate strategies to take advantage of its opportunities by building on its strengths, avoiding any threats, and compensating for its weaknesses. In the mid-1990s, **Dell Computer** used a SWOT analysis to create a business strategy that has helped it become a very strong competitor in its industry value chain. Dell identified its strengths in selling directly to customers and in designing its computers and other products to reduce manufacturing costs. It acknowledged the weakness of having no relationships with local computer dealers. Dell faced threats from competitors such as Compaq (now a part of Hewlett-Packard) and IBM, both of which had much stronger brand names and reputations for quality at that time. Dell identified an opportunity by noting that its customers were becoming more knowledgeable about computers and could specify exactly what they wanted without having Dell salespersons answer questions or develop configurations for them. It also saw the Internet as a potential marketing tool. The results of Dell's SWOT analysis appear in Figure 1-12.

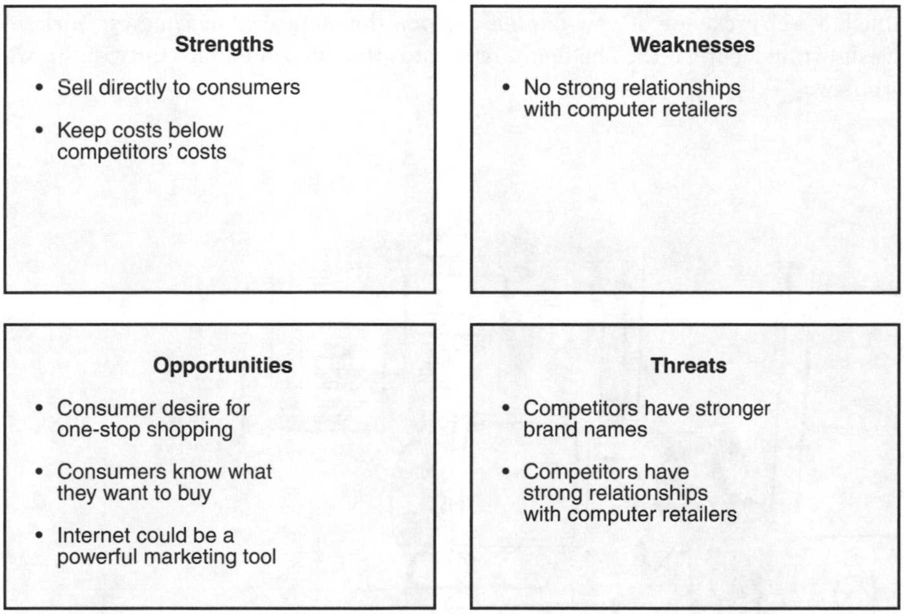

Strengths	**Weaknesses**
• Sell directly to consumers • Keep costs below competitors' costs	• No strong relationships with computer retailers
Opportunities	**Threats**
• Consumer desire for one-stop shopping • Consumers know what they want to buy • Internet could be a powerful marketing tool	• Competitors have stronger brand names • Competitors have strong relationships with computer retailers

FIGURE 1-12   Results of Dell's SWOT analysis

The strategy that Dell followed after doing the analysis took all four of the SWOT elements into consideration. Dell decided to offer customized computers built to order and sold over the phone, and eventually, over the Internet. Dell's strategy capitalized on its strengths and avoided relying on a dealer network. The brand and quality threats posed by Compaq and IBM were lessened by Dell's ability to deliver higher perceived quality because each computer was custom made for each buyer.

INTERNATIONAL NATURE OF ELECTRONIC COMMERCE

Because the Internet connects computers all over the world, any business that engages in electronic commerce instantly becomes an international business. When companies use the Web to improve a business process, they are automatically operating in a global environment. The key issues that any company faces when it conducts international commerce include trust and culture, language, and infrastructure. These topics are covered in the following sections. The related issues of international law and currency are covered in Chapter 7.

Trust Issues on the Web

It is important for all businesses to establish trusting relationships with their customers. Companies with established reputations in the physical world often create trust by ensuring that customers know who they are. These businesses can rely on their established brand names to create trust on the Web. New companies that want to establish online businesses face a more difficult challenge because a kind of anonymity exists for companies trying to establish a Web presence. A now-famous cartoon that appeared in *The New Yorker* magazine is shown in Figure 1-13. The figure illustrates the inherent anonymity of the Web in a humorous way.

"On the Internet, nobody knows you're a dog."

FIGURE 1-13   This classic cartoon from *The New Yorker* illustrates anonymity on the Web

For example, a U.S. bank can establish a Web site that offers services throughout the world. No potential customer visiting the site can determine just how large or well established the bank is simply by browsing through the site's pages. Because Web site visitors will not become customers unless they trust the company behind the site, a plan for establishing credibility is essential. Sellers on the Web cannot assume that visitors will know that the site is operated by a trustworthy business.

Customers' inherent lack of trust in "strangers" on the Web is logical and to be expected; after all, people have been doing business with their neighbors—not strangers—for thousands of years. When businesses grew to become large corporations with multinational operations, their reputations grew commensurately. Before a company could do business in dozens of countries, it had to prove its trustworthiness by satisfying customers for many years as it grew. Businesses on the Web must find ways to overcome this well-founded tradition of distrusting strangers, because today a company can incorporate one day and, through the Web, be doing business the next day with people in almost every country in the world. For businesses to succeed on the Web, they must find ways to generate quickly the trust that traditional businesses took years to develop.

Language Issues

Most companies realize that the only way to do business effectively in other cultures is to adapt to those cultures. The phrase "think globally, act locally" is often used to describe this approach. The first step that a Web business usually takes to reach potential customers in other countries, and thus in other cultures, is to provide local language versions of its Web site. This may mean translating the Web site into another language or regional dialect. Researchers have found that customers are far more likely to buy products and services from Web sites in their own language, even if they can read English well. Only 370 million of the world's 6 billion people learned English as their native language.

Researchers estimate that about 60 percent of the content available on the Internet today is in English, but more than 50 percent of current Internet users do not read English. International Data Corporation predicts that by 2007, more than 75 percent of Internet users will be outside the United States, and 60 percent of electronic commerce transactions will involve at least one party located outside the United States.

The non-English languages used most frequently by U.S. companies on their Web sites are Spanish, German, Japanese, and Chinese. Following closely behind is a second tier of languages that includes Italian, French, Korean, Portuguese, Dutch, Russian, and Swedish. In general, non-English languages used on the Internet have approximately the same levels of popularity. Some minor differences do exist, because "Internet use" includes activities other than electronic commerce and because some non-English speaking people do not conduct business with U.S. companies. The Web site of **Global Reach**, a consulting firm that offers Web site globalization services, maintains current information about language use on the Web.

Some languages require multiple translations for separate dialects. For example, the Spanish spoken in Spain is different from that spoken in Mexico, which is different from that spoken elsewhere in Latin America. People in parts of Argentina and Uruguay use yet a fourth dialect of Spanish. Many of these dialect differences are spoken inflections, which are not important for Web site designers (unless, of course, their sites include audio or video elements); however, a significant number of differences occur in word meanings and spellings. You might be familiar with these types of differences, since they occur in the U.S. and British dialects of English. The U.S. spelling of *gray* becomes *grey* in Great Britain, and the meaning of *bonnet* changes from a type of hat in the United States to an automobile hood in Great Britain. Chinese has two main systems of writing: one used in mainland China, and another used in Hong Kong and Taiwan.

Most companies that translate their Web sites translate all of their pages. However, as Web sites grow larger, companies are becoming more selective in their translation efforts. Some sites have thousands of pages with much targeted content; the businesses operating those sites can find the cost of translating all pages to be prohibitive.

The decision whether to translate a particular page should be made by the corporate department responsible for each page's content. The home page should have versions in all supported languages, as should all first-level links to the home page. Beyond that, pages that are devoted to marketing, product information, and establishing brand should be given a high translation priority. Some pages, especially those devoted to local interests, might be maintained only in the relevant language. For example, a weekly update on local news and employment opportunities at a company's plant in Frankfurt probably needs to be maintained only in German.

Firms that provide Web page translation services and translation software for companies include **Alis Technologies**, **Berlitz**, **Rubric, Ltd.**, **ScanSoft**, **Transparent Language**, and **Worldpoint Interactive**. These firms translate Web pages and maintain them for a fee that is usually between 25 and 90 cents per word for translations done by skilled human translators. Languages that are complex or that are spoken by relatively few people are generally more expensive to translate than other languages.

Different approaches can be appropriate for translating the different types of text that appear on an electronic commerce site. For key marketing messages, the touch of a human translator can be essential to capture subtle meanings. For more routine transaction-processing functions, automated software translation may be an acceptable alternative. Software translation, also called **machine translation**, can reach speeds of 400,000 words per hour, so even if the translation is not perfect, businesses might find it preferable to a human who can translate about 500 words per hour. Many of the companies in this field are working to develop software and databases of previously translated material that can help human translators work more efficiently and accurately.

The translation services and software manufacturers that work with electronic commerce sites do not generally use the term "translation" to describe what they do. They prefer the term **localization**, which means a translation that considers multiple elements of the local environment, such as business and cultural practices, in addition to local dialect variations in the language. The cultural element is very important because it can affect—and sometimes completely change—the user's interpretation of text.

Culture Issues

An important element of business trust is anticipating how the other party to a transaction will act in specific circumstances. That is one reason why companies with established brands can build online businesses more quickly and easily than a new company without a reputation. The brand conveys some expectations about how the company will behave. For example, a potential buyer might like to know how the seller would react to a claim by the buyer that the seller misrepresented the quality of the goods sold. Part of this knowledge derives from the buyer and seller sharing a common language and common customs. Business partners ideally have a common legal structure for resolving disputes. The combination of language and customs is often called **culture**. Most researchers agree that culture varies across national boundaries and, in many cases, varies across regions within nations. All companies must be aware of the differences in language and customs that make up the culture of any region in which they intend to do business. Under the heading **Trust and Culture**, this book's Online Companion includes links to Web sites that provide detailed information on culture issues for specific countries.

Managers at Virtual Vineyards (now a part of **Wine.com**), a company that sells wine and specialty food items on the Web, were perplexed. The company was getting an unusually high number of complaints from customers in Japan about short shipments. Virtual Vineyards sold most of its wine in case (12 bottles) or half-case quantities. Thus, to save on operating costs, it stocked shipping materials only in case, half-case, and two-bottle sizes. After investigation, the company determined that many of its Japanese customers ordered only one bottle of wine, which was shipped in a two-bottle container. To these Japanese customers, who consider packaging to be an important element of a high-quality product such as wine, it was inconceivable that anyone would ship one bottle of wine in a two-bottle container. They were e-mailing to ask where the other bottle was, notwithstanding the fact that they had ordered only one bottle.

Some errors stemming from subtle language and cultural standards have become classic examples that are regularly cited in international business courses and training sessions. For example, General Motors' choice of name for its Chevrolet Nova automobile amused people in Latin America—*no va* means "it will not go" in Spanish. Pepsi's "Come Alive" advertising campaign fizzled in China because its message came across as "Pepsi brings your ancestors back from their graves."

Another story that is widely used in international business training sessions is about a company that sold baby food in jars adorned with the picture of a very cute baby. The jars sold well everywhere they had been introduced except in parts of Africa. The mystery was solved when the manufacturer learned that food containers in those parts of Africa always carry a picture of their contents. This story is particularly interesting because it never happened. However, it illustrates a potential cultural issue so dramatically that it continues to appear in marketing textbooks and international business training materials.

Designers of Web sites for international commerce must be very careful when they choose icons to represent common actions. For example, in the United States, a shopping cart is a good symbol to use when building an electronic commerce site. However, many Europeans use shopping *baskets* when they go to a store and may never have seen a shopping *cart*. In Australia, people would recognize a shopping cart image but would be confused by the text "shopping cart" if it were used with the image. Australians call them shopping *trolleys*. In the United States, people often form a hand signal (the index finger touching the thumb to create a circle) that indicates "OK" or "everything is just fine." A Web designer might be tempted to use this hand signal as an icon to indicate that the transaction is completed or the credit card is approved, unaware that in countries such as Brazil, this hand signal is an obscene gesture.

The cultural overtones of simple design decisions can be dramatic. In India, for example, it is inappropriate to use the image of a cow in a cartoon or other comical setting.

Potential customers in Muslim countries can be offended by an image that shows human arms or legs uncovered. Even colors or Web page design elements can be troublesome. For example, white, which denotes purity in Europe and the Americas, is associated with death and mourning in China and many other Asian countries. A Web page that is divided into four segments can be offensive to a Japanese visitor because the number four is a symbol of death in that culture.

Bol.com is a company that resulted from the mergers of a number of online book-stores from 12 countries around the world, including China, Finland, Germany, Japan, Malaysia, Sweden, and Switzerland. If you explore the Bol.com site, you can see a number of different design approaches used in the home pages for each of the different countries. For example, the Bol.com Japan home page (a portion of which appears in Figure 1-14) has very few graphics and a large amount of text. Japanese culture values factual information over broad visual appeal. Also, the graphics are considerably smaller than those that appear on Web pages directed at U.S. consumers. Few Japanese consumers have broadband connections to the Internet, and this Web page is designed with that in mind.

Japanese shoppers have resisted the U.S. version of electronic commerce because they generally prefer to pay in cash or by cash transfer instead of by credit card, and they have a high level of apprehension about doing business online. **Softbank**, a major Japanese firm that invests in Internet companies, devised a way to introduce electronic commerce to a reluctant Japanese population. Softbank created a joint venture with 7-Eleven, Yahoo! Japan, and Tohan (a major Japanese book distributor) to sell books and CDs on the Web. This venture, called eS-Books, allows customers to order items on the Internet, and then pick them up and pay for them in cash at the local 7-Eleven convenience store. By adding an intermediary that satisfies the needs of the Japanese customer, Softbank has been highly successful in bringing business-to-consumer electronic commerce to Japan.

Nike, a major U.S.-based maker of sports products, realized that it had to create special Web pages to attract the millions of its customers who live outside the United States. One such effort is the **Nike Football** site shown in Figure 1-15.

This soccer imagery is not what most U.S. visitors would expect to see when visiting a "football" site! Since Nike already had a site that was designed for its U.S. audience (and that includes coverage of the U.S. game of "football"), it uses the Nike Football site to appeal to soccer fans throughout the world. The site allows the user to select from more than a dozen languages.

only a few, small graphics appear on page

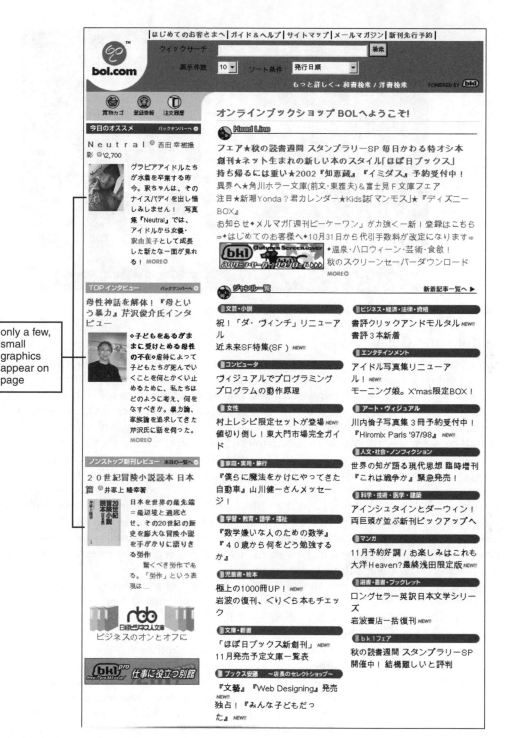

FIGURE 1-14 Bol.com Japan home page

links to various language versions of the site

FIGURE 1-15 Nike Football site

Culture and Government

Some parts of the world have cultural environments that are extremely inhospitable to the type of online discussion that occurs on the Internet. These cultural conditions, in some cases, lead to government controls that can limit electronic commerce development. The Internet is a very open form of communication. This type of unfettered communication is not desired or even considered acceptable in some cultures. For example, a **Human Rights Watch** report stated that many countries in the Middle East and North Africa do not allow their citizens unrestricted access to the Internet. The report notes that many governments in this part of the world regularly prevent free expression by their citizens and have taken specific steps to prevent the exchange of information outside of state controls. For instance, Saudi Arabia, Yemen, and the United Arab Emirates all make efforts to filter the Web content that is available in their countries. An organization devoted to the international promotion of democracy and civil liberties, **Freedom House**, offers a number of downloadable publications on its site, including in-depth reports on Internet censorship activities of governments throughout the world.

In most North African and Middle Eastern countries, officials have publicly denounced the Internet for carrying materials that are sexually explicit, anti-Islam, or that cast doubts on the traditional role of women in their societies. In many of these countries, uncontrolled use of Internet technologies is so at odds with existing traditions, cultures, and laws that electronic commerce is unlikely to exist in these countries at any significant level in the near future. In contrast, other Islamic countries in that part of the world, including Algeria, Morocco, and the Palestinian Authority, do not limit online access or content.

The censorship of Internet content and communications restricts electronic commerce because it prevents certain types of products and services from being sold or advertised. Further, it reduces the interest level of many potential participants in online activities. If large numbers of people in a country are not interested in being online, businesses that use the Internet as an information and product delivery channel will not develop in those countries.

Other countries, such as the People's Republic of China and Singapore, are wrestling with the issues presented by the growth of the Internet as a vehicle for doing business. These countries have a tradition of controlling their citizens' access to information from outside the country, but they want their economies to reap the benefits of electronic commerce. China created a complex set of registration requirements and regulations that govern any business that engages in electronic commerce. These regulations are enforced by the Public Security Bureau, which is a branch of the state police, not an independent administrative agency. For example, companies in China that sell Internet services must register all of their customers with the Public Security Bureau and must retain copies of all e-mail messages and chat room conversations for 60 days. Chinese citizens entering a chat room at **Sohu.com**, one of China's leading portal sites ("sohu" means "search fox" in Chinese), are greeted with a Web page containing the following text (translated here from the original Chinese):

> Warning! Please take note that the following issues are prohibited according to Chinese law: 1) Criticism of the People's Republic of China Constitution. 2) Revealing State secrets, and discussion about overthrowing the Communist government. 3) Topics which damage the reputation of the State.

The Chinese government regularly conducts reviews of ISPs and their records. Every year, the Chinese Public Security Bureau shuts down thousands of Internet cafes for failing to keep adequate records and requires many others to suspend operations while they implement required electronic record-keeping procedures. Operators of Web sites in China are required to monitor all content that appears on their sites. Blogbus was a Chinese site that allowed visitors to post essays and articles. The Chinese government shut down the site in March 2004 because one posting (out of 15,000) contained an essay that included what the government deemed to be "forbidden content." More than 50 people have been jailed in China for posting "subversive" content on Web pages. Singapore has also adopted a number of restrictive rules and policies. These countries will continue to face difficult policy choices as they maintain their attempts to control individuals' use of the Internet while at the same time trying to encourage increases in online business transaction activity.

Some countries, although they do not ban electronic commerce entirely, have strong cultural requirements that have found their way into the legal codes that govern business conduct. In France, an advertisement for a product or service must be in French. Thus, a business in the United States that advertises its products on the Web and is willing to ship goods to France must provide a French version of its pages if it intends to comply with French law. Many U.S. electronic commerce sites include in their Web pages a list of the countries from which they will accept orders through their Web sites.

The official language of the Canadian province of Quebec is French. Quebec provincial law requires street signs, billboards, directories, and advertising created by Quebec businesses to be in French. In 1999, the government of Quebec fined Quebec photographer Michael Calomiris and ordered him either to remove his English-language Web site or add a French translation of the pages to the site. Calomiris had been advertising his photographs for sale on his Quebec-based Web site and had targeted his ads to the U.S. market. He paid the fine and appealed the government's decision. He has had his Web site at **Michaels Photography Studio** in English during the six years his appeal has been pending.

Infrastructure Issues

Businesses that successfully meet the challenges posed by trust, language, and culture issues still face the challenges posed by variations and inadequacies in the infrastructure that supports the Internet throughout the world. Internet infrastructure includes the computers and software connected to the Internet and the communications networks over which the message packets travel. In many countries other than the United States, the telecommunications industry is either government owned or heavily regulated by the government. In many cases, regulations in these countries have inhibited the development of the telecommunications infrastructure or limited the expansion of that infrastructure to a size that cannot reliably support Internet data packet traffic.

Local connection costs through the existing telephone networks in many countries are very high compared to U.S. costs for similar access. This can have a profound effect on the behavior of electronic commerce participants. For example, in countries where Internet connection costs are high, few businesspeople would spend time surfing the Web to shop for a product. They would use a Web browser only to navigate to a specific site that they know offers the product they want to buy. Thus, to be successful in selling to businesses in such countries, a company would need to advertise its Web presence in traditional media instead of relying on Web search engines to deliver customers to their Web sites.

The Organization for Economic Cooperation and Development's (OECD) Directorate of Science, Technology, and Industry issued a number of **OECD Statements on Information and Communications Policy** that deal with telecommunications infrastructure development issues throughout the world. These OECD statements provide guidance for businesses and governments as they begin building the technological capabilities that will support international electronic commerce in the future.

Business and government leaders in an increasing number of European countries have been demanding that their countries' telecommunications providers offer flat-rate telephone line Internet access. Until recently, most Europeans paid for the amount of time they used the telephone line, including time for local calls. In a **flat-rate access** system, the consumer or business pays one monthly fee for unlimited telephone line usage. Activists in these countries argued that flat-rate access was a key to the success of electronic commerce in the United States. Although many factors contributed to the rapid rise of U.S. electronic commerce, many industry analysts agree that flat-rate access was one of the most important factors. As more European telecommunications providers have begun to offer flat-rate access, electronic commerce in those countries has increased dramatically.

More than half of all businesses on the Web turn away international orders because they do not have the processes in place to handle such orders. Some of these companies are losing millions of dollars' worth of international business each year. This problem is global; not only are U.S. businesses having difficulty reaching their international markets, but businesses in other countries are having similar difficulties reaching the U.S. market.

The paperwork and often convoluted processes that accompany international transactions are targets for technological solutions. Most firms that conduct business internationally rely on a complex array of freight forwarding companies, customs brokers, international freight carriers, and importers to navigate the maze of paperwork that must be completed at every step of the transaction to satisfy government and insurance requirements. The multiple flows of information and transfers of physical objects that occur in a typical international trade transaction are illustrated in Figure 1-16.

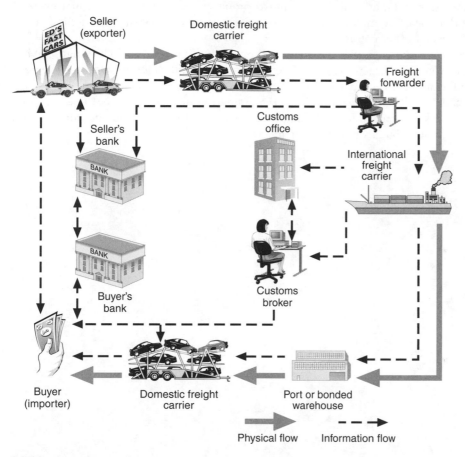

FIGURE 1-16 Parties involved in a typical international trade transaction

As you can see in Figure 1-16, the information flows can be complex. Domestic transactions usually include only the seller, the buyer, their respective banks, and one freight carrier. International transactions almost always require physical handling of goods by several freight carriers, storage in a freight forwarder's facility before international shipment, and storage in a port or bonded warehouse facility in the destination country. This handling and storage require monitoring by government customs offices in addition to the monitoring by seller and buyer that occurs in domestic transactions. International transactions usually require the coordinated efforts of customs brokers and freight forwarding agencies because the regulations and procedures governing international transactions are so complex. You will learn more about how businesses complete international transactions in Chapter 11.

The United Nations estimated that the annual cost of handling paperwork for international transactions is $600 billion, or approximately 6 percent of the total $10 trillion spent in worldwide international trade. Some companies sell software that can automate some of the paperwork, however, many countries have their own paper-based forms and procedures with which international shippers must comply. To further complicate matters, some countries that have automated some procedures use computer systems that are incompatible with those of other countries.

Some governments provide assistance to companies that want to do international business on the Web. The Argentine government operates the **Fundación Invertir** Web site to provide information to companies that want to do business in Argentina. The **U.S. Commercial Service** (an agency of the U.S. Department of Commerce) operates the **BuyUSA.com** site, a portal for U.S. companies that want to sell abroad and non-U.S. companies that want to buy from U.S. companies.

Summary

In this chapter, you learned that commerce, the negotiated exchange of goods or services, has been practiced in traditional ways for thousands of years. Electronic commerce is the application of new technologies, particularly Internet and Web technologies, to help individuals, businesses, and other organizations conduct business more effectively. As in the Industrial Revolution, electronic commerce will be adopted in waves of change. The first wave of electronic commerce ended in 2000. More recently, a second wave with new approaches to integrating Internet technologies into business processes has begun. In the second wave, businesses are focusing less on overall business models and more on improving specific business processes.

Not all activities lend themselves to improvement with these technologies, but many do. Using electronic commerce, some businesses have been able to create new products and services, and others have improved the promotion, marketing, and delivery of existing offerings. Firms have also found many ways to use electronic commerce to improve purchasing and supply activities; identify new customers; and operate their finance, administration, and human resource management activities more efficiently. You learned that electronic commerce can help businesses reduce transaction costs or create network economic effects that can lead to greater revenue opportunities.

You examined an overview of markets, hierarchies, and networks—the economic structures in which businesses operate—and learned how electronic commerce fits into those structures. Porter's ideas about value chains at the business unit and industry levels were presented, and you learned how to use value chains and SWOT analysis as ways to understand business processes and analyze their suitability for electronic commerce implementation.

The inherently global nature of electronic commerce leads to many opportunities and a few challenges. Businesses that want to use electronic commerce to sell across international borders must be careful to understand the trust, cultural, and language legal issues that arise in international business.

Key Terms

Activity

Business model

Business processes

Business unit

Business-to-business (B2B)

Business-to-consumer (B2C)

Business-to-government (B2G)

Commodity item

Company

Consumer-to-consumer (C2C)

Culture

E-procurement

Electronic business (e-business)

Electronic commerce (e-commerce)

Electronic data interchange (EDI)

Electronic funds transfer (EFT)

Firm

Flat-rate access

Hierarchical business organization

Industry

Industry value chain

Law of diminishing returns

Localization

Machine translation

Market

Merchandising

Network economic structure	Telecommuting
Network effect	Telework
Primary activities	Trading partners
Procurement	Transaction
Revenue model	Transaction costs
Shipping profile	Value-added network (VAN)
Strategic alliance	Value chain
Strategic business unit	Value system
Strategic partner	Vertical integration
Strategic partnership	Virtual community
Supply management	Virtual company
Supporting activities	Wire transfer
SWOT analysis	

Review Questions

RQ1. Describe three factors that would cause a company to continue doing business in traditional ways and avoid electronic commerce.

RQ2. Figure 1-5 lists roommate-matching services as a type of business that is well-suited to a combination of electronic and traditional commerce. In one paragraph, describe the elements of this service that would be best handled using traditional commerce and explain why.

RQ3. Choose one major difference between the first wave and the second wave of electronic commerce. Write a paragraph that describes this difference to a person who is not familiar with either business or Internet technologies.

RQ4. What are transaction costs and why are they important?

RQ5. Provide one example of how electronic commerce could help change an industry's economic structure from a hierarchy to a network.

RQ6. How might managers use SWOT analysis to identify new applications for electronic commerce in their strategic business units?

RQ7. In about 200 words, explain the difference between language translation and language localization.

RQ8. In a paragraph, describe the advantages of a flat-rate telecommunications access system for countries that want to encourage electronic commerce.

Exercises

E1. You have decided to buy a new color laser printer for your home office. List specific activities that you must undertake as you gather information about printer capabilities and features. Use the **CompUSA**, **HPshopping.com**, **Office Depot**, **OfficeMax**, and **Staples** Web sites to gather information. Write a short summary of the process you undertook so that others who plan to undertake a similar task can use your information.

E2.   Choose one of the Web sites listed in the previous question and identify three ways in which the company has reduced its transaction costs by using a Web site to provide information about printers. List these three transaction cost reduction elements and write a paragraph in which you discuss one transaction cost reduction opportunity that you believe the company missed.

E3.   Read the following business messages and come up with a list of words or phrases in each message that you believe might be troublesome for automated translation software. Then use either the **AltaVista Translation** Web site or the **FreeTranslation** Web site to translate the messages from English to one of the foreign languages available on that site. Translate each message back into English. Write a short memo that compares the problems you anticipated with those that occurred in the automated translation. The business messages are:

    a.   The flight has been delayed for several hours and your shipment of components will not arrive as scheduled.

    b.   We would be happy to bid on your proposal; however, we will need the drawings of subassembly #24 and the supervising mechanical engineer's quality control report by next Thursday.

    c.   Our company offers the latest and greatest hot deals on wheels. We would love to send you a brochure that explains why our brakes, wheels, and suspension components will do the job for you effectively and economically.

E4.   Create a diagram (similar to the diagram in Figure 1-10) that describes the industry value chain for the retail book business. You can use the Online Companion links for this exercise to examine the Web sites for **Amazon.com**, **Barnes & Noble**, **Books-A-Million**, **eCampus**, **Internet Bookshop,** and **Powell's Books**.

Cases

C1. Amazon.com

In 1994, a 29-year-old financial analyst and fund manager named Jeff Bezos became intrigued by the rapid growth of the Internet. Looking for a way to capitalize on this hot new marketing tool, he made a list of 20 products that might sell well on the Internet. After some intense analysis, he determined that books were at the top of that list. Although Bezos liked the name Abracadabra, he decided to call his online bookshop Amazon.com. Today, **Amazon.com** has more than 40 million customers and sells billions of dollars worth of all types of merchandise.

When he started, Bezos had no experience in the book-selling business, but he realized that books had an ideal shipping profile for online sales. He believed that many customers would be willing to buy books without inspecting them in person and that books could be impulse purchase items if properly promoted on a Web site. By accepting orders on its Web site, Bezos believed that Amazon.com could reduce transaction costs in the sale to the customer.

More than 4 million book titles are in print at any one time throughout the world, and more than 1 million of those are in English. However, the largest physical bookstore cannot stock more than 200,000 books and carries even fewer titles because bookstores stock more than one copy of each title. Having a wide selection was important because Bezos believed it would help create a network economic effect. People would visit Amazon.com whenever they wanted to buy a book because it would be the most likely store (physical or online) to have a particular title. After becoming satisfied customers, people would return to Amazon.com to buy more books and would eventually stop looking elsewhere.

The structure of the supply side of the book business was equally important to Amazon.com's success. Music CDs, which were second on Bezos' list, were produced by a few major recording companies who could easily control Amazon.com's supply. In contrast, there were a large number of book publishers, none of which held a dominant position in the book-selling marketplace. Thus, it was unlikely that a single supplier could restrict Bezos' supply of books or enter his market as a competitor. He decided to locate his firm in Seattle, close to a large pool of programming talent and near one of the largest book distribution warehouses in the world. These supply factors were important because Bezos wanted to develop efficiencies that would allow Amazon.com to reduce transaction costs for its purchases as well as its sales transactions.

Bezos encouraged early customers to submit reviews of books, which he posted with the publisher's information about the book and with reviews written by Amazon.com employees. This customer participation served as a substitute for the corner bookshop staff's friendly advice and recommendations. Bezos saw the power of the Internet in reaching small, highly focused market segments, but he realized that his comprehensive bookstore could not be all things to all people. Therefore, he created a sales associate program in which Web sites devoted to a particular topic, such as model railroading, could provide links to Amazon.com books that related to that topic. In return, Amazon.com remits a percentage of the referred sales to the owner of the referring site.

Although Bezos' original vision was to create an online bookstore with the world's best selection, Amazon has moved into other product lines where opportunities for network economic effects and transaction cost reductions looked promising. In 1998, Amazon.com began selling music CDs and videotapes. The Web site's software can track a customer's purchases and recommend similar book, CD, or video titles. In fact, the site can recommend related products in a variety of product categories now sold on Amazon.com. These product categories include consumer electronics, computers, toys, clothing, art, tools, hardware, housewares, furniture, and car parts.

By paying attention to every process involved in buying, promoting, selling, and shipping consumer goods, and by working to improve each process continually Bezos and Amazon.com have become one of the first highly visible success stories in electronic commerce. In fact, Amazon.com now generates significant revenue by supplying other sellers of consumer goods with the technology to sell those goods online. One of its first partnerships was with Toys R Us, a company that had experienced difficulties in selling online and making deliveries on time in the 1999 holiday shopping season. Toys R Us signed an agreement with Amazon.com in 2000 that placed Toys R Us products on the Amazon.com Web site. Amazon.com would accept the orders on its Web site and would ship products to customers for Toys R Us in exchange for a percentage of each sale. Amazon.com also agreed not to sell toys itself or on behalf of other partners for whom it might provide online sales services in the future. For example, when Amazon agreed to sell Target products online, it could not sell Target's toy lines on its Web site (Target is the third-largest toy retailer in the world, behind Wal-Mart and Toys R Us).

In addition to the online sales services Amazon.com provides to Toys R Us, Target, Borders, CDNow, and other large companies, it provides similar services to many smaller companies with its zShops offering. In zShops, small retailers become members of an online shopping mall on Amazon's site.

Toys R Us sells more than $300 million worth of toys each year through the Amazon.com site. Both Toys R Us and Amazon.com benefit from the network economics effect they obtain by having toys available for sale on Amazon.com's well-known electronic commerce site. Many small retailers in the zShops program who sell toys also benefit because shoppers visit the Amazon.com site looking for toys. When a site visitor searches for a toy, the zShops retailers' offerings are presented on the search results page along with results from Toys R Us, Amazon.com, and other companies for which Amazon.com provides online sales services.

Required:

1. In 2004, Toys R Us sued Amazon.com for violating terms of the agreement between the companies (specifically, Toys R Us objected to Amazon.com's permitting toys to be sold on its zShops Web pages). Amazon.com responded by filing a countersuit. Prepare a report of about 200 words in which you summarize the current state of the litigation.

2. Outline the advantages and disadvantages that Amazon.com would have considered before it made the agreement with Toys R Us to limit competing toy sales. In about 200 words, summarize these advantages and disadvantages, then evaluate Amazon.com's decision to enter such an agreement.

3. In about 200 words, outline specific recommendations to Amazon.com for negotiating a settlement with Toys R Us that would benefit both companies.

Note: Your instructor might assign you to a group to complete this case and might ask you to prepare a formal presentation of your results to your class.

C2. Hal's Hardware, Inc.

Hal Donovan is the president of Hal's Hardware, Inc. (HHI), a regional chain of 14 hardware stores located in Michigan, Ohio, and western Pennsylvania. HHI currently has a Web site that includes information about the company and some store information, such as locations and hours. Hal is thinking about expanding the HHI Web site to include online shopping. He believes that HHI customers might find the Web site to be a useful way to order items, see whether items are in stock at the nearest store, and comparison shop among different brands of a particular item. Hal is also hopeful that the Web site can reach customers who are not located near an HHI store. Many of the items sold at HHI are small and have high value-to-weight ratios, so they have good delivery service shipping profiles. Hal has decided that not all of HHI's inventory items should be available for sale on the Web site. Items such as wheelbarrows and live plants would probably be among the types of products that should be excluded. Hal does want customers to be able to order these items on the Web and pick them up in the store, however.

HHI enjoys an excellent reputation as a chain of friendly neighborhood stores. The store managers are all active in their communities and the stores regularly sponsor youth sports teams and support local charities. When hired, salespeople go through a comprehensive training program that includes skill training in the areas of the store in which they will work (plumbing, electrical, power tools, flooring, garden, and so on), and they are trained in customer service skills. As a result of HHI's focus on service, most of the stores have become community gathering places.

On Saturday afternoons, the stores are full of woodworking hobbyists, gardeners, and customers planning weekend projects of various kinds. On weekday mornings, electricians, plumbers, remodelers, and construction contractors stop by for the free coffee that the HHI stores offer when they open at 6:00 a.m. Each HHI store maintains a bulletin board next to the coffee urn in the contractors' area. Contractors can place help wanted or job wanted notices on the bulletin board. They can also place ads to buy and sell used equipment there. Many of HHI's regular customers obtained their current jobs through those bulletin boards.

HHI stores offer classes and workshops for the homeowner and hobbyist three evenings each month and regularly schedule seminars for professional customers on weekday mornings. Many of these workshops and seminars are underwritten and taught by manufacturers to promote their products, but an increasing number are being created by HHI staff members.

HHI's stores all face serious competition from national hardware chains such as **Home Depot** and **Lowe's**. These national chains have opened many new stores during the past few years, and they are larger, carry more items, and offer lower prices on some items. The competition is fierce; for example, all HHI stores have closed their lumber departments because of this competition. The national chains buy lumber in such large quantities that they can offer far lower prices. HHI was unable to earn a profit when matching the large competitors' prices, and the lumber operations consumed a large amount of store space. Hal is worried that this sort of problem could develop in other departments, so he is always looking for ways to add value to the HHI customer experience, especially ways that the national chains are not willing or able to do. For example, Hal believes that most people want to try out a new power tool in person before they spend hundreds of dollars on a purchase. Thus, every HHI store has a tool demonstration area that is always staffed with salespeople who are experts in power tool operation. For each major type of power tool (drills, power saws, joiners, grinding tools, and so on), HHI has created a small booklet of hints for using that type of tool. HHI gives these booklets to customers as free handouts. HHI also sells its own low-cost instructional videotapes and DVDs.

Hal is also concerned about competition from other sources as well. Some of the tool manufacturing companies are talking about selling directly to customers on their Web sites. None of HHI's major suppliers has done this yet, but Hal is worried that it could occur in the future. HHI also faces competition from companies such as **Outlet Tool Supply**, **Tool Crib**, **Southern Tool**, and **Tool Crib of the North**, which has formed an alliance with Amazon.com to appear on its Web site.

HHI buys most of its inventory directly from the manufacturers, but it does buy some items from distributors. Most items are shipped to one of HHI's three warehouses, but some items are shipped directly to the store locations. HHI has a new companywide inventory control system that was just installed last year at a cost of about $200,000. This information system monitors inventory in real time. When a new shipment arrives at an HHI store, it is entered into the system on the receiving dock. Each item is bar coded so it can be tracked as it moves from the receiving dock to the warehouse to the store shelf and, finally, out the door past a point-of-sale terminal (which Hal still calls a cash register). This inventory-tracking system is accessible through a Web browser and can be connected to a Web site, so HHI could sell inventory from its existing warehouses and stores through the Web. The cost for the software is $42,000, including installation and configuration.

Required:

1. Conduct a SWOT analysis for HHI's proposed electronic commerce Web site. You can use the information in the case narrative, your personal knowledge of the retail hardware industry, and information you obtain by following links in the Online Companion or doing independent searches of the Web as you conduct your analysis. You should create a diagram similar to Figure 1-12 to summarize your SWOT analysis results.

2. Based on your SWOT analysis, write a report of about 400 words that includes a summary of your assumptions and a list of recommendations for HHI. The recommendations should be specific and should address the content that HHI's Web site should include, the features that HHI should make available on the site, and how HHI might overcome any of the weaknesses or threats you identified in the SWOT analysis.

Note: Your instructor might assign you to a group to complete this case and might ask you to prepare a formal presentation of your results to your class.

For Further Study and Research

Agrawal, V., L. Arjona, and R. Lemmens. 2001. "E-Performance: The Path to Rational Exuberance," *The McKinsey Quarterly*, January, 31–43.

Al-Kibsi, G., K. de Boer, M. Mourshed, and N. Rea. 2001. "Putting Citizens On-Line, Not In-Line," *The McKinsey Quarterly*, April, 65–73.

Arthur, W. 2002. "Is the Information Revolution Dead? If History Is a Guide, It Is Not," *Business 2.0*, 3(3), March, 65–72.

Athitakis, M. 2003. "How to Make Money on the Net: The Second Internet Boom Is Quietly Taking Shape," *Business 2.0*, 4(4), May, 83–90.

Barker, P. 2002. "Swimming Lessons: Moving Your Business Online Requires More Than Just a Web Address and a Product to Sell," *Financial Post*, June 10, FP18.

Barlas, P. 2003. "Autobytel Survives Dot-com Crash, Looks to Grow," *Investor's Business Daily*, February 20.

Betts, M. 2001. "Report: Global E-Commerce Still Faces Big Challenges," *Computerworld*, May 3. (http://www.computerworld.com/cwi/story/0,1199,NAV47_STO60164.00.html)

Bingi, P., A. Mir, and J. Khamalah. 2000. "The Challenges Facing Global E-Commerce," *Information Systems Management*, 17(4), Fall, 26–34.

Bodeen, C. 2004. "China Shuts Down Internet Blogs," *Salon.com*, March 19. (http://www.salon.com/news/wire/2004/03/19/blogs2/index.html)

Brown, J., S. Durchslag, and J. Hagel. 2002. "Loosening Up: How Process Networks Unlock the Power of Specialization," *The McKinsey Quarterly*, Special Edition, 59–69.

Business Fleet Magazine. 2001. "Autobytel Generates More Vehicle Sales Online than Any Other Website for Third Year in a Row," November 29. (http://www.fleet-central.com/bf/bn_srch_c.cfm?rank=667)

Castells, M. 1996. *The Rise of the Network Society*. Cambridge, MA: Blackwell.

Chan, B. and S. Al-Hawamdeh. 2002. "The Development of E-Commerce in Singapore: The Impact of Government Initiatives," *Business Process Management Journal*, 8(3), 278-288.

Coase, R. 1937. "The Nature of the Firm," *Economica*, 4(4), November, 386–405.

Cohn, M. 2001. "China Seeks to Build the Great Firewall," *The Toronto Star*, July 21, A1.

Collett, S. 1999. "SWOT Analysis," *Computerworld*, 33(29), July 19, 58.

Computerworld. 2001. "Autopsy of a Dot Com," January 19. (http://www.computerworld.com/cwi/story/0,1199,NAV47_STO56616,00.html)

Cooper, L. 2004. "High Speed Access Rising Rapidly, FCC Says," *InternetWeek*, June 9. (http://www.internetweek.com/story/showArticle.jhtml?articleID=21600225)

DiLodovico, A., W. Lewis, V. Palmade, and S. Sankhe. 2001. "India—From Emerging to Surging," *The McKinsey Quarterly*, October, 28–65.

Drickhamer, D. 2003. "EDI Is Dead! Long Live EDI!" *Industry Week*, 252(4), April, 31–35.

Einhorn, B., A. Webb, and P. Engardio. 2000. "China's Tangled Web: Will Beijing Ruin the Net by Trying to Control It?" *Business Week*, July 17, 28–30.

Freeman, C. and F. Louçã, 2001. *As Time Goes By.* Oxford: Oxford University Press.

Friedman, M. 1999. "Photographer Fights Quebec Language Law," *Computing Canada*, 25(24), June 18, 1, 4.

Gantz, J. 2001. "Despite Crash, E-Commerce Will Still Flourish," *Computerworld*, 35(2), January 8, 31.

Glasner, J. 2001. "EToys Epitaph: 'End of an Error,'" *Wired News*, March 8. (http://www.wired.com/news/business/ 0,1367,42078,00.html)

Gold, J. 2004. "Amazon Countersues Toys R Us," *The Washington Post*, June 29, E5.

Goldstein, E. 1999. *The Internet in the Mideast and North Africa: Free Expression and Censorship.* Washington: Human Rights Watch.

Gosh, S. 1998. "Making Business Sense of the Internet," *Harvard Business Review*, 76(2), March–April, 126–135.

Hammer, M. and J. Champy. 1993. *Reengineering the Corporation: A Manifesto for Business Revolution.* New York: HarperBusiness.

Hansell, S. 2002. "Meg Whitman and eBay, Net Survivors," *The New York Times*, May 5, Section 3, 1.

Harrington, H., E. Esseling, and H. van Nimwegen. 1997. *Business Process Improvement Workbook: Documentation, Analysis, Design, and Management of Business Process Improvement.* New York: McGraw-Hill.

Harsany, J. 2004. "Web Grocer Hits Refresh: Online Grocer FreshDirect Takes the Hassle Out of City Shopping," *PC Magazine*, May 18, 76.

Hof, R. 2003. "Reprogramming Amazon," *Business Week*, December 22, 82.

Horrigan, J. and L. Rainie. 2002. *Getting Serious Online.* Washington: Pew Internet & American Life Project.

Jensen, M. 2002. *The African Internet: A Status Report.* Port St. Johns, South Africa: International Development Research Center. (http://www3.sn.apc.org/africa/afstat.htm)

Johnson, C. 2003. *U.S. E-Commerce: The Year in Review.* Cambridge, MA: Forrester Research.

Lapres, D. 2000. "Legal Do's and Don'ts of Web Use in China," *China Business Review*, 27(2), March–April, 26–28.

Le Seac'h, M. and A. Klotz. 1999. "Corporate Translating: Handle with Care," *Business and Economic Review*, 45(2), January–March, 12–14.

Leo, A. 2001. "The World Wide Translator," *Technology Review*, September 21. (http://www.techreview.com/web/leo/leo092101.asp)

Leon, M. 2003. "Online Sales Soared 48% in 2002," *CyberAtlas*, May 16. (http://cyberatlas.internet.com/markets/retailing/article/0,1323,6061_2208071,00.html)

Levaux, J. 2001. "Adapting Products and Services for Global E-Commerce: The Next Frontier is Beyond Localization," *World Trade*, 14(1), January, 52–54.

Lewis, S. 2002. "Online Lessons for Asia's SMEs," *Asian Business*, 38(1), January, 41.

Lico, N. 2001. "Can Cars Sell Via Web?" *Advertising Age*, 72(15), April 9, 20.

Mackey, C. 2003. "The Evolution of E-business, *Darwin*, May 1. (http://www.darwinmag.com/read/050103/ebiz.html)

Mearian, L. 2002. "Insurers Use IT to Fight Brokerage, Bank Rivals," *Computerworld*, 36(16), April 15, 12.

Morton, F., F. Zettelmeyer, and J. Risso. 2000. "Internet Car Retailing," *National Bureau of Economic Research Working Paper*, December. (http://faculty.haas.berkeley.edu/~florian/)

Murphy, C. 2003. "Five Internet Myths: An Interview with Jeff Bezos," *Information Week*, June 11. (http://www.informationweek.com/story/showArticle.jhtml?articleID=10300770)

Music Business International. 2001. "Losing the Golden Egg-Laying Goose," 11(6), December 1, 11.

Oakes, C. 2002. "Successful E-Commerce Means Going Back to the Basics," *International Herald Tribune*, June 24, 12.

Ouchi, M. 2004. "Dual Suits: Amazon.com, Toysrus.com cry 'Foul,'" *The Seattle Times*, July 11, E1.

Parker, P. 2002. "An Eye on the Multicultural Future," *ClickZ*, May 3. (http://www.clickz.com/feedback/uzz/article.php/1033911)

Perdue, L. 2001. "A Bright Future: After the Train Wreck," *Inc*, 23(4), March 15, 51–53.

Petzinger, T. 1999. *The New Pioneers: The Men and Women Who Are Transforming the Workplace and Marketplace*. New York: Simon & Schuster.

Porter, M. 1985. *Competitive Advantage*. New York: Free Press.

Porter, M. 1998. "Clusters and the New Economics of Competition," *Harvard Business Review*, 76(6), November–December, 77–90.

Porter, M. 2001. "Strategy and the Internet," *Harvard Business Review*, 79(3), March, 63–78.

Powell, W. 1990. "Neither Market nor Hierarchy: Network Forms of Organization," *Research in Organizational Behavior*, 12(3), 295–336.

Rainie, L. 2002. *Women Surpass Men as E-Shoppers During the Holidays: 2001 Sees More E-Commerce, and More Online Socializing*. Washington: Pew Internet & American Life Project.

Ramirez, C. 2001. "Disco Virtual Bills Four Times That of Offline Branch," *Business News Americas*, November 8. (http://www.bnamericas.com/story.xsql?id_noticia=78448&Tx_idioma=l&id_sector=1)

Rayport, J. and B. Jaworski. 2001. *E-Commerce*. New York: McGraw-Hill/Irwin.

Ring, R. and A. Van de Ven. 1992. "Structuring Cooperative Relationships Between Organizations," *Strategic Management Journal*, 13(4), 483–498.

Riquelme, H. 2002. "Commercial Internet Adoption in China: Comparing the Experience of Small, Medium, and Large Businesses," *Internet Research: Electronic Networking Applications and Policy*, 12(3), 276–286.

Rush, L. 2003. "U.S. E-commerce to See Significant Growth by 2008," *CyberAtlas*, August 7. (http://cyberatlas.internet.com/markets/retailing/article/0,1323,6061_2246041,00.html)

Schneider, G. 2005. "Digital Products on the Web: Pricing Issues and Revenue Models," 154–174. In Kehal, H. and V. Singh, eds., *Digital Economy: Impacts, Influences, and Challenges*. Hershey, PA: Idea Group.

Shannon, P. 2000. "Including Language In your Global Strategy for B2B E-Commerce," *World Trade*, 13(9), September, 66–68.

Shapiro, A. 1999. *The Control Revolution: How the Internet Is Putting Individuals in Charge and Changing the World We Know*. New York: The Century Foundation.

Shapiro, C. and H. Varian. 1999. *Information Rules: A Strategic Guide to the Network Economy*. Boston: Harvard Business School Press.

Shari, M. 2000. "Cutting Red Tape in Singapore," *Business Week*, September 18, 92.

Tapscott, D. 2001. "Rethinking Strategy in a Networked World: Or Why Michael Porter is Wrong About the Internet," *strategy+business*, 21(3), 1–8.

Taylor, D. and A. Terhune. 2001. *Doing E-Business: Strategies for Thriving in an Electronic Marketplace*. New York: John Wiley & Sons.

Tedeschi, B. 2004. "Broad Gains in Online Shopping," *The New York Times*, March 29, C4.

Totty, M. and A. Grimes. 2002. "If at First You Don't Succeed... Some Retailers Are Finding Success in Industries Long Thought Off-Limits to E-Commerce," *The Wall Street Journal*, February 11, R6.

U.S. Census Bureau. 2004. *Statistical Abstract of the United States*. Washington: U.S. Census Bureau.

The Wall Street Journal. 2001. "Autobytel Expects Profits in Second Half, Says CEO Lorimer," February 5, B15.

Wallraff, B. 2000. "What Global Language?" *The Atlantic Monthly*, 286(5), 52–66.

Weber, T. 2002. "Forget Dot-Com Bust: Net's Impact On the World Has Only Just Begun," *The Wall Street Journal Online*, May 13. (http://online.wsj.com/article/0,,SB1021238190524844800.djm,00.html)

Weernink, W. 2001. "Carmakers Target Online Middlemen," *Automotive News Europe*, April 9, 1–2.

Weintraub, A. 2001. "Make or Break for Autobytel," *Business Week*, July 9, EB30–EB32.

Williamson, O. 1975. *Markets and Hierarchies: Analysis and Antitrust Implications*. New York: Free Press.

Williamson, O. 1985. *The Economic Institutions of Capitalism*. New York: Free Press.

Willis, C. and S. Donahue. 1998. "Does Amazon.com Really Matter?" *Forbes*, 161(7), April 6, 55–58.

Wilson, T. 2001. "Spotty Infrastructure Impairs World View," *InternetWeek*, March 26, 1–3.

CHAPTER **2**

TECHNOLOGY INFRASTRUCTURE: THE INTERNET AND THE WORLD WIDE WEB

LEARNING OBJECTIVES

In this chapter, you will learn about:

- The origin, growth, and current structure of the Internet
- How packet-switched networks are combined to form the Internet
- How Internet protocols and Internet addressing work
- The history and use of markup languages on the Web, including SGML, HTML, and XML
- How HTML tags and links work on the World Wide Web
- The differences among internets, intranets, and extranets
- Options for connecting to the Internet, including cost and bandwidth factors
- Internet2 and the Semantic Web

INTRODUCTION

Many business executives made the statement "the Internet changes everything" during the late 1990s.

One of the first people to say those words publicly was John Chambers, CEO of **Cisco Systems**, in a

speech at a computer industry trade show in 1996. For his company, the Internet did indeed change everything. Cisco, founded in 1984, grew rapidly to become one of the largest and most profitable companies in the world by 2000.

Cisco designs, manufactures, and sells computer networking devices. In this chapter you will learn about these devices and how they make up the Internet. Cisco's earnings grew as telecommunications companies purchased the company's products to build the infrastructure of the Internet. Other companies also wanted to connect their business operations to the Internet; they became lucrative customers for Cisco, too. In its fiscal year ended July 2000, Cisco had sales of $19 billion and net income of $3 billion. Cisco was one of the true winners in the first wave of electronic commerce.

Because Cisco grew so rapidly during the first wave, it developed a strategy in which it built equipment before it received orders from customers. Cisco did not want to run short of equipment that customers were eager to get. In 2000, many of Cisco's most important telecommunications customers suddenly stopped expanding. Some of them even went out of business. Demand for Cisco's networking devices plummeted. The equipment that Cisco had built with the expectation of ever increasing orders sat in its warehouses, unwanted. In 2001, Cisco posted a loss of $1 billion on sales of $22 billion. Investors drove Cisco's stock, which had been trading at $80 per share, to $14 per share.

Chambers immediately undertook a series of belt-tightening steps. Between 2001 and 2004, Cisco wrote off $2 billion in inventory that it could not sell and laid off more than 10,000 of its 35,000 employees. The company also increased the efficiency of its operations by eliminating 20 percent of its 50 different product lines and reducing its number of suppliers by 60 percent. More than 75 percent of all corporate networks use Cisco equipment, and Chambers knew that market position would be important when business picked up again. All he needed to do was to keep the company operating through the difficult times the industry was facing.

Despite the sudden drop in demand for Cisco and other telecommunications companies' products,

the number of people using the Internet continued to increase rapidly. Businesses were finding new uses

for the Internet infrastructure that the telecommunications companies had built, especially now that

those financially troubled telecommunications companies were selling access to their infrastructure at

very low prices. In 2004, Cisco announced a new series of high-capacity networking equipment that

would handle the increased traffic on the Internet and within large companies. Four years after reporting

the $1 billion loss, Cisco was benefiting from the second wave of electronic commerce; its sales had

recovered and were growing again, although at a much slower rate. Cisco had weathered the storm and

was ready once again to make money by supplying technologies to a world in which the Internet had

changed everything.

THE INTERNET AND THE WORLD WIDE WEB

A **computer network** is any technology that allows people to connect computers to each
other. Computer networks and the Internet, which connects computer networks around the
world to each other, form the basic technology structure that underlies all electronic
commerce. This chapter introduces you to the hardware and software technologies that
make electronic commerce possible. First, you will learn how the Internet and the World
Wide Web work. Then, you will learn about other technologies that support the Internet, the
Web, and electronic commerce. In this chapter, you will be introduced to several complex networking technologies. If you are interested in learning more about how computer
networks operate, you can consult one of the computer networking books cited in the For
Further Study and Research section at the end of this chapter, or you can take courses in
data communications and networking.

Millions of people use the Internet every day, but only a small percentage of them really
understand how it works. The **Internet** is a large system of interconnected computer networks that spans the globe. Using the Internet, you can communicate with other people
throughout the world by means of electronic mail; read online versions of newspapers,
magazines, academic journals, and books; join discussion groups on almost any conceiv-
able topic; participate in games and simulations; and obtain free computer software.

In recent years, the Internet has allowed commercial enterprises to connect with one
another and with customers. Today, all kinds of businesses provide information about their
products and services on the Internet. Many of these businesses use the Internet to market and sell their products and services.

The part of the Internet known as the **World Wide Web**, or, more simply, the **Web**, is
a subset of the computers on the Internet that are connected to each other in a specific

way that makes them and their contents easily accessible to each other. The most important thing about the Web is that it includes an easy-to-use standard interface. This interface makes it possible for people who are not computer experts to use the Web to access a variety of Internet resources.

Origins of the Internet

In the early 1960s, the U.S. Department of Defense became concerned about the possible effects of nuclear attack on its computing facilities. The Defense Department realized that the weapons of the future would require powerful computers for coordination and control. The powerful computers of that time were all large mainframe computers, so the Defense Department began examining ways to connect these computers to each other and also connect them to weapons installations distributed all over the world. The Defense Department agency charged with this task hired many of the best communications technology researchers and, for many years, funded research at leading universities and institutes to explore the task of creating a worldwide network that could remain operational, even if parts of the network were destroyed by enemy military action or sabotage. These researchers worked to devise ways to build networks that could operate independently—that is, networks that did not require a central computer to control network operations.

Early computer networks used leased telephone company lines for their connections. Telephone company systems of that time established a single connection between sender and receiver for each telephone call, and that connection carried all data along a single path. When a company wanted to connect computers it owned at two different locations, the company placed a telephone call to establish the connection, and then connected one computer to each end of that single connection.

The Defense Department was concerned about the inherent risk of this single-channel method for connecting computers, and its researchers developed a different method of sending information through multiple channels. In this method, files and messages are broken into packets that are labeled electronically with codes for their origins, sequences, and destinations. You will learn more about how packet networks operate later in this chapter.

In 1969, Defense Department researchers in the Advanced Research Projects Agency (ARPA) used this network model to connect four computers—one each at the University of California at Los Angeles, SRI International, the University of California at Santa Barbara, and the University of Utah—into a network called the ARPANET. The ARPANET was the earliest of the networks that eventually combined to become what we now call the Internet. Throughout the 1970s and 1980s, many researchers in the academic community connected to the ARPANET and contributed to the technological developments that increased its speed and efficiency. At the same time, researchers at other universities were creating their own networks using similar technologies.

New Uses for the Internet

Although the goals of the Defense Department network were still to control weapons systems and transfer research files, other uses for this vast network began to appear in the early 1970s. E-mail was born in 1972 when a researcher wrote a program that could send and receive messages over the network. This new method of communicating became widely used very quickly. The number of network users in the military and education research communities continued to grow. Many of these new participants used the networking technology to transfer files and access computers remotely. You will learn about these file transfer tools in Chapter 8.

The first e-mail mailing lists also appeared on these networks. A **mailing list** is an e-mail address that forwards any message it receives to any user who has subscribed to the list. In 1979, a group of students and programmers at Duke University and the University of North Carolina started **Usenet**, an abbreviation for **User's News Network**. Usenet allows anyone who connects to the network to read and post articles on a variety of subjects. Usenet survives on the Internet today, with over 1000 different topic areas that are called **newsgroups**. Other researchers even created game-playing software for use on these interconnected networks.

Although the people using these networks were developing many creative applications, use of the networks was limited to those members of the research and academic communities who could access them. Between 1979 and 1989, these network applications were improved and tested by an increasing number of users. The Defense Department's networking software became more widely used in academic and research institutions as these organizations recognized the benefits of having a common communications network. As the number of people in different organizations using these networks increased, security problems were recognized. These problems have continued to become more important, and you will learn more about network security issues in Chapter 10. The explosion of personal computer use during the 1980s also helped more people become comfortable with computers. In the late 1980s, these independent academic and research networks merged into what we now call the Internet.

Commercial Use of the Internet

As personal computers became more powerful, affordable, and available during the 1980s, companies increasingly used them to construct their own internal networks. Although these networks included e-mail software that employees could use to send messages to each other, businesses wanted their employees to be able to communicate with people outside their corporate networks. The Defense Department network and most of the academic networks that had teamed up with it were receiving funding from the **National Science Foundation (NSF)**. The NSF prohibited commercial network traffic on its networks, so businesses turned to commercial e-mail service providers to handle their e-mail needs. Larger firms built their own networks that used leased telephone lines to connect field offices to corporate headquarters.

In 1989, the NSF permitted two commercial e-mail services, MCI Mail and CompuServe, to establish limited connections to the Internet for the sole purpose of exchanging e-mail transmissions with users of the Internet. These connections allowed commercial enterprises to send e-mail directly to Internet addresses, and allowed members of the research and education communities on the Internet to send e-mail directly to MCI Mail and CompuServe addresses. The NSF justified this limited commercial use of the Internet as a service that would primarily benefit the Internet's noncommercial users. As the 1990s began, people from all walks of life—not just scientists or academic researchers—started thinking of these networks as the global resource that we now know as the Internet. Although this network of networks had grown from four Defense Department computers in 1969 to more than 300,000 computers on many interconnected networks by 1990, the greatest growth of the Internet was yet to come.

Growth of the Internet

In 1991, the NSF further eased its restrictions on commercial Internet activity and began implementing plans to privatize the Internet. The privatization of the Internet was substantially completed in 1995, when the NSF turned over the operation of the main Internet connections to a group of privately owned companies. The new structure of the Internet was based on four **network access points (NAPs)** located in San Francisco, New York, Chicago, and Washington, D.C., each operated by a separate telecommunications company. As the Internet grew, more companies opened more NAPs in more locations. These companies, known as **network access providers**, sell Internet access rights directly to larger customers and indirectly to smaller firms and individuals through other companies, called **Internet service providers (ISPs)**.

The Internet was a phenomenon that had truly sneaked up on an unsuspecting world. The researchers who had been so involved in the creation and growth of the Internet just accepted it as part of their working environment. However, people outside the research community were largely unaware of the potential offered by a large interconnected set of computer networks. Figure 2-1 shows the growth in the number of **Internet hosts**, which are computers directly connected to the Internet. Except for a slight slowdown during 2001–2002, the Internet's growth has been consistently dramatic.

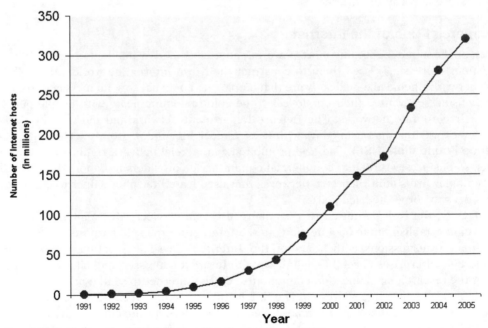

Source: Internet Software Consortium (http://www.isc.org/) and author's estimates

FIGURE 2-1   Growth of the Internet

In 30 years, the Internet has grown to become one of the most amazing technological and social accomplishments of the last century. Millions of people, most of whom are not computer researchers or experts, now use this complex, interconnected network of computers. These computers run thousands of different software packages. The computers are located in almost every country of the world. Every year, billions of dollars change hands over the Internet in exchange for all kinds of products and services. All of this activity occurs with no central coordination point or control, which is especially ironic given that the Internet began as a way for the military to maintain control while under attack.

The opening of the Internet to business activity helped dramatically increase its growth; however, there was another development that worked hand in hand with the commercialization of the Internet to spur its growth. That development was the World Wide Web.

Emergence of the World Wide Web

The Web is software that runs on computers that are connected to the Internet. The network traffic generated by Web software is the largest single category of traffic on the Internet today, outpacing e-mail, file transfers, and other data transmission traffic. But the Web is more a way of thinking about and organizing information storage and retrieval than it is a specific technology. As such, its history goes back many years. Two important innovations that became key elements of the Web are hypertext and graphical user interfaces.

The Development of Hypertext

In 1945, **Vannevar Bush**, who was director of the U.S. Office of Scientific Research and Development, wrote an article in *The Atlantic Monthly* about ways that scientists could apply the skills they learned during World War II to peacetime activities. The article included a number of visionary ideas about future uses of technology to organize and facilitate efficient access to information. Bush speculated that engineers would eventually build a machine that he called the Memex, a memory extension device that would store all of a person's books, records, letters, and research results on microfilm. Bush's Memex would include mechanical aids, such as microfilm readers and indexes, that would help users quickly and flexibly consult their collected knowledge.

In the 1960s, Ted Nelson described a similar system in which text on one page links to text on other pages. Nelson called his page-linking system **hypertext**. Douglas Engelbart, who also invented the computer mouse, created the first experimental hypertext system on one of the large computers of the 1960s. In 1987, Nelson published *Literary Machines*, a book in which he outlined project Xanadu, a global system for online hypertext publishing and commerce. Nelson used the term *hypertext* to describe a page-linking system that would interconnect related pages of information, regardless of where in the world they were stored.

In 1989, Tim Berners-Lee was trying to improve the laboratory research document-handling procedures for his employer, CERN: European Laboratory for Particle Physics. CERN had been connected to the Internet for two years, but its scientists wanted to find better ways to circulate their scientific papers and data among the high-energy physics research community throughout the world. Berners-Lee proposed a hypertext development project intended to provide this data-sharing functionality.

Over the next two years, Berners-Lee developed the code for a hypertext server program and made it available on the Internet. A **hypertext server** is a computer that stores files written in the hypertext markup language and lets other computers connect to it and read these files. Hypertext servers used on the Web today are usually called **Web servers**. **Hypertext Markup Language (HTML)**, which Berners-Lee developed from his original hypertext server program, is a language that includes a set of codes (or tags) attached to text. These codes describe the relationships among text elements. For example, HTML includes tags that indicate which text is part of a header element, which text is part of a paragraph element, and which text is part of a numbered list element. One important type of tag is the hypertext link tag. A **hypertext link**, or **hyperlink**, points to another location in the same or another HTML document.

Graphical Interfaces for Hypertext

Several different types of software are available to read HTML documents, but most people use a Web browser such as Netscape Navigator or Microsoft Internet Explorer. A **Web browser** is a software interface that lets users read (or browse) HTML documents and move from one HTML document to another through text formatted with hypertext link tags in each file. If the HTML documents are on computers connected to the Internet, you can use a Web browser to move from an HTML document on one computer to an HTML document on any other computer on the Internet.

An HTML document differs from a word-processing document in that it does not specify how a particular text element will appear. For example, you might use word-processing software to create a document heading by setting the heading text font to Arial, its font size to 14 points, and its position to centered. The document displays and prints these exact settings whenever you open the document in that word processor. In contrast, an HTML document simply includes a heading tag with the heading text. Many different browser programs can read an HTML document. Each program recognizes the heading tag and displays the text in whatever manner each program normally displays headings. Different Web browser programs might each display the text differently, but all of them display the text with the characteristics of a heading.

A Web browser presents an HTML document in an easy-to-read format in the browser's graphical user interface. A **graphical user interface (GUI)** is a way of presenting program control functions and program output to users. It uses pictures, icons, and other graphical elements instead of displaying just text. Almost all personal computers today use a GUI such as Microsoft Windows or the Macintosh user interface.

The World Wide Web

Berners-Lee called his system of hyperlinked HTML documents the World Wide Web. The Web caught on quickly in the scientific research community, but few people outside that community had software that could read the HTML documents. In 1993, a group of students led by Marc Andreessen at the University of Illinois wrote Mosaic, the first GUI program that could read HTML and use HTML hyperlinks to navigate from page to page on computers anywhere on the Internet. Mosaic was the first Web browser that became widely available for personal computers, and some Web surfers still use it today.

Programmers quickly realized that a functional system of pages connected by hypertext links would provide many new Internet users with an easy way to access information on the Internet. Businesses recognized the profit-making potential offered by a worldwide network of easy-to-use computers. In 1994, Andreessen and other members of the University of Illinois Mosaic team joined with James Clark of **Silicon Graphics** to found Netscape Communications (which is now owned by **Time Warner**). Its first product, the Netscape Navigator Web browser program based on Mosaic, was an instant success. Netscape became one of the fastest growing software companies ever. **Microsoft** created its Internet Explorer Web browser and entered the market soon after Netscape's success became apparent. A number of other Web browsers exist, but Internet Explorer dominates the market today.

The number of Web sites has grown even more rapidly than the Internet itself. The number of Web sites is currently estimated at about 60 million, and individual Web pages number in the billions because each Web site might include hundreds or even thousands of individual Web pages. Therefore, nobody really knows how many Web pages exist. For example, researchers at **BrightPlanet** estimate that the number of Web sites could be more than 500 million. Figure 2-2 shows how the growth rate of the Web increased dramatically between 1997 and 2000. After a brief consolidation period during 2001–2002, the Web is once again showing rapid growth.

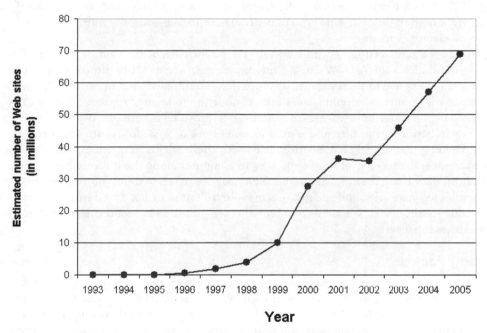

Adapted from Netcraft Computer Surveys (http://www.netcraft.com) and author's estimates

FIGURE 2-2   Growth of the World Wide Web

As more people gain access to the Web, commercial interest in using the Web to conduct business will continue to increase, and the variety of nonbusiness uses will become even greater. In the rest of this chapter, you will learn how Internet and Web technologies work to enable electronic commerce.

PACKET-SWITCHED NETWORKS

A network of computers that are located close together—for example, in the same building—is called a **local area network**, or a **LAN**. Networks of computers that are connected over greater distances are called **wide area networks**, or **WANs**.

The early models (dating back to the 1950s) for WANs were the circuits of the local and long-distance telephone companies of the time, because the first early WANs used leased telephone company lines for their connections. A telephone call establishes a single connection path between the caller and receiver. Once that connection is established, data travels along that single path. Telephone company equipment (originally mechanical, now electronic) selects specific telephone lines to connect to each other by closing switches. These switches work like the switches you use to turn lights on and off in your home, except that they open and close much faster, and are controlled by mechanical or electronic devices instead of human hands.

The combination of telephone lines and the closed switches that connect them to each other is called a **circuit**. This circuit forms a single electrical path between caller and receiver. This single path of connected circuits switched into each other is maintained for the entire length of the call. This type of centrally controlled, single-connection model is known as **circuit switching**.

Although circuit switching works well for telephone calls, it does not work as well for sending data across a large WAN or an interconnected network like the Internet. The Internet was designed to be resistant to failure. In a circuit-switched network, a failure in any one of the connected circuits causes the connection to be interrupted and data to be lost. Instead, the Internet uses packet switching to move data between two points. On a **packet-switched** network, files and e-mail messages are broken down into small pieces, called **packets**, that are labeled electronically with their origins, sequences, and destination addresses. Packets travel from computer to computer along the interconnected networks until they reach their destinations. Each packet can take a different path through the interconnected networks, and the packets may arrive out of order. The destination computer collects the packets and reassembles the original file or e-mail message from the pieces in each packet.

Routing Packets

As an individual packet travels from one network to another, the computers through which the packet travels determine the best route for getting the packet to its destination. The computers that decide how best to forward each packet are called **routing computers**, **router computers**, **routers**, **gateway computers** (because they act as the gateway from a LAN or WAN to the Internet), or **border routers** (because they are located at the border between the organization and the Internet). The programs on router computers that determine the best path on which to send each packet contain rules called **routing algorithms**. The programs apply their routing algorithms to information they have stored in **routing tables** or **configuration tables**. This information includes lists of connections that lead to particular groups of other routers, rules that specify which connections to use first, and rules for handling instances of heavy packet traffic and network congestion.

Individual LANs and WANs can use a variety of different rules and standards for creating packets within their networks. The network devices that move packets from one part of a network to another are called hubs, switches, and bridges. Routers are used to connect networks to other networks. As technologies have improved, many of the distinctions between these different types of network devices have become blurred. You can take a data communications and networking class to learn more about these network devices and how they work.

When packets leave a network to travel on the Internet, they must be translated into a standard format. Routers usually perform this translation function. As you can see, routers are an important part of the infrastructure of the Internet. When a company or organization becomes part of the Internet, it must connect at least one router to the other routers (owned by other companies or organizations) that make up the Internet. Figure 2-3 is a diagram of a small portion of the Internet that shows its router-based architecture. The figure shows only the routers that connect each organization's WANs and LANs to the Internet, not the other routers that are inside the WANs and LANs or that connect them to each other within the organization.

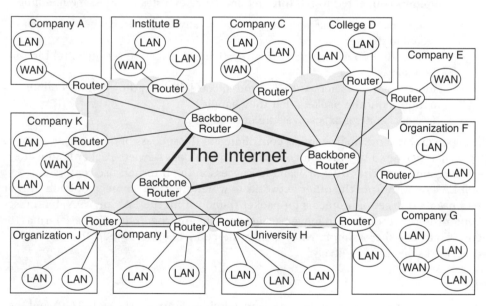

FIGURE 2-3 Router-based architecture of the Internet

The Internet also has routers that handle packet traffic along the Internet's main connecting points. These routers and the telecommunications lines connecting them are collectively referred to as the **Internet backbone**. These routers, sometimes called **backbone routers**, are very large computers that can each handle more than 50 million packets per second! You can see in Figure 2-3 that a router connected to the Internet always has more than one path to which it can direct a packet. By building in multiple packet paths, the designers of the Internet created a degree of redundancy in the system that allows it to keep moving packets, even if one or more of the routers or connecting lines fails.

INTERNET PROTOCOLS

As you learned earlier in this chapter, the first packet-switched network, the ARPANET, connected only a few universities and research centers. This experimental network grew over the next few years and used the **Network Control Protocol (NCP)**. A **protocol** is a collection of rules for formatting, ordering, and error-checking data sent across a network. For example, protocols determine how the sending device indicates that it has finished sending a message, and how the receiving device indicates that it has received (or not received) the message. A protocol also includes rules about what is allowed in a transmission and how it is formatted. Computers that communicate with each other must use the same protocol for data transmission. In the early days of computing, each computer manufacturer created its own protocol, so computers made by different manufacturers could not be connected to each other. This practice was called **proprietary architecture** or **closed architecture**. The **open architecture** philosophy developed for the evolving ARPANET, which later became the core of the Internet, included the use of a common protocol for all computers connected to the Internet and four key rules for message handling:

- Independent networks should not require any internal changes to be connected to the network.
- Packets that do not arrive at their destinations must be retransmitted from their source network.
- Router computers act as receive-and-forward devices; they do not retain information about the packets that they handle.
- No global control exists over the network.

The open architecture approach has contributed to the success of the Internet because computers manufactured by different companies (Apple, Dell, Hewlett-Packard, Sun, etc.) can be interconnected. The ARPANET and its successor, the Internet, use routers to isolate each LAN or WAN from the other networks to which they are connected. Each LAN or WAN can use its own set of protocols for packet traffic within the LAN or WAN, but must use a router (or similar device) to move packets onto the Internet in its standard format (or protocol). Following these simple rules makes the connections between the interconnected networks operate effectively.

TCP/IP

The Internet uses two main protocols: the **Transmission Control Protocol (TCP)** and the **Internet Protocol (IP)**. Developed by Internet pioneers Vincent Cerf and Robert Kahn, these protocols are the rules that govern how data moves through the Internet and how network connections are established and terminated. The acronym **TCP/IP** is commonly used to refer to the two protocols.

The TCP controls the disassembly of a message or a file into packets before it is transmitted over the Internet, and it controls the reassembly of those packets into their original formats when they reach their destinations. The IP specifies the addressing details for each packet, labeling each with the packet's origination and destination addresses. Soon after the new TCP/IP protocol set was developed, it replaced the NCP that ARPANET originally used.

In addition to its Internet function, TCP/IP is used today in many LANs. The TCP/IP protocol is provided in most personal computer operating systems commonly used today, including Linux, Macintosh, Microsoft Windows, and UNIX.

IP Addressing

The version of IP that has been in use for the past 20 years on the Internet is **Internet Protocol version 4**, abbreviated **IPv4**. It uses a 32-bit number to identify the computers connected to the Internet. This address is called an **IP address**. Computers do all of their internal calculations using a **base 2** (or **binary**) number system in which each digit is either a 0 or a 1, corresponding to a condition of either off or on. IPv4 uses a 32-bit binary number that allows over 4 billion different addresses ($2^{32} = 4,294,967,296$).

When a router breaks a message into packets before sending it onto the Internet, the router marks each packet with both the source IP address and the destination IP address of the message. To make them easier to read, IP numbers (addresses) appear as four numbers separated by periods. This notation system is called **dotted decimal** notation. An IPv4 address is a 32-bit number, so each of the four numbers is an 8-bit number (4 x 8 = 32). In most computer applications, an 8-bit number is called a **byte**; however, in networking applications, an 8-bit number is often called an **octet**. In binary, an octet can have values from 00000000 to 11111111; the decimal equivalents of these binary numbers are 0 and 255, respectively.

Because each of the four parts of a dotted decimal number can range from 0 to 255, IP addresses range from 0.0.0.0 (written in binary as 32 zeros) to 255.255.255.255 (written in binary as 32 ones). Although some people find dotted decimal notation to be confusing at first, most do agree that writing, reading, and remembering a computer's address as 216.115.108.245 is easier than 11011000011100110110110011110101, or its full decimal equivalent, which is 3,631,433,189.

Today, IP addresses are assigned by three not-for-profit organizations: the **American Registry for Internet Numbers (ARIN)**, the **Réseaux IP Européens (RIPE)**, and the **Asia-Pacific Network Information Center (APNIC)**. These registries assign and manage IP addresses for various parts of the world: ARIN for North America, South America, the Caribbean, and sub-Saharan Africa; RIPE for Europe, the Middle East, and the rest of Africa; and APNIC for countries in the Asia-Pacific area. These organizations took over IP address management tasks from the Internet Assigned Numbers Authority (IANA), which performed them under contract with the U.S. government when the Internet was an experimental research project.

You can use the **ARIN Whois** page at the ARIN Web site to search the IP addresses owned by organizations in North America. You can enter an organization name into the search box on the page, then click the Submit Query button, and the Whois server returns a list of the IP addresses owned by that organization. For example, performing a search on the word *Carnegie* displays the IP address blocks owned by Carnegie Bank, Carnegie Mellon University, and a number of other organizations whose names begin with Carnegie. You can also enter an IP address and find out who owns that IP address. If you enter "3.0.0.0" (without the quotation marks), you will find out that General Electric owns the entire block of IP addresses from 3.0.0.0 to 3.255.255.255. General Electric can use these addresses, which number approximately 16.7 million, for its own computers or it can lease them to other companies or individuals to whom it provides Internet access services.

In the early days of the Internet, the 4 billion addresses provided by the IPv4 rules certainly seemed to be more addresses than an experimental research network would ever need. However, about 2 billion of those addresses today are either in use or unavailable for use because of the way blocks of addresses were assigned to organizations. The new kinds of devices on the Internet's many networks, such as wireless personal digital assistants and cell phones that can access the Web, promise to keep demand high for IP addresses.

Network engineers have devised a number of stopgap techniques to stretch the supply of IP addresses. One of the most popular techniques is **subnetting**, which is the use of reserved private IP addresses within LANs and WANs to provide additional address space. **Private IP addresses** are a series of IP numbers that are not permitted on packets that travel on the Internet. In subnetting, a computer called a **Network Address Translation (NAT) device** converts those private IP addresses into normal IP addresses when it forwards packets from those computers to the Internet.

The **Internet Engineering Task Force (IETF)** worked on several new protocols that could solve the limited addressing capacity of IPv4, and in 1997, approved **Internet Protocol version 6 (IPv6)** as the protocol that will replace IPv4. The new IP is being implemented gradually because the two protocols are not directly compatible. The process of switching over to IPv6 will take at least another 10 years; however, network engineers have devised ways to run both protocols together on interconnected networks. The major advantage of IPv6 is that it uses a 128-bit number for addresses instead of the 32-bit number used in IPv4. The number of available addresses in IPv6 (2^{128}) is 34 followed by 37 zeros—billions of times larger than the address space of IPv4. The new IP also changes the format of the packet itself. Improvements in networking technologies over the past 20 years have made many of the fields in the IPv4 packet unnecessary. IPv6 eliminates those fields and adds fields for security and other optional information.

IPv6 has a shorthand notation system for expressing addresses, similar to the IPv4 dotted decimal notation system. However, because the IPv6 address space is much larger, its notation system is more complex. The IPv6 notation uses eight groups of 16 bits ($8 \times 16 = 128$). Each group is expressed as four hexadecimal digits and the groups are separated by colons; thus, the notation system is called **colon hexadecimal** or **colon hex**. A **hexadecimal (base 16)** numbering system uses 16 digits (0, 1, 2, 3, 4, 5, 6, 7, 8, 9, a, b, c, d, e, and f). An example of an IPv6 address expressed in this notation is: CD18:0000:0000: AF23:0000:FF9E:61B2:884D. To save space, the zeros can be omitted, which reduces this address to: CD18:::AF23::FF9E:61B2:884D.

Domain Names

The founders of the Internet were concerned that users might find the dotted decimal notation difficult to remember. To make the numbering system easier to use, they created an alternative addressing method that uses words. In this system, an address such as www. thomson.com is called a domain name. **Domain names** are sets of words that are assigned to specific IP addresses. Domain names can contain two or more word groups separated by periods. The rightmost part of a domain name is the most general. Each part of the domain name becomes more specific as you move to the left.

For example, the domain name www.business.sandiego.edu contains four parts separated by periods. Beginning at the right, the name "edu" indicates that the computer belongs to a four-year educational institution. The institution, University of San Diego, is identified by the name "sandiego", and "business" is the name of a specific computer or group of computers at the university. Finally, the "www" indicates that the computer is running software that makes it a part of the World Wide Web. Most, but not all, Web addresses follow this "www" naming convention. For an example of an exception, the group of computers that operate the **Yahoo! Games** service is named games.yahoo.com.

The rightmost part of a domain name is called a **top-level domain** (or **TLD**). For many years, these domains have included a group of general domains—such as .edu, .com, and .org—and a set of country domains. Since 1998, the **Internet Corporation for Assigned Names and Numbers (ICANN)** has had responsibility for managing domain names and coordinating them with the IP address registrars. ICANN is also responsible for setting standards for the router computers that make up the Internet. In 2000, ICANN added seven new TLDs to the general domain category. Although these new domain names were chosen after much deliberation and consideration of more than 100 possible new names, many people were highly critical of the selections (see, for example, the **ICANNWatch** Web site). In 2002, ICANN came under additional fire for acting in ways that many people thought violated the democratic principles on which the organization was founded. Figure 2-4 presents a list of the general TLDs, including the 2000 additions, and some of the more frequently used country TLDs.

Original general TLDs		Country TLDs		New (2000) general TLDs	
TLD	**Use**	**TLD**	**Country**	**TLD**	**Use**
.com	Commercial	.au	Australia	.aero	Air-transport industry
.edu	Four-year	.ca	Canada	.biz	Businesses
	educational institution	.de	Germany	.coop	Cooperatives
.gov	U.S. Federal government	.fi	Finland	.info	General use
.mil	U.S. Military	.fr	France	.museum	Museums
.net	General use	.jp	Japan	.name	Individual people
.org	Not-for-profit	.se	Sweden	.pro	Professionals
	organization	.uk	United Kingdom		(accountants, lawyers, physicians)

FIGURE 2-4 Top-level domain names

Web Page Request and Delivery Protocols

The Web is software that runs on computers that are connected to each other through the Internet. **Web client computers** run software called **Web client software** or **Web browser software**. Web client software sends requests for Web page files to other computers, which are called Web servers. A **Web server** computer runs software called **Web server software**. Web server software receives requests from many different Web clients and responds by sending files back to those Web client computers. Each Web client computer's Web client software renders those files into a Web page. Thus, the purpose of a Web server is to respond to requests for Web pages from Web clients. This combination of client computers running Web client software and server computers running Web server software is called a **client/server architecture**.

The set of rules for delivering Web page files over the Internet is in a protocol called the **Hypertext Transfer Protocol (HTTP)**, which was developed by Tim Berners-Lee in 1991. When a user types a domain name (for example, www.yahoo.com) into a Web browser's address bar, the browser sends an HTTP-formatted message to a Web server computer at Yahoo! that stores Web page files. The Web server computer at Yahoo! then responds by sending a set of files (one for the Web page and one for each graphic object, sound, or video clip included on the page) back to the client computer. These files are sent within a message that is HTTP formatted.

To initiate a Web page request using a Web browser, the user types the name of the protocol, followed by the characters "//:" before the domain name. Thus, a user would type http://www.yahoo.com to go to the Yahoo! Web site. Most Web browsers today automatically insert the http:// if the user does not include it. The combination of the protocol name and the domain name is called a **Uniform Resource Locator (URL)** because it lets the user locate a resource (the Web page) on another computer (the Web server).

Electronic Mail Protocols

Electronic mail, or **e-mail**, that is sent across the Internet must also be formatted according to a common set of rules. Most organizations use a client/server structure to handle e-mail. The organization has a computer called an **e-mail server** that is devoted to handling e-mail. The software on that computer stores and forwards e-mail messages. People in the organization might use a variety of programs, called **e-mail client software**, to read and send e-mail. These programs include **Microsoft Outlook**, **Mozilla Thunderbird**, **Netscape Messenger**, **Pegasus Mail**, **Qualcomm Eudora**, and many others. The e-mail client software communicates with the e-mail server software on the e-mail server computer to send and receive e-mail messages.

Many people also use e-mail on their computers at home. In most cases, the e-mail servers that handle their messages are operated by the companies that provide their connections to the Internet. An increasing number of people use e-mail services that are offered by Web sites such as Yahoo! Mail or Hotmail. In these cases, the e-mail servers and the e-mail clients are operated by the owners of the Web sites. The individual users only see the e-mail client software (and not the e-mail server software) in their Web browsers when they log on to the Web mail service.

With so many different e-mail client and server software choices, standardization and rules are very important. If e-mail messages did not follow standard rules, an e-mail message created by a person using one e-mail client program could not be read by a person using a different e-mail client program. As you have already learned in this chapter, rules for computer data transmission are called protocols.

SMTP and POP are two common protocols used for sending and retrieving e-mail. **Simple Mail Transfer Protocol (SMTP)** specifies the format of a mail message and describes how mail is to be administered on the e-mail server and transmitted on the Internet. An e-mail client program running on a user's computer can request mail from the organization's e-mail server using the **Post Office Protocol (POP)**. A POP message can tell the e-mail server to send mail to the user's computer and delete it from the e-mail server; send mail to the user's computer and not delete it; or simply ask whether new mail has arrived. The POP protocol provides support for **Multipurpose Internet Mail Extensions (MIME)**, which is a set of rules for handling binary files, such as word-processing documents, spreadsheets, photos, or sound clips, that are attached to e-mail messages.

IMAP, the **Interactive Mail Access Protocol**, is a newer e-mail protocol that performs the same basic functions as POP, but includes additional features. For example, IMAP can instruct the e-mail server to send only selected e-mail messages to the client instead of all messages. IMAP also allows the user to view only the header and the e-mail sender's name before deciding to download the entire message. POP allows users to search for and manipulate only those e-mail messages that they have downloaded to their computers. IMAP lets users create and manipulate mail folders (also called mailboxes), delete messages, and search for certain parts of a message while the e-mail is still on the e-mail server; that is, the user does not need to download the e-mail before working with it. The tools that IMAP provides are important to the increasing number of people who access their e-mail from different computers at different times. IMAP lets users manipulate and store their e-mail on the e-mail server and access it from any number of computers. POP allows users to access new e-mail messages from only one PC after they download the old messages to another PC. The main drawback to IMAP is that users' e-mail messages are stored on the e-mail server. As the number of users increases, the size of the e-mail server's disk drives must also increase. In general, server computers use faster (and thus, more expensive) disk drives than desktop computers. Therefore, it is more expensive to provide disk storage space for large quantities of e-mail on a server computer than to provide that same disk space on users' desktop computers. You can learn more about IMAP at the University of Washington's **IMAP Connection** Web site. You will learn more about specific electronic commerce uses of e-mail in Chapters 4 and 8.

MARKUP LANGUAGES AND THE WEB

Web pages can include many elements, such as graphics, photographs, sound clips, and even small programs that run in the Web browser. Each of these elements is stored on the Web server as a separate file. The most important parts of a Web page, however, are the structure of the page and the text that makes up the main part of the page. The page structure and text are stored in a text file that is formatted, or marked up, using a text markup language. A **text markup language** specifies a set of tags that are inserted into the text. These **markup tags**, also called **tags**, provide formatting instructions that Web client software can understand. The Web client software uses those instructions as it renders the text and page elements contained in the other files into the Web page that appears on the screen of the client computer.

The markup language most commonly used on the Web is HTML, which is a subset of a much older and far more complex text markup language called **Standard Generalized Markup Language (SGML)**. Figure 2-5 shows how HTML, XML, and XHTML have descended from the original SGML specification. SGML was used for many years by the publishing industry to create documents that needed to be printed in various formats and that were revised frequently. In addition to its role as a markup language, SGML is a **meta language**, which is a language that can be used to define other languages. Another markup language that was derived from SGML for use on the Web is **Extensible Markup Language (XML)**, which is increasingly used to mark up information that companies share with each other over the Internet. XML is also a meta language because users can create their own markup elements that extend the usefulness of XML (which is why it is called an "extensible" language).

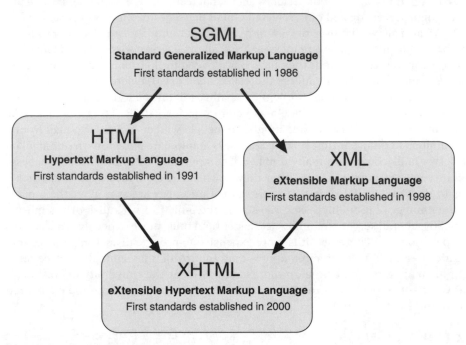

FIGURE 2-5 Development of markup languages

The **World Wide Web Consortium (W3C)**, a not-for-profit group that maintains standards for the Web, presented its first draft form of XML in 1996; the W3C issued its first formal version recommendation in 1998. Thus, it is a much newer markup language than HTML. In 2000, the W3C released the first version of a recommendation for a new markup language called **Extensible Hypertext Markup Language (XHTML)**, which is a reformulation of HTML version 4.0 as an XML application. The Online Companion includes a link to the **W3C XHTML Version 1.0 Specification**.

Standard Generalized Markup Language

Since the 1960s, publishers have used markup languages to create documents that can be formatted once, stored electronically, and then printed many times in various layouts that each interpret the formatting differently. U.S. Department of Defense contractors also used early markup languages to create manuals and parts lists for weapons systems. These documents contained many information elements that were often reprinted in different versions and formats. Using electronic document storage and programs that could interpret the formats to produce different layouts saved a tremendous amount of retyping time and cost.

A **Generalized Markup Language (GML)** emerged from these early efforts to create standard formatting styles for electronic documents. In 1986, after many elements of the standard had been in use for years, the **International Organization for Standardization (ISO)** adopted a version of GML called Standard Generalized Markup Language (SGML). SGML offers a system of marking up documents that is independent of any software application. Many organizations, such as the Association of American Publishers, Hewlett-Packard, and Kodak, use SGML because they have complex document management requirements.

SGML is nonproprietary and platform independent and offers user-defined tags. However, it is not well suited to certain tasks, such as the rapid development of Web pages. SGML is costly to set up and maintain, requires the use of expensive software tools, and is hard to learn. Creating document-type definitions in SGML can be expensive and time consuming.

Hypertext Markup Language (HTML)

HTML includes tags that define the format and style of text elements in an electronic document. HTML also has tags that can create relationships among text elements within one document or among several documents. The text elements that are related to each other are called **hypertext elements**.

HTML is easier to learn and use than SGML. HTML is the prevalent markup language used to create documents on the Web today. The early versions of HTML let Web page designers create text-based electronic documents with headings, title bar titles, bullets, lines, and ordered lists. As the use of HTML and the Web itself grew, HTML creator Berners-Lee turned over the job of maintaining versions of HTML to the W3C. Later versions of HTML included tags for tables, frames, and other features that helped Web designers create more complex page layouts. The W3C maintains detailed information about HTML versions and related topics on its **W3C HTML Page**.

The process for approval of new HTML features takes a long time, so Web browser software developers created some features, called **HTML extensions**, that would only work in their browsers. At various times during the history of HTML, both Microsoft and Netscape enabled their Web browsers to use these HTML extension tags before those tags were approved by the W3C. In some cases, these tags were enabled in one browser and not the other. In other cases, the tags used were never approved by the W3C or were approved in a different form than the one implemented in the Web browser software. Web page designers who wanted to use the latest available tags were often frustrated by this state of affairs. Many of these Web designers had to create separate sets of Web pages for the different types of browsers, which was inefficient and expensive. Most of these tag difference issues were resolved when the W3C issued the specification for HTML version 4.0 in 1997, although enough of them remain to cause regular problems for Web designers.

HTML Tags

An HTML document contains document text and elements. The tags in an HTML document are interpreted by the Web browser and used by it to format the display of the text enclosed by the tags. In HTML, the tags are enclosed in angle brackets (<>). Most HTML tags have an **opening tag** and a **closing tag** that format the text between them. The closing tag is preceded by a slash within the angle brackets. The general form of an HTML element is:

<tagname properties> Displayed information affected by tag </tagname>

Two good examples of HTML tag pairs are the boldface character-formatting tags and the italic character-formatting tags. For example, a Web browser reading the following line of text:

<B>A Review of the Book <I>HTML Is Fun!</I></B>

would recognize the <B> and </B> tags as instructions to display the entire line of text in bold and the <I> and </I> tags as instructions to display the text enclosed by those tags in italics. The Web browser would display the text as:

A Review of the Book *HTML Is Fun!*

Some Web browsers allow the user to customize the interpretations of the tags, so that different Web browsers might display the tagged text differently. For example, one Web browser might display text enclosed by bold tags in a blue color instead of displaying the text as bold. Tags can be written in either lowercase or uppercase letters; the tag <b> has the exact same meaning as the tag <B>. Although most tags are two-sided (they use both an opening and a closing tag), some are not. Tags that only require opening tags are known as one-sided tags. The tag that creates a line break (</br>) is a common one-sided tag. Some tags, such as the paragraph tag (<p> ... </p>), are two-sided tags for which the closing tag is optional. Designers often omit the optional closing tags, although many Web designers argue that this practice is poor markup style.

In a two-sided tag set, the closing tag position is very important. For example, if you were to omit the closing bold tag in the preceding example, any text that followed the line would be bolded. Sometimes an opening tag contains one or more property modifiers that further refine how the tag operates. A tag's property may modify a text display, or it may designate where to find a graphic element. Figure 2-6 shows some sample text marked up with HTML tags and Figure 2-7 shows this text as it appears in a Web browser. The tags in these two figures are among the most common HTML tags in use today on the Web.

Other frequently used HTML tags (not shown in the figures) let Web designers include graphics on Web pages and format text in the form of tables. The text and HTML tags that form a Web page can be viewed when the page is open in a Web browser by using the menu commands View, Source (in Internet Explorer) or View, Page Source (in Netscape Navigator). A number of good Web sources (such as the **W3C Getting Started with HTML** page) and textbooks are available that describe HTML tags and their uses, and you may wish to consult them for an in-depth look at HTML.

```
<html>

  <head>

    <title>HTML Tag Examples</title>

  </head>

  <body>

    <h1>This text is set in Heading One tags</h1>
    <h2>This text is set in Heading Two tags</h2>
    <h3>This text is set in Heading Three tags</h3>


    <p>
    This text is set within Paragraph tags. It will appear
    as one paragraph; the text will wrap at the end of each
    line that is rendered in the web browser no matter where
    the typed text ends. The text inside Paragraph tags is
    rendered without regard to extra spaces typed in the text,
    such as these:            Character formatting can also
    be applied within Paragraph tags. For example, <b>this
    text is set in bold</b>, <font face="Arial">this text is
    set in Arial</font>, and <i>this text is set in italics</i>.
    </p>

    <pre>
The Preformatted tag instructs the web browser to render the text
    exactly     the     way     it     is     typed,
as in this example.
    </pre>

    <p>
    HTML includes tags that instruct the web browser to place
    text in bulleted or numbered lists:
    </p>

    <ul>
      <li>Bulleted list item one</li>
      <li>Bulleted list item two</li>
      <li>Bulleted list item three</li>
    </ul>

    <ol>
      <li>Numbered list item one</li>
      <li>Numbered list item two</li>
      <li>Numbered list item three</li>
    </ol>

    <p>
    HTML also includes tags for left, center, and right justification:
    </p>

    <div align="left">Text aligned left</div>
    <div align="center">Text aligned center</div>
    <div align="right">Text aligned right</div>

    <p>
    The most important tag in HTML is the Anchor Hypertext
    Reference tag, which is the tag that provides a link to
    another web page (or another location in the same web page).
    For example, the underlined text
    <a href="http://www.course.com/">Course Technology</a>
    is a hyperlink to the web page of the publisher of this book.
    </p>

  </body>

</html>
```

FIGURE 2-6   Text marked up with HTML tags

This text is set in Heading One tags

This text is set in Heading Two tags

This text is set in Heading Three tags

This text is set within Paragraph tags. It will appear as one paragraph; the text will wrap at the end of each line that is rendered in the Web browser no matter where the typed text ends. The text inside Paragraph tags is rendered without regard to extra spaces typed in the text, such as these: Character formatting can also be applied within Paragraph tags. For example, **this text is set in bold**, this text is set in Arial, and *this text is set in italics*.

```
The Preformatted tag instructs the Web browser to render the text
        exactly    the    way    it    is    typed,
as in this example.
```

HTML includes tags that instruct the Web browser to place text in bulleted or numbered lists:

- Bulleted list item one
- Bulleted list item two
- Bulleted list item three

1. Numbered list item one
2. Numbered list item two
3. Numbered list item three

HTML also includes tags for left, center, and right justification:

Text aligned left

Text aligned center

Text aligned right

The most important tag in HTML is the Anchor Hypertext Reference tag, which is the tag that provides a link to another Web page (or another location in the same Web page). For example, the underlined text <u>Course Technology</u> is a hyperlink to the Web page of the publisher of this book.

FIGURE 2-7 Text marked up with HTML tags as it appears in a Web browser

HTML Links

The Web organizes interlinked pages of information residing on sites around the world. Hyperlinks on Web pages form a "web" of those pages. A user can traverse the interwoven pages by clicking hyperlinked text on one page to move to another page in the web of pages. Users can read Web pages in serial order or in whatever order they prefer by following hyperlinks. Figure 2-8 illustrates the differences between reading a paper catalog in a linear way and reading a hypertext catalog in a nonlinear way.

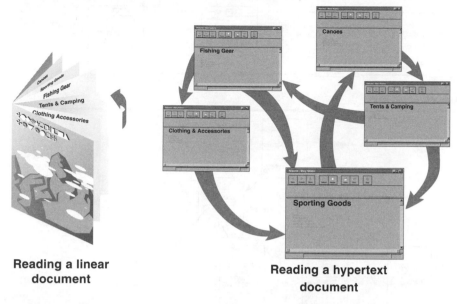

Reading a linear document

Reading a hypertext document

FIGURE 2-8   Linear vs. nonlinear paths through documents

An electronic commerce Web site can use links to direct customers to pages on the company's Web server. The way links lead customers through pages can affect the usefulness of the site and can play a major role in shaping customers' impressions of the company. Two commonly used link structures are linear and hierarchical. A **linear hyperlink structure** resembles conventional paper documents in that the reader begins on the first page and clicks the Next button to move to the next page in a serial fashion. This structure works well when customers fill out forms prior to a purchase or other agreement. In this case, the customer reads and responds to page one, and then moves on to the next page. This process continues until the entire form is completed. The only Web page navigation choices the user typically has are Back and Continue.

Another link arrangement is called a hierarchical structure. In a **hierarchical hyperlink structure**, the Web user opens an introductory page called a **home page** or **start page**. This page contains one or more links to other pages, and those pages, in turn, link to other pages. This hierarchical arrangement resembles an inverted tree in which the root is at the top and the branches are below it. Hierarchical structures are good for leading customers from general topics or products to specific product models and quantities. A company's home page might contain links to help, company history, company officers, order processing, frequently asked questions, and product catalogs. Many sites that use a hierarchical structure include a page on the Web site that contains a map or listing of the Web pages in their hierarchical order. This page is called a **site map**. Figure 2-9 illustrates the linear and hierarchical structures. Of course, pages combining linear and hierarchical structures are also possible.

In HTML, hyperlinks are created using the HTML **anchor tag**. Whether you are linking to text within the same document or to a document on a distant computer, the anchor tag has the same basic form:

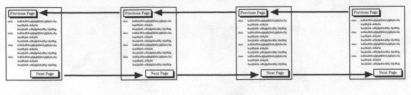

Linear structure

Hierarchical structure

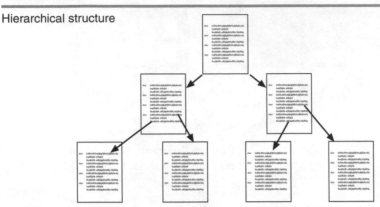

FIGURE 2-9 Two alternative hyperlink structures

<A HREF="address">Visible link text</A>

Anchor tags have opening and closing tags. The opening tag has a hypertext reference (HREF) property, which specifies the remote or local document's address. Clicking the text following the opening link transfers control to the HREF address, wherever that happens to be. A person creating an electronic résumé on the Web might want to make a university's name and address under the Education heading a hyperlink instead of plain text. Anyone viewing the résumé can click the link, which leads the reader to the university's home page. The following example shows the HTML code to create a hyperlink to another Web server:

<A HREF="http://www.sandiego.edu">University of San Diego</A>

Similarly, the résumé could include a local link to another part of the same document with the following marked up text:

<A HREF="#references">References are found here</A>

In both of these examples, the text between the anchors appears on the Web page as a hyperlink. Most browsers display the link in blue and underline it. In most browser software, the action of moving the mouse pointer over a hyperlink causes the mouse pointer to change from an arrow to a pointing hand.

Scripting Languages and Style Sheets

Versions of HTML released by the W3C after 1997 include an HTML tag called the object tag and include support for Cascading Style Sheets. Web designers can use the object tag to embed scripting language code on HTML pages. The most common scripting languages used on Web pages are JavaScript, JScript, Perl, and VBScript. Scripts written in these languages and embedded on Web pages can execute programs on computers that display those pages. You can learn more about embedding script languages in HTML documents (also called **client-side scripting**) by taking a course in Web programming or reading a book such as Kathleen Kalata's *Internet Programming* (a full reference for this book appears in the For Further Study and Research section at the end of this chapter).

Cascading Style Sheets (CSS) are sets of instructions that give Web developers more control over the format of displayed pages. Similar to document styles in word-processing programs, CSS let designers define formatting styles that can be applied to multiple Web pages. The set of instructions, called a **style sheet**, is usually stored in a separate file, and is referenced using the HTML style tag; however, it can be included as part of a Web page's HTML file. The term *cascading* means that designers can apply many style sheets to the same Web page, one on top of the other. For example, a three-stage cascade might include one style sheet with formatting instructions for text within heading 1 tags, a second style sheet with formatting instructions for text within heading 2 tags, and a third style sheet with formatting instructions for text within paragraph tags. A designer who later decides to change the formatting of heading 2 text can just replace the second style sheet with a different one.

Extensible Markup Language (XML)

As the Web grew, HTML continued to provide a useful tool for Web designers who wanted to create attractive layouts of text and graphics on their pages. However, as companies began to conduct electronic commerce on the Web, the need to present large amounts of data on Web pages also became important. Companies created Web sites that contained lists of inventory items, sales invoices, purchase orders, and other business data. The need to keep these lists updated was also important and posed a new challenge for many Web designers. The tool that had helped these Web designers create useful Web pages, HTML, was not such a good tool for presenting or maintaining information lists.

In the late 1990s, companies began turning to XML to help them maintain Web pages that contained large amounts of data. XML uses paired start and stop tags in much the same way as database software defines a record structure. For example, a company that sells products on the Web might have Web pages that contain descriptions and photos of the products it sells. The Web pages are marked up with HTML tags, but the product information elements themselves, such as prices, identification numbers, and quantities on hand, are marked up with XML tags. The XML document is embedded within the HTML document.

XML includes data management capabilities that HTML cannot provide. To better understand the strengths of XML and weaknesses of HTML in data management tasks, consider the simple example of a Web page that includes a list of countries and some basic information about each country. A Web designer might decide to use HTML tags to show each information item the same way for each country. Each information item would use a different tag. Assume that the Web designer in this case decided to use the HTML heading tags to present the data. Figure 2-10 shows the data and the HTML heading tags for four countries (this is only an example; the actual list would include more than 150 countries). The first item in the list provides the definitions for each tag. Figure 2-11 shows this HTML document as it appears in a Web browser.

```
<html>
    <head>
        <title>Countries</title>
    </head>
    <body>
        <h1>Countries</h1>

        <h2>CountryName</h2>
        <h3>CapitalCity</h3>
        <h4>AreaInSquareKilometers</h4>
        <h5>OfficialLanguage</h5>
        <h6>VotingAge</h6>

        <h2>Argentina</h2>
        <h3>Buenos Aires</h3>
        <h4>2,766,890</h4>
        <h5>Spanish</h5>
        <h6>18</h6>

        <h2>Austria</h2>
        <h3>Vienna</h3>
        <h4>83,858</h4>
        <h5>German</h5>
        <h6>19</h6>

        <h2>Barbados</h2>
        <h3>Bridgetown</h3>
        <h4>430</h4>
        <h5>English</h5>
        <h6>18</h6>

        <h2>Belarus</h2>
        <h3>Minsk</h3>
        <h4>207,600</h4>
        <h5>Byelorussian</h5>
        <h6>18</h6>

    </body>
</html>
```

FIGURE 2-10   Country list data marked up with HTML tags

These figures reveal some of the shortcomings of using HTML to present a list of items when the meaning of each item in the list is important. The Web designer in this case used HTML heading tags. HTML has only six levels of heading tags; thus, if the individual items had any more information elements than shown in this example (such as population and continent), this approach would not work at all. The Web designer could use various combinations of text attributes such as size, font, color, bold, or italics to distinguish among items, but none of these tags would convey the meaning of the individual data elements. The only information about the meaning of each country's listing appears in the first list

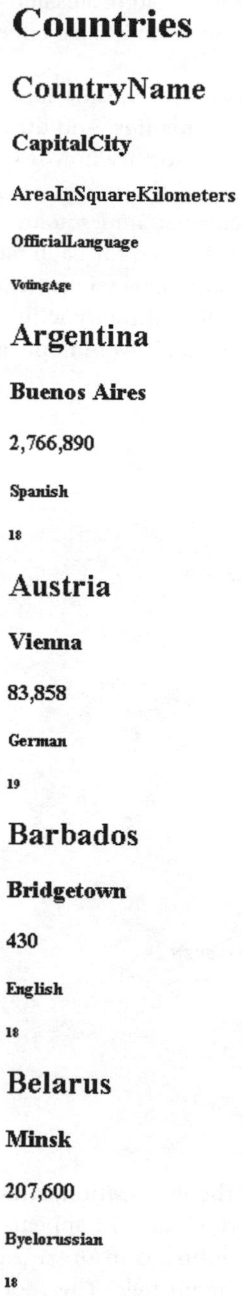

Countries

CountryName

CapitalCity

AreaInSquareKilometers

OfficialLanguage

VotingAge

Argentina

Buenos Aires

2,766,890

Spanish

18

Austria

Vienna

83,858

German

19

Barbados

Bridgetown

430

English

18

Belarus

Minsk

207,600

Byelorussian

18

FIGURE 2-11   Country list data as it appears in a Web browser

item, which includes the definitions for each element. In the late 1990s, Web professionals began to consider XML as a list-formatting alternative to HTML that would more effectively communicate the meaning of data.

XML differs from HTML in two important respects. First, XML is not a markup language with defined tags. It is a framework within which individuals, companies, and other organizations can create their own sets of tags. Second, XML tags do not specify how text appears on a Web page; the tags convey the meaning (the semantics) of the information included within them. To understand this distinction between appearance and semantics, consider the list of countries example again. In XML, tags can be created for each fact that define the meaning of the fact. Figure 2-12 shows the countries data marked up with XML tags. Some browsers, such as Internet Explorer, can render XML files directly without additional instructions. Figure 2-13 shows the country list XML file as it would appear in an Internet Explorer browser window.

```
declaration ──<?xml version="1.0"?>

            ┌─<CountriesList>
root        │  <Country Name = "Argentina">
element     │     <CapitalCity>Buenos Aires</CapitalCity>
            │     <AreaInSquareKilometers>2,766,890</AreaInSquareKilometers>
            │     <OfficialLanguage>Spanish</OfficialLanguage>
            │     <VotingAge>18</VotingAge>
            │  </Country>

               <Country Name = "Austria">
                  <CapitalCity>Vienna</CapitalCity>
                  <AreaInSquareKilometers>83,858</AreaInSquareKilometers>
                  <OfficialLanguage>German</OfficialLanguage>
                  <VotingAge>19</VotingAge>
               </Country>

               <Country Name = "Barbados">
                  <CapitalCity>Bridgetown</CapitalCity>
                  <AreaInSquareKilometers>430</AreaInSquareKilometers>
                  <OfficialLanguage>English</OfficialLanguage>
                  <VotingAge>18</VotingAge>
               </Country>

               <Country Name = "Belarus">
                  <CapitalCity>Minsk</CapitalCity>
                  <AreaInSquareKilometers>207,600</AreaInSquareKilometers>
                  <OfficialLanguage>Byelorussian</OfficialLanguage>
                  <VotingAge>18</VotingAge>
               </Country>

            </CountriesList>
```

FIGURE 2-12   Country list data marked up with XML tags

The first line in the XML file shown in Figures 2-12 and 2-13 is the declaration, which indicates that the file uses version 1.0 of XML. XML markup tags are similar in appearance to SGML markup tags, thus the declaration can help avoid confusion in organizations that use both. The second line and the last line are the root element tags. The root element of an XML file contains all of the other elements in that file and is usually assigned a name that describes the purpose or meaning of the file.

The other elements are called child elements; for example, Country is a child element of CountriesList. Each of the other attributes is, in turn, a child element of the Country element. The names of these child elements were created for use in this file. If

```
<?xml version="1.0" ?>
- <CountriesList>
  - <Country Name="Argentina">
      <CapitalCity>Buenos Aires</CapitalCity>
      <AreaInSquareKilometers>2,766,890</AreaInSquareKilometers>
      <OfficialLanguage>Spanish</OfficialLanguage>
      <VotingAge>18</VotingAge>
    </Country>
  - <Country Name="Austria">
      <CapitalCity>Vienna</CapitalCity>
      <AreaInSquareKilometers>83,858</AreaInSquareKilometers>
      <OfficialLanguage>German</OfficialLanguage>
      <VotingAge>19</VotingAge>
    </Country>
  - <Country Name="Barbados">
      <CapitalCity>Bridgetown</CapitalCity>
      <AreaInSquareKilometers>430</AreaInSquareKilometers>
      <OfficialLanguage>English</OfficialLanguage>
      <VotingAge>18</VotingAge>
    </Country>
  - <Country Name="Belarus">
      <CapitalCity>Minsk</CapitalCity>
      <AreaInSquareKilometers>207,600</AreaInSquareKilometers>
      <OfficialLanguage>Byelorussian</OfficialLanguage>
      <VotingAge>18</VotingAge>
    </Country>
</CountriesList>
```

FIGURE 2-13 Country list data marked up with XML tags as it would appear in Internet Explorer

programmers in another organization were to create a file with country information, they might use different names for these elements (for example, "Capital" instead of "Capital-City"), which would make it difficult for the two organizations to share information. Thus, the greatest strength of XML, that it allows users to define their own tags, is also its greatest weakness.

To overcome that weakness, many companies have agreed to follow common standards for XML tags. These standards, in the form of data type definitions (DTDs) or XML schemas, are available for a number of industries, including the **ebXML** initiative for electronic commerce standards, the **eXtensible Business Reporting Language (XBRL)** for accounting and financial information standards, **LegalXML** for information in the legal profession, and **MathML** for mathematical and scientific information. A number of industry groups have formed to create standard XML tag definitions that can be used by all companies in that industry. The **Open Buying on the Internet (OBI)** Consortium and **RosettaNet** are two examples of such industry groups. In 2001, the W3C released a set of rules for XML document interoperability that many researchers believe will help resolve incompatibilities between different sets of XML tag definitions. A set of XML tag definitions is sometimes called an **XML vocabulary**. Hundreds of publicly defined XML vocabularies are currently circulating, many of which are registered with the **XML Registry**. You can learn more about XML by reading the **W3C XML Pages**.

Although it is possible to display XML files in some Web browsers, XML files are not intended to be displayed in a Web browser. XML files are intended to be translated using another file that contains formatting instructions or to be read by a program. Formatting instructions are often written in the **Extensible Stylesheet Language (XSL)**, and the programs that read or transform XML files are usually written in the Java programming language. These programs, sometimes called **XML parsers**, can format an XML file so it can appear on the screen of a computer, a wireless PDA, a mobile phone, or some other device. A diagram showing one way that a Web server might process an HTTP request for an XML page appears in Figure 2-14.

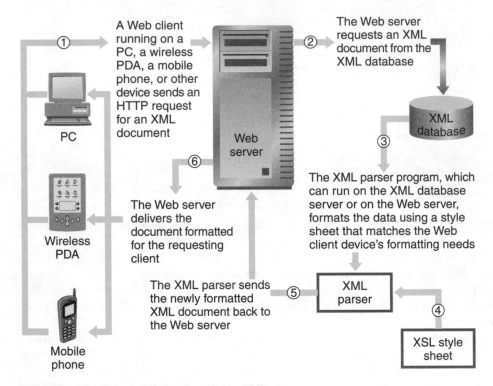

FIGURE 2-14 Processing a request for an XML page

HTML and XML Editors

Web designers can create HTML documents in any general-purpose text editor or word processor. However, one of the special-purpose HTML editors can help Web designers create Web pages much more easily. There are many freeware, shareware, and commercial HTML editors available for download on the Internet, including CoffeeCup, HotDog Professional, HomeSite, and CuteHTML. The Additional Resources section of the Online Companion for this chapter includes links to Web sites that offer downloads of these **HTML Editing Programs**.

HTML editors are also included as part of more sophisticated Web site design and creation programs that are sometimes called Web page builder software. With these programs, Web designers can create and manage complete Web sites, including features for database access, graphics, and fill-in forms. These programs display the Web page as it will appear in a Web browser in one window and display the HTML-tagged text in another window. The designer can edit in either window and changes are reflected in the other window. For example, the designer can drag and drop objects such as graphics onto the Web browser view page and the program automatically generates the HTML tags to position the graphics.

Web site design and creation software also provides maintenance tools that allow the designer to create a Web site on a PC and then upload the entire site (HTML documents, graphics files, and so on) to a Web server computer. When the site needs to be edited later, the designer can edit the copy of the site on the PC and instruct the program to synchronize those changes on the copy of the site that resides on the Web server. Examples of Web site design and creation programs include **Microsoft FrontPage** and **Macromedia Dreamweaver**. Figure 2-15 shows a Web page being edited in Dreamweaver.

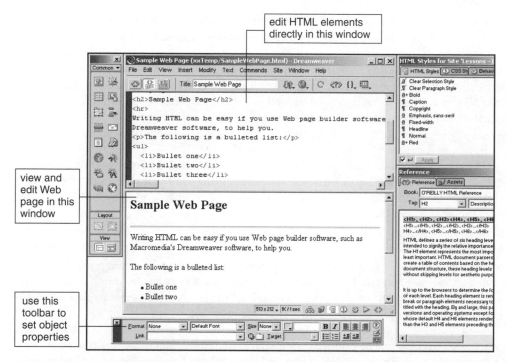

FIGURE 2-15 A Web page being edited in Dreamweaver

XML files, like HTML files, can be created in any text editor. However, programs designed to make the task of designing and managing XML files easier are also available. These programs include Epic Editor, TurboXML, XMetal, and XML Spy. These programs provide tag validation and XML creation capabilities in addition to making the job of marking up text with XML tags more efficient. You can find links to these programs' Web sites in the Additional Resources section of the Online Companion under the heading **XML Editing Programs**.

Not all TCP/IP networks connect to the Internet. Many companies build internets (small "i"), or interconnected networks, that do not extend beyond their organizational boundaries. An **intranet** is an interconnected network (or internet), usually one that uses the TCP/IP protocol set, and does not extend beyond the organization that created it. An **extranet** is an intranet that has been extended to include specific entities outside the boundaries of the organization, such as business partners, customers, or suppliers. Although fax, telephone, e-mail, and overnight express carriers have been the main communications tools for business for many years, extranets can replace many of them at a lower cost.

Intranets

Intranets are an excellent low-cost way to distribute internal corporate information. Based on the client-server model, intranet requests for files, documents, or schematic drawings work the same way they do on the Internet. An intranet uses Web browsers and Internet-based protocols, including TCP/IP, FTP, Telnet, HTML, and HTTP. Because intranets are compatible with the Internet, information from intranets can be shared among departments that use different technologies as well as among external consumers. Intranets are often the most efficient way to distribute internal corporate information, because producing and distributing paper is usually slower and more expensive than using Web-based communications.

Companies can also use intranets to reduce software maintenance and update costs for their employees' computer workstations. Computing staff can place software updates and patches on the intranet, and then provide a script to update employee workstations automatically the next time they log on.

Extranets

Extranets are networks that connect companies with suppliers, business partners, or other authorized users. Each participant in the extranet has access to the databases, files, or other information stored on computers connected to the extranet. An extranet can be set up through the Internet, or it can use a separate network.

Some extranets start out as intranets that eventually provide access of intranet data to select Internet users. For example, for many years, FedEx let customers track their packages by calling a FedEx toll-free number, and then giving the operator a tracking number. In the early 1990s, FedEx began giving package-tracking software to any customer who wanted it. Once it was installed on the customer's computer, the software dialed the FedEx computer using a modem, queried the status of the customer's package, and displayed the results on the customer's computer with no operator required. In the mid-1990s, FedEx eliminated client-machine software and made package tracking available on its Web site. Instead of having thousands of programs running on customers' computers, FedEx has its customers use their Web browsers to run one program running on its Web site. This Web-based system is called **FedEx Ship Manager**, and it gives customers Web access (from any browser on any computer on the Web) to package tracking, air bill creation, shipment logging, and FedEx supply shipments. Critical information, such as a package's location, is stored and made available to customers through the FedEx Ship Manager section of the FedEx extranet.

Public and Private Networks

A **public network** is any computer network or telecommunications network that is available to the public. The Internet is one example of a public network. Although a company can operate its extranet using a public network, very few do because of the high level of security risks. The Internet, as you will learn in later chapters, does not provide a high degree of security in its basic structure.

A **private network** is a private, leased-line connection between two companies that physically connects their intranets to one another. A **leased line** is a permanent telephone connection between two points. Unlike the normal telephone connection you create when you dial a telephone number, a leased line is always active. The advantage of a leased line is security. Only the two parties that lease the line to create the private network have access to the connection. The largest drawback to a private network is cost. Leased lines are expensive. Every pair of companies wanting a private network between them requires a separate line connecting them. For instance, if a company wants to set up an extranet connection over a private network with seven other companies, the company must pay the cost of seven leased lines, one for each company. If the extranet expands to 20 other companies, the extranet-sponsoring company must rent another 13 leased lines. As each new company is added, costs increase by the same amount and soon become prohibitive. Vendors refer to this as a **scaling problem**; that is, increasing the number of leased lines in private networks is difficult, costly, and time consuming. As the number of companies that need to join the extranet increases, other networking options become appealing.

Virtual Private Network (VPN)

A **virtual private network (VPN)** is an extranet that uses public networks and their protocols to send sensitive data to partners, customers, suppliers, and employees using a system called IP tunneling or encapsulation. **IP tunneling** effectively creates a private passageway through the public Internet that provides secure transmission from one computer to another. The virtual passageway is created by VPN software that encrypts the packet content and then places the encrypted packets inside another packet in a process called **encapsulation**. The outer packet is called an **IP wrapper**. The Web server sends the encapsulated packets to their destinations over the Internet. The computer that receives the packet unwraps it and decrypts the message using VPN software that is the same as, or compatible with, the VPN software used to encrypt and encapsulate the packet at the sending end. The "virtual" part of VPN means that the connection seems to be a permanent, internal network connection, but the connection is actually temporary. Each transaction between two intranets using a VPN is created, carries out its work over the Internet, and is then terminated.

VPN software must be installed on the computers at both ends of the transmission. A VPN provides security shells, with the most sensitive data under the tightest control. The

VPN is like a separate, covered commuter lane on a highway (the Internet) in which the passengers are protected from being seen by the vehicles traveling in the other lanes. Company employees in remote locations can send sensitive information to company computers using the VPN private tunnels established on the Internet.

Unlike private networks using leased lines, VPNs establish short-term logical connections in real time that are broken once the communication session ends. Establishing VPNs does not require leased lines. The only infrastructure required outside each company's intranet is the Internet. Companies such as **Aventail**, **Cisco**, **SonicWall**, and **V-ONE** are making VPNs simpler to install and maintain.

Extranets are sometimes confused with VPNs. Although a VPN is an extranet, not every extranet is a VPN. Figure 2-16 shows a diagram of a VPN. VPNs usually work as part of a firewall (the firewall-VPN combinations are shown as brick walls in the figure). A firewall is a program or hardware device that protects information inside an organization's network from attacks that originate outside the network. You will learn more about VPNs, firewalls, and other network security devices in Chapter 10.

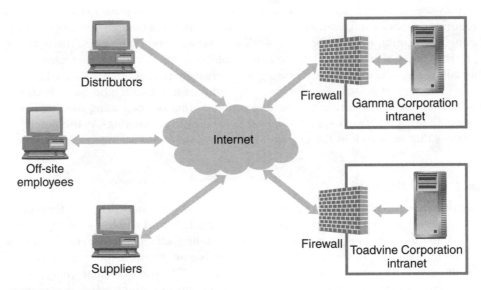

FIGURE 2-16 VPN architecture example

INTERNET CONNECTION OPTIONS

The Internet is a set of interconnected networks. A corporation or individual cannot become part of the Internet without a telephone connection or a connection to a LAN or intranet. Larger firms that provide Internet access to other businesses, called **Internet access providers (IAPs)** or ISPs, usually offer several connection options. This section briefly describes current connection choices and presents their advantages and disadvantages.

Connectivity Overview

ISPs offer several ways to connect to the Internet. The most common connection options are voice-grade telephone line, various types of broadband connections, leased line, and wireless. One of the major distinguishing factors between various ISPs and their connection options is the bandwidth they offer. **Bandwidth** is the amount of data that can travel through a communication line per unit of time. The higher the bandwidth, the faster data files travel and the faster Web pages appear on your screen. Each connection option offers different bandwidth, and each ISP offers varying bandwidth for each connection option. Traffic on the Internet and at your local service provider greatly affects **net bandwidth**, which is the actual speed that information travels. When few people are competing for service from an ISP, net bandwidth approaches the carrier's upper limit. On the other hand, users experience slowdowns during high-traffic periods.

Bandwidth can differ for data traveling to or from the ISP depending on the user's connection type. **Symmetric connections** provide the same bandwidth in both directions. **Asymmetric connections** provide different bandwidths for each direction. **Upstream bandwidth**, also called **upload bandwidth**, is a measure of the amount of information that can travel from the user to the Internet in a given amount of time. **Downstream bandwidth**, also called **download** or **downlink bandwidth**, is a measure of the amount of information that can travel from the Internet to a user in a given amount of time (for example, when a user receives a Web page from a Web server).

VOICE-GRADE TELEPHONE CONNECTIONS

The most common way to connect to an ISP is through a modem connected to your local telephone service provider. **POTS**, or **plain old telephone service**, uses existing telephone lines and an analog modem to provide a bandwidth of between 28 and 56 Kbps. Some telephone companies offer a higher grade of service called **Digital Subscriber Line (DSL)** protocol. DSL connection methods do not use a modem. They use a piece of networking equipment that is similar to a network switch, but most people call this piece of equipment (incorrectly) a "DSL modem." **Integrated Services Digital Network (ISDN)** was the first technology developed to use the DSL protocol suite and has been available in parts of the United States since 1984. ISDN is more expensive than regular telephone service and offers bandwidths of between 128 Kbps and 256 Kbps.

Broadband Connections

Connections that operate at speeds of greater than about 200 Kbps are called **broadband** services. One of the newest technologies that uses the DSL protocol to provide service in the broadband range is **asymmetric digital subscriber line (ADSL**, usually abbreviated **DSL)**. It provides transmission bandwidths from 100 to 640 Kbps upstream and from 1.5 to 9 Mbps (million bits per second) downstream. For businesses, a **high-speed DSL (HDSL)** connection service can provide more than 768 Kbps of symmetric bandwidth.

Cable modems—connected to the same broadband coaxial cable that serves a television—typically provide transmission speeds between 300 Kbps and 1 Mbps from the client to the server. The downstream transmission rate can be as high as 10 Mbps. In the United States alone, more than 100 million homes and organizations have broadband cable service available, and over 70 million homes subscribe to cable television. The latest estimates indicate that there are more than 9 million cable modem subscribers in the United States and another 3 million households that have broadband DSL or satellite connections.

DSL is a private line with no competing traffic. Unlike DSL, cable modem connection bandwidths vary with the number of other subscribers competing for the shared resource. Transmission speeds can decrease dramatically in heavily subscribed neighborhoods at prime times—in neighborhoods where many people are using cable modems simultaneously.

Connection options based on cable or telephone line connections are wonderful for urban and suburban Web users, but those living in rural areas often have very limited telephone service and no cable access at all. The telephone lines used to cover the vast distances between rural customers are usually **voice-grade lines**, which cost less than telephone lines designed to carry data, are made of lower grade copper, and were never intended to carry data. These lines can carry only limited bandwidth—usually less than 14 Kbps. Telephone companies have wired most urban and suburban areas with **data-grade lines** (made more carefully and of higher grade copper than voice-grade lines) because the short length of the lines in these areas makes it less expensive to install than in rural areas where connection distances are much longer. It is also likely that urban and suburban lines will someday be leased to companies willing to pay the higher fees charged for data-grade lines.

LEARNING FROM FAILURES

NORTHPOINT COMMUNICATIONS

In 1997, Michael Malaga was a successful telecommunications executive with an idea. He wanted to sell broadband Internet access to small businesses in urban areas. DSL technology was just gaining acceptance, and leased telephone lines were available from telephone companies. He wanted to avoid residential customers because they would soon have inexpensive cable modem access to meet their broadband needs. He also wanted to avoid suburban and rural businesses to keep the telephone line leasing costs low (lease charges are higher for longer distances). He and five friends started NorthPoint Communications with $500,000 of their combined savings and raised another $11 million within a few months. After six months, the company had raised more money from investors and had acquired 1500 customers, but it was posting a net loss of $30 million. On the strength of its number of customers, the company began the task of raising the $100 million that Malaga estimated it would need to create the network infrastructure.

continued

Independent DSL providers such as NorthPoint were pressed by customers to install service rapidly, but had to rely on local telephone companies to ensure that their lines would support DSL. In many cases, the telephone companies had to install switches and other equipment to make DSL work on a particular line. The telephone companies often were in no rush to do this because they also sold DSL service, and speedy service would be helping a competitor. The delays led to unpredictable installation holdups and many unhappy NorthPoint customers. Customers with problems after the service was installed often were bounced from the telephone company to NorthPoint, without obtaining satisfactory or timely resolutions of their problems.

Although NorthPoint was unable to make its relationship with each customer profitable, Malaga and his team were rapidly raising money in the hot capital markets of the time. The company raised $162 million before its first stock offering in 1999, which brought in an additional $387 million. At that time, the company had 13,000 customers, which means that NorthPoint had raised over $42,000 from outside investors for each customer. Considering that each customer would generate revenue of about $1000 per-year, the economics of the business did not look good. By the end of 1999, NorthPoint had spent $300 million of the cash it had raised to build its network infrastructure and reported an operating loss of $184 million. At this point, NorthPoint was operating in 28 cities.

During the next year, the company continued to raise additional funds, gain more customers, and lose money on each customer. In August 2000, the telephone company Verizon agreed to purchase 55 percent of the company for $800 million paid in installments. The total funding that NorthPoint had obtained by the end of 2000, including the partial payments received from Verizon, added up to $1.2 billion. By the end of the year, NorthPoint was in 109 cities and needed to spend $66 million in cash per month just to stay in business. Verizon withdrew from the purchase agreement, the stock plunged, and the layoffs began.

NorthPoint filed for bankruptcy in January 2001 and sold its networking hardware to AT&T in March for $135 million. AT&T was not interested in continuing the DSL business (it just wanted the hardware), so NorthPoint's 87,000 small business customers lost their Internet service overnight. In many of the cities that NorthPoint had served, there were no competitors to pick up the service. Because the capital markets of the late 1990s were so eager to invest in anything that appeared to be connected with the Internet, NorthPoint was able to raise incredible amounts of money. However, NorthPoint sold Internet access to customers for less than it cost to provide the service. No amount of investor money could overcome that basic business mistake.

Leased-Line Connections

Large firms with large amounts of Internet traffic can connect to an ISP using higher bandwidth connections that they can lease from telecommunications carriers. These connections use a variety of technologies and are usually classified by the equivalent number of telephone lines they include (the connection technologies they use were originally developed to carry large numbers of telephone calls).

A telephone line designed to carry one digital signal is called DS0 (digital signal zero, the name of the signaling format used on those lines) and has a bandwidth of 56 Kbps. A

T1 line (also called a DS1) carries 24 DS0 lines and operates at 1.544 Mbps. Some telecommunications companies offer **fractional T1**, which provides service speeds of 128 Kbps and upward in 128-Kbps increments. **T3** service (also called DS3) offers 44.736 Mbps (the equivalent of 30 T1 lines or 760 DS0 lines). All of these leased telephone line connections are much more expensive than POTS, ISDN, or DSL connections.

Large organizations that need to connect hundreds or thousands of individual users to the Internet require very high bandwidth. NAPs use T1 and T3 lines. NAPs and the computers that perform routing functions on the Internet backbone also use technologies such as **frame relay** and **asynchronous transfer mode (ATM)** connections and **optical fiber** (instead of copper wire) connections with bandwidths determined by the class of fiber-optic cable used. An OC3 (optical carrier 3) connection provides 156 Mbps, an OC12 provides 622 Mbps, an OC48 provides 2.5 Gbps (gigabits, or 1 billion bits per second), and an OC192 provides 10 Gbps.

Wireless Connections

For many people in rural areas, satellite microwave transmissions have made connections to the Internet possible for the first time. In the first satellite technologies, the customer placed a receiving dish antenna on the roof or in the yard and pointed it at the satellite. The satellite sent microwave transmissions to handle Internet downloads at speeds of around 500 Kbps. Uploads were handled by a POTS modem connection. For Web browsing, this was not too bad, since most of the uploaded messages were small text messages (e-mails and Web page requests). People who wanted to send large e-mail attachments or transfer files over the Internet found the slow upload speeds unsatisfactory.

In the past two years, companies such as **DirecPC**, **DIRECWAY**, and **StarBand** have begun offering satellite Internet connections that do not require a POTS modem connection for uploads. These connections use a microwave transmitter for Internet uploads. This transmitter provides upload speeds as high as 150 Kbps. Initially, the installation charges were much higher than for other residential Internet connection services because a professional installer was needed to carefully aim the transmitter's dish antenna at the satellite. Recently, the accuracy of the antennas improved, and some of these companies now offer a self-installation option that drastically reduces the initial cost. For installations in North America, the antennas must have a clear line of sight into the southwestern sky. This requirement can make these services unusable for many people living in large cities or on the wrong side of an apartment building.

No discussion of Internet connections would be complete without mention of the wireless devices that can be connected to the Internet. People today use mobile phones, wireless personal digital assistants (PDAs), tablet computers, and even laptops equipped with wireless network cards to connect to networks that, in turn, are connected to the Internet. Several wireless standards are in use today and more are being developed.

Bluetooth

One of the first wireless protocols, designed for personal use over short distances, is called **Bluetooth** (the protocol was developed in Norway and is named for Harald Bluetooth, a 10th century Scandinavian king). Bluetooth operates reliably over distances of up to 35 feet and can be a part of up to 10 networks of eight devices each. It is a low-bandwidth technology, with speeds of up to 722 Kbps. Bluetooth is useful for tasks such as wireless synchronization of laptop computers with desktop computers and wireless printing from laptops, PDAs, or mobile phones. These small Bluetooth networks are called **personal area networks (PANs)** or **piconets**.

One major advantage of Bluetooth technology is that it consumes very little power, which is an important consideration for mobile devices. Another advantage is that Bluetooth devices can discover each other and exchange information automatically. For example, a person using a laptop computer in a temporary office can print to a local Bluetooth-enabled printer without logging in to the network or installing software on either device. The printer and the laptop computer electronically recognize each other as Bluetooth devices and immediately can begin exchanging information.

Wireless Ethernet (Wi-Fi)

The most common wireless connection technology for use on LANs is called **Wi-Fi**, **wireless Ethernet**, or **802.11b** (802.11 is the number of the technology's **network specification**, which is the set of rules that equipment connected to the network must follow). A computer equipped with a Wi-Fi network card can communicate through a wireless access point connected to a LAN to become a part of that LAN. A **wireless access point (WAP)** is a device that transmits network packets between Wi-Fi-equipped computers and other devices that are within its range. The user must have authorization to connect to the LAN and might be required to perform a login procedure before the laptop can access the LAN through the WAP.

Wi-Fi has a potential bandwidth of 11 Mbps and a range of about 300 feet. In actual installations, the achieved bandwidth and range can be dramatically affected by the construction material of the objects (such as walls, floors, doors, and windows) through which the signals must pass. For example, reinforced concrete walls and certain types of tinted glass windows greatly reduce the effective range of Wi-Fi. Despite these limitations, organizations can make Wi-Fi a key element of their LAN structures by installing a number of WAPs throughout their premises. Wi-Fi devices are capable of **roaming**, that is, shifting from one WAP to another, without requiring intervention by the user. Increasingly, Wi-Fi is becoming available in public places such as airports, convention centers, hotels, and office lobbies. The users of these networks authorize a connection charge when they log in and then have access to the wireless LAN's resources, including access to the Internet.

In 2002, an improved version of Wi-Fi, called **802.11a** (the 802.11b protocol was easier to implement, thus it was introduced first) was introduced. The 802.11a protocol is capable of transmitting data at speeds up to 54 Mbps, but it is not compatible with 802.11b devices. Later in 2002, the **802.11g** protocol, which has the 54 Mbps speed of 802.11a and is compatible with 802.11b devices, was introduced. Because of its compatibility with the many 802.11b devices that were in use, 802.11g was an immediate success. The next protocol in this series, **802.11n**, is expected to offer even greater speeds (up to 320 Mbps). Most industry experts expect the 802.11n protocol to be released in 2006 or 2007.

Some organizations operate WAPs that are open to the public. These access points are called **hot spots**. Some organizations allow free access to their hot spots, others charge an access fee. A growing number of retail establishments, such as McDonalds, Panera, and Starbucks, offer hot spots. Hotels have found that installing a WAP is cheaper and easier than running network cable, especially in older buildings. Some hotels offer wireless access free, others charge a small fee. There are several Web sites that offer hot spot directories that show hot spots by location, but these sites tend to open and close frequently. The best way to find hot spots (or a hot spot directory) is to use your favorite search engine.

Some communities have installed wireless networks that can be accessed from anywhere in the area. For example, the city of Grand Haven, Michigan installed a metropolitan area Wi-Fi network. Grand Haven is a growing town on the shores of Lake Michigan. The company that built the network, Ottawa Wireless, sells network access to residents and businesses throughout the area. The company offers access not only on land, but on boats up to 15 miles out on Lake Michigan. Several small company owners use this network to conduct their online business while sailing!

Fixed-Point Wireless

In some areas, companies such as **Etherlinx** and **Getwireless.net** are beginning to offer fixed-point wireless service. One version of **fixed-point wireless** uses a system of repeaters to forward a radio signal from the ISP to customers. The **repeaters** are transmitter-receiver devices (also called **transceivers**) that receive the signal and then retransmit it toward users' roof-mounted antennas and to the next repeater, which receives the signal and passes it on to the next repeater, which can be up to 20 miles away. The users' antennas are connected to a device that converts the radio signals into Wi-Fi packets that are sent to the users' computers or wireless LANs. Another version of fixed-point wireless directly transmits Wi-Fi packets through hundreds, or even thousands, of short-range transceivers that are located close to each other. This approach is called **mesh routing**. As Wi-Fi technologies improve, the number and variety of options for wireless connections to the Internet should continue to increase.

Cellular Telephone Networks

In 2003, there were about 500 million mobile phones in the world and industry experts expect that number to grow by 50 million or more per year for at least the next several years. These phones are often called cellular (or cell) phones because they broadcast signals to (and receive signals from) antennas that are placed about three miles apart in a grid, and the hexagonal area that each antenna covers within this grid is called a cell.

Although cell phones were originally designed to handle voice communications, they have always been able to transmit data. However, their data transmission speeds were very low, ranging from 10 Kbps to 384 Kbps. Several changes in cell phone technology have increased those speeds in today's most capable cell phones to 2 Mbps. The devices that combine the latest technologies available today are called **third-generation (3G) cell phones**. Many cell phones have a small screen and can be used to send and receive short text messages using a protocol called **short message service (SMS)**. Some cell phones and combination phone-PDAs include tiny Web browsers. These devices can provide users with Web access, e-mail connections, and other Internet services in addition to short message service. In 2004, mobile telephone and PDA manufacturers introduced the first devices (telephones and PDAs) that could automatically switch their connections between the cellular network and a local Wi-Fi network. Many industry analysts expect these types of devices to become the standard within a few years.

Today, companies that sell mobile telephone services also sell Internet access through their cellular networks. These companies offer a variety of pricing plans and service levels, but most charge a fixed fee plus a charge for the amount of data traffic sent or received during the month.

Many companies have seen great business potential for these wireless networks and the devices connected to them. They use the term **mobile commerce** or **m-commerce** to describe the kinds of resources people might want to access (and pay for) using wireless devices. You will learn more about revenue models that use wireless technologies in Chapter 3 and cost reduction strategies that use wireless technologies in Chapter 5. In Chapter 11, you will learn how some companies are using these technologies to process online payments for goods and services. Figure 2-17 summarizes speed and cost information for the most commonly available wired and wireless options for connecting a home or business to the Internet.

Service	Upstream speed (Kbps)	Downstream speed (Kbps)	Capacity (number of simultaneous users)	One-time startup costs	Continuing monthly Costs
Residential and small business services					
Modem (POTS)	28–56	28–56	1	$0–$20	$12–$20
ISDN	128–256	128–256	1–3	$60–$300	$50–$90
ADSL	100–640	4,500–9,000	1–4	$50–$100	$40–$90
Cable modem	300–1,000	1,000–10,000	1–4	$0–$100	$40–$70
Satellite	125–150	400–500	1–3	$600–$1,200	$60–$70
Business services					
Leased digital line (DS0)	64	64	1–10	$50–$200	$40–$150
Fractional T1 leased line	128–1,544	128–1,544	5–180	$50–$800	$100–$1,000
HDSL	768–3,000	768–3,000	50–100	$300–$1,500	$400–$900
T1 leased line	1,544	1,544	100–200	$100–$2,000	$900–$1,600
T3 leased line	44,700	44,700	1,000–10,000	$1,000–$9,000	$5,000–$12,000
Large business, ISP, NAP, and Internet 2 services					
OC3 leased line	156,000	156,000	1,000–50,000	$3,000–$12,000	$9,000–$22,000
OC12 leased line	622,000	622,000	Backbone	Negotiated	$25,000–$100,000
OC48 leased line	2,500,000	2,500,000	Backbone	Negotiated	Negotiated
OC192 leased line	10,000,000	10,000,000	Backbone	Negotiated	Negotiated

FIGURE 2-17 Internet connection options

INTERNET2 AND THE SEMANTIC WEB

At the high end of the bandwidth spectrum, a group of network research scientists from nearly 200 universities and a number of major corporations joined together in 1996 to recapture the original enthusiasm of the ARPANET with an advanced research network called **Internet2**. When the National Science Foundation turned over the Internet backbone to commercial interests in 1995, many scientists felt that they had lost a large, living laboratory. **Internet2** is the replacement for that laboratory. An experimental test bed for new networking technologies that is separate from the original Internet, Internet2 has achieved bandwidths of 10 Gbps and more on parts of its network.

Internet2 is also used by universities to conduct large collaborative research projects that require several supercomputers connected at very fast speeds, or that use multiple video feeds—things that would be impossible on the Internet given its lower bandwidth limits. Internet2 promises to be the proving ground for new technologies and applications of those technologies that will eventually find their way to the Internet. One of the most important elements of Internet2 today is the **Abilene** project. Abilene is a high-bandwidth backbone built by the Internet2 organization with Indiana University and three commercial partners: Juniper Networks, Nortel Networks, and Qwest .

The Internet2 project is focused mainly on technology development. In contrast, Tim Berners-Lee has announced a project that will blend technologies and information to create a next-generation Web, which he calls the Semantic Web. The **Semantic Web** project, if successful, would result in words on Web pages being tagged (using XML) with their meanings. The Web would become a huge machine-readable database. People could use intelligent programs called **software agents** to read the XML tags to determine the meaning of the words in their contexts. For example, a software agent given the instruction to find an airline ticket with certain terms (date, cities, cost limit) would launch a search on the Web and return with an electronic ticket that meets the criteria. Instead of a user having to visit several Web sites to gather information, compare prices and itineraries, and make a decision, the software agent would automatically do the searching, comparing, and purchasing.

The key elements that must be added to Web standards so that software agents can perform these functions include XML, a resource description framework, and an ontology. You have already seen how XML tags can describe the semantics of data elements. A **resource description framework (RDF)** is a set of standards for XML syntax. It would function as a dictionary for all XML tags used on the Web. An **ontology** is a set of standards that defines, in detail, the relationships among RDF standards and specific XML tags within a particular knowledge domain. For example, the ontology for cooking would include concepts such as ingredients, utensils, and ovens; however, it would also include rules and behavioral expectations, such as that ingredients can be mixed using utensils, that the resulting product can be eaten by people, and that ovens generate heat within a confined area. Ontologies and the RDF would provide the intelligence about the knowledge domain so that software agents could make decisions as humans would.

Most observers agree that Berners-Lee and the researchers at the W3C who are defining the elements of the Semantic Web have a great deal of complex work to do before the results are usable. You can learn more about this project by following the link in the Online Companion to the **W3C Semantic Web** pages.

Summary

In this chapter, you learned about the history of the Internet and the Web, including how these technologies emerged from research projects and grew to be the supporting infrastructure for electronic commerce today. You also learned about the protocols, programs, languages, and architectures that support the Internet and the World Wide Web. TCP/IP is the protocol suite used to create and transport information packets across the Internet. IP addresses identify computers on the Internet. Domain names such as www.amazon.com also identify computers on the Internet, but those names are translated into IP addresses by the routing computers on the Internet. HTTP is the set of rules for transferring Web pages and requests for those Web pages on the Internet. POP, SMTP, and IMAP are protocols that help manage e-mail.

Hypertext Markup Language, or HTML, was derived from the more generic meta language SGML. HTML defines the structure and content of Web pages using markup symbols called tags. Over time, HTML has evolved to include a large number of tags that accommodate graphics, Cascading Style Sheets, and other Web page elements. Hyperlinks are HTML tags that contain a URL. The URL can be a local or remote computer. The better HTML editors facilitate Web page construction with helpful tools and drag-and-drop capabilities.

Extensible Markup Language, or XML, is also derived from SGML. However, unlike HTML, XML uses markup tags to describe the meaning, or semantics, of the text, rather than its display characteristics. XML offers businesses hope for a common language that they will be able to use to describe products, services, and even business processes to each other in common, shared databases. XML could help companies dramatically reduce the costs of handling intercompany information flows.

Intranets are private internal networks that use the same protocols as the Internet. Employees can access the intranet and find, view, or print information just as they would Internet-based material. When companies want to collaborate with suppliers, partners, or customers, they can connect their intranets to each other and form an extranet. The three types of extranets are: public network, private network, and virtual private network. Virtual private networks, or VPNs, provide security at a low cost, whereas public network extranets have no security at all.

Internet service providers offer many different types of connections to the Internet. Basic telephone connections are the most economical and easiest to install, but they are the slowest. Broadband cable, satellite microwave transmission, and DSL services provide Internet access at relatively high speeds. Other, more expensive options provide the bandwidth that larger businesses need. A variety of wireless connection options are becoming available for businesses and homes. The wireless connection options available through cell phones show promise in creating new opportunities for revenue generation, cost reduction, and payment-processing applications.

Internet2 is an experimental network built by a consortium of research universities and businesses that provides a test bed for creating and perfecting the networking technologies of tomorrow. The W3C Semantic Web project holds out hope that in the future, many mundane user interactions with the Web will be handled by intelligent software agents.

Key Terms

802.11a, 802.11b, 802.11g, 802.11n

Anchor tag

Asymmetric connection

Asymmetric digital subscriber line (ADSL or DSL)

Asynchronous transfer mode (ATM)

Backbone router

Bandwidth

Base 2 (binary)

Bluetooth

Border router

Broadband

Byte

Cascading Style Sheets (CSS)

Circuit

Circuit switching

Client/server architecture

Client-side scripting

Closed architecture

Closing tag

Colon hexadecimal (colon hex)

Computer network

Configuration table

Data-grade lines

Digital Subscriber Line (DSL)

Domain name

Dotted decimal

Download

Downstream bandwidth (downlink bandwidth)

Electronic mail (e-mail)

E-mail client software

E-mail server

Encapsulation

Extensible Hypertext Markup Language (XHTML)

Extensible Markup Language (XML)

Extensible Stylesheet Language (XSL)

Extranet

Fixed-point wireless

Fractional T1

Frame relay

Gateway computer

Generalized Markup Language (GML)

Graphical user interface

Hexadecimal (base 16)

Hierarchical hyperlink structure

High-speed DSL (HDSL)

Home page

Hot spot

HTML extensions

Hypertext

Hypertext element

Hypertext link (hyperlink)

Hypertext Markup Language (HTML)

Hypertext server

Hypertext Transfer Protocol (HTTP)

Integrated Services Digital Network (ISDN)

Interactive Mail Access Protocol (IMAP)

Internet

Internet access provider (IAP)

Internet backbone

Internet host

Internet Protocol (IP)

Internet Protocol version 4 (IPv4)

Internet Protocol version 6 (IPv6)

Internet service provider (ISP)

Internet2

Intranet

IP address

IP tunneling

IP wrapper

Leased line

Linear hyperlink structure

Local area network (LAN)

Mailing list

Markup tags (tags)

Mesh routing

Meta language

Mobile commerce (m-commerce)

Multipurpose Internet Mail Extensions (MIME)

Net bandwidth

Network access points (NAPs)

Network access provider

Network Address Translation (NAT) device

Network Control Protocol (NCP)

Network specification

Newsgroup

Octet

Ontology

Open architecture

Opening tag

Optical fiber

Packet

Packet-switched

Personal area network (PAN)

Piconet

Plain old telephone service (POTS)

Post Office Protocol (POP)

Private IP address

Private network

Proprietary architecture

Protocol

Public network

Repeater

Resource description framework

Roaming

Router

Router computer (routing computer)

Routing algorithm

Routing table

Scaling problem

Semantic Web

Short message service (SMS)

Simple Mail Transfer Protocol (SMTP)

Site map

Software agents

Standard Generalized Markup Language (SGML)

Start page

Style sheet

Subnetting

Symmetric connection

T1

T3

TCP/IP

Text markup language

Third-generation (3G) cell phones

Top-level domain (TLD)

Transceiver

Transmission Control Protocol (TCP)

Uniform Resource Locator (URL)

Upload bandwidth

Upstream bandwidth

Usenet (User's News Network)

Virtual private network (VPN)

Voice-grade lines

Web

Web browser

Web browser software

Web client computer

Web client software

Web server

Web server software

Wi-Fi (wireless ethernet, 802.11b, 802.11a, 802.11g, 802.11n)

Wide area network (WAN)

Wireless access point (WAP)

World Wide Web (WWW)

World Wide Web Consortium (W3C)

XML parser

XML vocabulary

Review Questions

RQ1. What were the main forces that led to the commercialization of the Internet? Summarize your answer in about 100 words.

RQ2. Describe in two paragraphs the origins of HTML. Explain how markup tags work in HTML, and describe the role of at least one person involved with HTML's development.

RQ3. In about 200 words, compare the POP e-mail protocol to the IMAP e-mail protocol. Describe situations in which you would prefer to use one protocol or the other and explain the reasons for your preference.

RQ4. In about 400 words, describe the similarities and differences between XML and HTML. Provide examples of at least two situations in which you would use XML and two situations in which you would use HTML.

RQ5. Use your favorite search engine and the links in the Online Companion (under the heading **Internet Connection Options**) to search for more information about broadband satellite connections, DSL connections, wireless connections, and cable connections. Prepare a four-column table (one column for each technology) in which you list the advantages and disadvantages of each connection method. Include at least two advantages and two disadvantages for each connection method.

Exercises

E1. You are the assistant to Julie Davidson, the sales manager of Old Reliable Life Insurance Company. Julie is interested in equipping her sales force with the technology they need to sell Old Reliable's insurance products. Most of her salespeople visit customers in their homes or offices. Today, the salespeople carry a laptop computer to show value projections and cash flow summaries for various policies. Many of them also carry a PDA for appointments and a mobile phone. Julie would like to ensure that salespeople have access to the home office server computers while they are making their sales presentations to customers. This access will let salespeople download the latest product information and obtain online assistance from office staff and in-house experts when the salespeople get a question from a customer that they are not able to answer. A correct and quick answer to a customer's question can often help close a difficult sale. Julie asks you to investigate various options for giving salespeople remote access to the home office server computers. She wants you to consider both wireless (directly to the laptop computers or through salespeoples' cell phones or PDAs) and wired options. Prepare a report for Julie in which you briefly review at least four options, writing no more than three paragraphs for each option. Then choose the best wired option and the best wireless option and write a one-page evaluation of strengths and weaknesses for each of them. Use the Online Companion links and your favorite Web search engines to do your research.

E2. Bridgewater Engineering Company (BECO), a privately held machine shop, makes industrial-quality, heavy-duty machinery for assembly lines in other factories. It sells its presses, grinders, and milling equipment using a few inside salespeople and telephones. This traditional approach worked well during the company's start-up years, but BECO is getting a lot of competition from abroad. Because you worked for the company during the summers of your college years, BECO's president, Tom Dalton, knows you and realizes that you are Web savvy. He wants to form close relationships with the steel companies and small parts manufacturers that are BECO's suppliers so that he can tap into their ordering systems and request supplies when he needs them. Tom wants you to investigate how he can use the Internet to set up such electronic relationships. Use the Web and the links in the Online Companion to locate information about extranets and VPNs. Write a report that briefly describes how companies use extranets to link their systems with those of their suppliers, then write an evaluation of at least two companies (using information you have gathered in your Web searches) that could help develop an extranet that would work for Tom. Close the report with an overview of how BECO could use VPN technologies in this type of extranet. The three parts of your report should total about 700 words.

E3.   Frieda Bannister is the IT manager for the State of Iowa's Department of Transportation (DOT). She is interested in finding ways to reduce the costs of operating the DOT's vehicle repair facilities. These facilities purchase replacement parts and repair supplies for all of the state's cars, trucks, construction machinery, and road maintenance equipment. Frieda has read about XML and thinks that it might help the DOT send orders to its many suppliers throughout the country more efficiently. Use the Online Companion links, the Web, and your library to conduct research on the use of XML in state, local, and federal government operations. Provide Frieda with a report of about 1000 words that includes sections that discuss what XML is and explain why XML shows promise for the ordering application Frieda envisions. Your report should also identify other DOT business processes or activities that might benefit from using XML. The report should also include a summary of the main disadvantages of using XML today for integrating business transactions. End the report with a brief summary of how the W3C Semantic Web project results might help the DOT operate more efficiently in the future.

Cases

C1. Covad

In 1996, three enterprising executives decided to leave their jobs at Intel and form a company that would take advantage of an opportunity provided by the recently enacted Telecommunications Act of 1996. The law eliminated the monopoly that local telephone companies had held and allowed other companies to offer telecommunications services to businesses and individuals in what had been the local telephone companies' protected service areas. Since the goal of the company was to offer converged voice and data services, the founders named the company **Covad**. During its first two years, the company became a solid regional company in the San Francisco Bay and Silicon Valley areas that sold Internet access to businesses and ISPs. Their ISP customers provided DSL access to smaller businesses and residential customers. But the Internet boom was in full swing, and in 1998, Covad hired U.S. West senior vice president Robert Knowling to take the company to the next level. Over the next two years, Covad raised more than $2 billion from stock and bond offerings and expanded into 98 metropolitan areas throughout the country. By the end of 1999, it had more than 200,000 customers, including AOL, MCI, and some of the fastest growing regional ISPs in the country. It was following the lead of its main competitor, NorthPoint Communications (featured in this chapter's Learning from Failures feature), and pursuing a strategy that included rapid expansion using external funds.

In 2000, Covad's largest customers, the ISPs that sold Internet access to smaller companies and residential users, stopped paying their Covad bills because their customers were disappearing. The Internet bubble had burst. Covad had expanded too fast and was in serious trouble. It had put all of its investors' money into equipment and infrastructure during its rapid growth and had no cash reserves to take it through a period of slower growth.

In 2001, the company brought in a new manager, Charlie Hoffman, to take the company through a Chapter 11 bankruptcy reorganization. Covad's bondholders received 19 cents on each dollar after the reorganization. This gave the company a much lower debt payment load and allowed it to focus on rebuilding its business. Hoffman changed the basic strategy of the company by decreasing its emphasis on sales to ISPs who would resell Internet DSL access to small businesses and residential customers. Instead, Covad began selling these access services directly to those customers. By 2004, Covad had more than 700,000 DSL and T1 customers and was getting close to making a profit.

Although the company's original plans were to sell both voice and data services, Covad grew rapidly selling only data services. In 2004, Covad began offering telephone voice services over its data lines (this service is called voice-over-IP, or VoIP). In the direct market for Internet access and VoIP services, Covad faces serious competition from cable companies, who have offered Internet access for many years and who are now also offering VoIP services. Covad's Internet access services also face competition from independent DSL ISPs (including many of Covad's former customers) and from satellite and fixed-point wireless access providers. Finally, Covad's VoIP services also face competition from local telephone companies and the large national mobile telephone service providers.

Required:

1. Prepare a 200-word analysis in which you describe why Covad's recent strategy of selling Internet access directly rather than through ISPs has been successful.

2. Use the links in the Online Companion for this chapter, your favorite search engine, and resources in your library to learn more about Covad and its current competitors in the cable industry (primarily **Cox Cable, Comcast,** and **Time Warner Cable**, but there are others). Prepare a report of about 600 words in which you outline and analyze the strengths and weaknesses that Covad has with respect to each of these competitors.

3. Use the links in the Online Companion for this chapter, your favorite search engine, and resources in your library to learn more about Covad and its current competitors in the local telephone company and national wireless industries. Prepare a report of about 600 words in which you outline and analyze the strengths and weaknesses that Covad has with respect to each of these competitors.

Note: Your instructor might assign you to a group to complete this case, and might ask you to prepare a formal presentation of your results to your class.

C2. Portable Fun Instruments

Yash Gupta is the founder and president of Portable Fun Instruments (PFI), a company that has had great success in the handheld game market. Its first products were dedicated handheld devices that each offered a specific game, such as backgammon, checkers, or chess. As the power of microprocessors grew, and the size and cost of those microprocessors shrank, PFI was able to build better and more complex games into its devices.

Today, PFI offers a wide variety of dedicated devices on which users can play card games, adventure games, and sports simulations, and solve various kinds of puzzles. Most of the elements in the game displays are graphics, not words. This helps PFI sell the devices in many different markets around the world without having to build separate interfaces for each language. PFI's game devices have retail prices that range between $30 and $70, but the retailers and distributors buy them from PFI for prices that range between $12 and $25.

PFI is profitable because Yash has worked hard to keep development and production costs low. Most of the programming is done in Bangalore, India, and the devices are built in production facilities located in Xixiang, China and Penang, Malaysia. Although Yash has been successful in controlling production costs, he worries about continuing to operate the company with a long-term strategy that requires PFI to build a new physical device for each sale. The large retail chains that have become PFI's main customers are always asking for discounts and reduced prices on new orders, and production costs are creeping upward even though the facilities are located in some of the lowest-cost areas in the world.

Yash wants to explore the potential PFI has for moving its games to other platforms. PFI has translated some of its games to PDA platforms, such as the Palm Operating System and the Microsoft Windows CE Operating System (sometimes called the Pocket PC system), but the results have been disappointing. Most PDA users are businesspeople who use their PDAs for appointments, address books, travel expenses, and other data management functions. These users are not avid game players, and sales of PFI's games for these platforms have also not been strong.

Some of PFI's marketing team members have been telling Yash about the success that Japan's DoCoMo has had offering a variety of entertainment products that are downloaded from the Internet and that are displayed on mobile phones. DoCoMo charges users for their use of these products, which include games, and shares those fees with the providers of the products. Yash has also heard that a U.S. mobile phone technology company, Qualcomm, is offering a similar service called Binary Runtime Environment for Wireless (BREW). Yash believes that these markets might be worth pursuing.

Yash has hired you as a consultant to investigate DoCoMo, BREW, and any similar delivery systems for selling online access to PFI's games to mobile phone users. He is also interested in learning more about the programming and markup languages used by these delivery systems. The programming teams in Bangalore are becoming familiar with XML because they are doing contract programming projects for other companies, and Yash wants to know if these skills will help them adapt PFI's games to mobile phone delivery systems.

Required:

1. Use the links in the Online Companion for this case, your favorite search engine, and resources in your library to learn more about DoCoMo, BREW, and similar content delivery systems for mobile phones. Prepare a 400-word executive summary for Yash that describes each delivery system, identifies the company or companies behind each system, and outlines the current availability of each system for content providers such as PFI.

2. Prepare a report for Yash and the PFI executive team in which you outline and analyze the strengths and weaknesses of each content delivery system. Your report should conclude with a specific recommendation regarding the suitability of each content delivery system for PFI's games. This report should be 1000 to 2000 words in length.

3. Prepare a 300-word report that describes the programming and markup languages used in each delivery system. In this report, discuss whether the PFI programming team's experience with XML will help it adapt PFI's games to mobile phone content delivery systems.

Note: Your instructor might assign you to a group to complete this case, and might ask you to prepare a formal presentation of your results to your class.

For Further Study and Research

Alschuler, L. 2001. "Getting the Tags In: Vendors Grapple with XML-Authoring, Editing and Cleanup," *Seybold Report on Internet Publishing*, 5(6), February, 5–10.

Angwin, J. 2002. "ICANN Leader Seeks Big Changes in How Internet Is Governed," *The Wall Street Journal*, February 26, B6.

Babcock, C. 2001. "XML Databases Offer Greater Search Capabilities," *Interactive Week*, 8(18), May 7, 11–13.

Bannan, K. 2002. "Satellite: The Only Game Out of Town," *PC Magazine*, 21(3), February 21, 99.

Benner, J. 2002. "Getting a Lock on Broadband," *Salon.com*, June 7. (http://www.salon.com/tech/feature/2002/06/07/broadband/print.html)

Bergman, M. 2001. *The Deep Web: Surfacing Hidden Value.* Sioux Falls, SD: BrightPlanet.com. (http://brightplanet.com/technology/deepweb.asp)

Bonner, P. 2002. "The Semantic Web," *PC Magazine*, 21(13), July, IP01–IP02.

Bosak, J. and T. Bray. 1999. "How XML Will Fix the Web: Tags Categorizing Facts, Not Formats, Speed Up Transactions," *Scientific American*, 280(5), May, 89.

Boyle, M. 2002. "The Shiniest Reputations in Tarnished Times," *Fortune*, 145(5), March 4, 70–72.

Brewin, B. 2004. "Michigan City Turns on Citywide Wi-Fi," *Computerworld*, July 30. (http://www.computerworld.com/mobiletopics/mobile/wifi/story/0,10801,94928,00.html)

Campbell, T. 1998. "The First E-Mail," *Pretext Magazine*, March. (http://www.pretext.com/mar98/features/story2.htm)

Caulfield, B. 2002. "Wi-Fi Goes to Work," *Business 2.0*, 3(6), June, 122–123.

Comer, D. 2001. *Computer Networks and Internets With Internet Applications*, 3rd Edition. Upper Saddle River, NJ: Prentice Hall.

Computergram Weekly. 2003. "Cisco Sees Flat Sales But Chambers More Cheerful," May 7, 7–8.

Cope, J. 2001. "IPv6: Is It Inevitable?" *Computerworld*, 35(22), May 28, 58–59.

Costa, D. 2001. "Cable: This Technology Is the Simplest and Most Popular Option," *PC Magazine*, 20(3), February 6, 149–151.

Davis, K. and E. Burt. 2001. "Mad as Hell about DSL," *Kiplinger's Personal Finance Magazine*, 55(7), July 2001, 84–85.

Dornan, A. 2003. "Unwiring the Last Mile," Network Magazine, 18(1), January, 34–37.

Duffy, J. 2003. "RBOCs and Cable Wage Turf War," Network World, August 18, 11–14.

Dyck, T. 2002. "Going Native: XML Databases," *PC Magazine*, 21(12), June 30, 136–139.

The Economist. 2002. "ICANNOT," March 2, 59.

Eisenzopf, J. 2002. "Is VoiceXML Right for Your Customer Service Strategy?" *New Architect*, 7(3), March, 20–21.

Elgin, B. and M. Roman. 2002. "Cisco Rides Again," *Business Week*, May 20, 50.

Fensel, D., J. Hendler, H. Lieberman, and W. Wahlster. 2002. *Spinning the Semantic Web: Bringing the World Wide Web to Its Full Potential.* Cambridge, MA: MIT Press.

Fitchard, K. 2004. "Covad's Quiet Authority," *Telephony*, 245(12), June 7, 34–39.

Fixmer, R. 2002. "Broadband Homeland," *eWeek*, March 4, 41–43.

Floyd, M. 2002. "XML Exposed," *PC Magazine*, 21(12), June 30, 132–140.

Fruhlinger, J. 2002. "Broadband and the New User Experience," *Web Techniques*, 7(2), February, 23–25.

Garcia, J. and J. Wilkins. 2001. "Cable Is Too Much Better to Lose," *The McKinsey Quarterly*, January, 185–188.

Goldfarb, C. 1981. "A Generalized Approach to Document Markup," *ACM Sigplan Notices*, (16)6, June, 68–73.

Hardy, Q. 2001. "Cisco Kidding?" *Forbes*, 167(11), May 14, 52–53.

Hardy, Q. 2002. "The Great Wi-Fi Hope," *Forbes*, 169(6), March 18, 56–64.

Hawn, C. 2001. "Management By Stock Market: NorthPoint Rode the Web Wave," *Forbes*, 167(10), April 30, 52–53.

Henderson, T. and T. Ritchey. 2002. "Bring in the (802.11) A Team," *Network World*, 19(4), January 28, 55–56.

Hochmuth, P. 2002. "What Customers Want From Cisco," *Network World*, 19(2), June 3, 1.

Huston, J. 2001. "Scaling the Internet," *Satellite Broadband: The Cutting Edge of Satellite Communications,* 2(3), March, 18–22.

Kalata, K. 2001. *Internet Programming with VBScript and JavaScript.* Boston: Course Technology.

Kaven, O. 2004. "Wired Ethernet and 802.11g Outpace the Rest," *PC Magazine*, 23(6), April 6, 104.

LaBarba, L. 2001. "DSL Pains Reach End Users," *Telephony*, 240(15), April 9, 14–15.

Lawson, S., K. Miyake, and J. Evers. 2002. "IPv6 Enters the Real World," *InfoWorld*, 24(7), February 18, 35–36.

Liebman, L. 2001. "XML's Tower Of Babel," *InternetWeek*, April 30, 25–26.

Lowry, T. 2004. "Satellite's Hot Pursuit of Cable, *Business Week*, May 24, 46.

Markoff, J. 2002. "Two Tinkerers Say They've Found a Cheap Way to Broadband," *The New York Times*, June 10, C1.

Marsan, C. 2004. "It's a New Domain-Name Game," *Network World*, March 1, 1, 14.

Martin, M. 2002. "Not Part of the Cable and DSL Boom?" *Network World*, February 11, 21.

Metz, C. 2002. "The All Mail Revue," *PC Magazine*, 21(9), May 7, 85–99.

Milstein, S. 2003. "Be Your Own Wireless Network," *The New York Times,* February 27, 4.

Mitchell, R. 2002. "Wireless at Full Throttle," *Computerworld*, 36(20), May 13, 63.

Nelson, T. 1987. *Literary Machines*. Swarthmore, PA: Nelson.

Nielsen, J. 2003. "Mobile Devices: One Generation From Useful," *Alertbox*, August 18. (http://www.useit.com/alertbox/20030818.html)

Nolle, T. 2002. "Are Cable Companies the Key to Local Access?" *Network Magazine*, April, 104.

Nolle, T. 2002. "Why Is Cisco Making Money?" *Network World*, 19(22), June 3, 51.

O'Connor, R. 2000. "Under Construction: Two Research Groups Work to Build a Better Internet," *Interactive Week*, 7(46), November 13, 44–48.

Olivia, R. 2001. "The Promise of XML," *Marketing Management*, 10(1), Spring, 46–49.

Panko, R. 2002. *Business Data Networks and Telecommunications*. 4th Edition. Upper Saddle River, NJ: Prentice Hall.

PC Magazine. 2001. "Turn XML into HTML," 20(11), IP01–IP04.

Pimm, F. 2001. "Boeing Shows How XML Can Help Business," *Computerworld*, 35(11), March 12, 28–29.

Port, O. 2002. "The Next Web," *Business Week*, March 4, 96–102.

Rendleman, J. 2002. "Cisco Positioned to Profit From Changing Market," *Information Week*, May 13, 32.

Richtel, M. 2004. "Where Entrepreneurs Go and the Internet Is Free," *The New York Times*, June 7. (http://www.nytimes.com/2004/06/07/technology/07wifi.html)

Robertson, D. 2001. "Tweaking Protocols," *Satellite Broadband: The Cutting Edge of Satellite Communications*, 2(2), February, 26–28.

Schonfeld, E. 2002. "Unwiring the Masses," *Business 2.0*, 3(6), June, 18–20.

Schrick, B. and M. Riezenman. 2002. "Wireless Broadband in a Box," *IEEE Spectrum*, June. (http://www.spectrum.ieee.org/WEBONLY/publicfeature/jun02/wire.html)

Shafer, S. 2002. "Secure Extranets Arrive," *InfoWorld*, 24(10), March 11, 27.

Simon, B. 2003. "Some Bet the Future of Broadband Belongs to Regional Bells, Not Cable," *The New York Times,* July 21. (http://www.nytimes.com/2003/07/21/technology/21BROA.html)

Spangler, T. 2002. "Crossing the Broadband Divide," *PC Magazine*, 21(3), February 21, 92–101.

Swanson, S. 2002. "Post Office In—and Out—of Wireless Early," *Information Week*, April 8, 24.

Thurm, S. 2002. "Cisco Profit Exceeds Expectations," *The Wall Street Journal*, May 8, A3.

Vogelstein, F. 2004. "The Cisco Kid Rides Again," *Fortune*, 150(2), July 26, 132–137.

Weber, T. 2002. "Wanted: A Peace Envoy to End Net's Bickering Over Address System," *The Wall Street Journal*, March 25, B1.

White, C. 2002. *Data Communications and Computer Networks: A Business User's Approach,* 2nd Edition. Boston: Course Technology.

Wildstrom, S. 2002. "Broadband Needs Home Improvement," *Business Week*, April 1, 19.

Witte, G. 2003. "Bringing Broadband Over the Mountain: Roadstar Puts Wireless Technology to the Test," *The Washington Post*, September 15, E1.

Zhang, M. and R. Wolff. 2004. "Crossing the Digital Divide: Cost-Effective Broadband Wireless Access for Rural and Remote Areas," *IEEE Communications Magazine*, 42(2), February, 99–105.

PART **2**

BUSINESS STRATEGIES FOR ELECTRONIC COMMERCE

CHAPTER **3**

SELLING ON THE WEB: REVENUE MODELS AND BUILDING A WEB PRESENCE

LEARNING OBJECTIVES

In this chapter, you will learn about:

- Revenue models
- How some companies move from one revenue model to another to achieve success
- Revenue strategy issues that companies face when selling on the Web
- Creating an effective business presence on the Web
- Web site usability
- Communicating effectively with customers on the Web

INTRODUCTION

The **Vanguard Group** manages a variety of pooled investment accounts for individuals and institutions. These pooled accounts are called mutual funds. Vanguard earns revenue by charging an annual management fee that is based on the size of each mutual fund. Thus, Vanguard is interested in attracting new investors to its mutual funds and managing the funds' holdings so they grow over time. To sell its investment products effectively, Vanguard must maintain good communications with both current and potential customers. In the past, Vanguard used the telephone and mail to develop and maintain contact with its customers and prospects. As more and more investors began using the Internet, Vanguard realized that the Web would give it another good way to stay in touch with customers and prospects. To that end, Vanguard spent more than $100 million to develop and refine its Web site. In its current version, Vanguard's Web site allows customers to obtain account information, manage current investments, and make further investments in Vanguard mutual funds. Many other mutual fund management companies have Web sites and most of those sites are focused on promoting the investment products those companies offer. Vanguard has chosen not to focus its Web site on direct product promotion. Instead, Vanguard's strategy has been to use its Web site to build customer loyalty and to promote its products indirectly.

The stated purpose of Vanguard's Web site is to educate its customers and provide them with a high level of service. Many times, information on the site discourages customers from buying particular mutual fund shares if those shares are inappropriate for their investment goals. Vanguard's corporate policy is to help its customers make good investment decisions rather than to pick up a quick profit by selling customers on particular investments. Although Vanguard risks losing sales in the short run, it believes that by providing information and educating its potential customers, it will achieve better long-term growth. The company's Web site is a consistent representation of that basic corporate strategy.

REVENUE MODELS

As you learned in Chapter 1, a useful way to think about electronic commerce implementations is to consider how they can generate revenue. Not all electronic commerce initiatives have the goal of providing revenue; some are undertaken to reduce costs or improve customer service. You will learn about cost reduction initiatives in Chapter 5. In this chapter, you will learn about various models for generating revenue used by Web businesses today, including Web catalog, advertising-supported, advertising-subscription mixed, and fee-based models. These approaches can work for both business-to-consumer (B2C) and business-to-business (B2B) electronic commerce. Many companies create one Web site to handle both B2C and B2B sales. Even when companies create separate sites (or separate pages within one site), they often use the same revenue model for both types of sales.

Web Catalog Revenue Model

Many companies sell goods and services on the Web using an adaptation of a mail order catalog revenue model that is over 100 years old. In 1872, a traveling salesman named Aaron Montgomery Ward started selling dry goods to farmers through a one-page list. Richard Sears and Alvah Roebuck began mailing catalogs to farmers and small town residents in 1895. Both Montgomery Ward (which closed in 2001) and Sears, Roebuck & Company grew to become dominant retailers in the United States by the 1950s, with retail stores serving urban markets in addition to the catalog business that served their rural and small-town markets.

In this traditional catalog-based retail revenue model, the seller establishes a brand image, and then uses the strength of that image to sell through printed information mailed to prospective buyers. Buyers place orders by mail or by calling the seller's toll-free telephone number. This revenue model, which is often called the **mail order** or **catalog model**, has proven to be successful for a wide variety of consumer items, including apparel, computers, electronics, housewares, and gifts.

Companies can take this catalog model online by replacing or supplementing their print catalogs with information on their Web sites. When the catalog model is expanded this way, it is often called the **Web catalog revenue model**. Customers can place orders through the Web site or by telephone. This flexibility is important because many consumers are still reluctant to buy on the Web. In the first few years of consumer electronic commerce, most shoppers used the Web to obtain information about products and compare prices and features, but then made their purchases by telephone. These shoppers found early Web sites hard to use and were often afraid to send their credit card numbers over the Internet. Although these fears are less prevalent today, most companies that use the Web catalog revenue model do give customers a way to complete the payment part of the transaction by telephone or by mail.

Many of the most successful Web catalog sales businesses are firms that were already operating in the mail order business and simply expanded their operations to the Web. Other companies that use the Web catalog revenue model adopted it after realizing that the products they sold in their physical stores could also be sold on the Web. This additional sales outlet did not require them to build additional stores, yet provided access to customers throughout the world. Types of businesses using the Web catalog revenue model include sellers of computers and consumer electronics; books, music, and videos; luxury goods; clothing; flowers and gifts; and general discount merchandise. In the next sections, you will learn how these types of businesses have applied the Web catalog revenue model to their operations.

Computers and Consumer Electronics

Leading computer manufacturers such as **Apple**, **Dell**, **Gateway**, and **Sun Microsystems** have had great success selling on the Web. All four of these companies sell a full range of products—from small desktop computers to large server computers—to individuals, businesses, and other organizations through their Web sites.

Dell has been a leader in allowing customers to specify exactly the configuration of computers they order on the Web. Dell created value by designing its entire business around offering this high degree of configuration flexibility to its customers. Other personal computer manufacturers that sell directly to customers on the Web have followed Dell's lead by offering visitors different ways to access product information. These sites usually offer links to specific products and pages designed for specific categories of customers, such as home, small business, education, or government users.

Retailers of consumer electronics products have also been active in undertaking electronic commerce using the Web catalog revenue model. Companies such as **Crutchfield** and **The Sharper Image** expanded their successful mail order catalog operations to include Web sites such as the Sharper Image site shown in Figure 3-1. Other companies that had strong retail presences in their physical stores, such as **Best Buy**, **Circuit City**, **The Good Guys!**, **J&R Music World**, and **Radio Shack**, also opened Web sites to sell the same products that they had been selling in their stores.

FIGURE 3-1   Sharper Image home page

Books, Music, and Videos

Retailers using the Web catalog model to sell books, music, and videos have been among the most visible examples of electronic commerce. In 1994, a 29-year-old Wall Street financial analyst named Jeff Bezos became intrigued by the rapid growth of the Internet. Looking for a way to capitalize on this new marketing tool, he made a list of 20 products that he thought would sell well on the Internet. After some intense analysis, he determined that books were at the top of that list. Bezos had no experience in the book-selling business, but he realized that books were small-ticket commodity items and were easy and inexpensive to ship. He knew many customers would be willing to buy books without inspecting them in person and that books could be impulse purchase items if properly promoted. Over 4 million book titles are in print at any one time throughout the world; however, even the largest physical bookstore cannot stock more than 200,000 books. Bezos had identified a strategic opportunity for selling online. Ten years later, **Amazon.com**, the company Bezos formed to sell books on the Internet, has annual sales of over $6 billion and more than 60 million customers. Amazon.com has evolved to become a general retailer that sells books, music, videos, consumer electronics, housewares, tools, and many other items.

The rapid growth of Amazon.com inspired many booksellers to undertake electronic commerce. A number of well-established companies that operated physical bookstores, such as **Barnes & Noble**, **Blackwell's**, **Books-A-Million**, and **Powell's Books**, all adopted the Web catalog model in their online sales endeavors. Borders eventually decided to close its site and have Amazon.com handle its online business, but the other companies continue to operate their own sites successfully.

In 1994, the same year that Jeff Bezos started his online bookstore, 24-year-old twin brothers Jason and Matthew Olim began an online music store they called **CDnow** that used the Web catalog revenue model. By 1997, CDnow had one-third of the online music business. Its success attracted many competitors. Companies such as **Tower Records** and **Sam Goody**, which had been selling music in their retail stores for years, opened Web sites to compete with CDnow. Web-only retailers such as **CD Universe** copied CDnow's approach exactly. CDnow's founders sold their company to German music conglomerate Bertelsmann AG, which created an alliance with Amazon.com to use Amazon.com's electronic commerce platform for its sales. The CDnow site appears in Figure 3-2.

FIGURE 3-2 CDnow site running on the Amazon.com platform

Luxury Goods

For some types of products, people are still reluctant to buy through a Web site. This is particularly true for luxury goods and high-fashion clothing items. The Web sites of couturiers **Vera Wang** and **Versace**, for example, were not constructed to generate revenue directly, but to provide information to shoppers who would visit the physical stores to examine items they had seen on the sites. Such sites tend to make heavy use of graphics and animation. **Evian**, the purveyor of premium-priced bottled water, went so far as to create a Web site that works well only on computers that are connected to the Internet by a broadband connection. Evian intentionally designed its site for a select, affluent group of customers. The Flash animation takes a long time to download to computers that have an inexpensive dial-up modem connection.

Another company that has a Web site created entirely in Flash is upscale jewelry and gift retailer **Tiffany & Co.** The Flash version of the Tiffany site loads in its own browser window, which covers the browser window that gives users the option of choosing an HTML-only site. By loading the Flash version of the site as the default, the company is catering to customers who can afford a broadband connection.

Clothing Retailers

A number of apparel sellers have adapted their catalog sales model to the Web, including **bobo**, **Gap**, **Lands' End**, **L. L. Bean**, **Talbots**, and **Wet Seal**. Unlike sellers in the high-fashion clothing category discussed above, these Web stores display photos of casual and business clothing with prices, sizes, colors, and tailoring details. Their intent is to have customers examine the clothing and place orders through the Web site. Lands' End pioneered the idea of online Web shopping assistance with its Lands' End Live feature in 1999. A Web customer with a question can initiate a text chat with a customer service representative or click a button on the Web page to have the representative call. In addition to answering questions, the representative can offer suggestions by pushing Web pages to the customer's browser.

Many of Lands' End's competitors (including Eddie Bauer, L.L. Bean, and Talbots) added similar text chat and call-back features to their sites. More recently, Lands' End added personal shopper and virtual model features to its site. The **personal shopper** is an intelligent agent program that learns the customer's preferences and makes suggestions. The **virtual model** is a graphic image built from customer measurements on which customers can try clothes. About 15 percent of visitors to the site use the virtual model and, on average, dress the model 40 times during a visit. Lands' End has found that the dollar amount of orders placed by customers who use the virtual model is about 10 percent larger than other orders. The Canadian company that developed this Web site feature, **My Virtual Model**, has sold the technology to a number of other clothing retailers. A person who constructs a virtual model on one of those sites can use that same model on the other sites. Figure 3-3 shows the Lands' End virtual model in action.

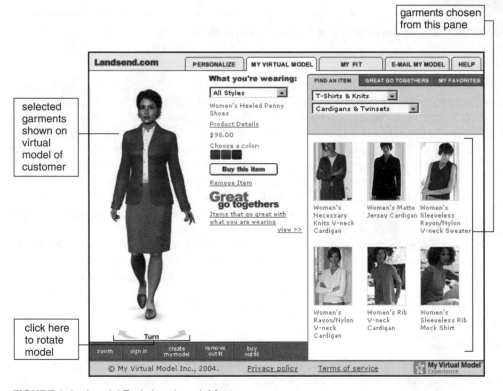

FIGURE 3-3 Lands' End virtual model feature

Lands' End also has a feature that allows two shoppers to browse the Web site together from different computers. Only one of the shoppers can purchase items, but either shopper can select items to view. These items appear in both Web browsers.

In the fast-changing clothing business, retailers have always had to deal with the problem of overstocks—products that did not sell as well as hoped. Many retailers use outlet stores to sell their overstocks. Lands' End found that its overstocks Web page worked so well that it has closed some of its physical outlet stores. An online overstocks store works well because it reaches more people than a physical store and it can be updated more frequently than a printed overstocks catalog.

In addition to general apparel retailers, a number of specialty retailers opened stores on the Web. For example, women's shoe retailers such as **Steve Madden** and **Nine West** use the Web catalog model to sell directly to consumers on their sites.

One problem that the Web presents for clothing retailers of all types is that the color settings on computer monitors vary widely. It is difficult for customers to get an accurate idea of what the product's color will look like when it arrives. Until technology solves this problem, most online clothing stores will send a fabric swatch on request. The swatch also gives the customer a sense of the fabric's texture—an added benefit not provided by catalogs. Most Web catalog retailers also have generous return policies that allow customers to return unused merchandise for any reason.

Flowers and Gifts

Gift retailers also use the Web catalog revenue model. Florist **1-800-Flowers** created an online extension to its highly successful telephone order business to compete with online-only florists such as **Calyx & Corolla** and **Proflowers.com**. Chocolatier **Godiva** offers business gift plans on its site. For gift shoppers who want a familiar brand name, shopping mall mainstays **Hickory Farms** and **Mrs. Fields Cookies** both have created Web catalog sites. **Harry and David**, famous for its trademarked "Fruit-of-the-Month" club, opened an informational Web site to promote its existing catalog business. The company was surprised by the volume of sales leads that the site generated, and quickly added online ordering features to the site, which appears in Figure 3-4.

General Discounters

A number of new companies have started retail operations on the Web. Some of these completely new businesses, such as **Buy.com**, operate as Web-based deep discounters. Borrowing a concept from the physical world's Wal-Marts and discount club stores, these discounters sell merchandise such as computer equipment, software, consumer electronics, books, music CDs, and sports equipment at extremely low prices.

Some of these Web discount retailers originally sold advertising on their sites to subsidize their low product prices. Beyond.com closed its retail operation and now sells the software it created for operating a Web catalog site. Buy.com changed its approach because advertising revenues were not sufficient subsidies. Buy.com now relies on the same volume-purchasing strategy as physical world retailers to keep prices low. As in the physical world, the online discount retail business is fiercely competitive and many of these companies operate on thin margins—and consequently earn little profit. Cyberian Outpost began business in 1995 as one of the first retailers on the Web. In 2001, after six years of winning

FIGURE 3-4 Harry and David home page

awards for customer service, it ran out of cash and was purchased by Fry's Electronics, which continues to operate the **Outpost.com** Web site as a subsidiary.

Traditional discount retailers, such as **Costco**, **Kmart**, **Target**, and **Wal-Mart**, were slow to introduce electronic commerce on their Web sites. Many industry observers criticized these traditional retailers for their slow entry into online sales; however, those same industry observers now expect the traditional retailers to do very well competing against the retailers that started on the Web.

LEARNING FROM FAILURES

WALMART.COM

Wal-Mart is the world's largest retailer, with more than 4000 stores and annual sales of $280 billion. Founded in 1962 by retailing legend Sam Walton, the company has won numerous awards for business innovation. However, Wal-Mart's move into online retailing has been troubled, to say the least.

Wal-Mart launched its first Web site in July 1996. Like most company sites of that time, it contained some information about the company, but did not offer any products for sale. Wal-Mart did little to develop the Web site over the next three years, but it did add a Web store—just in time to participate in the disastrous 1999 holiday shopping season.

Wal-Mart was not the only Web retailer to have trouble in 1999. Many companies found that they were ill prepared for the large number of customers who decided to try electronic commerce in that year's holiday season. Lost orders, unfilled orders, and shipments that failed to arrive until January 2000 were common for many Web retailers that year. Wal-Mart was noted as an industry leader in shipping and logistics management; however, the announcement on its Web site that it could not promise Christmas delivery for items ordered after December 14 was particularly embarrassing.

To make matters worse, Wal-Mart was in the middle of developing a new Web site that it had hoped to launch before the holiday season. The project, which industry analysts estimate cost over $100 million, ran months late and did not operate until January 2000.

After eight months of operating the new Web site, Wal-Mart found itself with low levels of customer traffic (well below those of its major rivals J.C. Penney, Sears, Kmart, and Target) and high levels of criticism from Web site design experts who found the site slow, difficult to use, and lacking customer service features.

In October 2000, Wal-Mart closed the site completely for four weeks. Earlier in the year, it had created Walmart.com, a joint venture with Accel Partners to develop a new Web site, but the new site was not ready to launch until November. Industry analysts widely criticized Wal-Mart's decision to completely shut down its Web operations for such a long time period at the beginning of the holiday shopping season.

The new Web site is a vast improvement over the old site. It is much better organized and offers improved browsing and search functions. The new site offers about the same number of items as the old site did (about 500,000—several times more than what the physical stores carry); however, the new site has more offerings of consumer electronics, toys, and sporting goods, and fewer offerings of consumable products. Behind the scenes is a new distribution center that serves Walmart.com exclusively.

Walmart.com's experience is a testament to how difficult it can be to get Web retailing right. Success eluded the largest retailer in the world for years. Wal-Mart is estimated to have spent more than $150 million on its various Web implementations before it was able to present a truly usable site to its customers.

Digital Content Revenue Models

Firms that own intellectual property or rights to that property have embraced the Web as a new and highly efficient distribution mechanism. **LexisNexis** began as a legal research tool, and it has been available as an online product for years. Today, LexisNexis offers a variety of information services, including legal information, corporate information, government information, news, and resources for academic libraries. The original legal information product exists on the Web today as **Lexis.com** and provides full-text search of court cases, laws, patent databases, and tax regulations. In the past, law firms had to subscribe to and install expensive dedicated computer systems to obtain access to this information.

The Web has given LexisNexis customers much more flexibility in how they purchase information. Through the Lexis.com Web site, law firms can subscribe to several versions of the service that are customized for different firm sizes and usage patterns. The Web site even offers a credit card charge option for infrequent users who do not want a subscription. LexisNexis has used the Web to improve the delivery and variety of its existing product line and has been able to devise new products that take advantage of the Web's features.

ProQuest, a Web site that sells digital copies of published documents, has its roots in two businesses: the former Bell and Howell learning materials business and University Microfilms International (UMI). These firms acquired reproduction rights to a variety of published and unpublished materials. For example, UMI had contracts with most North American universities to publish all doctoral dissertations and masters theses on demand. ProQuest offers digital versions of these documents for sale, along with a number of newspapers, journals, and other specialized academic publications. Many schools and libraries have subscriptions to ProQuest. **Ovid** and **EBSCO Information Services** also sell subscriptions to digital versions of journals and books to corporate and university libraries. These companies sell access to bibliographic databases and electronic journals to individuals, schools, companies, and libraries as well. The EBSCO Information Services home page that appears in Figure 3-5 shows the types of services it offers.

Dow Jones, a business-focused publisher of newspapers such as *The Wall Street Journal* and *Barron's*, was one of the first publishers to create a Web site for selling subscriptions to digitized newspaper, magazine, and journal content. The Dow Jones Interactive site offered a customized digital clipping service that provided subscribers with a daily e-mail message of news on topics of interest to them. In 2002, Dow Jones and Reuters, a British company, joined to create an online content management and integration service called **Factiva**. In addition to the content and services previously offered on the Dow Jones Interactive site, Factiva gives companies the ability to integrate their existing content (such as a corporate library) with Dow Jones and Reuters news sources.

One of the first academic organizations to make the transition to electronic distribution on the Web was (not surprisingly) the Association for Computer Machinery (ACM). The **ACM Digital Library** offers subscriptions to electronic versions of its journals to its members and to library and institutional subscribers. Academic publishing has always been a difficult business in which to make a profit because the base of potential subscribers is so small. Even the most highly regarded academic journals often have fewer than 2000 subscribers. To break even, academic journals often must charge each subscriber hundreds or even thousands of dollars per year. Electronic publishing eliminates the high costs

FIGURE 3-5   EBSCO Information Services home page

of paper, printing, and delivery, and makes dissemination of research results less expensive and more timely.

As was the case for other technologies, such as VCRs and subscription cable television, many of the early commercial users of Web technology were dealers in adult-themed entertainment material. Many of the first profitable sites on the Web were sellers of adult digital content. These sites pioneered the online processing of credit card payment transactions (about which you will learn in Chapter 11) and many different digital video technologies that are now used by all types of businesses on the Web.

Advertising-Supported Revenue Models

The **advertising-supported revenue model** is the one used by network television in the United States. Broadcasters provide free programming to an audience along with advertising messages. The advertising revenue is sufficient to support the operations of the network and the creation or purchase of the programs. Many observers of the Web in its early growth period believed that the potential for Internet advertising was tremendous. Web advertising grew from essentially zero in 1994 to $2 billion in 1998. However, Web advertising was flat or declining in the years 2000 through 2002. Since then, Web advertising has once again started to grow, but at much lower rates than in the early years of the Web. After trying to develop profitable advertising-supported revenue models on the Web, most companies today are considerably less optimistic about the general potential of these revenue models. However, a few information sites, such as **About.com**, **HowStuffWorks**, and the ***Drudge Report***, are successful in using advertising-supported revenue models. The sites that have been successful tend to be sites that attract a specific group of visitors to which advertisers can direct specific messages. For example, About.com and HowStuffWorks both provide pages of information that are directed at visitors with highly focused interests. A visitor looking for an explanation of how heating stoves work on either of these sites would be a good prospect for advertisers that sell heating stoves. The site would not need to obtain any specific information from the visitor, the fact that the visitor is viewing the heating stoves information page is enough justification for charging an advertiser a higher rate for ads placed on those pages.

The overall success of online advertising has been hampered by two major problems. First, no consensus has emerged on how to measure and charge for site visitor views. Since the Web allows multiple measurements, such as number of visitors, number of unique visitors, number of click-throughs, and other attributes of visitor behavior, it has been difficult for Web advertisers to develop a standard for advertising charges. In addition to the number of visitors or page views, stickiness is a critical element in creating a presence that attracts advertisers. The **stickiness** of a Web site is its ability to keep visitors at the site and attract repeat visitors. People spend more time at a **sticky** Web site and are thus exposed to more advertising.

The second problem is that very few Web sites have sufficient numbers of visitors to interest large advertisers. Most successful advertising on the Web is targeted to very specific groups. The set of characteristics that marketers use to group visitors is called **demographic information**, which includes such things as address, age, gender, income level, type of job held, hobbies, and religion. It can be difficult to determine whether a given Web site is attracting a specific market segment unless that site collects demographic information from its visitors—information that visitors are increasingly reluctant to provide because of privacy concerns.

Web Portals

Few general-interest sites have generated sufficient traffic to be profitable based on advertising revenue alone. The drop in advertising rates and spending that occurred between 2000 and 2002 created difficulties for even the largest advertising-supported sites. One of the leading general-interest sites is **Yahoo!**, which was one of the first Web directories. A **Web directory** is a listing of hyperlinks to Web pages. Because so many people use Yahoo! as a starting point for searching the Web, it has always attracted a large number of visitors. This large number of visitors made it possible for Yahoo! to expand its Web directory into one of the first portal sites. A **portal** or **Web portal** is a site that people use as a launching point to enter the Web (the word "portal" means "doorway"). A portal almost always includes a Web directory and search engine, but it also includes other features that help visitors find what they are looking for on the Web and thus make the Web more useful. Most portals include features such as shopping directories, white pages and yellow pages searchable databases, free e-mail, chat rooms, file storage services, games, and personal and group calendar tools.

Because the Yahoo! portal's search engine presents visitors' search results on separate pages, it can include advertising on each results page that is triggered by the terms in the search. For example, when the Yahoo! search engine detects that a visitor has searched on the term "new car deals," it can place a Ford ad at the top of the search results page. Ford is willing to pay more for this ad because it is directed only at visitors who have expressed interest in new cars. This example demonstrates one attractive option for identifying a target market audience without collecting demographic information from site visitors. Unfortunately, only a few high-traffic sites are able to generate significant advertising revenues this way.

Besides Yahoo!, the main portal sites using the advertising-supported revenue model today are **AOL**, **AltaVista**, **Excite**, **Google**, **Lycos**, **Netscape NetCenter**, and **MSN**. Smaller general-interest sites, such as the Web directory **refdesk.com**, have had more difficulty attracting advertisers than the larger search engine sites. This may change in the future as more people use the Web. Another type of portal that may be able to earn a profit with smaller numbers of visitors is the portal that offers items of interest to a specialized interest group. The technology portal **C-NET** is one example of this type of portal. You will learn more about portal strategies in Chapter 6.

Newspaper Publishers

Many newspapers publish all or part of their print content on the Web. The **Internet Public Library Online Newspapers** page includes links to hundreds of newspaper sites around the world. It is unclear whether a newspaper's presence on the Web helps or hurts the newspaper's business as a whole. Although it provides greater exposure for the newspaper's name and a larger audience for advertising that the paper carries, it also can take away sales from the print edition. Like retailers or distributors whose online sales lead to the loss of their brick-and-mortar sales, publishers also experience sales losses as a result of online distribution. Newspapers and other publishers worry about these sales losses because they are very difficult to measure. Some publishers have conducted surveys in which they ask people whether they do not buy the newspaper because the content they want to see is available online, but the results of such surveys are not very reliable.

In addition to the concern about lost sales of print editions, most newspaper publishers have found that the cost of operating their Web sites cannot be covered by the revenue they generate from selling advertising on the sites. Thus, many newspaper publishers are currently experimenting with various other ways of generating revenue from their Web sites. You will learn about these alternative revenue models later in this chapter. Because newspapers are now using several different online revenue models, you will see newspapers mentioned in the discussions of several different revenue models.

Targeted Classified Advertising Sites

Although attempts to create general-interest Web sites that generate sufficient advertising revenue to be profitable have met with mixed results, sites that target niche markets have been more successful. For newspapers, classified advertising is very profitable; thus, Web sites that specialize in providing only classified advertising do have profit potential. This is especially true if they can reach a narrow target market and charge higher rates because the advertising reaches the right audience.

One implementation of the advertising-supported revenue model that is successful is Web employment advertising. Industry analysts estimate that e-recruiting revenues will exceed $80 billion by 2007. Companies such as **CareerSite** offer international distribution of employment ads. As the number of people using the Web increases, these businesses will be able to move beyond their current focus on technology and higher-level jobs and include advertising for all kinds of positions. These sites can use the same approach that search engine sites use to offer advertisers target markets. When a visitor specifies an interest in, for example, engineering jobs in Dallas, the results page can include a targeted banner ad for which an advertiser will pay more because it is directed at a specific segment of the audience.

Employment ad sites can also target specific categories of job seekers by including short articles on topics of interest. These articles increase the site's stickiness and attract people who are not necessarily looking for a job. This is a good tactic because people who are not looking for a job are often the candidates most highly sought by employers. The **Monster.com** page directed at management-level job applicants (as distinguished from entry-level job applicants or executive-level job applicants) appears in Figure 3-6. This page offers links to articles, reports, a message board, and chat sessions that might interest a mid-career manager. It also offers a subscription to a newsletter for managers.

Another type of classified advertising Web site that can generate sufficient revenue to be profitable is the used vehicle site. Trader Publishing has printed advertising newspapers for many years and now operates the **AutoTrader.com**, **CycleTrader.com**, and **BoatTrader.com** sites. These sites accept paid advertising from individuals and companies that want to sell cars, motorcycles, and boats. Trader Publishing charges a fee for each listing and gives the seller the option of running the ad on the Web site only or on the Web and in the print version of the advertising newspaper. If the product has a dedicated following, this type of site can be successful by catering to small audiences. For example, the **VetteFinders** site sells classified ads for Corvette automobiles only.

Any product that is likely to be useful after the original buyer uses it provides the potential for a classified advertising site. People who want to sell used musical instruments can place ads on the **Musicians Buy-Line** site. Comic book collectors will find classified ads directed to them at **ComicLink.com**. Golfers who have given up the game or moved on to better clubs can place classified ads for their old equipment on **The Golf Classifieds**.

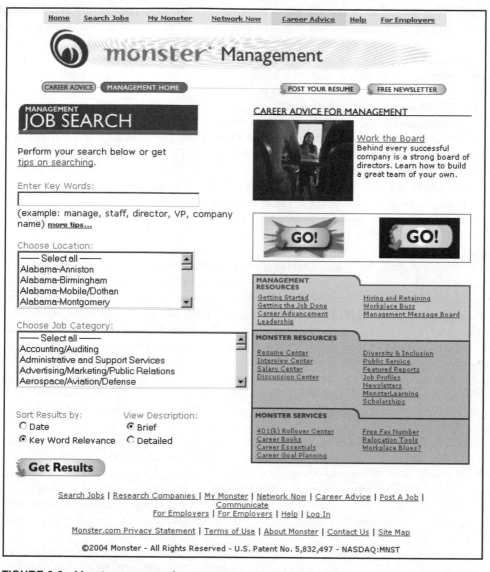

FIGURE 3-6   Monster.com page for management-level job candidates

Advertising-Subscription Mixed Revenue Models

In an **advertising-subscription mixed revenue model**, which has been used for many years by traditional print newspapers and magazines, subscribers pay a fee and accept some level of advertising. On Web sites that use the advertising-subscription revenue model, subscribers are typically subjected to much less advertising than they are on advertising-supported sites. Firms have had varying levels of success in applying this model and a number of companies have moved to or from this model over their lifetimes.

Two of the world's most distinguished newspapers, **The New York Times** and **The Wall Street Journal**, use a mixed advertising-subscription model. *The New York Times* version is mostly advertising supported, with a $35 annual subscription fee for visitors who want online access to the newspaper's premium crossword puzzle pages. The paid subscription service originally included the bridge and chess columns, but those are now free to any registered site visitor (registration is free). *The New York Times* also provides a searchable archive of articles dating back to 1996 and charges a small fee for viewing any article older than one week. *The Wall Street Journal's* mixed model is weighted more heavily to subscription revenue. The site allows nonsubscriber visitors to view the classified ads and certain stories from the newspaper, but most of the content is reserved for subscribers who pay an annual fee for access to the site. Visitors who already subscribe to the print edition are offered a reduced rate on subscriptions to the online edition.

Note that both of these newspapers use one version of this revenue model for their print editions and another version for their online editions. More and more newspapers and magazines are finding that they need to use different revenue models for their print and online editions.

Some newspapers, including **The Washington Post** and the **Los Angeles Times**, use another variation of the mixed revenue model. These newspapers do not charge any subscription fees for access to their Web sites. Instead, they offer current stories free of charge on their Web sites, but require visitors to pay for articles retrieved from their archives.

Business Week offers yet another variation on the mixed model theme. It offers some free content at its **Business Week online** site, but requires visitors to buy a subscription to the *Business Week* print magazine if they want to gain access to the entire site. Subscribers who want to read archived articles that are more than five years old are levied an additional charge per article. *Business Week* does place content in the subscriber section of its Web site before the magazine appears on the newsstands or is delivered to subscribers.

Sports fans visit the **ESPN** site for all types of sports-related information. Leveraging its brand name from its cable television businesses, ESPN is one of the most visited sites on the Web. It sells advertising and offers a vast amount of free information, but die-hard fans can subscribe to its Insider service to obtain access to even more sports information. Thus, ESPN uses a mixed model that includes advertising and subscription revenue, but it only collects the subscription revenue from Insider subscribers, who make up only a small portion of site visitors.

Consumers Union, the publisher of product evaluations and ratings monthly magazine *Consumer Reports*, operates a Web site, **ConsumerReports.org**, that relies heavily on subscriptions. Consumers Union is a not-for-profit organization that does not accept advertising as a matter of policy because it might appear to influence its research results. Thus, the site is supported by a combination of subscriptions and a small amount of charitable donations. The Web site does offer some free information as a way to attract subscribers and fulfill its organizational mission of encouraging improvements in product safety.

Fee-for-Transaction Revenue Models

In the **fee-for-transaction revenue model**, businesses offer services for which they charge a fee that is based on the number or size of transactions they process. Some of these services lend themselves well to operating on the Web. To the extent that companies can offer

Web site visitors the information they need about the transaction, companies can offer much of the personal service formerly provided by human agents. If customers are willing to enter transaction information into Web site forms, these sites can provide options and execute transactions much less expensively than traditional transaction service providers. The removal of an intermediary, such as a human agent, from a value chain is called **disintermediation**. The introduction of a new intermediary, such as a fee-for-transaction Web site, into a value chain is called **reintermediation**.

Travel Agents

Travel agents earn commissions on each airplane ticket, hotel reservation, auto rental, or vacation that they book. These commissions are paid to the travel agent by the transportation or lodging provider. The travel agency revenue model involves receiving a fee for facilitating a transaction. The value added by a travel agent is that of information consolidation and filtering. A good travel agent knows many things about the traveler's destination and knows enough about the traveler to select the information elements that are useful and valuable to the traveler. Computers, particularly computers networked to large databases, are very good at information consolidation and filtering. In fact, travel agents have used networked computers, such as the **Sabre** system, for many years to make reservations for their customers.

When the Internet emerged as a new way to network computers and then became available to commercial users, a number of online travel agencies began doing business on the Web. Existing travel agencies did not, in general, rush to the new medium. They believed that the key value they added, personal customer service, could not be replaced with a Web site. Therefore, the first Web-based travel agencies were new entrants. One of these sites, **Travelocity**, is based on the same Sabre system that traditional travel agents use (Travelocity is also owned by Sabre).

Microsoft also established a position in the online travel agency business with its **Expedia** subsidiary. **Travelocity**, **Expedia**, **Hotels.com**, and **Hotel Discount Reservations** are regularly listed among the top electronic commerce sites in surveys and industry analyst rankings. All four are profitable. In 2001, a consortium of five major U.S. airlines (American, Continental, Delta, Northwest, and United) launched a new Web travel site, **Orbitz**. A number of consumer groups and the attorneys general of 20 states expressed concern over possible antitrust issues that could arise with a site such as Orbitz, which is sponsored by competing airlines. For example, Orbitz offers a monetary incentive to airlines that agree to offer Orbitz customers their lowest fares at all times. That means that the airline cannot offer a special low fare on its own Web site (or on another site such as Travelocity or Expedia) unless it also makes that fare available through Orbitz. The site launch was met with mixed reviews in surveys and criticism from industry analysts. The site encountered some technical difficulties; however, the site did immediately generate significant amounts of visitor traffic. Within a year, Orbitz had become one of the three most visited travel sites on the Web. The Orbitz home page appears in Figure 3-7.

In addition to earning commissions from the transportation and lodging providers, these sites generate advertising revenue from ads placed on travel information pages. These ads are similar to those on search engine results pages because advertisers can target them without obtaining demographic details about the site visitor. For example, if you are

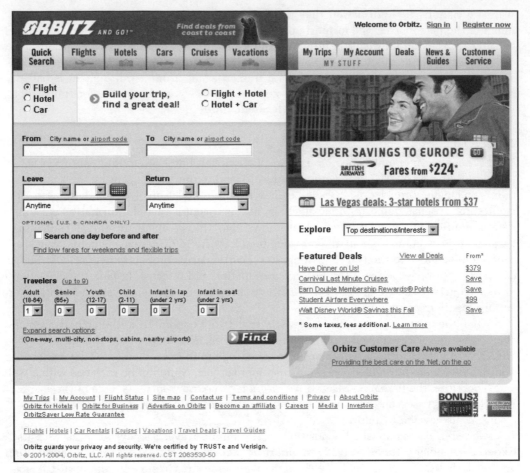

FIGURE 3-7 Orbitz home page

booking a flight to Chicago, the page that lists airline ticket options may also carry a banner ad for a hotel in Chicago or a car rental company that is running a promotion in the Chicago area.

Many traditional travel agents are finding themselves squeezed out of the business today. Airlines are reducing or even eliminating the commissions they pay to travel agents on each ticket. Many travel agents (including the online travel sites) now charge their customers a flat fee for processing a ticket on an airline that has reduced or eliminated the fees it pays to travel agents.

Although these changes have hurt all travel agents, some industry observers believe that the large online travel agencies will have a better chance of surviving any shakeout that occurs in the business. Other industry observers note that smaller traditional travel agents often specialize in cruises or finding specialized hotel accommodations. Both cruise lines and hotels still view travel agents as an important part of their selling strategy and continue to pay reasonable commissions to travel agents on the sales that they make.

Some travel agents have been successful by following a reintermediation strategy with a focus on specific groups of travelers. These travel agents identify a group of travelers with specific needs and create travel packages designed for that group. For example, surf vacations have become increasingly popular in recent years. The stereotypical surfer of years gone by (a young unemployed male) has been replaced by a much broader demographic. Today's surfers often have significant financial resources and enjoy surfing in exotic locations. Web sites such as **WaveHunters.com** and **WannaSurf** have followed a reintermediation strategy and cater to this highly specialized market in ways that generalist travel agents have not.

Automobile Sales

Auto dealers buy cars from the manufacturer and sell them to consumers. They provide showrooms and salespeople to help customers learn about product features, arrange financing, and make a purchase decision. Most auto dealers negotiate the prices at which they sell their cars; thus, the salesperson's job also includes extracting the highest possible price from the consumer. Many people do not like negotiating car prices, especially if they have taken the time to learn about car features, arrange financing, and are ready to purchase a car without further assistance from a salesperson.

As you learned in Chapter 1, Autobytel and other firms offer knowledgeable consumers an option that removes the salesperson from the process. Autobytel and similar firms, such as **MSN Autos** and **CarsDirect.com**, provide an information service to car buyers. Each of these firms implements the fee-for-transaction revenue model in a slightly different way. For example, CarsDirect.com offers customers the ability to select a specific car (model, color, options) at a price it determines. CarsDirect.com then finds a local dealer that has such a car and is willing to sell it for the CarsDirect.com price. Alternatively, **Autoweb.com** and **Autobytel** locate dealers in the buyer's area that are willing to sell the car specified by the buyer (including make, model, options, and color) for a small premium over the dealer's nominal cost. The buyer can purchase the car from the dealer without negotiating with a salesperson. Autobytel and Autoweb.com charge participating dealers a fee for this service. In effect, firms such as Autobytel, Autoweb.com, and CarsDirect.com are taking the salesperson out of the value chain. To the extent that the salesperson provides little or no value to the consumer, these firms are reducing the transaction costs in the process. The car salesperson is disintermediated and the Web site becomes the new intermediary in the transaction, which is an example of reintermediation.

Stockbrokers

Stock brokerage firms also use a fee-for-transaction model. They charge their customers a commission for each trade executed. In the past, stockbrokers offered investment advice and made specific buy and sell recommendations to customers. They did not charge for this advice, but they did charge relatively high commissions on the trades they handled for their customers. After the U.S. government deregulated the securities trading business in the early 1970s, a number of discount brokers opened. These discount brokers distinguished themselves from the established "full-line" brokerage houses by not offering any investment advice and charging very low commissions. Because the full-line brokers had failed to provide value to some of their customers, those customers were very happy to move their business to the discount brokers.

The Web made it possible for firms such as **E*TRADE** and Datek (later purchased by **Ameritrade**) to offer investment advice (posted on Web pages) similar to that offered by a full-line broker, without incurring many of the costs of distributing the advice (such as stockbroker salaries, overhead, and the costs of printing and mailing newsletters). Web-based brokerage firms could also offer fast execution of trades that customers entered into Web page forms. Thus, in the 1990s, discount brokers who had taken business away from full-line brokers for 15 years faced new competition from online firms. Of course, the full-line brokers found that they were losing business to both the discount brokers and the online brokers. In response, both discount brokers (such as **Charles Schwab** and **Ameritrade**) and full-line brokers (such as **Merrill Lynch** and **Smith Barney**) opened new stock trading and information Web sites.

The online brokers are offering customers the same kind of transaction cost reductions as the online auto buying sites. Stockbrokers are finding themselves disintermediated in the same way as car salespeople. Online brokers are offering an alternative service that has greater perceived value for many investors today.

Insurance Brokers

Other sales agency businesses are moving to the Web. Although insurance companies themselves were slow to offer policies and investments for sale on the Web, a number of intermediaries that sell insurance policies from a variety of companies have been online since the early days of the Web. Quotesmith, which began business in 1984 as a policy-quoting service for independent insurance brokers, decided in 1996 to offer its policy price quotes directly to the public over the Internet. By quoting policies and accepting applications directly, Quotesmith is disintermediating the independent insurance agents with whom it formerly worked. Quotesmith operates the **Insure.com** Web site, which appears in Figure 3-8.

Other Web sites that offer insurance policy information, comparisons, and sales include **InsWeb**, **Answer Financial**, **Insurance.com**, and **YouDecide.com**, which was created by the human resources software development company **ProAct Technologies**. In response to the appearance of independent Web sites that offered customers a way to compare prices from various insurance companies, **Progressive Insurance** decided to offer quotes on its Web site with an interesting twist. The company provides quotes for its insurance products and also for its competitors' products. If a site visitor finds that one of the competitor's products is less expensive, Progressive provides a link to that company's site so site visitors can buy their insurance elsewhere. Progressive has always promoted itself as offering the lowest-priced insurance, and this is a way that the company reinforces its image. If it cannot offer the lowest price, the company invites potential customers to buy elsewhere. Progressive's well-advertised strategy encourages many insurance shoppers to visit the Progressive site instead of an independent comparison site. Today, many major insurance companies, such as **Allstate**, **GEICO**, and **State Farm Insurance**, offer information or policies for sale on their Web sites.

tabs link to information about specific types of insurance policies

FIGURE 3-8 Quotesmith's Insure.com Web site

Event Tickets

Obtaining tickets for concerts, shows, and sporting events can be a challenge. Some venues only offer tickets for sale at their own box offices, and others sell tickets through ticket agencies that can be difficult for patrons to find. The Web offers event promoters an ability to sell tickets from one virtual location to customers practically anywhere in the world. Traditional ticket agencies such as **Ticketmaster** have opened shop online. Other new entrants, such as **Tickets.com** and **TicketWeb**, also offer a wide variety of tickets for events in many different locations. These electronic commerce initiatives reduce transaction costs for both buyers and sellers of tickets.

Real Estate and Mortgage Loan Brokers

Other fee-for-transaction businesses are also starting to open electronic commerce Web sites, including real estate brokers and mortgage loan brokers. Online real estate brokers provide all of the services that a traditional broker might provide—except that online brokers provide these services through their Web sites. Leading real estate sites include Web pioneers **eRealty** and **zipRealty.com**. Industry observers agree that these new online brokers do a much better job selling on the Web than traditional real estate brokers that have opened Web sites, such as **Coldwell Banker** and **Prudential**. The industry's trade association, the National Association of Realtors, sponsors a Web site, **Realtor.com**, that carries ads for houses listed by its member companies.

IndyMac Bank Home Lending offers mortgage loan seekers an online credit review and decision in minutes. Approved customers can then print an approval letter from their own computers and take it with them the same day to shop for a new house. This rapid decision-making ability and other customer service features have helped IndyMac become one of the leading mortgage loan sites on the Web, funding more than $12 billion in home loans each year. Other successful mortgage brokers on the Web include **Ditech** and **E-LOAN**. These sites often provide helpful features such as the IndyMac Bank Quick Pricer, a loan payment calculator that appears in Figure 3-9.

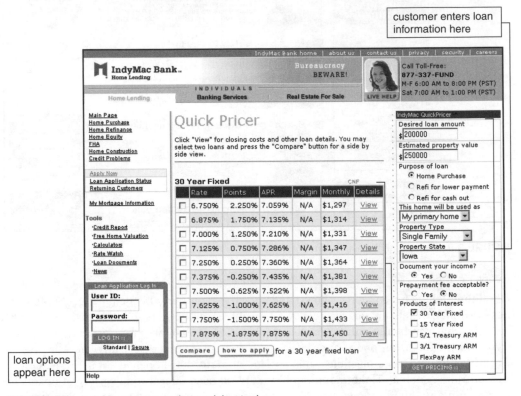

FIGURE 3-9   IndyMac mortgage loan pricing tool

Online Banking and Financial Services

Because financial services do not involve a physical product, they are easy to offer on the Web. The greatest concerns that most people have when they consider moving their financial transactions to the Web are security and the reliability of the financial institution—the same concerns that exist in the physical world. However, on the Web, it is much more difficult for a firm to establish its reputation for security and trust than it is in the physical world, where massive buildings and clearly visible room-sized safes can help create the necessary image. Many people who are willing to buy products and services online are unwilling to trust a Web site for their banking services. Fewer than 15 percent of all people who have made a purchase online also conduct their banking business online. You will learn more about the security of Web payments and other Internet financial transactions in Chapter 11.

Some existing banks opened online "branches" that carry the identification and reputation of the physical world bank's brand (such as Citibank's **Citibank Online**). Other firms started online banks that are not affiliated with an existing bank (such as the **First Internet Bank of Indiana**). Bank One opened an online bank under the name Wingspan in 1999. Bank One's expressed intent at the time was to present Wingspan as a new and separate entity, in the spirit of the dot-com boom that was then under way. After operating Wingspan separately for about two years, Bank One decided to close Wingspan and merge it with the **BankOne.com** site. Industry observers today agree that an online bank can benefit from using the name of an established traditional bank to help create its reputation and provide customers with a sense of trustworthiness.

Online banks handle only a tiny portion of the world's financial transactions today, but as the reputation and reliability of online banks grow, more customers will accept them as a good way to conduct their banking business. Banks are interested in serving their customers online because it costs the bank less to provide services to a customer online than to provide services through a personal interaction with bank employees in a branch office. In addition to customers' concerns about trust and security online, two other significant barriers are preventing a more rapid rate of growth in the online banking business: a lack of bill presentment features and a lack of account aggregation tools.

Today's online banks give customers a way to pay their bills electronically, but the customers still receive most of the bills in the mail. The bills that they can access on the Web are at sites with **bill presentment** features. Unfortunately, most people must visit a different Web site to view each online bill. As online banks add bill presentment features that allow their customers to view all of their bills on the bank's Web site (and pay each of them with a single click), those banks will find more customers willing to do their banking on the Web.

Another important feature that few online banks currently offer is **account aggregation**, which is the ability to obtain bank, investment, loan, and other financial account information from multiple Web sites and display it all in one location at the bank's Web site. Many of a bank's best customers have credit card, loan, investment, and brokerage accounts with multiple financial institutions. Having all of this information in one place would be very useful for these customers. Some nonbank sites, including **Yodlee** and **MSN Money**, are offering account aggregation tools. These companies offer account aggregation services and, in some cases, license their technology to other Web sites. For example, both the **Quicken Banking Center** and the **Yahoo! Finance** page use Yodlee technology. Some banks have also begun to add bill presentment features to their online banking services.

Fee-for-Service Revenue Models

Companies are offering an increasing variety of services on the Web for which they charge a fee. These are neither broker services nor services for which the charge is based on the number or size of transactions processed. The fee is based on the value of the service provided. These **fee-for-service revenue models** range from games and entertainment to financial advice and the professional services of accountants, lawyers, and physicians.

Online Games

Computer and video games are a huge industry. In the United States alone, more than $10 billion per year is spent on these types of games. An increasing portion of that revenue is generated online. Although many sites that offer games relied on advertising revenue in the past, a growing number now include premium games in their offerings. Site visitors must pay to play these premium games, either by buying and downloading software to install on their computers, or by paying a subscription fee to enter the premium games area on the site. Microsoft's **MSN Games by Zone.com**, Sony's **Station.com**, RealNetworks' **RealArcade**, and **Electronic Arts** are among the leading game sites that include subscription game services. For example, Sony's EverQuest adventure game draws more than 400,000 players who have purchased a $40 software package and pay $10 per month to continue playing the game. Most of the game sites charge a monthly subscription of between $5 and $20 for access to all their fee-based games offerings. The **Entertainment Software Association** is an industry group that tracks computer and video game use. Its Web site includes a number of interesting statistics about computer game sales and demographics of game players. For example, more than 40 percent of frequent computer game players are over the age of 35!

Concerts and Films

As more households obtain broadband access to the Internet, an increasing number of companies provide streaming video of concerts and films to paying subscribers. With a revenue model patterned after cable television companies, **Intertainer** began selling subscriptions for delivery of video content to computers and other devices through cable modem and DSL connections in 1999. Intertainer had built its business to more than 140,000 subscribers in 2002 when it closed and filed a lawsuit against several major media companies alleging that they were illegally controlling the market. Despite the ongoing lawsuit, MGM Studios, Paramount Pictures, Sony Pictures Entertainment, Universal Studios, and Warner Brothers Studios formed a joint venture to build the **Movielink** site. Movielink offers downloadable movies drawn from the content owned by the joint venture partners and licensed from other content owners such as Walt Disney Pictures and Miramax. Movielink sells a 24-hour window of access to the downloaded movies for a $2 to $5 fee. RealNetwork's **RealOne SuperPass** subscription includes sporting events, music videos, comedy, and other entertainment offerings for $13 per month.

The main technological limitation these companies face is that each additional customer who downloads a video stream requires that the provider purchase additional bandwidth from its ISP. Television broadcasters, on the other hand, need only pay the fixed cost of a transmitter—the airwaves are free and carry the transmission to an unlimited number of viewers at no additional cost. In contrast, as the number of an Internet-based provider's subscribers increases, the cost of the provider's Internet connection increases. However, if these Web entertainment companies can charge a high enough monthly fee, they should be able to cover the additional costs of technology upgrades and still make a profit.

Professional Services

State laws have been one of the main forces preventing U.S. professionals (such as physicians, lawyers, accountants, and engineers) from extending their practices to the Web. Since most professionals are licensed by individual states, state laws can effectively prevent them from practicing their professions on the Web because their patients or clients could be located in other states. If they were to offer their services over the Web, professionals could be charged with unlicensed practice in those other states. State laws that address the location of services are vague—it is difficult to determine where a service provided over the Internet actually occurred. This uncertainty arises because most state professional practice laws were written long before the Internet existed.

Although some medical, legal, and other professional practices are using online services such as **MyDocOnline** to allow patients to make appointments, most practices are reluctant to do even that limited amount of business activity on the Web. The major concern expressed by physicians regards the protection of patients' privacy. Until the Web is perceived as more secure, most people will continue to be reluctant to make medical appointments, or even refill prescriptions, on the Web.

The **Law on the Web** site offers legal consultations on a variety of matters for residents of the United Kingdom. Accounting professionals in the United States can be located through the **CPA Directory**, and a number of legal referral sites can direct site visitors to local attorneys. Although a number of Web sites such as **RealAge**, **Dr. Andrew Weil's Self Healing**, **pain.com**, and **WebMD** (shown in Figure 3-10) offer general health information, physicians and other health care professionals have been reluctant to sell specific advice to specific patients over the Internet. The difficulty of diagnosing medical problems without a physical examination of the patient is a significant barrier to providing many types of health care services on the Web.

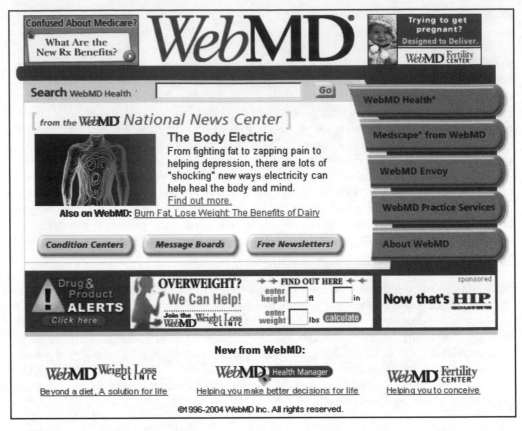

FIGURE 3-10 WebMD home page

REVENUE MODELS IN TRANSITION

Many companies have gone through transitions in their revenue models as they learn how to do business successfully on the Web. As more people use the Web to buy goods and services, and as the behavior of those Web users changes, companies often find that they must change their revenue models to meet the needs of those new and changing Web users. Some companies created electronic commerce Web sites that needed many years to grow large enough to become profitable. This is not unusual; both CNN and ESPN took more than 10 years to become profitable and they had both created new businesses in television, which was an existing and well-established medium. After the investment community became reluctant to continue funding most Web businesses in 2000, many Web companies that were counting on additional investments to support them during their unprofitable growth phases were forced to either change their revenue models or go out of business.

This section describes the revenue model transitions undertaken by five different companies as they gained experience in the online world and faced the changes that occurred in that world. As the world embarks on the second wave of electronic commerce, these and other companies might well face the need to make further adjustments to their revenue models.

Subscription to Advertising-Supported Model

Microsoft founded its **Slate** **magazine** Web site as an upscale news and current events publication. Because *Slate* included experienced writers and editors on its staff, many people expected the online magazine to be a success. Microsoft believed that the magazine had a high value, too. At a time when most online magazines (also called **e-zines**, or electronic magazines) were using an advertising-supported revenue model, *Slate* began charging an annual subscription fee after a limited free introductory period.

Although *Slate* drew a wide readership and received acclaim for its incisive reporting and excellent writing, it was unable to draw a sufficient number of paid subscribers. At its peak, *Slate* had about 27,000 subscribers generating annual revenue of $500,000, which was far less than the cost of creating the content and maintaining the Web site. *Slate* is now operated as an advertising-supported site. Because it is a part of Microsoft, *Slate* does not report its own profit numbers, but most industry observers believe that the site does not earn a profit. Microsoft maintains the *Slate* site as part of its MSN portal, so it is likely that *Slate* increases the stickiness of the portal.

Advertising-Supported to Advertising-Subscription Mixed Model

Another upscale online magazine, **Salon.com**, that has also received acclaim for its innovative content, has moved its revenue model in the direction opposite to *Slate*'s transition. After operating for several years as an advertising-supported site, *Salon.com* now offers an optional subscription version of its site. The subscription offering was motivated by the company's inability to raise the additional money from investors that it needed to continue operations.

Subscribers pay $30 per year to view a version of the magazine called *Salon Premium*, which is free of advertising and can be downloaded for storage and later offline reading on the subscriber's computer. Premium subscribers also gain access to additional content such as downloadable music, e-books, and audio books.

Advertising-Supported to Fee-for-Services Model

Xdrive Technologies opened its original advertising-supported Web site in 1999. Xdrive offered free disk storage space online to users. The users saw advertising on each page and had to provide personal information that allowed Xdrive to send targeted e-mail advertising to them. Its offering was very attractive to Web users who had begun to accumulate large files, such as MP3 music files, and wanted to access those files from several computers in different locations.

After two years of offering free disk storage space, Xdrive found that it was unable to pay the costs of providing the service with the advertising revenue it had been able to generate. It switched to a subscription-supported model and began selling the service to business users as well as individuals. The amount of the monthly subscription is based on the amount of disk space reserved for the user. In recent years, disk drive costs have been dropping and Xdrive has frequently adjusted its monthly fee downward. Currently, Xdrive offers 10 GB of storage for about $20 per month. Other companies, such as **IBackup** and **Kela** have followed Xdrive's lead in offering online storage for a monthly fee.

Advertising-Supported to Subscription Model

Northern Light was founded in August 1997 as a search engine with a twist. In addition to searching the Web, it searched its own database of journal articles and other publications to which it had acquired reproduction rights. When a user ran a search, Northern Light returned a results page that included links to Web sites and abstracts of the items in its own database. Users could then follow the links to Web sites, which were free, or purchase access to the database items.

Thus, Northern Light's revenue model was a combination of the advertising-supported model used by most other Web search engines plus a fee-based information access service, similar to the subscription services offered by ProQuest and EBSCO that you learned about earlier in this chapter. The difference in the Northern Light model was that users could pay for just one or two articles (the cost was typically $1–5 per article) instead of paying a large amount of money for unlimited access to its database on an annual subscription basis. Northern Light also offered subscription access to most of its database to companies, schools, and libraries, however.

In January 2002, Northern Light decided that the advertising revenue it was earning from the ads it sold on search results pages was insufficient to justify continuing to offer that service. It stopped offering public access to its search engine and converted to a new revenue model that was primarily subscription supported. Northern Light's main revenue source in its new model is from annual subscriptions to large corporate clients. It still offers an individual account option, however. A person interested in having the ability to search the Northern Light database can open an account, supply a credit card number, and be billed monthly for the articles accessed. In late 2003, new owners took over the operation of Northern Light and announced an intention to reopen the Web search engine site.

Multiple Transitions

Encyclopædia Britannica is an excellent example of a company that transferred its existing reputation for high quality to the Web. Encyclopædia Britannica has developed one of the most respected brand names in research and education over its many years in print publishing. It is particularly interesting that Encyclopædia Britannica began in 1768 as a sort of precomputer-age frequently asked questions (FAQ) list. A group of academics collected notes they had made while conducting research and decided to publish them as a series of articles.

Encyclopædia Britannica began its online expansion with two Web-based offerings. The Britannica Internet Guide was a free Web navigation aid that classified and rated information-laden Web sites. It featured reviews written by Britannica editors who also selected and indexed the sites. The company's other Web site, Encyclopædia Britannica Online, was available for a subscription fee or as part of the Encyclopædia Britannica CD package. Britannica used the free site to attract users to the paid subscription site.

In 1999, disappointed by low subscription sales, Britannica converted to a free, advertiser-supported site. The first day the new site, **Britannica.com**, became available at no cost to the public, it had over 15 million visitors, forcing Britannica to shut down for two weeks to upgrade its servers.

The Britannica.com site then offered the full content of the print edition in searchable form, plus access to the *Merriam-Webster's Collegiate Dictionary* and the *Britannica Book of the Year*. One of the most successful aspects of the site was the way it integrated the Britannica Internet Guide Web-rating service with its print content. The Britannica Store sold the CD version of the encyclopedia along with other educational and scientific products to help generate revenue.

After two years of trying to generate a profit using this advertising-supported model, Britannica faced declining advertising revenues. In 2001, Britannica returned to a mixed model in which it offered free summaries of encyclopedia articles and free access to the *Merriam-Webster's Collegiate Dictionary* on the Web, with the full text of the encyclopedia available for a subscription fee of $50 per year or $5 per month.

Britannica has gone from being a print publisher to a seller of information on the Web to an advertising-supported Web site to a mixed advertising subscription model—three major revenue model transitions—in just a few short years. The main value that Britannica has to sell is its reputation and the expertise of its editors, contributors, and advisors. For now, Britannica has decided that the best way to capitalize on that reputation and expertise is through a combined format of subscriptions and advertising support. Figure 3-11 shows the Britannica search page available to subscribers. The clean look of the advertising-free page is noticeable. The page offers direct links to all of Britannica's subscriber-only services.

FIGURE 3-11   Britannica paid subscriber search page

REVENUE STRATEGY ISSUES

In the first part of this chapter, you learned about the revenue models that companies are using on the Web today. In this section, you will learn about some issues that arise when companies implement those models. You will also learn how companies deal with those issues.

Channel Conflict and Cannibalization

Companies that have existing sales outlets and distribution networks often worry that their Web sites will take away sales from those outlets and networks. For example, **Levi Strauss & Company** sells its Levi's jeans and other clothing products through department stores and other retail outlets. The company began selling jeans to consumers on its Web site in mid-1998. Many of the department stores and retail outlets that had been selling Levi's products for many years complained to the company that the Web site was now competing with them. In January 2000, Levi Strauss decided to stop selling products on its own Web site.

Such a **channel conflict** can occur whenever sales activities on a company's Web site interfere with its existing sales outlets. The problem is also called **cannibalization** because the Web site's sales consume sales that would be made in the company's other sales channels. The **Levi's** Web site now provides product information, but directs customers who want to buy its products to online stores that carry those products. Levi's product pages also include links that lead to a store finder page, so that customers who want to shop for Levi's products in a physical store can find stores near them.

Maytag, the manufacturer of home appliances, found itself in the same position as Levi Strauss. It created a Web site that allowed customers to order directly from Maytag. After less than two years of operating its direct sales outlet and receiving many complaints from its authorized distributors and resellers, Maytag decided to incorporate online partners into its Web site store design. Now, after searching and gathering information about specific products from the Maytag Web site, a customer can select a retailer who will deliver and install the appliance. These retail store partners are authorized Maytag distributors. The customer can complete the transaction on the Maytag site or can choose to complete the transaction on the retailer's site.

Both Levi's and Maytag faced channel conflict and cannibalization issues with their retail distribution partners. Their established retailers sold many times the dollar volume that either company ever hoped they could sell on their own Web sites. Thus, to avoid angering their retailers, who could always sell competing products, both Levi's and Maytag decided that it would be best to work with their retail partners. Similar issues can also arise within a company if that company has established sales channels that would compete with direct sales on the company's Web site.

Eddie Bauer, a retailer of clothing and outdoor gear, was selling through a catalog and retail stores located primarily in major shopping malls when it decided to begin selling products on its Web site. The company believed that it could make online sales more attractive by allowing customers to return unwanted products that they had purchased online at the retail store locations. The managers of these stores were concerned about the time it would take for their sales associates to process these returns and about having to add the items to their stores' inventories. In a retail store operation, managing labor costs and inventory are very important in achieving store profitability. The managers at the company's catalog division were also worried. They feared that sales through the Web site would cannibalize sales through the catalog.

By making adjustments in the managers' compensation and bonus plans, Eddie Bauer was able to convince all of the managers to support the Web site. The retail store managers were credited with an inventory and labor cost allowance for each Web site return they handled. The catalog division managers were given a credit for existing catalog customers who purchased goods from the Web site. By giving their customers access to the company's products through a coordinated presence in all three distribution channels, Eddie Bauer was able to increase overall sales to those customers. This type of solution is called **channel cooperation**.

Strategic Alliances and Channel Distribution Management

When two or more companies join forces to undertake an activity over a long period of time, they are said to create a strategic alliance. When companies form a strategic alliance, they are operating in the network form of organization that you learned about in Chapter 1. Companies form strategic alliances for many purposes. An increasing number of businesses are forming strategic alliances to sell on the Web. For example, the relationships that Levi's created with its retail partners by giving them space on the Levi's Web site to sell Levi's products is an example of a strategic alliance.

Earlier in this chapter you learned about Yodlee, the account aggregation services provider, and the Web portal sites that offer these services to consumers. The relationship between Yodlee and its portal site clients is another example of a strategic alliance. Yodlee can concentrate on developing the technology and services while the Web portals provide the customers. Because account aggregation services increase the propensity of customers to return to the site, they add to the portal sites' stickiness. Thus, both parties benefit from the strategic alliance.

As you learned earlier in this chapter, Amazon.com has added many product lines to its original offering of books. In some cases, Amazon.com built these businesses from the ground up. In other cases, it forged strategic alliances with existing firms. Amazon has joined with Target to sell that discount retailer's products on a Web site devoted to Target products. The Target site is housed within the Amazon.com site. Its partner in the tools and hardware category is **Tool Crib of the North**. Amazon.com agreed to operate the online sales function for **Borders** after that retailer decided it did not want to continue in the online books and music business. Amazon.com has teamed up with **ToysRUs** to sell toys and with **CDnow** to sell music and video products on the Amazon.com Web site. The ToysR Us page appears in Figure 3-12.

Another type of strategic alliance that industry observers expect to see taking place in the near future is the joining of Web sites with channel distribution management companies. **Channel distribution managers**, also called **fulfillment managers** or **category managers**, are companies that take over responsibility for a particular product line within a retail store. For example, the **Handleman Company** is a channel distribution manager that has specialized in retail music sales for more than 50 years. Wal-Mart, Kmart, and other large retailers have Handleman manage their inventories of music CDs in their physical stores. Companies such as Handleman monitor inventory levels, order CDs, maintain in-store product displays, and coordinate marketing and advertising of music CDs for their partners.

The idea behind this type of affiliation is that the channel distribution manager can develop and maintain more knowledge about the specific product line, or category, than its partners because it specializes in that product category. Even after taking its fee, which is usually a percentage of the sales volume of the products it manages, the channel distribution manager's efforts can yield a greater profit for the retailer than if the retailer managed the product category itself.

Handleman and other music product channel distribution managers such as **Alliance Entertainment** have extended their businesses, offering assistance to online retailers. As these companies offer Web site support in addition to their physical retailer management programs, they compete with companies such as Amazon.com because they offer an alternative to the strategic alliance approach.

ToysRUs.com loads as a tabbed section within the Amazon.com Web page

FIGURE 3-12 ToysRUs.com store within Amazon.com

Mobile Commerce

In Chapter 2, you learned about a number of technologies that use wireless data transmission technologies to link laptop computers, PDAs, cell phones, and other devices to the Internet. Beginning in 1997, industry experts would predict each year that mobile commerce would present important growth opportunities for online businesses. Each year, those experts have been wrong. Although people do obtain stock quotes, directions, weather forecasts, airline flight schedules, and other information on their cell phones and PDAs, few companies have been successful in generating any significant revenues from the sale of this information to consumers.

One exception is NTT's **DoCoMo I-Mode** service, which is offered in Japan. DoCoMo is a successful cell phone that people avidly use to send short messages, play games, and obtain weather forecasts and other information. Many of these services are not free, and NTT, which is a telephone company, charges by the minute for connect time to the Internet. After several years of rapid growth, however, DoCoMo's sales have flattened and NTT is struggling to find new services to sell on the network to generate future revenue growth.

In the United States, **AvantGo** offers consumers several thousand channels of information as PDA downloads. The content of most of these channels is provided by online magazines and newspapers, who are permitted to include advertising with the content. AvantGo also sells a small amount of advertising that it includes with the content downloads. Most of AvantGo's revenue, however, is generated by other services it offers to businesses. You will learn more about business uses of mobile technology for cost reduction in Chapter 5.

Although few companies have made money in mobile commerce to date, many industry observers still believe that a company with the right service could be successful. As you learned in Chapter 2, the bandwidth of wireless devices is increasing. A service that could capitalize on the increased bandwidth, such as the delivery of short movies or news reports, could be successful if the audience would be willing to pay a subscription fee or view ads. A few companies have introduced services that allow people to pay for items with their cell phones and wireless PDAs. You will learn about these payment services in Chapter 11.

CREATING AN EFFECTIVE WEB PRESENCE

Businesses have always created a presence in the physical world by building stores, factories, warehouses, and office buildings. An organization's **presence** is the public image it conveys to its stakeholders. The **stakeholders** of a firm include its customers, suppliers, employees, stockholders, neighbors, and the general public. Most companies tend not to worry much about the image they project until they grow to a significant size—until then, they are too focused on just surviving to spare the effort. On the Web, presence can be much more important. Many customers and other stakeholders of a Web business know the company only through its Web presence. Creating an effective Web presence can be critical even for the smallest and newest firms operating on the Web.

Identifying Web Presence Goals

When a business creates a physical space in which to conduct its activities, its managers focus on very specific objectives. Few of these objectives are image driven. The new company must find a location that will be convenient for its customers, with sufficient floor space and features to allow the selling activity to occur. A new business must balance its needs for inventory storage space and employee work space with the costs of obtaining that space. The presence of a physical business location results from satisfying these many other objectives and is rarely a main goal of designing the space.

On the Web, businesses and other organizations have the luxury of building their Web sites intentionally to create distinctive presences. A firm's physical location must satisfy so many other business needs that it often fails to convey a good presence. A good Web site design can provide many image-creation and image-enhancing features very effectively—it can serve as a sales brochure, a product showroom, a financial report, an employment ad, and a customer contact point. Each entity that establishes a Web presence should decide which features the Web site can provide and which of those features are the most important to include.

Making Web Presence Consistent with Brand Image

Different firms, even those in the same industry, might establish different Web presence goals. For example, **Coca Cola** and **Pepsi** are two companies that have established powerful brand images in the same business, but they have developed significantly different Web presences. These two companies frequently change their Web pages, but the Coca Cola page usually includes a trusted corporate image such as the Coke bottle. Alternatively, the Pepsi page is usually filled with hyperlinks to a variety of activities and product-related promotions.

These Web presences convey the images each company wishes to project. Each presence is consistent with other elements of the marketing efforts of these companies—Coca Cola's traditional position as a trusted classic, and Pepsi's position as the upstart product favored by a younger generation.

Achieving Web Presence Goals

An effective site is one that creates an attractive presence that meets the objectives of the business or organization. These objectives include:

- Attracting visitors to the Web site
- Making the site interesting enough that visitors stay and explore
- Convincing visitors to follow the site's links to obtain information
- Creating an impression consistent with the organization's desired image
- Building a trusting relationship with visitors
- Reinforcing positive images that the visitor might already have about the organization
- Encouraging visitors to return to the site

Profit-Driven Organizations

The **Toyota** site that appears in Figure 3-13 is a good example of an effective Web presence. The site provides links to detailed information about each vehicle model, links to a dealer locator page, links to information about the company and the financing services it offers, and a link to a site search/help feature. The presence that Toyota created with its home page is consistent with its corporate goal of providing cars and trucks that meet the needs of a wide variety of customers.

In contrast, **Quaker Oats** created a Web site that did not offer a particularly strong sense of corporate presence, although it provided a good selection of information about the firm. Figure 3-14 (on page 143) shows the Quaker Oats Web site as it appeared until 1999.

FIGURE 3-13 Toyota U.S. home page

The site was a straightforward presentation of links to information about the firm. This page included 24 links to financial information, employment opportunities, current press releases, and other information about the company. It included links to contact information for the firm. It even had the corporate "Quaker man" logo. Although this Quaker Oats site provided access to useful information about the company, it offered the visitor a completely different experience and impression than that provided by the Toyota site. In 1999, Quaker changed its Web page to include some pictures of its products and improve its general appearance and user-friendliness. The overall impression of this new site is much more lively and fun than the original site. The redesigned Quaker Oats home page appears in Figure 3-15.

Although the redesigned site is more colorful and interesting, the basic information offered is essentially the same as that offered by the previous version. The new site offers this information more efficiently, using fewer links. On both the old and new sites, Quaker Oats uses versions of its "Quaker man" logo to convey a highly recognizable image that is strongly associated with the company. Even though the new site is the result of a major overhaul, Quaker decided to include this image because it conveys the Quaker brand so well. This Web page remained essentially unchanged when Quaker was merged into Pepsi in 2001, which indicates that Pepsi was likely satisfied with the presence conveyed by the site for the Quaker brand products it acquired in the merger. The current Quaker Oats page has been updated somewhat, but it has not been substantially redesigned since 1999.

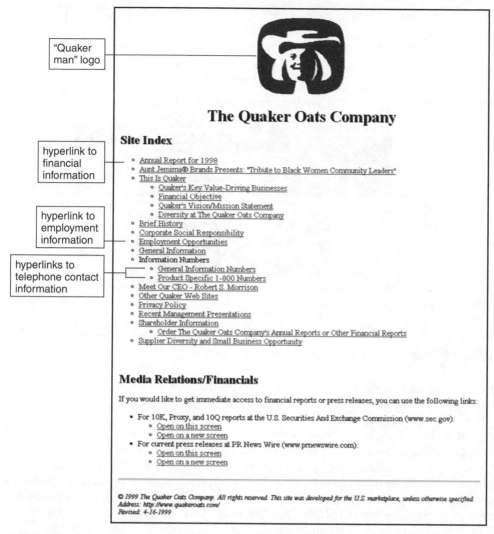

FIGURE 3-14 Quaker Oats home page before 1999 redesign

Not-for-Profit Organizations

The Toyota and Quaker Oats examples show how companies can enhance their images by providing information. For some organizations, this image-enhancement capability is a key goal of their Web presence efforts. Not-for-profit organizations are an excellent example of this. They can use their Web sites as a central resource for communications with their varied and often geographically dispersed constituencies.

A key goal for the Web sites of many not-for-profit organizations is information dissemination. The Web allows these groups to integrate information dissemination with fund-raising in one location. Visitors who become engaged in the issues presented are

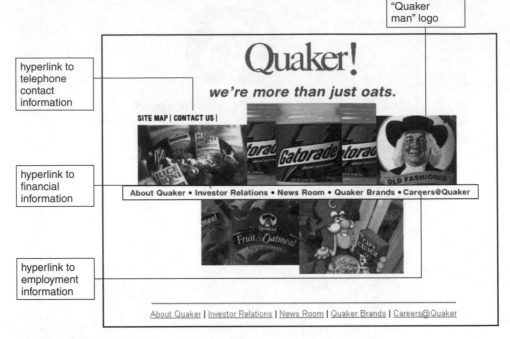

hyperlink to telephone contact information

hyperlink to financial information

hyperlink to employment information

"Quaker man" logo

FIGURE 3-15 Quaker Oats home page after 1999 redesign

usually just one or two clicks away from a page offering memberships or other opportunities to donate using a credit card. Web pages also provide a two-way contact channel for people who are engaged in the organization's efforts but who do not work directly for the organization—for example, many not-for-profits rely on volunteers and coordination with other organizations to accomplish their goals. This combination of information dissemination and a two-way contact channel is a key element on any successful electronic commerce Web site. Interestingly, not-for-profit organizations are ahead of many businesses in accomplishing this combination of elements in their Web presences. Figure 3-16 shows the home page of the **American Civil Liberties Union (ACLU)**, which is devoted to the advocacy of individual rights in the United States.

This page allows interested visitors to learn more about the ACLU and join the organization if their interests are piqued by what they see. The Feedback link at the bottom of the page leads to a form that visitors can use to report a civil liberties violation, obtain assistance with legal research, ask questions about ACLU membership, or request permission to reprint ACLU publications. The page contains three links (the Join button, the Donate Now link at the top of the page, and the Support Us link at the bottom of the page) to a page that allows individuals to join the ACLU (and thus make a financial contribution to the organization).

The ACLU's use of a Web site is especially valuable because the organization serves many different constituencies, not all of whom agree with the ACLU or with each other on specific issues. If the ACLU were to create a print newsletter that contained interesting information for some of its supporters, that same information might offend other supporters.

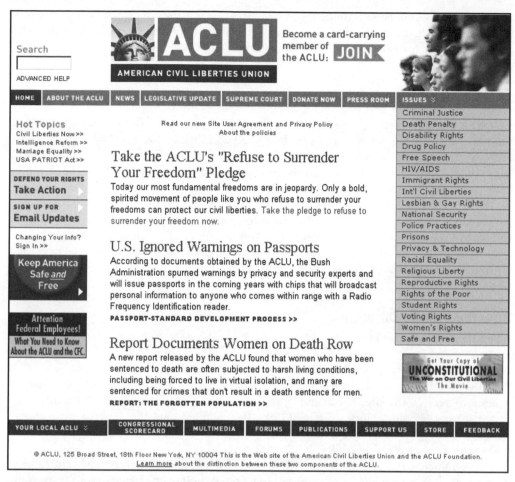

FIGURE 3-16   American Civil Liberties Union home page

The Issues links at the right side of the ACLU home page allow site visitors to select the issues in which they are interested—and only those issues.

Not-for-profit organizations can use the Web to stay in touch with existing stakeholders and identify new opportunities for serving them. Organizations as diverse as the **American Red Cross**, the **Public Broadcasting System**, and the **Union of Concerned Scientists** have created effective Web presences. Organizations such as **Amnesty International** and the **United Nations** also use the Web to create international communities of interested persons.

Political parties want to offer information about party positions on issues, recruit members, keep existing members informed, and provide communication links to visitors who have questions about the party. All the major United States political parties have Web sites and each year candidates running for public office set up their own Web sites. In addition, political organizations that are not affiliated with a specific party, such as the nonpartisan **Center for Responsive Politics**, also accomplish similar goals with their Web presences.

WEB SITE USABILITY

Research indicates that few businesses accomplish all of their goals for their Web sites in their current Web presences. Even sites that succeed in achieving most of these goals often fail to provide sufficient interactive contact opportunities for site visitors. Most firms' Web sites give the general impression that the firm is too important and its employees are too busy to respond to inquiries. This is no way to encourage visitors to become customers!

In this section, you will learn how the Web is different from other ways in which companies have communicated with their customers, suppliers, and employees in the past. You will learn how companies can improve their Web presences by making their sites accessible to more people and easier to use, and by making sure that their sites encourage visitors to trust and even develop feelings of loyalty toward the organization behind the Web site.

How the Web Is Different

When firms first started creating Web sites in the mid-1990s, they often built simple sites that conveyed basic information about their businesses. Few firms conducted any market research to see what kinds of things potential visitors might want to obtain from these Web sites, and even fewer considered what business infrastructure adjustments would be needed to service the site. For example, few firms had e-mail address links on their sites. Those firms that did include an e-mail link often understaffed the department responsible for answering visitors' e-mail messages. Thus, many site visitors sent e-mail messages that were never answered.

This failure to understand how the Web is different from other presence-building media is one reason that so many businesses fail to achieve their Web objectives. To learn more about this issue, see Jakob Nielsen's **Failure of Corporate Websites** page in the Online Companion; the article was written in 1998, but still accurately describes far too many Web sites.

Most Web sites that are designed to create an organization's presence in the Web medium include links to a fairly standard information set. The site should give the visitor easy access to the organization's history, a statement of objectives or mission statement, information about products or services, financial information, and a way to communicate with the organization. Sites achieve varying levels of success based largely on how they offer this information. Presentation is important, but so is realizing that the Web is an interactive medium.

A number of Web designers and consultants have taken firms to task for their uninspired use of the Web's interactive nature. Some of these criticisms appear in the print media, but many appear in online newsletters or e-zines. For example, Christopher

Locke argues that large corporations should encourage their employees to engage in unrestricted online dialog with the firm's customers, suppliers, and other stakeholders. He believes that this type of dialog is in the spirit of the Internet and helps companies create more honest and real personalities as part of their Web presences (see the references to his work in the For Further Study and Research section at the end of this chapter). David Weinberger presents similar arguments in his online **Journal of the Hyperlinked Organization**.

The main point these consultants make is that large firms must acknowledge and use the Web's capability for two-way, meaningful communication with their customers. They further argue that use of this communication process is not optional; companies that fail to communicate effectively through this channel will lose customers to competitors that do.

Meeting the Needs of Web Site Visitors

Businesses that are successful on the Web realize that every visitor to their Web sites is a potential customer. Thus, an important concern for businesses crafting Web presences is the variation in visitor characteristics. People who visit a Web site seldom arrive by accident; they are there for a reason.

Many Motivations of Web Site Visitors

Unfortunately for the Web designer trying to make a site that is useful for everyone, visitors arrive for many different reasons, including these:

- Learning about products or services that the company offers
- Buying products or services that the company offers
- Obtaining information about warranty, service, or repair policies for products they purchased
- Obtaining general information about the company or organization
- Obtaining financial information for making an investment or credit-granting decision
- Identifying the people who manage the company or organization
- Obtaining contact information for a person or department in the organization

Creating a Web site that meets the needs of visitors with such a wide range of motivations can be challenging. Not only do Web site visitors arrive with different needs, they arrive with different experience and expectation levels. In addition to the problems posed by the diversity of visitor characteristics, technology issues can also arise. These Web site visitors are connected to the Internet through a variety of communication channels that provide different bandwidths and data transmission speeds. They will also be using several different Web browsers. Even those using the same browser can have a variety of configurations. The wide array of browser add-in and plug-in software adds yet another dimension to visitor variability. Considering and addressing the implications of these many variations in visitor characteristics when building a Web site can help convert these visitors into customers.

Making Web Sites Accessible

One of the best ways to accommodate a broad range of visitor needs is to build flexibility into the Web site's interface. Many sites offer separate versions with and without frames and give visitors the option of choosing either one. Some sites offer a text-only version. As researchers at the **Trace Center** note, this can be an especially important feature for visually impaired visitors who use special browser software, such as the **IBM Home Page Reader**, to access Web site content. The **W3C Web Accessibility Initiative** site includes a number of useful links to information regarding these issues.

If the site uses graphics, it can give the visitor the option to select smaller versions of the images so that the page loads on a low-bandwidth connection in a reasonable amount of time. If the site includes streaming audio or video clips, it can give the visitor the option to specify a connection type so that the streaming media adjusts itself to the bandwidth for that connection.

A good site design lets visitors choose among information attributes, such as level of detail, forms of aggregation, viewing format, and downloading format. Many electronic commerce Web sites give visitors a selectable level of detail by presenting product information by product line. The site presents one page for each line of products. A product line page contains pictures of each item in that product line accompanied by a brief description. By using hyperlinked graphics for the product pictures, the site offers visitors the option of clicking the product picture, which opens a page of detailed specifications for that product.

One of the more controversial developments in Web site design is the use of animated graphics software, such as **Macromedia Flash**, to create Web pages. These pages (or large portions of the pages) are not rendered in HTML and are often very large files that take considerable time to download, especially for visitors who do not have broadband connections. Many Web site designers love these products because they provide them with an exciting creative design tool, but few major electronic commerce sites use these types of animated graphics pages. For interesting discussions of the disadvantages of Flash and similar tools, see WebWord.com's **Flash Usability Challenge** pages or Jakob Nielsen's **Flash: 99% Bad** commentary. A good solution to the controversy is to give visitors an option to select a Flash or non-Flash version of the site on its home page.

Some specific tasks that customers want to perform do lend themselves to animated Web pages. For example, the **Lee FitFinder** is a series of Flash animation pages that can help Lee customers find the right size and style of jeans. One of the Lee Jeans FitFinder animation pages is shown in Figure 3-17.

Web sites can also offer visitors multiple information formats by including links to files in those formats. For example, a page offering financial information could include links to an HTML file, an Adobe PDF file, and an Excel spreadsheet file. Each of these files would contain the same financial information in different formats; visitors can then choose the format that best suits their immediate needs. Visitors looking for a specific financial fact might choose the HTML file so that the information appears in their Web browsers. Other visitors who want a copy of the entire annual report as it was printed would select the PDF file and either view it in their browsers or download and print the file. Visitors who want to conduct analyses on the financial data would download the spreadsheet file and perform calculations using the data in their own spreadsheet software.

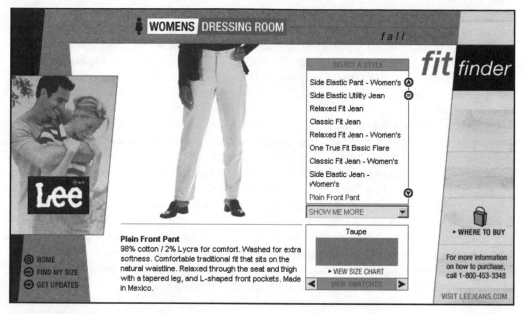

FIGURE 3-17   Lee Jeans FitFinder Flash animation

To be successful in conveying an integrated image and offering information to potential customers, businesses should try to meet the following goals when constructing their Web sites:

- Offer easily accessible facts about the organization.
- Allow visitors to experience the site in different ways and at different levels.
- Provide visitors with a meaningful, two-way (interactive) communication link with the organization.
- Sustain visitor attention and encourage return visits.
- Offer easily accessible information about products and services and how to use them.

Trust and Loyalty

When companies first started selling on the Web, many of them believed that their customers would use the abundance of information to find the best prices and disregard other aspects of the buying experience. For some products, this may be true; however, most products include an element of service. When customers buy a product, they are also buying that service element. A seller can create value in a relationship with a customer by nurturing customers' trust and developing it into loyalty. Recent studies by business researchers have found that a 5 percent increase in customer loyalty measures (such as proportion of returning customers) can yield profit increases ranging from 25 percent to 80 percent.

Even when products are commodity items, the service element can be a powerful differentiating factor for which customers will pay extra. These services include such things as delivery, order handling, help with selecting a product, and after-sale support. Because many of these services are things that a potential customer cannot evaluate before purchasing a product, the customer must trust the seller to provide an acceptable level of service.

When a customer has an experience with a seller who provides good service, that customer begins to trust the seller. When a customer has multiple good experiences with a seller, that customer feels loyal to the seller. Thus, the repetition of satisfactory service can build customer loyalty that can prevent a customer from seeking alternative sellers who offer lower prices.

Many companies doing business on the Web spend large amounts of money to obtain customers. If they do not provide levels of customer service that lead customers to develop trust in and loyalty to the firm, the companies are unlikely to recover the money they spend to attract the customers in the first place, much less earn a profit.

Customer service is a problem for many electronic commerce sites. Recent research indicates that customers rate most retail electronic commerce sites to be average or low in customer service. A common weak spot for many sites is the lack of integration between the companies' call centers and their Web sites. As a result, when a customer calls with a complaint or problem with a Web purchase, the customer service representative does not have information about Web transactions and is unable to resolve the caller's problem.

A number of studies show that the e-mail responsiveness of electronic commerce sites has also been disappointing. Many major companies are slow to respond to e-mail inquiries about product information, order status, or after-sale problems. A significant number of companies in these studies never acknowledged or responded to the e-mail queries.

Rating Electronic Commerce Web Sites

Two companies routinely review electronic commerce Web sites for usability, customer service, and other factors. Many people have found these review sites to be useful as they decide which sites to patronize. Unfortunately, one of the sites, **Gomez.com**, no longer publishes most of its scorecards for electronic commerce sites. It now sells the information it gathers to the companies that operate the Web sites and offers suggestions for improvements. **BizRate.com** provides a comparison shopping service and offers links to sites with low prices and good service ratings for specific products. BizRate.com compiles its ratings by conducting surveys of sites' customers. When the customer places an order, a pop-up window appears asking for a rating of various aspects of his or her experience with the site. The customer is offered a chance at a prize in exchange for providing this information.

Usability Testing

Only a few companies perform any **usability testing** on their Web sites; however, more and more companies are realizing its importance and are doing some usability testing. As the practice of usability testing becomes more common, more Web sites will meet the goals outlined previously in this chapter.

Experts estimate that average electronic commerce Web sites frustrate up to 70 percent of their customers to the point that they leave without buying anything. Even the best sites lose half of their customers because the sites are confusing or difficult to use. Simple changes in site usability can increase customer satisfaction and sales.

Companies that have done usability tests, such as **Eastman Kodak**, **T. Rowe Price**, and **Maytag**, have found that they can learn a great deal about meeting visitor needs by conducting focus groups and watching how different customers navigate through a series of Web site test designs. Industry analysts agree that the cost of usability testing is so low compared to the total cost of a Web site design or overhaul that it should almost always be included in such projects. Two pioneers of usability testing are Ben Shneiderman and Jakob Nielsen. Dr. Shneiderman founded the **University of Maryland Human-Computer Interaction Lab** and has published a number of books on interface design. Dr. Nielsen's **Alertbox** Web site includes much information about how to conduct usability testing and use its results to improve Web site design and operation.

Customer-Centric Web Site Design

An important part of a successful electronic business operation is a Web site that meets the needs of potential customers. In the list of goals for constructing Web sites that you learned about earlier in the chapter, the focus was on meeting the needs of all site visitors. Putting the customer at the center of all site designs is called a **customer-centric** approach to Web site design. A customer-centric approach leads to some guidelines that Web designers can follow when creating a Web site that is intended to meet the specific needs of *customers*, as opposed to all Web site visitors. These guidelines include the following:

- Design the site around how visitors will navigate the links, not around the company's organizational structure.
- Allow visitors to access information quickly.
- Avoid using inflated marketing statements in product or service descriptions.
- Avoid using business jargon and terms that visitors might not understand.
- Build the site to work for visitors who are using the oldest browser software on the oldest computer connected through the lowest bandwidth connection—even if this means creating multiple versions of Web pages.
- Be consistent in use of design features and colors.
- Make sure that navigation controls are clearly labeled or otherwise recognizable.
- Test text visibility on smaller monitors.
- Check to make sure that color combinations do not impair viewing clarity for color-blind visitors.
- Conduct usability tests by having potential site users navigate through several versions of the site.

Web marketing consultant Kristin Zhivago of **Zhivago Marketing Partners** has a number of recommendations for Web sites that are designed specifically to meet the needs of online customers. She encourages Web designers to create sites focused on the customer's buying process rather than the company's perspective and organization. For example, she suggests that companies examine how much information their Web sites provide and how useful that information is for customers. If the site does not provide substantial "content for your click" to visitors, they will not become customers.

Using these guidelines when you create your site can help make visitors' Web experiences more efficient, effective, and memorable. Usability is an important element of creating an effective Web presence. For an interesting look at Web design issues, you can visit the **Webby Awards** site. The Webby Awards are given to sites that "exemplify the kinds of sites that Internet users should visit every day for information and entertainment," as judged by a panel of Web designers, journalists, and industry leaders.

CONNECTING WITH CUSTOMERS

An important element of a corporate Web presence is communicating with site visitors who are customers or potential customers. In this section, you will learn how Web sites can help firms identify and reach out to customers.

The Nature of Communication on the Web

Most businesses are familiar with two general ways of identifying and reaching customers: personal contact and mass media. These two approaches are often called **communication modes** because they each involve a characteristic way (or mode) of conveying information from one person to another (or communicating). In the **personal contact** model, the firm's employees individually search for, qualify, and contact potential customers. This personal contact approach to identifying and reaching customers is sometimes called **prospecting**. In the **mass media** approach, firms prepare advertising and promotional materials about the firm and its products or services. They then deliver these messages to potential customers by broadcasting them on television or radio, printing them in newspapers or magazines, posting them on highway billboards, or mailing them.

Some experts distinguish between broadcast media and addressable media. **Addressable media** are advertising efforts directed to a known addressee and include direct mail, telephone calls, and e-mail. Since few users of addressable media actually use address information in their advertising strategies, in this book, we consider addressable media to be mass media. Many businesses use a combination of mass media and personal contact to identify and reach customers. For example, Prudential uses mass media to create and maintain the public's general awareness of its insurance products and reputation, while its salespeople use prospecting techniques to identify potential customers. Once an individual becomes a customer, Prudential maintains contact through a combination of personal contact and mailings.

The Internet is not a mass medium, even though a large number of people now use it and many companies seem to view their Web sites as billboards or broadcasts. Nor is the Internet a personal contact tool, although it can provide individuals the convenience of making personal contacts through e-mail and newsgroups. Jeff Bezos, founder of Amazon.com, described the Web as the ideal tool for reaching what he calls "the hard middle"—markets that are too small to justify a mass media campaign, yet too large to cover using personal contact. Figure 3-18 illustrates the position of the Web as a customer contact medium, between the large markets addressed by mass media and the highly focused markets addressed by personal contact selling and promotion techniques.

To help you better understand the differences shown in Figure 3-18, read the following scenario. The scenario assumes that you have heard about a new book, but would like to learn more about it before buying it. Consider how your information acquisition process would vary, depending on the medium you used to gather the information.

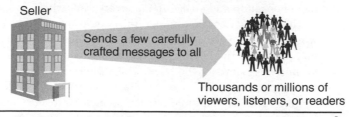

Mass media
One-to-many

Seller

Sends a few carefully crafted messages to all

Thousands or millions of viewers, listeners, or readers

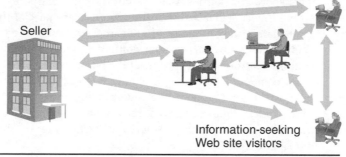

The Web
Many-to-one
and
many-to-many

Seller

Information-seeking
Web site visitors

Personal contact
One-to-one

Salesperson

Customer or prospect

FIGURE 3-18 Business communication modes

- *Mass media*: You might have been exposed to general promotional messages from book publishers that have created impressions about quality associated with particular book brands. If your existing knowledge includes a brand identity for the book's publisher, these messages may influence your perceptions of the book. You may have been exposed to an ad for the title on television or radio, or in print. You might have heard the book's author interviewed on a radio program or read a review of the book in a publication such as *The New York Times Book Review* or *Booklist* magazine. Notice that most of these process elements involve you as a passive recipient of information. This communication channel is labeled "Mass media" and appears at the top of Figure 3-18. Communication in this model flows from one advertiser to many potential buyers and thus is called a **one-to-many communication model**. The defining characteristic of the mass media promotion process is that the seller is active and the buyer is passive.

- *Personal contact*: Small-value items are not frequently sold through this medium because the costs of devoting a salesperson's efforts to a small sale are prohibitive. However, in the case of books, local bookshop owners and employees often devote considerable time and resources to developing close relationships with their customers. Although each individual book sale is a small-value transaction, people who frequent local bookshops tend to buy large numbers of books over time. Thus, the bookseller's investment in developing personal contacts is often rewarded. In this scenario, you may visit your local bookshop and strike up a conversation with a knowledgeable bookseller. In the personal contact model, this would most likely be a bookseller with whom you have already established a relationship. The bookseller would offer an opinion on the book based on having read that book, books by the same author, or reviews of the book. This opinion would be expressed as part of a two-way conversational interchange. This interchange usually includes a number of conversational elements, such as discussions about the weather, local sports, or politics, that are not directly related to the transaction you are considering. These other interchanges are part of the trust-building and trust-maintaining activities that businesses undertake to develop the relationship element of the personal contact model. The underlying **one-to-one communication model** appears at the bottom of Figure 3-18 and is labeled "Personal contact." The defining characteristic of information gathering in the personal contact model is the wide-ranging interchange that occurs within the framework of an existing trust relationship. Both the buyer and the seller (or the seller's representative) actively participate in this exchange of information.

- *The Web*: To obtain information about a book on the Web, you could search for Web site references to the book, the author, or the subject of the book. You would likely identify a number of Web sites that offer such information. These sites might include those of the book's publisher, firms that sell books on the Web, independent book reviews, or discussion groups focused on the book's author or genre. **The New York Review of Books** and **Booklist** magazine, both staples of mass media book promotion, now have online Web editions. Book review sites that did not originate in a print edition, such as **BookBrowser**, also appear on the Web. Most Web-based booksellers maintain searchable space on their sites for readers to post reviews and comments about specific titles. If the author of the book is famous, there might even be independent Web fan sites devoted to him or her. If the book is about a notable person, incident, or time period, you might find Web sites devoted to those notable topics that include reviews of books related to the topic. You could examine any number of these resources to any extent you desired. You might encounter some advertising material created by the publisher while searching the Web. However, if you choose not to view the publisher's ads, you will find it as easy to click the Back button on your Web browser as it is to surf television channels with your remote control. The Web affords you many communication channels. Figure 3-18 shows only one of the communication models that can occur when using the Web to search for product information. The model labeled "The Web" in Figure 3-18 is the **many-to-one communication model**. The Web gives you the flexibility to use a one-to-one model (as in the personal contact model) in which you communicate over the Web with an individual working for the seller, or engage in **many-to-many communications** with other potential buyers. The defining characteristic of a product information search on the Web is that the buyer actively participates in the search and controls the length, depth, and scope of the search.

Summary

In this chapter, you learned that businesses are using six main approaches to generate revenue on the Web: the Web catalog, digital content sales, advertising-supported, advertising-subscription mixed, fee-for-transaction, and fee-for-service models. You learned how these models work and what kinds of businesses use which models. You also learned that some companies have changed models as they learned more about their customers and the business environment in which their Web sites operate.

Companies sometimes face the challenges of channel conflict and cannibalization either within their own organizations or with the companies that have traditionally provided sales distribution to consumers for them. In accordance with the network model of organization that you learned about in Chapter 1, companies undertaking electronic commerce initiatives sometimes form strategic alliances with other companies or contract with channel distribution managers to obtain their skills in Web site operation or their product category knowledge. You learned that business-to-consumer mobile commerce has not yet been widely successful; however, increasing bandwidth could make it possible for new services to emerge that will be successful.

By understanding how the Web differs from other media and by designing a Web site to capitalize on those differences, companies can create an effective Web presence that delivers value to visitors. Every organization must realize that visitors to its Web site arrive with a variety of expectations, prior knowledge, and skill levels, and are connected to the Internet through different technologies. Knowing how these factors can affect the visitor's ability to navigate the site and extract information from the site can help organizations design better, more usable Web sites. Enlisting the help of users when building test versions of the Web site is also a good way to create a Web site that represents the organization well.

Firms must understand the nature of communication on the Web so they can use it to identify and reach the largest possible number of customers and qualified prospects. Using a many-to-one communications model enables Web sites to effectively reach potential customers.

Key Terms

Account aggregation

Addressable media

Advertising-subscription mixed revenue model

Advertising-supported revenue model

Bill presentment

Cannibalization

Catalog model

Category managers

Channel conflict

Channel cooperation

Channel distribution managers

Communication modes

Customer-centric

Demographic information

Disintermediation

E-zine

Fee-for-service revenue model

Fee-for-transaction revenue model

Fulfillment managers

Mail order model

Many-to-many communications

Many-to-one communication model

Mass media

One-to-many communication model

One-to-one communication model

Personal contact

Personal shopper   Stickiness
Portal (Web portal)   Sticky
Presence   Usability testing

Prospecting   Virtual model
Reintermediation   Web catalog revenue model
Stakeholders   Web directory

Review Questions

RQ1. Write a paragraph in which you describe the conditions under which a Web site could become profitable by relying exclusively on advertising revenue. In a second paragraph, provide an example of a company not mentioned in the chapter that is using the advertising-supported model and that is likely to be successful in the long run. Explain why you think it will succeed.

RQ2. Describe two possible service-for-fee offerings that might become available to users of Internet-enabled wireless devices (such as PDAs or mobile phones) in the near future. Write one paragraph for each service in which you outline the profit potential and risk of losses for each.

RQ3. In two paragraphs, explain why a customer-centric Web site design is so important, yet is so difficult to accomplish.

RQ4. In one paragraph, define the term "presence." Write an additional paragraph in which you explain why firms that do business on the Web should be more concerned about presence than firms that operate only in the physical world.

RQ5. Many real estate agents today have Web sites that list the properties they have for sale. These agents also advertise the properties in classified newspaper ads and sometimes in television ads. Write three paragraphs in which you briefly describe the things that real estate agents can best accomplish through (1) their Web sites, (2) mass media advertising, and (3) personal contact.

RQ6. Promoting products on the Web is different from using mass media promotion or personal contact. Assume that you want to explain these differences to a person who is planning to open a Web site that will sell snowboarding vacation packages. Write one paragraph about each approach (mass media, personal contact, the Web). In each paragraph, explain the advantages and disadvantages of the approach for the snowboarding vacation package Web business.

Exercises

E1. Page 151 includes a list of things that Web sites can do to meet the needs of visitors. Create a table in which the first column is a list of the five needs of Web site visitors. Find three Web sites that meet three or more of the needs. Create a column for each Web site next to the first column (columns two through four) and rate how well each site meets each need in the first column. You may want to use the **Webby Awards** site as a starting point in your search, but do not use any of the award nominees or winners in your table.

E2. Evaluate the usability of two Web sites that sell large-screen televisions. A list of links to companies that sell this product is included in the Online Companion for this exercise, but you may use other sites if you wish. In your evaluation, compare the sites on how easy it is to learn about the product and purchase the product. Your report should include a section of about 200 words in which you describe the criteria you used in your evaluation, a section of about 300 words that summarizes your findings, and a section of about 100 words in which you present your conclusion.

E3. You have been employed by Bob Drudge, the owner of **refdesk.com**, to explore revenue-generating alternatives. Currently, the site is using an advertising-supported revenue model. Bob wants you to consider each of the other revenue models and the potential of strategic alliances that might make sense for his site. Write a report of about 400 words in which you summarize your research and state your recommendations. Your instructor might assign you to a group to complete this exercise and might also ask you to present your recommendations in class.

E4. High-end jewelry retailers such as **Harry Winston** and **Tiffany & Co.** often use Macromedia's Flash software to create their Web sites. Present three arguments for and three arguments against the use of Flash animations in sites such as these. Consider the retailers' objectives, the characteristics of the products being sold, and the type of customers who visit these sites. Limit your answer to 400 words.

Cases

C1. Lonely Planet

In 1972, Tony and Maureen Wheeler were newlyweds who decided to have one last adventure travel experience before settling down. Their trip was an overland trek from London to Australia through Asia. So many other travelers asked them about their experiences that they sat down at their kitchen table and wrote a book titled *Across Asia on the Cheap*. They published the book themselves and were surprised by how many copies they sold. Three decades and 60 million books later, their publishing enterprise has turned out to be one of the most successful in history.

The Wheelers' publishing company, Lonely Planet, has grown rapidly, with typical annual sales increases of 20 percent or more. The company is privately held and does not release sales figures, but industry analysts estimate current annual revenues to be $50 million. Lonely Planet publishes more than 650 titles in 17 languages and holds a 13 percent share of the travel guide market. The company has more than 400 employees in its United Kingdom, United States, French, and Australian offices who perform editorial, production, graphic design, and marketing tasks. Travel guide content is written by a network of more than 150 contract authors in 20 countries. These authors are knowledgeable about everything from visa regulations to hotel prices to the names of the hottest new entertainment spots. The combined expertise of the in-house staff and the in-country authors has kept Lonely Planet ahead of its competitors for many years.

In recent years, Lonely Planet has expanded its business beyond the publication of travel guides. The company offers travel services that include a phonecard, hotel and hostel room-booking, airplane tickets, European rail travel reservations and tickets, package tours, and travel insurance. These services are offered by telephone and on the Lonely Planet Web site.

The Web site has won numerous awards, including the Society of American Travel Writers 2003 Silver Award and a spot on *Time* magazine's 2003 "Fifty Best Web Sites" list. It has also won the best travel site Webby three times, most recently in 2004. The site was launched in 1994 and includes an online store in which Lonely Planet publications are sold. However, the site's main draws are its comprehensive collection of information about travel destinations and its online bulletin board, the Thorn Tree, which has over 220,000 registered users and more than 400,000 message posts each year. Another section of the Web site, Lonely Planet Images, includes 250,000 digital photos and other graphics and is used by more than 25,000 registered users.

Lonely Planet is always looking for ways to expand its market and brand image through new technologies. For example, it has formed a joint venture with Nokia to provide city guides on mobile telephones in more than 40 cities worldwide. The company has also sold its content for use on portal sites such as Yahoo! and has created a B2B division that provides customized content to large corporate customers for their internal use.

Despite its excellent Web site and its use of new technologies, most of Lonely Planet's revenues are still generated by book sales. The typical production cycle of a travel guide is about eight months long. This is the time it takes to commission authors, conduct research, work through several drafts of writing and editing, select photos, create the physical book, and print it. This production cycle causes new books to be almost a year out of date by the time they are published. Only the most popular titles are revised annually. Other titles are on two, three, or four year revision cycles. The time delay in publication means that many details in the guides are outdated or wrong; restaurants and hotels close (or move), exchange rates and visa regulations change, and once-hot night spots are abandoned by fickle clientele.

Lonely Planet publications are well researched and of high quality, but the writers do not work continuously because the books are not published continuously. The Web site often has information that is more current than the published travel guides. Lonely Planet has adopted new technologies, but has not used them to revise its revenue model or to make basic changes in the production of its main product, the travel guides.

Required:

1. Prepare a report in which you analyze the marketing channel conflicts and cannibalization issues that Lonely Planet faces as it is currently operating. Suggest solutions that might reduce the revenue losses or operational frictions that result from these issues.

2. Prepare a list of new products that Lonely Planet might introduce to take advantage of Internet technologies (including wireless technologies) and address customers' concerns about the timeliness and currency of information in the printed travel guides. Briefly describe any problems that Lonely Planet will face as it introduces these new products.

3. Many loyal Lonely Planet customers carry their travel guides (which can be several hundred pages thick) with them as they travel around the world. In many cases, these customers do not use large portions of the travel guides. Also, Internet access can be a problem for many of these customers while they are traveling. Describe a product (or products) that might address this customer concern and also yield additional revenue for Lonely Planet. Your answer here may build on ideas that you developed in your solution to part 2.

Note: Your instructor might assign you to a group to complete this case and might ask you to prepare a formal presentation of your results to your class.

C2. Association for the Study of International Business

The Association for the Study of International Business (ASIB) is an organization of researchers, professors, and business executives interested in the study, analysis, and promotion of business activities beyond domestic borders. Mario DiPonetti, ASIB's executive director, hires you as a consultant to help him map out a future Web revenue strategy for the association.

The ASIB has about 3000 members located in countries throughout the world; however, about half of its members are in the United States. Each member pays an annual membership fee of $100, so ASIB's dues revenue totals about $300,000 per year. ASIB sponsors several conferences each year; it also publishes a monthly newsletter and two journals. The conferences break even; that is, the conference and exhibitor fees cover the costs of running the conference, but they do not yield any profit that can be applied to other ASIB operating costs.

One of the journals, *Annals of International Business*, has an academic focus and is read by researchers interested in international business topics. All ASIB members receive a copy of this journal and ASIB sells about 600 subscriptions to the journal at $300 per year. Most of the subscribers are university libraries. This journal is published four times each year. The second journal, *International Business Today*, is written for business executives. It includes articles and features that report on current trends in international business. All ASIB members receive a copy of this journal and ASIB sells about 1000 subscriptions to the journal at $60 per year. Most of the subscribers are business executives. This journal is published 12 times each year. The total subscription revenue from the two journals is $240,000 per year. The business journal also sells advertising that yields about $20,000 per year. ASIB uses that total revenue of $260,000 to cover the costs of producing and mailing both journals. The cost of producing one issue of either journal, which includes proofreading, editing, and typesetting costs, is about $5000. The printing and mailing costs, which have been increasing rapidly over the past several years, average about $2 per journal (the mailing costs to some members are much higher than others because they are located in distant countries). Each year, ASIB produces 16 issues (four of the academic journal and 12 of the business journal) and mails 62,400 journals (14,400 of the academic journal and 48,000 of the business journal) to members and subscribers at a total cost of $204,800 (16 x $5000 plus 62,400 x $2). Thus, ASIB's current journal operations yield a net profit of $55,200 ($260,000 – $204,800) that can help support other ASIB activities.

ASIB has a Web site that it constructed at a cost of $56,000 three years ago. One of ASIB's staff members spends approximately half of her time managing the site. One-half of her salary along with other recurring expenses, such as software licenses and computer upgrades for the Web site, total about $35,000 per year. Mario explains to you that one of the ASIB's greatest cost reduction successes was last year's decision to offer the newsletter by e-mail. About half of the members chose to receive the newsletter by e-mail. The paper newsletters cost 50 cents each to print and mail, but creating and sending the e-mails took less than $50 worth of staff members' time. Thus, ASIB realized an immediate savings of about $700 each month. The newsletters are also placed on the Web site so that members can check there if they happen to miss the e-mailed newsletter. This success prompted Mario to think about ways to reduce the cost of distributing the journals. He wants to make sure, however, that ASIB continues to receive as much of the journal revenue as possible under any new revenue model.

One of the companies you learned about in the chapter, EBSCO, approached Mario with an offer to handle electronic distribution of the academic journal. EBSCO will take a copy of the journal when it is published, scan each article into Adobe Portable Document Format (PDF) and into HTML format, index the articles, and place them into several of EBSCO's databases. Many university and research libraries subscribe to EBSCO databases. The EBSCO representative explained to Mario that most of the libraries would continue their print subscriptions to the journal, but that about 30 percent of the libraries would stop subscribing and rely on their electronic access to the journal through the EBSCO database. Mario called some of his friends who are executive directors of other associations and confirmed that this percentage was correct in their experience. EBSCO would pay ASIB $10,000 per year for access to the journal plus $50 per year for every library that subscribed to an EBSCO database that included the journal. The EBSCO representative estimated that the number of subscribing libraries would be about 1000.

Mario outlined an alternative to the EBSCO contract. In this alternative, ASIB would itself scan the journals into PDF files and make them available on the ASIB Web site for a subscription fee. Mario estimated that it would cost about $1000 to create the PDF files for one issue and place them on the Web site. He also estimated that managing the accounts and passwords would consume about $500 per month of staff time and costs.

EBSCO was not interested in purchasing access to the business journal, but Mario is considering ways to make some or all of the content from that journal available on the ASIB Web site. He is considering offering reduced rate "Web access only" subscriptions to business executives. He is also thinking about offering some of the best stories from the print edition on the Web and including ads offering full subscriptions on each page. He is even considering placing the first part of the best stories on the Web site and offering readers a chance to subscribe so they can read the rest of the story.

Several companies that sell products and services to businesses that sell internationally currently run ads in the business journal. These companies expressed an interest in placing banner ads on ASIB Web pages that contain content (such as stories from the business journal). Mario estimates that ASIB could earn between $3000 and $9000 per month from these banner ads, but he is concerned that having the best content from the business journal on the Web site might convince some business executives to drop their subscriptions to the print edition.

Required:

Prepare a comprehensive report for Mario in which you outline and analyze the possible revenue models that ASIB might use for its Web site. You should address the two journals as separate issues. Your report should provide the basis for a presentation to the ASIB executive board and should include specific recommendations where possible.

Note: Your instructor might assign you to a group to complete this case and might ask you to prepare a formal presentation of your results to your class.

For Further Study and Research

Anthes, G. 2002. "What Lies Ahead for Web Merchants? Advanced Search, 3-D Images, Real-Time Inventory—and Credibility," *Computerworld*, June 16, 26.

Barron, K. 1999. "Realtors 1, Web 0," *Forbes*, 163(3), February 8, 64–65.

Berchtold, J., J. Grass, B. Johnson, and E. Stephenson. 2002. "Can Broadband Save Internet Media?" *The McKinsey Quarterly*, June. (http://www.mckinseyquarterly.com/article_page.asp?ar=1189&L2=17&L3=104)

Brown, S., A. Tilton, and D. Woodside. 2002. "The Case for Online Communities," *The McKinsey Quarterly*, January. (http://www.mckinseyquarterly.com/article_page.asp?ar=1143&L2=24&L3=45)

Buckman, R. 1999. "Wall Street Is Rocked by Merrill's Online Plans," *The Wall Street Journal*, June 2, C1.

Carr, D. 2003. "Slate Sets a Web Magazine First: Making Money," *The New York Times*, April 28, C1.

Cavillini, S. 2001. "Stocks That Work: Job-Hunting Sites Are Among the Select Dot-Coms That Haven't Plunged," *The Industry Standard*, August 20–27, 78.

Champy, J. 1998. "Direct Sales or Electronic Channels? Give the Customer a Choice," *Computerworld*, 32(47), November 23, 64.

Chan, S. 2001. "Usability Guru Philosophizes on Web Subjects," *The Seattle Times*, April 7, E1.

Charski, M. 2001. "Games Could Be Good Bet for Advertisers," *Interactive Week*, 7(52), December 25, 14.

Christensen, C. and M. Overdorf. 2000. "Meeting the Challenge of Disruptive Change," *Harvard Business Review*, 78(2), March-April, 66–75.

Coursaris, C. and K. Hassanein. 2002. "Understanding M-Commerce," *Quarterly Journal of Electronic Commerce*, 3(3), 247–271.

Cox, B. 2003. "Bluefly Gets Another Cash Infusion," *Internet News*, January 29. (http://www.internetnews.com/ec-news/article.php/1576331)

Crawford, W. 2004. "Keeping the Faith: Playing Fair with Your Visitors," *EContent*, 27(4), September, 42–43.

Dahm, T. 2001. "Five Steps to Accessible Web Sites," *E-Commerce News*, June 29. (http://www.zdnet.com/ecommerce/stories/main/0,10475,2781418,00.html)

Daly, J. 2000. "Sage Advice: Interview with Peter Drucker," *Business 2.0*, August 22, 134–144.

Dano, M. 2003. "U.S. Still Largely Left Out of M-Commerce," *RCR Wireless News*, 22(38), September 22, 8–9.

Dayal, S., T. French, and V. Sankaran. 2002. "The E-Tailer's Secret Weapon: Category Management," *The McKinsey Quarterly*, June, 72–79.

Doonar, J. 2004. "It's Not Such a Lonely Planet," *Brand Strategy*, January, 24–25.

Duff, A. 2003. "Lonely at the Top," *Director*, 57(3), October, 78–82.

Eisenmann, T. 2002. *Internet Business Models*. New York: McGraw-Hill.

English, J., K. Garret, and S. Pearson. 2001. TravelLite Project. University of California, Berkeley. (http://www.sims.berkeley.edu/courses/final-projects/travelite/index.htm)

Evans, P. and T. Wurster. 1997. "Strategy and the New Economics of Information," *Harvard Business Review*, 75(5), September-October, 71–83.

Garrity, B. 2001. "Amazon, Handleman Plant Seeds In Crowded Field Of Online Fulfillers," *Billboard*, 113(33), August 18, 47.

Greco, S. 2001. "Fanatics!" *Inc*, 23(5), April 1, 36–43.

Hamilton, A. 2001. "Net Addictions," *Time*, 157(12), March 26, 76.

Hellweg, E. 2003. "Do Investor-relations Web Sites Work? A New Study Examines Problems With Information Found on Corporate Web Sites," *CNN-Money*, February 25. (http://money.cnn.com/2003/02/25/technology/techinvestor/hellweg/index.htm)

Hill, A. 2001. "Stop Shopping Cart Abandonment: Top Five Reasons Customers Abandon Shopping Carts," *Smart Business*, February 13. (http://www.zdnet.com/smartbusinessmag/stories/all/ 0,6605,2677306,00.html)

Keighley, G. 2003. "The Secrets of Drudge, Inc." *Business 2.0*, April. (http://www.business2.com/articles/mag/0,1640,47762,00.html)

Kemp, T. 2000. "Wal-Mart No Web Mart," *InternetWeek*, October 9, 1–2.

Kooser, A. 2003. "Signs of Life?" *Entrepreneur*, 31(10), October, 28.

Kotha, S., S. Rajgopal, and M. Venkatachalam. 2004. "The Role of Online Buying Experience as a Competitive Advantage: Evidence from Third-Party Ratings for E-Commerce Firms," *Journal of Business*, 77(Supplement), April, S109–S134.

Lais, S. 2002. "How to Stop Web Shopper Flight," *Computerworld*, June 17, 44.

Levine, R., C. Locke, D. Searle, and D. Weinberger. 2000. *The Cluetrain Manifesto: The End of Business as Usual*. Cambridge, MA: Perseus.

Livingston, B. 2002. "Factiva: 8,000 Publications, 1.5 million Subscribers," *InfoWorld*, May 28. (http://www.infoworld.com/articles/op/xml/02/05/30/020530opsecrets.xml)

Locke, C. 2000. "Smart Customers, Dumb Companies," *Harvard Business Review*, 78(6), November–December, 187–191.

Lorek, L. 2001. "Orbitz Takes Off," *Interactive Week*, 8(21), May 28, 16–19.

Luna, L. 2002. "The M-Commerce M-Plosion: Is There Still Hope?" *Telephony*, 243(12), October 14, 48–51.

Magura, B. 2003. "What Hooks M-Commerce Customers?" *MIT Sloan Management Review*, 44(3), Spring, 9–10.

McLaughlin, K. 2001. "Real Makes Gaming Play: Hopes Its Customer Base Will Pay for Online Diversions," *Business 2.0*, March 21. (http://www.business2.com/ebusiness/2001/03/28783.htm)

Merrick, A. 2002. "Sears' Deal to Buy Lands' End Reunites Stores With Big Catalogs," *The Wall Street Journal*, May 14, A1.

Mullaney, T. 2001. "Orbitz Doesn't Soar," *Business Week*, July 9. (http://www.businessweek.com/magazine/ content/01_29_/b3740621.htm)

Neuborne, E. 2001. "Bridging the Loyalty Gap," *Business Week*, January 22, EB10.

Nielsen, J. 1999. *Designing Websites With Authority: Secrets of an Information Architect*. Indianapolis, IN: New Riders.

Nielsen, J. 2000. "End of Web Design," *Alertbox*, July 23. (http://www.useit.com/alertbox/20000723.html)

Nielsen, J. 2000. "Flash: 99% Bad," *Alertbox*, October 29. (http://www.useit.com/alertbox/20001029.html)

Nielsen, J. 2001. "Usability Metrics," *Alertbox*, January 21. (http://www.useit.com/alertbox/20010121.html)

Nielsen, J. and M. Tahir. 2002. *Homepage Usability: 50 Websites Deconstructed*. Indianapolis, IN: New Riders.

Nielsen, J., K. Coyne, and M. Tahir. 2001. "Make It Usable," *PC Magazine*, 20(3), February 6, IPO1-IPO6.

Nunes, P., D. Wilson, and A. Kambil. 2000. "The All-in-One Market," *Harvard Business Review*, 78(3), May-June, 19–20.

Oliva, R. 2003. "Going Mobile: Quietly, B2B Marketers Are Finding New Applications for Mobile Platforms," *Marketing Management*, 12(4), July-August, 46–48.

Opdyke, J. 2001. "Tools to Trade By: Technology Is Giving Investors the Ability to Do All Kinds of New Things," *The Wall Street Journal*, June 11, R8.

Oreskovic, A. 2001. "Testing 1-2-3," *The Industry Standard*, March 5. (http://www.thestandard.com/article/ 0,1902,22338,00.html).

Pomerantz, D. 2001. "Shop 'Til You Flop," *Forbes*, 167(2), January 22, 56.

Rayport, J. and J. Sviokla. 1995. "Exploiting the Virtual Value Chain," *Harvard Business Review*, 73(6), November-December, 75–85.

Rich, M. 2002. "Book 'Em: The Internet Introduced a Middleman to the Hotel Industry—Now Hotels Want That Business Back," *The Wall Street Journal*, June 10, R10.

Roth, A. 2001. "Dimon Kills Wingspan; Cost of Brand Is Cited," *American Banker*, 166(125), June 29, 1–2.

Sanderfoot, A. and C. Jenkins. 2001. "Content Sites Pursue Fee-Based Model," *Folio: The Magazine for Magazine Management*, 30(6), 15–16.

Schneider, I. 2002. "Yodlee Adds Business Intelligence to Aggregation," *Bank Systems & Technology*, May 31. (http://www.banktech.com/story/BSTeNews/BNK20020531S0008)

Schroeder, C. 2001. "Content Sites Will Work on the Web," *The Wall Street Journal*, April 16, A18.

Schwartz, E. 1997. *Webonomics*. New York: Broadway Books.

Schwartz, E. 1999. *Digital Darwinism*. New York: Broadway Books.

Shneiderman, B. 1997. *Designing the User Interface: Strategies for Effective Human-Computer Interaction*. Reading, MA: Addison-Wesley.

Sklar, J. 2002. *Principles of Web Design,* Second Edition. Boston, MA: Course Technology.

Slatalla, M. 2000. "Wonder How You'd Look in Yellow Eye Shadow?" *The New York Times*, August 18, G-4.

Sliwa, C. 2002. "Traditional Retailers Debate Pulling Plug on E-Commerce," *Computerworld*, February 13. (http://www.computerworld.com/managementtopics/ebusiness/story/0,10801,68278,00.html)

Surmacz, J. 2001. "Big Pipe Dreams: Evian Uses Broadband to Create a Web Site That Fits Its Image," *CIO E-Business Research Center*, January 31. (http://www.cio.com/forums/ec/edit/013101_evian.html)

Tedeschi, B. 2000. "Charitable Groups Discover New Revenue in Retailing Goods via Their Own Web Sites," *The New York Times*, March 27, C-11.

Tedeschi, B. 2000. "Retailers Are Letting their Right Hand Know What the Left Hand Is Up To for Better Customer Service," *The New York Times*, July 24, C-9.

Totty, M. 2001. "The Consumers' Choice: Why Do Certain Sites Become Favorites? We Went Straight to the Source," *The Wall Street Journal*, December 10, R6.

Vickery, L. 2001. "Keeping the Cachet: Luxury-Goods Makers Want to Jump into the Chaos of Cyberspace Without Sullying Their Images," *The Wall Street Journal*, April 23, R28.

Waldman, A. 2000. "Vanguard Preps Site Overhaul," *Financial NetNews*, 5(46), November 13, 1–2.

Weinberger, D. 1999. "0:1 Marketing," *Journal of Hyperlinked Organizations*, May 20. (http://www.hyperorg.com/backissues/joho-may20-99.html#01)

Weingarten, M. 2001. "Flash Backlash," *The Industry Standard*, March 5. (http://www.thestandard.com/article/ 0,1902,22330,00.html)

Weiss, T. 2000. "Walmart.com Back Online After Four-Week Overhaul," *Computerworld*, 34(45), November 6, 24.

Wood, C. 2001. "Collusion in the Air," *PC Magazine*, 20(9), May 8, 199.

Yudkin, M. 2001. "The Web's Pipsqueaks Stand Tall," *Business 2.0*, May 1. (http://www.business2.com/magazine/ 2001/05/webs_pipsqueaks.htm)

Zabbal, C. 2002. "Playing Games With Broadband," *The McKinsey Quarterly*, June. (http://www.mckinseyquarterly.com/article_page.asp?ar=1196&L2=17&L3=65)

Zeitchik, S. 2003. "New Worlds at Lonely Planet," *Publishers Weekly*, 250(25), June 23, 12.

Zellner, W. 2001. "Where the Net Delivers: Travel," *Business Week*, June 11. (http://www.businessweek.com/ magazine/content/01_24/b3736111.htm)

Zimmerman, A. 2000. "Wal-Mart Launches Web Site for a Third Time, This Time Emphasizing Speed and Ease," *The Wall Street Journal*, October 31, B12.

CHAPTER 4

MARKETING ON THE WEB

INTRODUCTION

In September 1997, a new gift shop opened for business on the Web. There were already many gift shops on the Web at that time; however, this store, named 911Gifts.com, carried items that were chosen specifically to meet the needs of last-minute gift shoppers. Using 911—the emergency telephone number used in most parts of the United States—in the store's name was intended to convey the impression of crisis-solving urgency. The company's two major strengths were its promise of next-day delivery on all items and its site layout, in which gift selections were organized by holiday rather than by product type. Thus, a harried shopper could simply click the Mother's Day gifts link and view a set of gift choices appropriate for that holiday that were ready for immediate delivery. The site also included a reminder service, called GiftAlert, to help its customers avoid another emergency gift situation on the next holiday.

By 1999, the company had 90,000 customers signed up for GiftAlert and was doing about $1 million in annual sales. It carried about 500 products, and each of the products was chosen to yield a gross margin of at least 40 percent. 911Gifts.com was a successful business, but the company's founders realized they would need to build wider awareness of their brand. They also realized that building a brand would require a substantial investment of funds. The company hired Hilary Billings, a retail marketing executive whose experience included building the Pottery Barn catalog business at Williams-Sonoma, to create a brand-building strategy and obtain financing to implement that strategy. Billings undertook a complete reevaluation of the 911Gifts.com marketing plan and, after revising it, took it to investors who committed over $30 million for a rebranding and complete overhaul of the company's Web site. In October 1999, the new brand was born as **RedEnvelope**. In many Asian countries, gifts of money are enclosed in a simple red envelope. The new brand was designed to create a sense of elegant simplicity to replace the sense of panic and emergency solutions conveyed by the old brand name.

The product line was revamped to fit the new image as well. About 300 products were dropped and replaced with different products that focus groups had judged to be more appealing. The new product line had an even higher average gross margin than the old line. Billings launched a massive brand-awareness campaign that included online advertising, buses in seven major cities painted red and festooned with large red bows, and print advertising in upscale publications. The most important change in advertising strategy was the launch of a print catalog. RedEnvelope catalogs are mailed to customers to coincide with major gift-giving holidays and serve as additional reminders. Because RedEnvelope sells a small set of products that are chosen for their visual appeal and for the status they are intended to convey, the full-color, lushly-illustrated print catalogs are a powerful selling tool.

One year later, the results of this extensive makeover and substantial monetary investment were clear. RedEnvelope had tripled its number of customers and had increased sales by over 400 percent.

The company chose a specific part of the gifts market and targeted its offerings to meet the needs and desires of those customers. The company created a brand, a marketing plan, and a set of advertising and promotion strategies that would expose the company to the largest portion of that market it could afford to reach. The most important point is that RedEnvelope matched its inventory selection, delivery methods, and marketing efforts to each other and to the needs of its customers. Today, the company continues to use print catalogs and a focus on upscale product lines to keep its sales increasing each year.

WEB MARKETING STRATEGIES

In this chapter, you will learn how companies are using the Web in their marketing strategies to advertise their products and services and promote their reputations. Increasingly, companies are classifying customers into groups and creating targeted messages for each group. The sizes of these targeted groups can be smaller when companies are using the Web—in some cases, just one customer at a time can be targeted. New research into the behavior of Web site visitors has even suggested ways in which Web sites can respond to visitors who arrive at a site with different needs at different times. This chapter will also introduce you to some of the ways companies are making money by selling advertising on their Web sites.

Most companies use the term **marketing mix** to describe the combination of elements that they use to achieve their goals for selling and promoting their products and services. When a company decides which elements it will use, it calls that particular marketing mix its **marketing strategy**. As you learned in Chapter 3, companies—even those in the same industry—try to create unique presences in their markets. A company's marketing strategy is an important tool that works with its Web presence to get the company's message across to both its current and prospective customers.

Most marketing classes organize the essential issues of marketing into the **four Ps of marketing**: product, price, promotion, and place. **Product** is the physical item or service that a company is selling. The intrinsic characteristics of the product are important, but customers' perceptions of the product, called the product's **brand**, can be as important as the actual characteristics of the product.

The **price** element of the marketing mix is the amount the customer pays for the product. In recent years, marketing experts have argued that companies should think of price in a broader sense, that is, the total of all financial costs that the customer pays (including transaction costs) to obtain the product. This total cost is subtracted from the benefits that a customer derives from the product to yield an estimate of the **customer value** obtained in the transaction. Later in this book, you will learn how the Web can create new opportunities for creative pricing and price negotiations through online auctions, reverse auctions, and group buying strategies. These Web-based opportunities are helping companies find new ways to create increased customer value.

Promotion includes any means of spreading the word about the product. On the Internet, new possibilities abound for communicating with existing and potential customers. In Chapter 2, you learned how companies are using the Internet to engage in meaningful dialogs with their customers using e-mail and other means. In this chapter, you will learn even more communication techniques that companies are using to promote their products.

For years, marketing managers dreamed of a world in which instant deliveries would give all customers exactly what they wanted when they wanted it. The issue of **place** (also called **distribution**) is the need to have products or services available in many different locations. The problem of getting the right products to the right places at the best time to sell them has plagued companies since commerce began. Although the Internet does not solve all of these logistics and distribution problems, it can certainly help. For example, digital products (such as information, news, software, music, video, and e-books) can be delivered almost instantly on demand through the Internet. Companies that sell products that must be shipped have found that the Internet gives them much better shipment tracking and control than did previous information technologies.

Product-Based Marketing Strategies

In Chapter 3, you learned about the importance of a company's Web presence and how this presence must integrate with the brand or other established images the company uses in its promotional activities. Most companies offer a variety of products that appeal to different groups. When creating a marketing strategy, managers must consider both the nature of their products and the nature of their potential customers.

Managers at many companies think of their businesses in terms of the products and services they sell. This is a logical way to think of a business because companies spend a great deal of effort, time, and money to design and create those products and services. If you ask managers to describe what their companies are selling, they usually provide you with a detailed list of the physical objects they sell or use to create a service. When customers are likely to buy items from particular product categories, or are likely to think of their needs in terms of product categories, this type of product-based organization makes sense. Most office supplies stores on the Web believe their customers organize their needs into product categories. The **Staples** home page, shown in Figure 4-1, uses product categories as a very strong organizing theme.

The Staples page has tabbed headings near the top of the page that are links to product categories. More detailed product category links fill the center of the page. Staples designed its page to meet the needs of the customer who has a specific product category in mind. Even the search box near the top of the page includes a drop-down list of categories so that customers can narrow their searches within categories.

A company that sells to a different market, but that uses a similar product-based marketing strategy, is **Sears**. Sears sold its products through catalogs and later in physical stores for many years before opening its Web site. Most companies that used print catalogs in the past organized them by product category. As you can see in Figure 4-2, Sears has carried over its product-focused marketing strategy to its Web site.

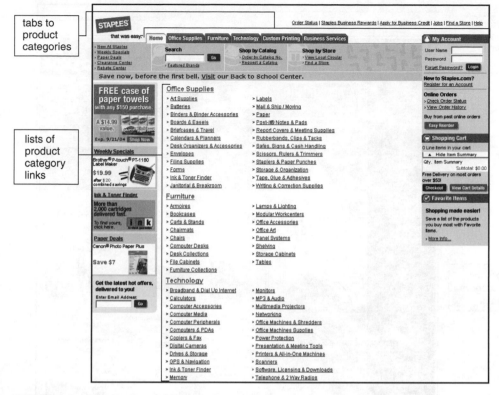

FIGURE 4-1   Staples home page

Both of these companies are using a product-based strategy. They organized their Web sites from an internal viewpoint, that is, according to the way that they arranged their product design and manufacturing processes. If customers arrive at these Web sites looking for a specific type of product, this approach works well. Alternatively, customers who are looking to fulfill a specific need, such as outfitting a new sales office or choosing a graduation gift, might not find these Web sites as useful. Many marketing researchers and consultants advise companies to think as if they were their own customers and to design their Web sites so that customers find them to be enabling experiences that can help customers meet their individual needs.

FIGURE 4-2   Sears home page

Customer-Based Marketing Strategies

In Chapter 3, you learned that the Web creates an environment that allows buyers and sellers to engage in complex communications modes. The communication structures on the Web can become much more complex than those in traditional mass media outlets such as broadcast and print advertising. When a company takes its business to the Web, it can create a Web site that is flexible enough to meet the needs of many different users. Instead of thinking of their Web sites as collections of products, companies can build their sites to meet the specific needs of various types of customers.

A good first step in building a customer-based marketing strategy is to identify groups of customers who share common characteristics. **Sabre Holdings** is a company that sells marketing services and technology to support those services to the travel industry. Its customers include travel agencies, airlines, large companies that have in-house travel departments, and travel consolidators (companies that buy blocks of airline seats and hotel rooms, then resell them as vacation packages). Sabre also operates the Travelocity B2C travel site that you learned about in Chapter 3. The Sabre Holdings home page, which appears in Figure 4-3, includes links to separate sections of its site that are designed to meet the needs of each of its major customer groups. By following these links, Sabre's different customers can find specific products and services targeted to each of their needs.

FIGURE 4-3   Sabre home page

Although Sabre's approach of breaking customers into four main groups is a good first step, subgroups probably exist within each of those groups. Marketers can use their experience with selling in their industries to identify those subgroups and then develop marketing strategies and tactics that will effectively reach customers in each subgroup. The use of customer-based marketing approaches was pioneered on B2B sites. B2B sellers were more aware of the need to customize product and service offerings to match their customers' needs than were the operators of B2C Web sites. In recent years, B2C sites have increasingly added customer-based marketing elements to their Web sites. One of the most noticeable trends in this direction is in university Web sites. In the early days of the Web, university sites were usually organized around the internal elements of the school (such as departments, colleges, and programs). Today, most university home pages include links to separate sections of the Web site designed for specific stakeholders, such as current students, prospective students, parents of students, potential donors, and faculty.

COMMUNICATING WITH DIFFERENT MARKET SEGMENTS

Identifying groups of potential customers is just the first step in selling to those customers. An equally important component of any marketing strategy is the selection of communication media to carry the marketing message.

In the physical world, companies can convey large parts of their messages by the way they construct buildings and design their floor spaces. For example, banks have traditionally been housed in large, solid-looking buildings that provide passersby an ample view of the main safe and its thick, sturdy door. Banks use these physical manifestations of reliability and strength to convey an important part of their service offerings—that a customer's money is safe and secure with the bank.

Media selection can be critical for an online firm because it does not have a physical presence. The only contact a potential customer might have with an online firm could well be the image it projects through the media and through its Web site. The challenge for online businesses is to convince customers to trust them even though they do not have an immediate physical presence.

Trust and Media Choice

The Web is an intermediate step between mass media and personal contact, but it is a very broad step. Using the Web to communicate with potential customers offers many of the advantages of personal contact selling and many of the cost savings of mass media. Figure 4-4 shows how these three information dissemination models compare on another important dimension: trust.

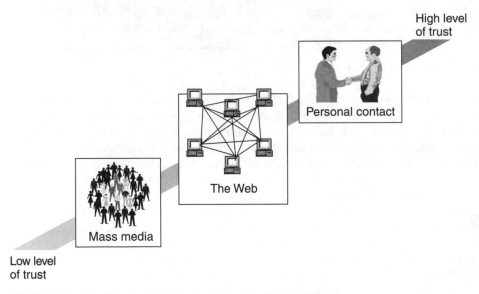

FIGURE 4-4 Trust in three information dissemination models

Although mass media offers the lowest level of trust, many companies continue to use it successfully. The cost of mass media advertising can be spread over the many people in its large audiences. For example, the cost of creating a television ad can be several hundred thousand dollars, but that ad will be viewed by millions of people. Thus, the cost of advertising per viewer is very low. Its low cost makes mass media advertising attractive to many companies.

After years of being barraged by television and radio commercials, many people have developed a resistance to the messages conveyed in the mass media. The impact on an audience of the shouted expression "New and improved!" is very low. The overuse of superlatives has caused many people to distrust or ignore much mass media. Television remote controls have mute buttons and make channel surfing easy for a reason. Attempts to re-create mass media advertising on the Web are likely to fail for the same reasons—many people ignore or resist messages that lack content of any specific personal interest to them.

Mass media advertising campaigns that are successful often rely on the passive nature of the media consumption experience. People watching television or listening to radio are usually in a passive and receptive state of mind. Thus, advertisers can include messages in mass media advertising that recipients might not consider valid or convincing if they were actively evaluating those statements. The messages are accepted by recipients because they are in a nonquestioning and passive state of mind. In contrast, Web users are actively engaged in the medium, with hands on the keyboard and mouse, as they view Web pages. This active state of mind makes Web users far more likely to actively evaluate advertising messages they see and less likely to accept the content of those messages.

Companies can use the Web to capture some of the benefits of personal contact, yet avoid some of the costs inherent in that approach. Most experts agree that it is better to make the trust-based model of personal contact selling work on the Web than to adopt the mass marketing approach on the Web. In 1996, when companies were beginning to do business online, rising consumer expectations and reduced product differentiation led to increased competition and a splintering of mass markets. Both of these results were reducing the effectiveness of mass media advertising. Thus, the Internet provided a new vehicle for achieving high levels of customer-focused marketing strategies.

Market Segmentation

Advertisers' response to this decrease in effectiveness was to identify specific portions of their markets and target them with specific advertising messages. This practice, called **market segmentation**, divides the pool of potential customers into **segments**. Segments are usually defined in terms of demographic characteristics such as age, gender, marital status, income level, and geographic location. Thus, for example, unmarried men between the ages of 19 and 25 might be one market segment.

In the early 1990s, firms began identifying smaller and smaller market segments for specific advertising and promotion efforts. This practice of targeting very small market segments is called **micromarketing**. However, the low cost per viewer of traditional mass media advertising campaigns becomes much higher when those methods are used to target very small market segments. This cost increase hampered the success of micromarketing strategies. Even though micromarketing was an improvement over mass media advertising, it still used the same basic approach and suffered from the weaknesses of that model.

Marketers have traditionally used three categories of variables to identify market segments. One variable is location. Firms divide their customers into groups by where they live or work. In this type of segmentation, called **geographic segmentation**, companies create different combinations of marketing efforts for each geographical group of customers. The grouping can be by nation, state (or province), city, or even by neighborhood. Alternatively, companies can develop one marketing strategy for urban customers, another for suburban customers, and yet a third for rural customers.

The second category uses information about age, gender, family size, income, education, religion, or ethnicity to group customers. This type of segmentation is called **demographic segmentation**. Demographic variables are frequently used by traditional marketers because research has shown that customers' need for and usage of products are strongly related to these types of variables. Demographic segmentation also exists on the Web. For example, a number of sites are devoted to women's issues or directed at specific age groups (such as teenagers) whose members tend to purchase music CDs and trendy clothing. Often, demographic and geographic segmenting strategies are combined. For example, an airline might target middle-income families living in Wisconsin and Michigan with mid-winter advertising for vacation trips to Florida.

In **psychographic segmentation**, marketers try to group customers by variables such as social class, personality, or their approach to life. For example, an auto company might direct advertising for a sports car to customers who are gregarious and have a high need for achievement. The use of psychographic segmentation has increased dramatically in recent years as marketers attempt to identify characteristic lifestyles and then design advertising to reach people who see themselves as having a particular lifestyle.

Companies that advertise on television often create messages designed to reach the likely audiences of various types of programs. These audiences represent one or more market segments. The market segments can be geographic, demographic, psychographic, or a combination of these. Figure 4-5 presents some examples from the television medium that show how companies do this.

Type of television program	Type of advertising
Children's cartoons	Children's toys and games
Daytime dramas	Household and laundry goods, pet foods
Late-night talk shows	Snack foods and nonprescription drugs
Golf tournaments	Golf equipment, investment services, and life insurance
Baseball and football games	Snack foods, beer, autos
Documentary films	Books, CDs, educational videotapes

FIGURE 4-5   Television advertising messages tailored to program audience

Children's television shows are likely to feature advertising for products that appeal to children. Ads on daytime dramas are directed at people who are home during the day and who thus might be interested in household and laundry care products. These people are more likely than others to own pets, so they also will see ads for pet foods. Advertisers on late-night talk shows often direct their ads at people who might have trouble falling asleep. Advertisers also believe that this late-night audience is receptive to promotions for snack foods to eat while watching these programs or for nonprescription medications for ailments that might be keeping them up so late.

Advertisers use sports programming as a vehicle for two different market segments. Some sports shows, such as golf tournaments or tennis matches, appeal to higher-income viewers. Other sports shows, such as baseball or football game broadcasts, appeal to viewers with more moderate incomes. As a result, programs that cover golf or tennis are more likely to include ads for investment and insurance products and luxury automobiles than are baseball or football programs. Also, because viewers of golf tournaments and tennis matches are likely to play the sport, these programs often include ads for game equipment. Baseball or football games rarely include ads for game equipment because few viewers of these games are participants in the sport themselves.

Programs that feature documentaries (such as those on the History Channel or the Discovery Channel) often carry ads for books, book clubs, CDs, and educational videos. Advertisers have found that these types of products appeal to the intellectual, arts-loving audiences of these programs.

Companies do much more than just match advertising messages to market segments. They also build a sales environment for their product or service that corresponds to the market segment they are trying to reach. In the physical world, store design and layout are often directed at specific market segments. If you walk through a shopping mall, you can observe that colors, displays, lighting, background music, and even the clothes worn by sales clerks vary with the targeted segment. For example, a clothing store for young women presents a completely different experience to its customers than a clothing store that sells expensive, conservative attire targeted toward more mature women with larger incomes.

Market Segmentation on the Web

The Web gives companies an opportunity to present different store environments online. Figures 4-6 and 4-7 show the home pages of **Steve Madden** and **Talbots**, respectively.

Both pages are well designed and functional. However, you can see that they are each directed to different market segments. The Steve Madden site is targeted at young, fashion-conscious buyers. The site uses a wide variety of typefaces, bold graphics, and photos of brightly colored products to convey its tone. The emphasis on the page is to make a bold fashion statement and, presumably, become the envy of your friends. In contrast, the Talbots page is rendered in a more muted, conservative style. The page is designed for older, more established buyers. The messages emphasized on the page are stability, home life, and the trademark Talbots red doors. These images appeal to a market segment of people looking for classics instead of the latest trends.

In the physical world, retail stores have limited floor and display space. These limitations often force physical stores to decide on one particular message to convey. Exceptions do exist, such as a music store that has a separate room for classical recordings (with different background music than the rest of the store) or a large department store that can use lighting and display space differently in each department; however, smaller retail stores usually choose the one image that appeals to most of their customers. On the Web, retailers can provide separate virtual spaces for different market segments. Some Web retailers provide the ultimate in targeted marketing—they allow their customers to create their own stores, as you will learn in the next section.

FIGURE 4-6 Steve Madden home page

FIGURE 4-7 Talbots home page

Offering Customers a Choice on the Web

Dell has done many things well in its online business. Its Web site offers customers a number of different ways to do business with the company. Its USA home page, shown in Figure 4-8, includes links for each major group of customers it has identified and also includes links to specific product categories.

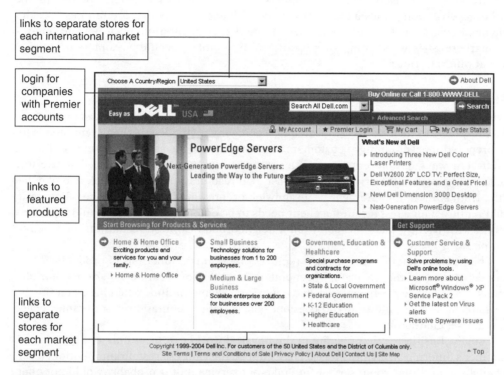

links to separate stores for each international market segment

login for companies with Premier accounts

links to featured products

links to separate stores for each market segment

FIGURE 4-8   Dell USA home page

Two of Dell's major customer groups, businesses and public organizations (the second and third columns in the figure), are divided into subgroups. Dell has found that, for example, its Education customers have different interests and needs than its Healthcare customers do, and that K-12 Education and Higher Education are distinct segments within the general Education category.

Dell Premier accounts give users the ultimate level of customer-based market segmentation. In these accounts, Dell offers each customer its own Dell Web site. Dell can customize a company's Premier account pages to show product selections for which price and terms have already been negotiated. Dell even allows individual employees of its customers to create their own personalized pages within their companies' Premier pages. This highly customized approach to offering products and services that match the needs of a particular customer is called **one-to-one marketing**. The Internet gives marketers the best opportunity for highly customized interactions with customers that they have had since the heyday of the door-to-door salesperson in the 1940s and 1950s.

BEYOND MARKET SEGMENTATION: CUSTOMER BEHAVIOR AND RELATIONSHIP INTENSITY

In the previous sections, you learned how companies can target groups of customers that are similar to each other as market segments. You also learned how one-to-one marketing gives companies a chance to create Web experiences that are unique to each individual customer. The next step—beyond market segmentation, even beyond one-to-one marketing—is when companies use the Web to target specific customers in different ways at different times.

Segmentation Using Customer Behavior

In the physical world, businesses can sometimes create different experiences for customers in response to their needs. For example, a company might decide that its mission is to sell prepared meals to hungry customers. A given potential customer responds to hunger in different ways at different times. If a person is hungry in the morning, but late for work, that person might drive through a fast food restaurant or grab a quick cup of coffee at the train station. Lunch might be a sandwich ordered and delivered to the office, or it could require a nice restaurant if a client needs to be entertained. Dinner could be at a restaurant with friends, take-out food from a neighborhood Chinese restaurant, or a delivered pizza.

The point is that the same person requires different combinations of products and services depending on the occasion. In general, the creation of separate experiences for customers based on their behavior is called **behavioral segmentation**. When based on things that happen at a specific time or occasion, behavioral segmentation is sometimes called **occasion segmentation**.

Usually, businesses that operate in the physical world can meet only one or a few of a customer's differing behavioral needs. For example, the Chinese restaurant mentioned earlier might offer dining room service and take-out service, but it probably would not offer a drive-through window or a morning coffee kiosk. Very few restaurants are able to offer everything from fast food through a five-course dinner. In the online world, it is much easier to design a single Web site that meets the needs of visitors who arrive in different behavioral modes. Thus, a Web site design can include elements that appeal to different behavioral segments.

Marketing researchers are just beginning to study how and why people prefer different combinations of products, services, and Web site features and how these preferences are affected by their modes of interaction with the site. Market researchers are finding that people want Web sites that offer a range of interaction possibilities from which they can select to meet their needs. Remember that a particular person might visit a particular Web site at different times and might search for different interactions each time. Customizing visitor experiences to match the site usage behavior patterns of each visitor or type of visitor is called **usage-based market segmentation**. Researchers have begun to identify common patterns of behavior and to categorize those behavior patterns. One set of categories that marketers use today includes browsers, buyers, and shoppers.

Browsers

Some visitors to a company's Web site are just surfing or browsing. Web sites intended to appeal to potential customers in this mode must offer them something that piques their interest. The site should include words that are likely to jog the memories of visitors and remind them of something they want to buy on the site.

These key words are often called **trigger words** because they prompt a visitor to stay and investigate the products or services offered on the site. Links to explanations about the site or instructions for using the site can be particularly helpful to this type of customer. A site should include extra content related to the product or service the site sells. For example, a Web site that sells camping gear might offer reviews of popular camping destinations with photos and online maps. Such content can keep a visitor who is in browser mode interested long enough to stay at the site and develop a favorable impression of the company. Once visitors have developed this favorable impression, they are more likely to buy on this visit or bookmark the site for a return visit.

Buyers

Visitors who arrive in buyer mode are ready to make a purchase right away. The best thing a site can offer a buyer is certainty that nothing will get in the way of the purchase transaction. For visitors who first choose a product from a printed catalog, many Web sites include a text box on their home pages that allows visitors to enter the catalog item number. This places that item in the site's shopping cart and takes the buyer directly to the shopping cart page. A **shopping cart** is the part of a Web site that keeps track of selected items for purchase and automates the purchasing process.

The shopping cart page should offer a link that takes the visitor back into the shopping area of the site, but the primary goal is to get the buyer to the shopping cart as quickly as possible, even if the buyer is at the site for the first time. The shopping cart should allow the buyer to create an account and log in after placing the item into the cart. To avoid placing barriers in the way of customers who want to buy, the site should not ask visitors to log in until they near the end of the shopping cart procedure. You will learn more about shopping carts in Chapter 9.

Perhaps the ultimate in shopping cart convenience is the 1-Click feature offered by **Amazon.com**, which allows customers to purchase an item with a single click. Any items that a customer purchases using the 1-Click feature within a 90-minute time period are aggregated into one shipment. Amazon.com has a patent on the 1-Click feature. You will learn more about such business process patents and other legal issues in Chapter 7.

Shoppers

Some customers arrive at a Web site knowing that it offers items they are interested in buying. These visitors are motivated to buy, but they are looking for more information before they make a purchase decision. For the visitor who is in shopper mode, a site should offer comparison tools, product reviews, and lists of features. One site that offers many useful features for shopper-mode visitors is the **Crutchfield** consumer electronics Web store. Crutchfield's site includes features that allow customers to specify the level of detail presented for each product, sort products by brand or price, and compare products with each other side by side. A page from the Crutchfield site showing some of the shopper-friendly features appears in Figure 4-9.

link to customer service pages with telephone number for immediate help

short description of product category

controls that allow the visitor to manipulate information

product list

FIGURE 4-9 Crutchfield's shopper-friendly features

Remember that a person might visit a Web site one day as a browser, and then return later as a shopper or a buyer. People do not retain behavioral categories from one visit to the next—even for the same Web site.

Although many companies work with these three visitor modes, other researchers are exploring alternative models. Much of Web site visitor behavior is not yet well understood. One study conducted by major consulting firm **McKinsey & Company** examined the online behavior of 50,000 active Internet users and identified six different groups. Following are the six behavior-based categories and their characteristic traits:

- *Simplifiers* are users who like convenience. They are attracted by sites that make doing business easier, faster, or otherwise more efficient than is possible in the physical world.
- *Surfers* use the Web to find information, explore new ideas, and shop. They like to be entertained, and they spend far more time on the Web than other people. To attract Surfers, sites must offer a wide variety of content that is attractive, well displayed, and constantly updated.

- *Bargainers* are in search of a good deal. Although they make up less than 10 percent of the online population, they make up over half of all visitors to the eBay auction site. They enjoy searching for the best price or shipping terms and are willing to visit many sites to do that.

- *Connectors* use the Web to stay in touch with other people. They are intensive users of chat rooms, instant messaging services, electronic greeting card sites, and Web-based e-mail. Connectors tend to be new to the Web, less likely than other people to purchase on the Web, and actively trying to learn what the Web has to offer them.
- *Routiners* return to the same sites over and over again. They use the Web to obtain news, stock quotes, and other financial information. Routiners like the comfort of working with a user interface that they know well.
- *Sportsters* are similar to Routiners, but they tend to spend time on sports and entertainment sites rather than news and financial information sites. Since they view the Web as an entertainment vehicle, Sportsters are attracted by sites that are interactive and attractive.

Other research studies have identified similar sets of characteristics and groupings. Companies in different industries or lines of business identify somewhat different sets of characteristics and group their Web site visitors using different names. The challenge for Web businesses is to identify which groups are visiting their sites and formulate ways of generating revenue from each segment. For example, some of these groups (such as Simplifiers and Bargainers) are ready to buy and would be interested in seeing specific product or service offerings. Other groups (such as Surfers, Routiners, and Sportsters) would be good targets for specific types of advertising messages. As more researchers study Web site visitor behavior, perhaps the industry will learn how to recognize the various modes in which visitors arrive and then channel them into the appropriate sections of the site. Until then, many Web sites use Dell's approach, in which visitors are asked to identify themselves as belonging to a particular category of customer when they enter the sites.

Customer Relationship Intensity and Life-Cycle Segmentation

One goal of marketing is to create strong relationships between a company and its customers. The reason that one-to-one marketing and usage-based segmentation are so valuable is that they help to strengthen companies' relationships with their customers. Good customer experiences can help create an intense feeling of loyalty toward the company and its products or services. Researchers have identified several stages of loyalty as customer relationships develop over time. A five-stage model of customer loyalty that is typical of these models appears in Figure 4-10.

This model shows the increase in intensity of the relationship as the customer moves through the first four stages: awareness, exploration, familiarity, and commitment. In the fifth stage, separation, a decline occurs and the relationship terminates. Not all customers go through the full five stages; some stop at a stage and continue the relationship at that level of intensity, or terminate the relationship at that point. Some customers in a particular stage might have contact with the company online while other customers in the same stage encounter the company offline. Companies should strive for a consistent customer experience at a particular life-cycle stage. That is, customers should experience the

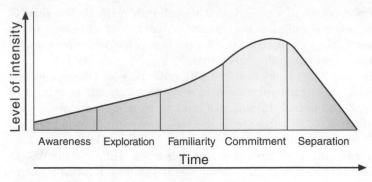

FIGURE 4-10 Five stages of customer loyalty

same level and quality of service whether they encounter the company online or offline. Online and offline customer contact points are often called **touchpoints**, and the goal of providing similar levels and quality of service at all touchpoints is called **touchpoint consistency**.

As the figure shows, changes in the nature of the relationship do not occur suddenly as a customer moves from one stage to the next. Within each stage, the level of intensity changes gradually as the customer moves through that stage. The characteristics of the five stages are outlined in the next sections.

Awareness

Customers who recognize the name of the company or one of its products are in the awareness stage of customer loyalty. They know that the company or product exists, but have not had any interaction with the company. Advertising a brand or a company name is a common way for companies to achieve this level of relationship with potential customers.

Exploration

In the exploration stage, potential customers learn more about the company or its products. The potential customer might visit the company's Web site to learn more, and the two parties will often communicate by telephone or e-mail. A large amount of information interchange can occur between the parties at this stage.

Familiarity

Customers who have completed several transactions and are aware of the company's policies regarding returns, credits, and pricing flexibility are in the familiarity stage of their relationship with the company. In this stage, they are as likely to shop and buy from competitors as they are from the company.

Commitment

After experiencing a considerable number of highly satisfactory encounters with a company, some customers develop a fierce loyalty or strong preference for the products or brands of that company. These customers have reached the commitment stage and are often willing to tell others about how happy they are with their interactions. To lure customers from the familiarity stage to the commitment stage, companies sometimes make concessions on price or terms. Usually, the value of the strong relationship is worth more to the company than the costs of these concessions.

Separation

Over time, the conditions that made the relationship valuable might change. The customer might be severely disappointed by changes in the level of service (either as provided by the company or as perceived by the customer) or product quality. The company can also evaluate the relationship and conclude that the loyal, committed customer is costing too much to maintain. As the intensity of the relationship fades, the parties enter a separation stage.

An important goal of any marketing strategy should be to move customers into the commitment stage as rapidly as possible and keep them there as long as possible. Companies want to see customers move into the separation stage only if they are costing more to serve than they are worth.

Life-Cycle Segmentation

Analyzing how customers' behavior changes as they move through the five stages can yield information about how they interact with the company and its products in each stage. The five stages are sometimes called the **customer life cycle**, and using these stages to create groups of customers that are in each stage is called **life-cycle segmentation**. Two companies that undertake continuing research into market segmentation and how companies can use segment information to develop better relationships with their customers are **Claritas** and **Donnelley Marketing**.

Claritas created one of the first segment marketing databases, named PRIZM, in the early 1970s. Claritas built PRIZM to take advantage of people's tendency to live near other people with similar tastes and preferences. Thus, PRIZM identifies the demographic characteristics of people by neighborhood. Claritas developed a number of other products that offer marketers databases with specific demographic, financial, and psychographic characteristics. Donnelley Marketing offers similar products, such as its Buyer Behavior Indicator and Affluence Models databases. Both Donnelley and Claritas extended their research from traditional direct marketing to help firms sell online. You can learn more about these companies and their products by following their links in the Online Companion for this chapter.

Acquisition, Conversion, and Retention of Customers

One goal of the strategies and tactics you will learn about in the rest of this chapter is to attract new visitors to a Web site. The benefits of acquiring new visitors are different for Web businesses with different revenue models. For example, an advertising-supported site is interested in attracting as many visitors as possible to the site and then keeping those visitors at the site as long as possible. That way, the site can display more advertising messages to more visitors, which is how the site earns a profit. For sites that operate a Web catalog, charge a fee for services, or are supported by subscriptions, attracting visitors to the site is only the first step in the process of turning those visitors into customers.

The total amount of money that a site spends, on average, to draw one visitor to the site is called the **acquisition cost**.

The second step that a Web business wants to take is to convert the first-time visitor into a customer. This is called a **conversion**. For advertising-supported sites, the conversion is usually considered to happen when the visitor registers at the site, or, in some cases, when a registered visitor returns to a site several times. For sites with other revenue models, the conversion occurs when the site visitor buys a good or service or subscribes to the site's content. The total amount of money that a site spends, on average, to induce one visitor to make a purchase, sign up for a subscription, or (on an advertising-supported site) register, is called the **conversion cost**. Most managers use a cumulative definition for conversion cost; that is, conversion cost includes acquisition cost.

For many Web businesses, the conversion cost is greater than the profit earned on the average sale (or the average first sale). In such cases, the Web business must induce the customer to return to the site and buy again (or renew the subscription, or view more advertising). Customers who return to the site one or more times after making their first purchases are called **retained customers**. Different businesses use different measures for determining when a customer is a retained customer. Some companies consider a customer retained if he or she returns just once and purchases again. Others use some number of subsequent purchases or some number of subsequent purchases within a specific time frame. The costs of inducing customers to return to a Web site and buy again are called **retention costs**.

Companies have found that measuring acquisition, conversion, and retention costs is important because it gives them an idea of which advertising and promotion strategies are successful. These measurements are more precise than classifying customers into the five stages of loyalty in the customer life-cycle model. It is much easier to determine, for example, whether a customer has been converted or retained than it is to determine whether that customer is in the familiarity stage or the commitment stage. For example, a company that is evaluating its promotion campaign can measure the conversion costs and compare them to the profit generated by the average first-time sale. Most companies are very interested in retaining customers, because the cost of acquiring a new customer is between 3 and 15 times (depending on the type of business) the cost of retaining an existing customer.

In the rest of this chapter, you will learn some specific techniques that can be elements of successful Web marketing strategies. Remember that each of these techniques makes sense only when used in concert with another. Not all techniques work well in all situations. For example, in the chapter's opening case, RedEnvelope found that a print catalog could be an integral part of promoting its online sales. RedEnvelope's success does not mean that printing catalogs is a good idea for all Web businesses (see the Kozmo Learning from Failures feature on the following page). It is only a good idea if it provides customers with recognizable value and augments the rest of the company's marketing strategy.

LEARNING FROM FAILURES

KOZMO

Throughout New York City, people were in their homes late at night craving videos and snack foods. Kozmo was launched in 1998 to meet those needs. With its orange-jacketed delivery people riding bicycles or motor scooters, Kozmo promised delivery of most items within an hour of ordering. Kozmo did not offer as wide a range of items as most convenience stores, so its main competitive advantage was its delivery service. Kozmo attempted to become profitable by adding high-margin items, such as DVD players and Sony PlayStations, and expanding its delivery areas to include higher-income neighborhoods. In addition to Manhattan, Kozmo opened operations for a short time in Houston and San Diego. In these cities, the higher average distances between deliveries made it even more difficult to cover costs.

Despite its best efforts, Kozmo was unable to create an image that was much different from that of a convenience store on wheels. Kozmo found it difficult to convince customers that delivered snack food items and videos were significantly more valuable than snack food items and videos on the shelves of nearby convenience stores. Most of Kozmo's product line consisted of items for which most people were accustomed to paying low prices.

In March 2001, just one month before closing operations, Kozmo announced a marketing plan that included spending $2.5 million to print and circulate 400,000 catalogs. The plan was a last-ditch attempt to increase brand awareness, gain new customers, and convince people who did not have an Internet connection to use Kozmo's phone order service. Unlike RedEnvelope, however, the Kozmo catalog was not a part of an integrated business plan and did not provide the same kind of added value that RedEnvelope's catalog provides—a bag of potato chips does not gain much appeal by appearing in a full-color catalog photo.

The lesson from Kozmo's experience is that using one element from a marketing strategy that worked for one company is no guarantee that it will work for every company. Marketing techniques are effective only when implemented as part of an integrated strategy that fits the company's products and gives customers a compelling reason to buy.

Customer Acquisition, Conversion, and Retention: The Funnel Model

Marketing managers need to have a good sense of how their companies acquire and retain customers. They often must evaluate competing marketing strategies to determine which are the most effective ways to attract and retain customers. The funnel model is used as a conceptual tool to understand the overall nature of a marketing strategy, but it also provides a clear structure for evaluating specific strategy elements.

The funnel model is very similar to the customer life-cycle model you learned about earlier in this chapter; however, the funnel model is less abstract and does a better job of showing the effectiveness of two or more specific strategies. The funnel is a good analogy for the operation of a marketing strategy because almost every marketing strategy starts with a large number of prospects and converts fewer and fewer of those prospects into serious prospects, customers, and finally, loyal customers. One example of a funnel model appears in Figure 4-11.

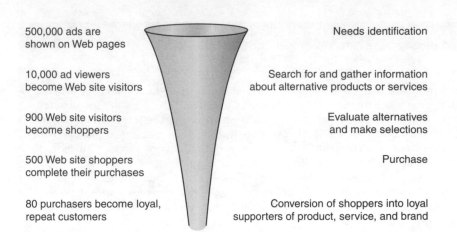

500,000 ads are shown on Web pages — Needs identification

10,000 ad viewers become Web site visitors — Search for and gather information about alternative products or services

900 Web site visitors become shoppers — Evaluate alternatives and make selections

500 Web site shoppers complete their purchases — Purchase

80 purchasers become loyal, repeat customers — Conversion of shoppers into loyal supporters of product, service, and brand

FIGURE 4-11   Funnel model of customer acquisition, conversion, and retention

In this funnel model, the steps that potential customers take as they become loyal, repeat customers are listed on the left side of the figure. The right side of the figure explains the increasing level of commitment that occurs in each step. Using market research and past history as a guide, the marketing manager develops the numbers that show the effectiveness of the planned strategy. The wider the bottom of the funnel, the better the strategy; that is, the more prospects are converted into loyal customers. The funnel model can be used in planning marketing strategies by comparing the projected results shown in the diagram with the results for alternative strategies shown in separate diagrams. The funnel model can also be used to show results that can then be compared with the costs of running the marketing campaign. Either way, the model gives marketing managers a tool for conceptualizing and evaluating alternative strategies.

ADVERTISING ON THE WEB

Advertising is all about communication. The communication might be between a company and its current customers, potential customers, or even former customers that the company would like to regain. To be effective, firms should send different messages to each of these audiences.

The five-stage customer loyalty model shown in Figure 4-10 (in the previous section) can be helpful in creating the messages to convey to each of these audiences. In the awareness stage, the advertising message should inform. The message could describe a new product, suggest new uses for existing products, or describe specific improvements to a product. Audiences in the exploration stage should receive messages that explain how a product or service works and encourage switching to that brand. In the familiarity stage, the advertising message should be persuasive—convincing customers to purchase specific products or request that a salesperson call. Customers in the commitment stage should be sent reminder messages. These ads should reinforce customers' good feelings about the brand and remind them to buy products or services. Companies generally do not target ads at customers who are in the separation stage.

Most companies that launch electronic commerce initiatives already have advertising programs in place. Online advertising should always be coordinated with existing advertising efforts. For example, print ads should include the company's URL. Banner ads are the dominant advertising format in use on the Web. Other online ad formats include pop-up ads, pop-behind ads, interstitial ads, and active ads.

Banner Ads

Most advertising on the Web uses banner ads. A **banner ad** is a small rectangular object on a Web page that displays a stationary or moving graphic and includes a hyperlink to the advertiser's Web site. Banner ads are versatile advertising vehicles—their graphic images can help increase awareness, and users can click them to open the advertiser's Web site and learn more about the product. Thus, banner ads can serve both informative and persuasive functions.

Early banner ads used a simple graphic, usually in GIF format, that loaded with the Web page and remained on the page until the user moved to another page or closed the browser. Today, a variety of **animated GIFs** and **rich media objects** created using Shockwave, Java, or Flash are used to make attention-grabbing banner ads. These ads can be rotated so that each time the Web page is loaded into a browser, the ad changes.

Although Web sites can create banner ads in any dimensions, advertisers decided early in the life of electronic commerce that it would be easier to standardize the sizes. The standard banner sizes that most Web sites have voluntarily agreed to use are called **interactive marketing unit (IMU) ad formats**. The **Interactive Advertising Bureau (IAB)** is a not-for-profit organization that promotes the use of Internet advertising and encourages effective Internet advertising (in 2001, the IAB changed its name from the Internet Advertising Bureau). The IAB has established voluntary standards for IMUs. As the Web grew, so did the creativity of Web advertisers. They were using an increasing number of IMU ad formats. By 2003, advertisers were using more than 15 different IMU ad formats. The IAB decided to encourage its members to agree to use only four standard formats. The four formats appear in Figure 4-12. You can learn more about banner ads, including examples of the latest IAB-approved sizes, by following the Online Companion link to the IAB Web site.

Most advertising agencies that work with online clients can create banner ads as part of their services. Web site design firms also can create banner ads. Charges for creating banner ads range from about $100 to over $1500, depending on the complexity of the ad. Companies can make their own banner ads by using a graphics program or the tools provided by some Web sites. **AdDesigner.com** is an advertising-supported Web site that offers free downloadable graphics.

Banner Ad Placement

Companies have three different ways to arrange for other Web sites to display their banner ads. The first is to use a banner exchange network. A **banner exchange network** coordinates ad sharing so that other sites run one company's ad while that company's site runs other exchange members' ads. Usually, the exchange requires each member site to accept two ads on its site for every one of its ads that appears on another member's site. The exchange then makes its profit by selling the extra ad space to other businesses. Since

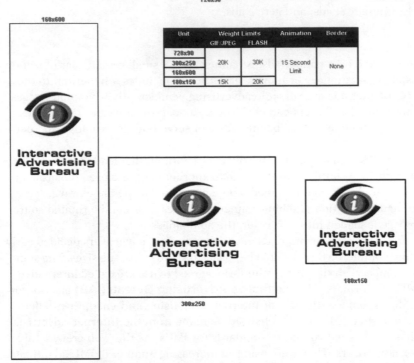

Unit	Weight Limits		Animation	Border
	GIF/JPEG	FLASH		
728x90				
300x250	20K	30K	15 Second Limit	None
160x600				
180x150	15K	20K		

FIGURE 4-12   IAB Universal Ad Package guidelines

banner exchanges are free, many Web businesses do use them; however, it is often difficult to find a group of other Web sites that have formed an exchange or that belong to an
exchange that are not direct competitors. This limitation prevents many businesses from
using banner exchange networks. Not-for-profit information Web sites are more likely to find
a banner exchange network suitable.

The second way that businesses can place their banner advertising is to find Web sites
that appeal to one of the company's market segments and then pay those sites to carry the
ads. This can take considerable time and effort. Smaller sites may not have an established pricing policy for advertising. Larger sites usually have high standard rates that they
discount for larger customers. Smaller customers generally pay the standard rates. A company can hire an advertising agency to negotiate lower rates and help with ad placement.
A full-service advertising agency can help design the ads, create the banners, and identify
appropriate Web sites on which to display them. Agencies that do a lot of Internet work can
often negotiate lower advertising rates with sites because the agencies can consolidate their
clients' budgets and buy large blocks of advertising space at one time.

A third way to place banner advertising is to use a banner advertising network. A **banner
advertising network** acts as a broker between advertisers and Web sites that carry ads. The

larger banner advertising networks, such as **DoubleClick**, LinkExchange (now a part of **Microsoft bCentral**), and **ValueClick**, offer many of the same services as comprehensive ad agencies and often broker space primarily on larger Web sites (such as Yahoo!) that have high traffic rates and are, thus, more expensive. The smaller firms, on the other hand, often sell only leftover discounted space. Many of these smaller firms have fallen on hard times with the recent decline in advertising purchases, and many have gone out of business.

Measuring Banner Ad Cost and Effectiveness

As more companies rely on their Web sites to make a favorable impression on potential customers, the issue of measuring Web site effectiveness has become important. Mass media efforts are measured by estimates of audience size, circulation, or number of addressees. When a company purchases mass media advertising, it pays a dollar amount for every thousand people in the estimated audience. This pricing metric is called **cost per thousand** and is often abbreviated **CPM** (the "M" is from the Roman numeral for "thousand").

Measuring Web audiences is more complicated because of the Web's interactivity and because the value of a visitor to an advertiser depends on how much information the site gathers from the visitor (for example, name, address, e-mail address, telephone number, and other demographic data). Because each visitor voluntarily chooses whether to provide these bits of information, all visitors are not of equal value. Internet advertisers have developed some Web-specific metrics for site activity, but these are not generally accepted and are currently the subject of considerable debate.

A **visit** occurs when a visitor requests a page from the Web site. Further page loads from the same site are counted as part of the visit for a specified period of time. This period of time is chosen by the administrators of the site and depends on the type of site. A site that features stock quotes might use a short time period because visitors may load the page to check the price of one stock and reload the page 15 minutes later to check another stock's price. A museum site would expect a visitor to load multiple pages over a longer time period during a visit and would use a longer visit time window. The first time that a particular visitor loads a Web site page is called a **trial visit**; subsequent page loads are called **repeat visits**. Each page loaded by a visitor counts as a **page view**. If the page contains an ad, the page load is called an **ad view**.

Some Web pages have banner ads that continue to load and reload as long as the page is open in the visitor's Web browser. Each time the banner ad loads is an **impression**. If the visitor clicks the banner ad to open the advertiser's page, that action is called a **click** or **click-through**. Banner ads are often sold on a CPM basis where the "thousand" is 1000 impressions. Rates vary greatly and depend on how much demographic information the Web site obtains about its visitors and what kinds of visitors the site attracts, but most rates range between $1 and $50 CPM. As recently as 1999, the range of online advertising rates was much higher, from about $5 to $100. Figure 4-13 shows a comparison of CPM rates for banner ads and other Web advertising media to CPM rates for advertising placed in traditional media outlets.

One of the most difficult things for companies to do as they move onto the Web is gauge the costs and benefits of advertising on the Web. Many companies have developed new metrics to evaluate the number of desired outcomes their advertising yields. For example, instead of comparing the number of click-throughs that companies obtain per dollar of advertising, they measure the number of new visitors to their site who buy for the first time

Medium	Description	Total cost	Audience size	Cost per thousand (CPM)
Network television	30-second commercial	$80,000 – $600,000	10 million – 20 million	$5 – $30
Local television station	30-second commercial	$1000 – $50,000	50,000 – 2 million	$3 – $25
Cable television	30-second commercial	$3000 – $10,000	100,000 – 500,000	$8 – $20
Radio	60-second commercial	$200 – $1000	50,000 – 2 million	$1 – $18
Major metro newspaper	Full-page ad, single insertion	$20,000 – $80,000	100,000 – 600,000	$80 – $130
Regional edition of a national magazine	Full-page ad, single insertion	$5000 – $50,000	50,000 – 900,000	$40 – $80
Local magazine	Full-page ad, single insertion	$2000 – $10,000	3000 – 80,000	$100 – $140
Direct mail – coupon pack	Mailed in letter-sized envelope	$100 – $3000	10,000 – 200,000	$15 – $20
Billboard	Highway billboard (one – three months)	$5000 – $25,000	100,000 – 3 million	$2 – $5
World Wide Web	Banner ad (one month)	$200 – $30,000	10,000 – 50 million	$1 – $50
World Wide Web	Rich media ad (one month)	$200 – $1 million	10,000 – 50 million	$18 – $30
World Wide Web	Site sponsorship (one month)	$300 – $2 million	10,000 – 50 million	$30 – $75
Targeted e-mail	Single mailing	$50 – $150,000	10,000 – 10 million	$5 – $15

FIGURE 4-13   CPM rates for advertising in various media

after arriving at the site by way of a click-through. They can then calculate the advertising cost of acquiring one customer on the Web and compare that to how much it costs them to acquire one customer through traditional channels.

When banner ads first appeared on the Web in the mid-1990s, they provided a new experience for Web surfers. As users saw more ads, however, the ads lost their ability to attract attention. Click-through rates, which had been as high as 2 percent when banner ads were first introduced, have steadily dropped and now range from .3 percent to .5 percent, depending on the site's content.

To battle this decrease, banner ad designers first introduced animated GIFs with moving elements in the hopes that they might be more attractive to the user's eye than stationary graphics. When animated GIFs failed to halt the decline in click-through rates, designers created ads that displayed rich media effects, such as movie clips. They added interactive effects by writing Java programs that could respond to a user's click with some action (other than simply loading the advertiser's page into the browser). Some of these interactive ads even act like miniature video games. Designers also created banner ads that appear to be dialog boxes in the hope that confused users would click them. Several examples of this type of banner ad are shown in Figure 4-14. These ads are designed to induce users to click a button in the ad to fix the "error", but the banners actually link to Web sites or begin installing a program on the user's computer.

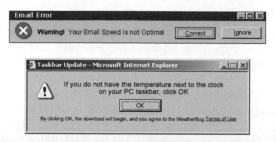

FIGURE 4-14   Disguised banner ads

Periodically, advertisers have tried new banner ad sizes and have placed the ads in page locations other than at the top or bottom. For example, some sites now use a large banner ad called a **skyscraper ad** that is designed to be placed on the side of a Web page and remain visible as the user scrolls down through the page.

Unfortunately for advertisers, none of these efforts has prevented the inexorable decline in click-through rates. In what some observers see as a last-ditch attempt to make advertising work on the Web, banner ad designers have turned to the alternative ad formats described in the next section.

Other Web Ad Formats

The steady decline in the effectiveness of banner ads has prompted advertisers to explore other formats for Web ads. One of these formats is the pop-up ad. A **pop-up ad** is an ad that appears in its own window when the user opens or closes a Web page. The window in which the ad appears does not include the usual browser controls. The only way to dismiss the ad is to click the small close button in the upper-right corner of the window's frame. Many users find pop-up ads extremely annoying. A particularly irritating variation on the pop-up ad technique occurs at Web sites that open more than one pop-up ad when a user leaves the site or closes the browser. If the user does not act quickly enough, the browser spawns multiple windows and can even crash the computer.

Another type of pop-up ad is called the pop-behind ad. A **pop-behind ad** is a pop-up ad that is followed very quickly by a command that returns the focus to the original browser window. The result is an ad that is parked behind the user's browser, waiting to appear when the browser is closed.

Marketing on the Web

Despite user objections to pop-up ads (in all their variations), an increasing number of Web sites are using them as a way of delivering a larger advertising image in a more forceful way. Some users have responded by using **ad-blocking software** that prevents banner ads and pop-up ads from loading (see the Online Companion's Additional Information section for links to Web sites that distribute or sell ad-blocking software). An increasing number of Web browsers can be configured not to display many of these ads; however, any site that uses methods for navigation that are similar to those used to deliver ads (such as pop-up information windows) cannot operate as intended in the reconfigured browser. Research conducted in 2004 by British interactive media consulting firm **Bunnyfoot Universality** found that pop-up ads not only annoy users, they actually create lasting bad will among users toward the company whose products are depicted in the ads. Despite these findings, many advertisers find pop-up ads to be effective tools for drawing customers to their sites and continue to use them.

Another intrusive ad format is the **interstitial ad**. When a user clicks a link to load a page, the interstitial ad opens in its own browser window, instead of the page that the user intended to load (the general meaning of the word "interstitial" is something that comes between two other things). Many interstitial ads close automatically, allowing the intended page to open in the existing browser window. Other interstitials require the user to click a button before they close. Because they open in a full-size browser window, interstitial ads offer the advertiser even more space than the pop-up ad format. These ads also completely cover the Web page that the user was trying to see. Many users find interstitials even more annoying than pop-up ads because they are larger and a more forceful interruption of the Web-browsing experience.

A fourth ad format is the rich media ad. **Rich media ads**, also called **active ads**, generate graphical activity that "floats" over the Web page itself instead of opening in a separate window. These ads always contain moving graphics and usually include audio and video elements. One of the first rich media ads featured the figure of a little man who walked into the displayed Web page, unrolled a movie poster, and then pasted the poster onto the Web page (covering up part of the Web page content—content that a user might have been reading!). After about 10 seconds, the figure walked off the page and the poster disappeared. While it was open on the page, the poster was an active link to the movie's Web site. Another early rich media ad showed a Ford Explorer driving into the Web page. The Web page appeared to shake with the vibrations of the Explorer as it drove through. Rich media ads are certainly attention getting and are even more intrusive than pop-ups or interstitials because they occur on the Web page itself and offer users no obvious way to dismiss them. Many industry observers believe that advertisers will create new ad formats as users become accustomed to seeing active ads and they lose their effectiveness.

Site Sponsorships

Some Web sites offer advertisers the opportunity to sponsor all or parts of their sites. These **site sponsorships** give advertisers a chance to promote their products, services, or brands in a more subtle way than by placing banner or pop-up ads on the sites (although some sponsorship packages include a certain number of banner and pop-up ads).

Companies that buy Web site sponsorships have goals that are similar to those of sporting event sponsors or television program sponsors; that is, they want to tie the company or product name to an event or a set of information. The idea is that the quality of the event or information set will carry over to the company's products, services, or brands. In general, sponsorships are used to build brand images and develop reputation rather than to generate immediate sales.

In some cases, the sponsor is given the right to create content for the site or to weave its advertising message into the site's content. This practice can raise ethical concerns if not done carefully. Sites that offer content spots to sponsors should always identify the content as an advertisement or as provided by the sponsor. Unfortunately, many sites do not use clear labels for sponsored content. This can confuse site visitors who are unable to distinguish between editorial content and advertising. Sites that offer medical information, for example, should be especially careful to distinguish between information that is generated by the site's reporters or editorial staff and information that is provided by pharmaceutical companies or medical device manufacturers.

Effectiveness of Online Advertising

After years of experimenting with a variety of online advertising formats, the effectiveness of online advertising remains difficult to measure. A major problem is the lack of a single industry standard measuring service, such as the service that the Nielsen ratings provide for television broadcasting or the Audit Bureau of Circulations procedures provide for the print media. In 2003, the Interactive Advertising Bureau (IAB) and the Institute of Practitioners in Advertising (IPA) created a joint task force to review four media measurement systems (Nielsen//NetRatings, ComScore, Hitwise, and RedSheriff) and recommend one as the single standard or devise an alternative measurement system. The task force has announced that it is currently considering only ComScore and Nielsen//NetRatings.

Part of the difficulty in measuring the effectiveness of online advertising arises from the ways in which site visitors change their Web surfing behaviors and habits. For example, as people using the Web gain experience, they change their behavior. An experienced Web user is far less likely than a new Web user to click a banner ad. Declining click-through rates might not be a good indicator of the success of online advertising, however. Many companies are finding that online advertising can be an important element in a comprehensive marketing strategy that uses several different media to deliver messages to potential customers. In 2003, Ad Age reported survey results that showed more potential car purchasers would be influenced by an online ad than by a television ad. Very few people would buy a car based on an online ad, but online ads might prove to be an effective way of building brand recognition and conveying information about cars to potential buyers. You can learn more about current developments in online advertising effectiveness by visiting the **AdAge.com**, **eMarketer**, and **Online Publishers Association** Web sites.

Most marketing analysts do agree that online advertising is much more effective if it is properly targeted. Online ads that reach site visitors who are looking for something specific that is related to the ad's message are much more successful than ads viewed by a general population. Thus, market segmentation is an important element in online advertising success. One useful marketing tool that uses market segmentation successfully is e-mail marketing, the subject of the next section.

Sociologists and cultural anthropologists have proclaimed e-mail to be one of the greatest tools for human communication to be developed in the 20th century. Because advertising is a process of communication, it is easy to see that e-mail can be a very powerful element in any company's advertising strategy. Many businesses would like to send e-mail messages to their customers and potential customers to announce new products, new product features, or sales on existing products. However, industry analysts have severely criticized some companies for sending e-mail messages to customers or potential customers. Some companies have even faced legal action after sending out mass e-mailings. You will learn more about unsolicited commercial e-mail (often called "spam") issues in Chapter 8. However, sending e-mail messages to Web site visitors who expressly request the e-mail messages is a completely different story. A key element in any e-mail marketing strategy is to obtain customers' approvals before sending them any e-mail that includes a marketing or promotional message.

Permission Marketing

Many businesses are finding that they can maintain an effective dialog with their customers by using automated e-mail communications. Sending one e-mail message to a customer can cost less than one cent if the company already has the customer's e-mail address. Purchasing the e-mail addresses of people who ask to receive specific kinds of e-mail messages adds between a few cents and a dollar to the cost of each message sent. Another factor to consider is the conversion rate. The **conversion rate** of an advertising method is the percentage of recipients who respond to an ad or promotion. Conversion rates on requested e-mail messages range from 10 percent to over 30 percent. These are much higher than the click-through rates on banner ads, which are currently under .5 percent and decreasing.

The practice of sending e-mail messages to people who request information on a particular topic or about a specific product is called **opt-in e-mail** and is part of a marketing strategy called **permission marketing**. Seth Godin, the founder of YoYoDyne and later the vice president for direct marketing at Yahoo!, developed this marketing strategy and publicized it in a book he wrote with Don Peppers titled *Permission Marketing*. Godin argues that, as the pace of modern life quickens, time becomes a valuable commodity. Most marketing efforts that traditional businesses use to promote their products or services depend on potential customers having enough time to listen to sales pitches and pay attention to the best ones. As time becomes more precious to everyone, people no longer wish to hear and evaluate advertising and promotional appeals for products and services in which they have no interest.

Thus, a marketing strategy that sends specific information only to people who have indicated an interest in receiving information about the product or service being promoted should be more successful than a marketing strategy that sends general promotional messages through the mass media. Two companies that offer opt-in e-mail services are **PostMasterDirect** and **yesmail.com**. These services provide the e-mail addresses to advertisers at rates that vary depending on the type and price of the product being promoted, but range from a minimum of about $1 to a maximum of 25–30 percent of the selling price of the product.

Combining Content and Advertising

One strategy for getting e-mail accepted by customers and prospects that many companies have found successful is to combine content with an advertising e-mail message. Articles and news stories that would interest specific market segments are good ways to increase acceptance of e-mail.

E-mail messages that include large articles or large attachments (such as graphics, audio, or video files) can fill up recipients' in-boxes very quickly, so many advertisers send content by inserting hyperlinks into e-mail messages. The hyperlinks should take customers to the content, which is stored on the company's Web site. Once customers are viewing pages on the Web site, it is easier to induce them to stay on the site and consider making purchases. Using hyperlinks that lead to a Web page instead of embedding content in e-mail messages is especially important if the content requires a browser plug-in to play (as many audio and video files do). The Web page can provide a link to the needed plug-in software.

An important element in any marketing strategy is coordination across media outlets. If a company is using e-mail to promote its products or services, it should make sure that any other marketing efforts it is undertaking at the same time, such as press releases, print media ads, or broadcast media ads, are delivering a message that is consistent with the e-mail campaign's message.

Outsourcing E-Mail Processing

Many companies find that the number of customers who opt-in to information-laden e-mails can grow rapidly. The job of handling e-mail lists and mass-mailing software can quickly outgrow the capacity of the company's information technology staff. A number of companies offer e-mail management services, and most small or medium-size companies outsource their e-mail-processing operations. The Additional Information section of the Online Companion pages for this chapter includes links to several companies that offer e-mail processing and management services. These companies will manage an e-mail campaign for a cost of between one and two cents per valid e-mail address. Many of these companies will also help their clients purchase lists of e-mail addresses from companies that compile such lists. You will learn more about the ethical issues faced by companies that compile and sell e-mail address lists in Chapter 7.

TECHNOLOGY-ENABLED CUSTOMER RELATIONSHIP MANAGEMENT

The nature of the Web, with its two-way communication features and traceable connection technology, allows firms to gather much more information about customer behavior and preferences than they can gather using micromarketing approaches. Now, companies can measure a large number of things that are happening as customers and potential customers gather information and make purchasing decisions. The information that a Web site can gather about its visitors (which pages were viewed, how long each page was viewed, the sequence, and similar data) is called a **clickstream**.

The idea of technology-enabled relationship management has become possible when promoting and selling on the Web. **Technology-enabled relationship management** occurs when a firm obtains detailed information about a customer's behavior, preferences, needs, and buying patterns, *and* uses that information to set prices, negotiate terms, tailor promotions, add product features, and otherwise customize its entire relationship with that customer.

Although companies can use technology-enabled relationship management concepts to help manage relationships with vendors, employees, and other stakeholders, most companies currently use these concepts to manage customer relationships. Thus, technology-enabled relationship management is often called **customer relationship management (CRM)**, **technology-enabled customer relationship management**, or **electronic customer relationship management (eCRM)**. Figure 4-15 lists seven dimensions of the customer interaction experience and shows how technology-enabled customer relationship management differs from traditional seller-customer interactions in each of those dimensions.

Dimensions	Technology-enabled customer relationship management	Traditional relationships with customers
Advertising	Provide information in response to specific customer inquiries	"Push and sell" a uniform message to all customers
Targeting	Identify and respond to specific customer behaviors and preferences	Market segmentation
Promotions and discounts offered	Individually tailor to customer	Same for all customers
Distribution channels	Direct or through intermediaries; customer's choice	Through intermediaries chosen by the seller
Pricing of products or services	Negotiated with each customer	Set by the seller for all customers
New product features	Created in response to customer demands	Determined by the seller based on research and development
Measurements used to manage the customer relationship	Customer retention; total value of the individual customer relationship	Market share; profit

FIGURE 4-15   Technology-enabled relationship management and traditional customer relationships

CRM as a Source of Value in the Marketspace

Harvard Business School researchers Jeffrey Rayport and John Sviolka observed that firms today do business in both a physical world and a virtual, information world. Rayport and Sviolka distinguish between commerce in the physical world, or marketplace, and commerce in the information world, which they term the **marketspace**. In the information world's marketspace, digital products and services can be delivered through electronic communication channels, such as the Internet.

In Chapter 1, you learned that the value chain model described the primary and support activities that firms use to create value. This value chain model is valid for activities in the physical world and in the marketspace. However, value creation requires different processes in the marketspace. By understanding that value creation in the marketspace is different, firms can identify value opportunities effectively in both the physical and information worlds.

For years, businesses have viewed information as a part of the value chain's supporting activities, but they have not considered how information itself might be a source of value. In the marketspace, firms can use information to create new value for customers. Many electronic commerce Web sites today offer customers the convenience of an online order history, recommendations based on previous purchases, and show current information about products in which the customer might be interested.

Successful Web marketing approaches all involve enabling the potential customer to find information easily and customizing the depth and nature of that information; such approaches should encourage the customer to buy. Firms should track and examine the behaviors of their Web site visitors, and then use that information to provide customized, value-added digital products and services in the marketspace. Companies that use these technology-enabled relationship management tools to improve their contact with customers are more successful on the Web than firms that adapt advertising and promotion strategies that were successful in the physical world, but are less effective in the virtual world. You can obtain updates on current developments in CRM at the **IntelligentCRM** Web site. In Chapter 9, you will learn more about the specific software tools and other technologies that companies are using to implement CRM.

CREATING AND MAINTAINING BRANDS ON THE WEB

A known and respected brand name can present to potential customers a powerful statement of quality, value, and other desirable qualities in one recognizable element. Branded products are easier to advertise and promote, because each product carries the reputation of the brand name. Companies have developed and nurtured their branding programs in the physical marketplace for many years. Consumer brands such as Ivory soap, Walt Disney entertainment, Maytag appliances, and Ford automobiles have been developed over many years with the expenditure of tremendous amounts of money. However, the value of these and other trusted major brands far exceeds the cost of creating them.

Elements of Branding

The key elements of a brand, according to researchers at advertising agency Young & Rubicam, are differentiation, relevance, and perceived value. Product differentiation is the first condition that must be met to create a product or service brand. The company must clearly distinguish its product from all others in the market. This makes branding difficult for commodity products such as salt, nails, or plywood—difficult, but not impossible.

A classic example of branding a near-commodity product is Procter & Gamble's creation of the Ivory brand over 100 years ago. The company was experimenting with manufacturing processes and had accidentally created a bar soap that contained a high percentage of air. When one of the workers noted that the soap floated in water, the company decided to sell the soap using this differentiating characteristic in packaging and advertising by claiming "it floats". Thus was the Ivory soap brand born. **Procter & Gamble** maintains this brand differentiation on its Web site even today by listing the link to its **Ivory Soap** site under the heading "Beauty and Skin Care Products".

The second element of branding—relevance—is the degree to which the product offers utility to a potential customer. The brand only has meaning to customers if they can visualize its place in their lives. Many people understand that **Tiffany & Co.** creates a highly differentiated line of jewelry and gift products, but very few people can see themselves purchasing and using such goods.

The third branding component—perceived value—is a key element in creating a brand that has value. Even if your product is different from others on the market and potential customers can see themselves using this product, they will not buy it unless they perceive value. Some large fast food outlets have well-established brands that actually work against them. People recognize these brands and avoid eating at these restaurants because of negative associations—such as low overall quality and high-fat-content menu items. Figure 4-16 summarizes the elements of a brand.

Element	Meaning to customer
Differentiation	In what significant ways is this product or service unlike its competitors?
Relevance	How does this product or service fit into my life?
Perceived value	Is this product or service good?

FIGURE 4-16   Elements of a brand

If a brand has established that it is different from competing brands and that it is relevant and inspires a perception of value to potential purchasers, those purchasers will buy the product and become familiar with how it provides value. Brands become established only when they reach this level of purchaser understanding and acceptance.

Unfortunately, brands can lose their value if the environment in which they have become successful changes. A dramatic example is Digital Equipment Corporation (DEC). For years, DEC was a leading manufacturer of midrange computers. When the market for computing shifted to personal computers, DEC found that its branding did not transfer to the personal computers that it produced. The consumers in that market did not see the same perceived value or differentiation in DEC's personal computers that the buyers of midrange systems had seen for years. This is an important element of branding for Web-based firms to remember, because the Web is still evolving and changing at a rapid pace.

Emotional Branding vs. Rational Branding

Companies have traditionally used emotional appeals in their advertising and promotion efforts to establish and maintain brands. One branding expert, Ted Leonhardt, has described "brand" as "an emotional shortcut between a company and its customer". These emotional appeals work well on television, radio, billboards, and in print media, because the ad targets are in a passive mode of information acceptance. However, emotional appeals are difficult to convey on the Web because it is an active medium controlled to a great extent by the customer. Many Web users are actively engaged in such activities as finding information, buying airline tickets, making hotel reservations, and obtaining weather forecasts. These users are busy people who will rapidly click away from emotional appeals.

Marketers are attempting to create and maintain brands on the Web by using **rational branding**. Companies that use rational branding offer to help Web users in some way in exchange for their viewing an ad. Rational branding relies on the cognitive appeal of the specific help offered, not on a broad emotional appeal. For example, Web e-mail services such as **Excite Mail**, **HotMail**, or **Yahoo! Mail** give users a valuable service—an e-mail account and storage space for messages. In exchange for this service, users see an ad on each page that provides this e-mail service.

Similarly, **MasterCard** promotes its brand name online through its **Shop Smart!** program. Shop Smart! is a third-party assurance mechanism. MasterCard ensures that any Web site displaying the Shop Smart! emblem (which happens to include a large Master-Card logo) is using what MasterCard defines as a "safe" method of processing transactions. In exchange for this assurance on a Web shopping site, the Web user sees the Master-Card logo.

Brand Leveraging Strategies

Rational branding is not the only way to build brands on the Web. One method that is working for well-established Web sites is to extend their dominant positions to other products and services, a strategy called **brand leveraging**. **Yahoo!** is an excellent example of a company that has used brand-leveraging strategies. Yahoo! was one of the first directories on the Web. It added a search engine function early in its development and has continued to parlay its leading position by acquiring other Web businesses and expanding its existing offerings. Yahoo! acquired GeoCities and Broadcast.com, and entered into an extensive cross-promotion partnership with a number of **Fox** entertainment and media companies. Yahoo! continues to lead its two nearest competitors, **Excite** and **Go.com**, in ad revenue by adding features that Web users find useful and that increase the site's value to advertisers. Amazon.com's expansion from its original book business into CDs, videos, and auctions is another example of a Web site leveraging its dominant position by adding features that are useful to existing customers.

Brand Consolidation Strategies

Another way to leverage the established brands of existing Web sites was pioneered by Della & James, an online bridal registry that is now doing business as part of **WeddingChannel.com**. Although a number of national department store chains, such as **Macy's**, have established online registries for their own stores, Della & James created a single registry that connects to several local and national department and gift stores, including **Crate&Barrel**, **Gump's**, **Neiman Marcus**, **Tiffany & Co.**, and **Williams-Sonoma**. The logo and branding of each participating store are featured prominently on the WeddingChannel.com site. The founders identified an opening for a market intermediary because the average engaged couple registers at three stores. Thus, WeddingChannel.com provides a valuable consolidating activity for registering couples and their wedding guests that no store operating alone could provide. WeddingChannel.com also provides wedding planning services and access to every item that a bride and groom might need—from the bridal gown to the cake—all in one convenient Web location.

Costs of Branding

Transferring existing brands to the Web or using the Web to maintain an existing brand is much easier and less expensive than creating an entirely new brand on the Web. In 1998, a large number of companies began spending significant amounts of money to build new brands on the Web. According to studies by the **Intermarket Group**, the top 100 electronic commerce sites each spent an average of $8 million that year to create and build their online brands. Two of the top spenders included the battling Web sites **Amazon.com**, which spent $133 million, and **BarnesandNoble.com**, which spent $70 million. Most of this spending was for television, radio, and print media—not for online advertising. Online brokerages E*TRADE and Ameritrade Holding were also among the top five in that first year of major brand building on the Web, spending $71 million and $44 million, respectively.

Brand-building activity continued on the Web through 1999 and into the first months of 2000. In March 2000, the supply of money from lenders and venture capitalists began drying up, which resulted in smaller advertising expenditures for most firms. By 2001, the peak of brand-building spending was over for new companies on the Web. Traditional firms realized that an opportunity had opened for them to move their offline brands to the Web.

Promoting any company's Web presence should be an integral part of brand development and maintenance. The company's URL should always be included on product packaging and in mass media advertising on radio, television, and in print. Integrating the URL with the company logo on brochures can also be helpful in getting the word out about the Web site. Ensuring that the site appears in search engine listings is also very important, as you will learn in the next section.

Affiliate Marketing Strategies

Of course, this leveraging approach works only for firms that already have Web sites that dominate a particular market. As the Web matures, it will be increasingly difficult for new entrants to identify unserved market segments and attain dominance. A tool that many new, low-budget Web sites are using to generate revenue is affiliate marketing. In **affiliate marketing**, one firm's Web site—the affiliate firm's—includes descriptions, reviews, ratings, or other information about a product that is linked to another firm's site that offers the item for sale. For every visitor who follows a link from the affiliate's site to the seller's site, the affiliate site receives a commission. The affiliate site also obtains the benefit of the selling site's brand in exchange for the referral.

The affiliate saves the expense of handling inventory, advertising and promoting the product, and processing the transaction. In fact, the affiliate risks no funds whatsoever.

CDnow and Amazon.com were two of the first companies to create successful affiliate programs on the Web. CDnow's Web Buy program, which included more than 250,000 affiliates before the company entered into its joint marketing agreement with Amazon.com, was one of CDnow's main sources for new customers. The Amazon.com program (which now includes the CDnow program) has over 800,000 affiliate sites. Most of these affiliate sites are devoted to a specific issue, hobby, or other interest. Affiliate sites choose books or other items that are related to their visitors' interests and include links to the seller's site on their Web pages. Books and CDs are a natural for this type of shared promotional activity, but sellers of other products and services also use affiliate marketing programs to attract new customers to their Web sites.

One of the more interesting marketing tactics made possible by the Web is **cause marketing**, which is an affiliate marketing program that benefits a charitable organization (and, thus, supports a "cause"). In cause marketing, the affiliate site is created to benefit the charitable organization. When visitors click a link on the affiliate's Web page, a donation is made by a sponsoring company. The page that loads after the visitor clicks the donation link carries advertising for the sponsoring companies. Many companies have found that the click-through rates on these ads are much higher than the typical banner ad click-through rates.

Affiliate Commissions

Affiliate commissions can be based on several variables. In the **pay-per-click model**, the affiliate earns a commission each time a site visitor clicks the link and loads the seller's page. This is similar to the click-through model of charging for banner advertising, and the rates paid per thousand click-throughs are similar to those paid for banner ads.

In the **pay-per-conversion model**, the affiliate earns a commission each time a site visitor is converted from a visitor into either a qualified prospect or a customer. An example of a seller that might use the qualified prospect definition is a credit card-issuing bank. The bank might decide that its best strategy is to pay affiliates only when the visitor turns out to be a good credit risk. Alternatively, the bank may decide it wants to pay the affiliate only if the visitor is approved for the card and then accepts the card (completes the sale). A site that pays its affiliates on completed sales usually pays a percentage of the sale amount rather than a fixed amount per conversion. Some sites use a combination of these methods to pay their affiliates. Commissions on completed sales range from 5 percent to 20 percent of the sale amount, depending on variables such as the type of product, the strength of the product's brand, how profitable the product is, and the size of an average order.

You can learn more about affiliate programs offered by different sites by going to each Web site and looking for a link to information about its affiliate program. Figure 4-17 shows the affiliate program information page for **Proflowers.com**, a top-rated online florist that has more than 40,000 affiliates sending business to its site.

Alternatively, you can visit an affiliate program broker site that offers affiliate program opportunities for a number of Web sites. An **affiliate program broker** is a company that serves as a clearinghouse or marketplace for sites that run affiliate programs and sites that want to become affiliates. These brokers also often provide software, management consulting, and brokerage services to affiliate program operators. For example, Proflowers.com uses affiliate program broker **Be Free** to manage its affiliate program. Be Free tracks

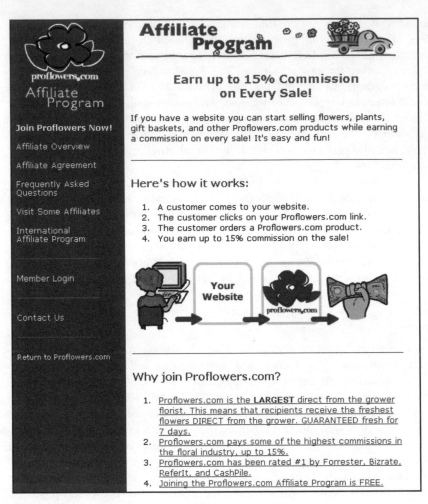

FIGURE 4-17   Proflowers.com affiliate program information page

affiliates' sales, calculates and pays affiliates' commissions, and handles any problems that arise. **Commission Junction** and **LinkShare** are two other popular affiliate program brokers. Other companies, such as **Performics**, offer affiliate program brokering along with other marketing services.

Viral Marketing Strategies

Traditional marketing strategies have always been developed with an assumption that the company would communicate with potential customers directly or through an intermediary acting on behalf of the company, such as a distributor, retailer, or independent sales organization. Because the Web expands the types of communication channels available, including customer-to-customer communication, another marketing approach, viral marketing, has become popular on the Web. **Viral marketing** relies on existing customers to tell other people—the company's prospective customers—about the products or services they have enjoyed using. Much as affiliate marketing uses Web sites to spread the word about a company, viral marketing approaches use individual customers to do the same thing. The number of customers increases the way a virus multiplies, thus the name.

Blue Mountain Arts, an electronic greeting card company, purchased very little advertising, but grew rapidly. Electronic greeting cards are e-mail messages that include a link to the greeting card site. When people received Blue Mountain Arts electronic greeting cards in their e-mail, they clicked a link in the e-mail message that opened the Blue Mountain Arts Web site in their browser. Once at the Blue Mountain Arts site, they were likely to search for cards that they might like to send to other friends. A greeting card recipient might send electronic greeting cards to several friends, who could then send greetings to their friends. Each new visitor to the site could spread the "virus," which in this case was the knowledge of Blue Mountain Arts. By late 1999, when the company was sold to At Home Corporation for $780 million, Blue Mountain had more than 10 million people visiting its site each month. Blue Mountain Arts built a large following using its approach to viral marketing. Today, the site requires visitors to pay for a subscription before they can send electronic greeting cards. However, the site's original strategy of offering free greetings combined with a viral marketing strategy helped it build a large customer base very quickly.

SEARCH ENGINE POSITIONING AND DOMAIN NAMES

Potential customers find Web sites in many different ways. Some site visitors are referred by a friend. Others are referred by an affiliate marketing partner of the site. Some see the site's URL in a print advertisement or on television. Others arrive after typing a URL that is similar to the company's name. But many site visitors are directed to the site by a search engine or directory Web site.

Search Engines and Web Directories

A **search engine** is a Web site that helps people find things on the Web. Search engines contain three major parts. The first part, called a **spider**, a **crawler**, or a **robot** (or simply **bot**), is a program that automatically searches the Web to find Web pages that might be interesting to people. When the spider finds Web pages that might interest search engine site visitors, it collects the URL of the page and information contained on the page. This information might include the page's title, key words included in the page's text, and information about other pages on that Web site. In addition to words that appear on the Web page, Web site designers can specify additional key words in the page that are hidden from the view of Web site visitors, but that are visible to spiders. These key words are enclosed in an HTML tag set called meta tags. The word "meta" is used for this tag set to indicate that the key words describe the content of a Web page and are not themselves part of the content.

The spider returns this information to the second part of the search engine to be stored. The storage element of a search engine is called its **index** or **database**. The index checks to see if information about the Web page is already stored. If it is, it compares the stored information to the new information and determines whether to update the page information. The index is designed to allow fast searches of its very large amount of stored information.

The third part of the search engine is the search utility. Visitors to the search engine site provide search terms, and the **search utility** takes those terms and finds entries for Web pages in its index that match those search terms. The search utility is a program that creates a Web page that is a list of links to URLs that the search engine has found in its index that match the site visitor's search terms. The visitor can then click the links to visit those sites. You will learn more about the technologies used in search engines in later chapters of this book.

Some search engine sites also provide classified hierarchical lists of categories into which they have organized commonly searched URLs. Although these sites are technically called Web directories, most people refer to them as search engines. The most popular of these sites, such as Yahoo!, include a Web directory and a search engine. They give users the option of using the search engine to find categories of URLs as well as the URLs themselves. This combination of Web directory and search engine can be a powerful tool for finding things on the Web. **Nielsen//NetRatings**, the online audience-measurement and analytics consulting firm, frequently issues press releases that list the most frequently visited Web sites. The search engine and Web directory sites **AltaVista**, **AOL**, **Excite**, **Google**, **Lycos**, **MSN**, and **Yahoo!** regularly appear on these lists.

Marketers want to make sure that when a potential customer enters search terms that relate to their products or services, their companies' Web site URLs appear among the first 10 returned listings. The weighting of the factors that search engines use to decide which URLs appear first on searches for a particular search term is called a **search engine ranking**. For example, if a site is near the top of the list of links returned for the search term "auto", that site is said to have a high search engine ranking for "auto". The combined art and science of having a particular URL listed near the top of search engine results is called **search engine positioning**, **search engine optimization**, or **search engine placement**. For sites that obtain most of their visitors from search engines, a high ranking that places their URL near the top of the list of links returned by the search engine is extremely important.

Paid Search Engine Inclusion and Placement

An increasing number of search engine sites have started making the task easier—but for a price. These search engine sites offer companies a **paid placement** (also called a **sponsorship** or a **search term sponsorship**; however, note that these search term sponsorships are not the same thing as the general site sponsorships you learned about earlier in this chapter), which is the option of purchasing a top listing on results pages for a particular set of search terms. The rates charged vary tremendously depending on the desirability of the search terms to potential sponsors.

Another option for companies is to buy banner ad space at the top of search results pages that include certain terms. For example, Chevrolet might want to buy banner ad space at the top of all search results pages that are generated by queries containing the words "new" and "car". Most search engine sites sell banner ad space on this basis. An increasing number sell space on results pages for the most desirable terms only to companies that agree to package deals that include paid placement and banner ad purchases.

Search engine positioning is a complex subject. A number of consulting firms do nothing but advise companies on positioning strategy. Entire books have been written on the

204

subject (one of the best currently available is Frederick Marckini's book, which is referenced in the For Further Study and Research section at the end of this chapter), and several major conferences are devoted to the subject each year.

The large drop in Internet advertising that took place in 2001-2002 is over and online ad spending is up. Spending exceeded $8.5 billion in 2004, and the rate of increase is expected to continue at more than 20 percent per year. Most of this increase is due to a boom in paid placement advertising. In 2002, paid placement ads were 15 percent of all online advertising. Today, they are more than 40 percent of the total. Banner ad spending has dropped steadily in recent years and currently is less than 20 percent of total online advertising.

The business of selling search engine inclusions and placements is complex because many search engines do not sell inclusion and placement rights on their pages directly to advertisers. They use **search engine placement brokers**, which are companies that aggregate inclusion and placement rights on multiple search engines and then sell those combination packages to advertisers. Two large search engine placement brokers are **LookSmart** and **Overture**. Another reason for the complexity in this business is that recent years have brought a flurry of mergers and acquisitions. For example, in 2003, Yahoo! purchased Overture. This put Yahoo! in the business of selling advertising for several of its major competitors (who were using Overture as their search engine placement broker), including AltaVista, InfoSpace, Lycos, and MSN. An excellent resource for keeping up with the rapid changes in this business is Danny Sullivan's **Search Engine Watch** Web site. Although some of the content on the site is limited to paid subscribers, the site does include many free resources and explanations that are useful for learning about search engines, placement brokers, and search engine optimization in general.

The most popular search engine, Google, does not use a placement broker to sell search term inclusion and placement for its site. Google sells these services directly through its **Google AdWords** program. The home page for Google's AdWords program appears in Figure 4-18.

Web sites that offer content can also participate in paid placement. Google offers its **AdSense** program to sites that want to carry ads that match the content offered on the site. Other companies, such as **Kanoodle** and Yahoo!'s **Overture** division, offer similar ad brokerage services, but Google is the leader in this market. The content site receives a placement fee from the broker in exchange for the ad placement and the broker sells the placement slots to interested advertisers. These techniques in which ads are placed in proximity to related content is sometimes called **contextual advertising**.

Of course, this approach is not without its flaws. In 2003, the *New York Post* ran a sensational story that described a gruesome murder. The murder victim's body had been cut into pieces, which the murderer hid in a suitcase. When the newspaper's Web site ran the story, it appeared with a paid placement ad for luggage. The ad broker's software had noted the word "suitcase" in the story and decided that it would be the perfect place for a luggage ad. Today, ad brokers use more sophisticated software and human reviewers to prevent this type of error; however, some industry analysts believe that contextual advertising on content sites will never be as successful as paid placement on search engine pages. They argue that search engine pages are provided to site visitors looking for something specific, often as part of a purchasing process. Content sites are used to explore and learn

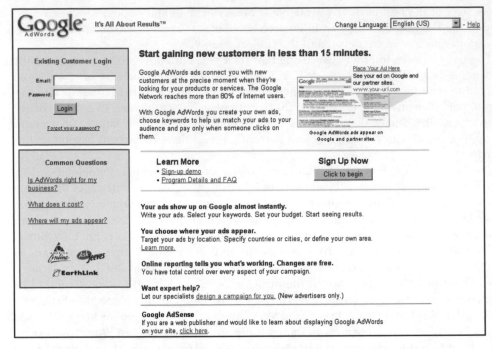

FIGURE 4-18   Google's AdWords program home page

about more general things. Thus, an ad on a search engine results page will always be more effective than an ad on a content site page.

Another variation of paid placement ads uses search engine results pages that are generated in response to a search for products or services in a specific local area. This technique, called **localized advertising**, places ads related to the location on the search results page. In 2004, Google launched the beta of its **Google Local** search service that lets users search by ZIP code or local address. The local advertising market (in outlets such as the Yellow Pages), is estimated to be more than $14 billion, a very attractive market for online advertisers. This promises to be an interesting area for the expansion of online advertising in the future.

Web Site Naming Issues

Companies that have a well-established brand name or reputation in a particular line of business usually want the URLs for their Web sites to reflect that name or reputation. Obtaining identifiable names to use on the Web can be an important part of establishing a Web presence that is consistent with the company's existing image in the physical world.

Two airlines that started their online businesses with troublesome domain names have both purchased more suitable domain names. Southwest Airlines' domain name was www.iflyswa.com until it purchased www.southwest.com. Delta Air Lines' original domain name was www.delta-air.com. After several years of complaints from confused customers who could never remember to include the hyphen, the company purchased the domain name www.delta.com.

Companies often buy more than one domain name. Some companies buy additional domain names to ensure that potential site visitors who misspell the URL will still be redirected (through the misspelled URL) to the intended site. For example, Yahoo! owns the name Yahow.com. Other companies own many URLs because they have many different names or forms of names associated with them. For example, General Motors' main URL is GM.com, but the company also owns GeneralMotors.com, Chevrolet.com, Chevy.com, GMC.com, and many others. In 1995, Procter & Gamble purchased hundreds of domain names that included the names of its products, such as Crisco.com, Folgers.com, Jif.com, and Pampers.com. It also bought names related to its products such as Flu.com, BadBreath.com, Disinfect.com, and Stains.com. Procter & Gamble hoped that people searching the Web for information about stains, for example, would find the Stains.com site, which featured links to the company's cleaning products. Procter & Gamble even purchased Pimples.com and Underarms.com (the company has since sold many of these domain names).

Buying, Selling, and Leasing Domain Names

In 1998, a poster art and framing company named Artuframe opened for business on the Web. With quality products and an appealing site design, the company was doing well, but it was concerned about its domain name, which was www.artuframe.com. After searching for a more appropriate domain name, the company's president found the Web site of Advanced Rotocraft Technology, an aerospace firm, at the URL www.art.com. After finding out that Advanced Rotocraft Technology's site was drawing 150,000 visitors each month who were looking for something art related, Artuframe offered to buy the URL. The aerospace firm agreed to sell the URL to Artuframe for $450,000. Artuframe immediately relaunched as **Art.com** and experienced a 30 percent increase in site traffic the day after implementing the name change.

The newly named site did not rely on the name change alone, however. It entered a joint marketing agreement with Yahoo! that placed ads for Art.com on art-related search results pages. Art.com also created an affiliate program with businesses that sell art-related products and not-for-profit art organizations. Although Art.com was ultimately unsuccessful in building a profitable business on the Web and liquidated in mid-2001, the domain name was snapped up immediately by already profitable Allwall.com for an undisclosed amount. The new Allwall.com site, relaunched with the Art.com domain name, experienced a 100 percent increase in site visitors within the first month.

Another company that invested in an appropriate domain name was **Cars.com**. The firm paid $100,000 to the speculator who had originally purchased the rights to the name. Cars.com is a themed portal site that displays ads for new cars, used cars, financing, leasing, and other car-related products and services. The major investors in this firm are newspaper publishers that wanted to retain an interest in automobile-related advertising as it moved online. As you learned in Chapter 3, classified ads are an important revenue source for many newspapers.

More recently, higher prices have prevailed in the market for domain names. Names such as Fruits.com, Question.com, Speaker.com, Tower.com, and Wisdom.com have each sold for prices over $100,000. Other names, including Cinema.com, Drugs.com, and ForSaleByOwner.com, have sold for more than $500,000 each. Not long ago, eCompanies paid $7.5 million for the domain name Business.com. Although most domains that have high value are in the .com top-level domain, the name engineering.org sold at auction to the American Society of Mechanical Engineers, a not-for-profit organization, for just under $200,000. Figure 4-19 lists domain names that sold for more than $1 million each.

Domain name	Price
Business.com	$ 7.5 million
Altavista.com	$ 3.3 million
Loans.com	$ 3.0 million
Wine.com	$ 3.0 million
Autos.com	$ 2.2 million
Express.com	$ 2.0 million
WallStreet.com	$ 1.0 million

FIGURE 4-19   Domain names that sold for more than $1 million

Some companies and individuals invested their money in the purchase of highly desirable domain names. Instead of selling these names to the highest bidder, some of these domain name owners decided to retain ownership of the domain names and lease the rights to the names to companies for a fixed time period. Usually, these domain name lessors rent their domain names through URL brokers.

URL Brokers and Registrars

Several legitimate online businesses, known as **URL brokers**, are in the business of selling, leasing, or auctioning domain names that they believe others will find valuable. Companies selling "good" (short and easily remembered) domain names include **BuyDomains.com**, and **GreatDomains**. The **Domain Notes** Web site provides links to URL brokers along with current information about the domain name market.

Companies can also obtain domain names that have never been issued, or that are currently unused, from a domain name registrar. **ICANN** (the Internet Corporation for Assigned Names and Numbers, about which you learned in Chapter 2) maintains a list of accredited registrars. Many of these registrars offer domain name search tools on their Web sites. A company can use these tools to search for available domain names that might meet their needs. Another service offered by domain name registrars such as **DirectNIC.com** is domain name parking. **Domain name parking**, also called **domain name hosting**, is a service that permits the purchaser of a domain name to maintain a simple Web site (usually one page) so that the domain name remains in use. The fees charged for this service are usually much lower than those for hosting an active Web site.

Summary

In this chapter, you learned how companies can use the principles of marketing strategy and the four Ps of marketing to achieve their goals for selling products and offering services on the Web. Some companies use a product-based marketing strategy and some use a customer-based strategy. The Web enables companies to mix these strategies and give customers a choice about which approach they prefer.

Market segmentation using geographic, demographic, and psychographic information can work as well on the Web as it does in the physical world. The Web gives companies the powerful added ability to segment markets by customer behavior and life-cycle stage, even when the same customer exhibits different behavior during different visits to the company's site.

Online advertising has become more intrusive since it was introduced in the mid-1990s. You learned how companies are using various types of online ads, including banners, pop-ups, pop-behinds, and interstitials to promote their sites to potential customers. Permission marketing and opt-in e-mail show promise as viable alternatives to decreasingly effective Web page ads.

Many companies are using the Web to manage their relationships with customers in new and interesting ways. By understanding the nature of communication on the Web, companies can use it to identify and reach the largest possible number of qualified customers. Technology-enabled customer relationship management can provide better returns for businesses on the Web than the traditional unaided approaches of market segmentation and micromarketing.

Firms on the Web can use rational branding instead of the emotional branding techniques that work well in mass media advertising. Some businesses on the Web are sharing and transferring brand benefits through affiliate marketing and cooperative efforts among brand owners. Others are using brand leveraging and viral marketing to increase their appeal and their customer bases.

Successful search engine positioning and domain name selection can be critical for many businesses in their quests for new online customers. A growing number of advertisers are paying for inclusion and placement services to guarantee that their sites' URLs appear among the top results provided to potential customers by search engines. They are also paying for placement of advertising messages in those pages and on other sites, such as content sites and local information sites. These paid advertising placements are the most rapidly growing forms of online advertising and are driving the overall increase in online ad spending that has occurred in the past two years. The most important theme in this chapter is that companies must integrate the Web marketing tools they use into a cohesive and customer-sensitive overall marketing strategy.

Key Terms

Acquisition cost	Banner ad
Active ad	Banner advertising network
Ad view	Banner exchange network
Ad-blocking software	Behavioral segmentation
Affiliate marketing	Brand
Affiliate program broker	Brand leveraging
Animated GIF	Cause marketing

Click

Clickstream

Click-through

Contextual advertising

Conversion

Conversion cost

Conversion rate

Cost per thousand (CPM)

Crawler

Customer life cycle

Customer relationship management (CRM)

Customer value

Database

Demographic segmentation

Domain name hosting

Domain name parking

Electronic customer relationship management (eCRM)

Four Ps of marketing

Geographic segmentation

Impression

Index

Interactive marketing unit (IMU) ad format

Interstitial ad

Life-cycle segmentation

Localized advertising

Market segmentation

Marketing mix

Marketing strategy

Marketspace

Micromarketing

Occasion segmentation

One-to-one marketing

Opt-in e-mail

Page view

Paid placement (sponsorship)

Pay-per-click model

Pay-per-conversion model

Permission marketing

Place (distribution)

Pop-behind ad

Pop-up ad

Price

Product

Promotion

Psychographic segmentation

Rational branding

Repeat visit

Retained customer

Retention cost

Rich media ad

Rich media object

Robot (bot)

Search engine

Search engine optimization

Search engine placement

Search engine placement broker

Search engine positioning

Search engine ranking

Search term sponsorship

Search utility

Segments

Shopping cart

Site sponsorship

Skyscraper ad

Spider

Technology-enabled customer relationship management

Technology-enabled relationship management

Touchpoint

Touchpoint consistency

Trial visit

Trigger words

URL brokers

Usage-based market segmentation

Viral marketing

Visit

Review Questions

RQ1. Assume you are a consultant to Fred's Sticks, a golf club manufacturer that sells its clubs directly to customers on the Web. Review Figure 4-5 in this chapter, which describes how advertisers select television programs that would be good hosts for their ads. Present a list of four magazines (other than golf magazines) in which Fred's Sticks should consider placing print advertising to support its Web sales effort. For each magazine, write one paragraph in which you explain why that magazine would be a good advertising outlet to reach potential customers of an online golf club store.

RQ2. In about 600 words, explain the differences between customer acquisition and retention and outline two marketing strategies that would help a company accomplish each of these two objectives. Be sure to present facts and logical arguments that support the use of each strategy for each objective.

RQ3. Select a retail store with which you are familiar that has a Web site on which it sells products or services similar to those it sells in its physical retail stores. Explore the Web site and examine it carefully for features that indicate the level of service it provides. Using your experience in the physical store and your review of the Web site, write a 200-word evaluation of the company's touchpoint consistency.

RQ4. Many people have strong negative reactions to pop-up, pop-behind, interstitial, and rich media ads. Write a 200-word letter to the editor of an Internet industry magazine in which you explain, from the advertiser's viewpoint, why these ads can be effective advertising media.

RQ5. In about 300 words, describe the key elements of technology-enabled customer relationship management and outline the advantages that technology-enabled customer relationship management has over traditional seller-customer interactions.

RQ6. In about 400 words, describe what a search engine inclusion and placement broker does and explain why an advertiser might use such a broker rather than working directly with a search engine site.

Exercises

E1. Visit the **RedEnvelope** Web site to examine how that company implements occasion segmentation. Write a report of approximately 200 words in which you describe two clear examples of occasion segmentation on the site.

E2. You are the new online advertising manager for the *Midland Daily Courier*, a local newspaper. The newspaper wants to sell banner advertising on its site in a variety of sizes to meet the needs of its advertisers. Examine the **IAB** Web site and other online resources of your choosing, then prepare a memo of about 500 words to the newspaper's advertising manager that outlines the current state of standards for banner ads. Include a specific recommendation regarding how many different sizes the newspaper should offer and support your recommendation with factual and logical arguments.

E3. You have been employed by **Switchboard.com** to sell space on its site to advertisers. Create a promotional press release of approximately 300 words in which you describe the advantages of advertising on Switchboard.com. You may decide to promote space on the main page, other specific pages, or all pages. Be prepared to explain why your promotional strategy should work. You may find the **Art of Web Site Promotion**, **Promotion World**, **Sitelaunch**, or Seltzer's **How to Publicize Your Web Site over the Internet** Online Companion links helpful in your task.

E4.   Marti Baron operates a small Web business, The Cannonball, that sells parts, repair kits, books, and accessories to hobbyists who restore antique model trains. Many model train hobbyists and collectors have created Web sites on which they share photos and other information about model trains. Marti is interested in creating an affiliate marketing program that would allow those hobbyists to place links on their sites to The Cannonball and be rewarded with commissions on sales that result from visitors following those links. Examine the services offered by **Be Free**, **Commission Junction**, **LinkShare**, and any other affiliate program brokers you can find on the Web. Recommend at least one affiliate program broker that would be a good fit for Marti's business. In about 500 words, explain your recommendation. Be sure to consider the characteristics of Marti's business in your analysis.

Cases

C1. Oxfam

For more than 60 years, Oxfam has worked through and with its donors, staff, project partners, and project participants to overcome poverty and injustice around the world. Early in World War II, Greece was occupied by German Nazis. Allied forces created a naval blockade around Greece to prevent further German expansion; however, the blockade created severe shortages of food and medicine among Greek civilian communities. In 1942, a number of Famine Relief Committees were established in Great Britain to ship emergency supplies through the Allied blockade. Although most of these committees ceased operations after the war ended, the Oxford Committee for Famine Relief saw a continuing need and enlarged its operations to provide aid throughout postwar Europe, and in later years, the rest of the world. The Committee eventually became known by its abbreviated telegraph address, Oxfam, and the name was formally adopted in 1965.

Oxfam's success and growth was due to many dedicated volunteers and donors who continued and expanded their financial support of the organization. In the 1960s, Oxfam began to generate significant revenues from its retail stores. These shops, located throughout Great Britain, accept donations of goods and handcrafted items from overseas for resale. Today, those stores number more than 800 and are staffed by more than 20,000 volunteers.

Oxfam often deals with humanitarian disasters that are beyond the scope of its resources. In these cases, the organization provides aid by mobilizing an international lobbying staff that has contacts with key aid agencies based in other countries, governments in the affected area, and the United Nations.

In 1996, Oxfam opened a Web site to provide information about its efforts to supporters and potential donors. The Web site included detailed reports on Oxfam's work, past and present, and allows site visitors to make donations to the organization. Although Oxfam gladly accepts any donations, it encourages supporters to commit to a continuing relationship by making regular donations. In exchange, it provides regular updates about its activities on the Web site and through an e-mailed monthly newsletter. The Web site includes a sign-up page for the e-mail newsletter, which goes out to more than 200,000 supporters.

Oxfam has been involved in relief work in Sudan since the 1970s, when it provided help to Ugandan refugees in the southern part of the country. More recently, Oxfam was an early responder to the 2004 crisis in that country. Oxfam set up sanitary facilities and provided clean drinking water in camps set up for thousands of displaced people fleeing pro-government Arab militias. The need in Sudan rapidly exceeded Oxfam's capacity and it decided to use e-mail to mobilize support for the project.

Oxfam planned an e-mail campaign that would send three e-mails in HTML format to supporters on its existing e-mail list over a six-week period. The first e-mail included a photo of children in one of the camps. The text of the e-mail message described Oxfam's efforts to provide clean water to the displaced people living in the Sudanese camps. The e-mail included links in two places that took recipients to a Web page that had been created specifically to receive visitors responding to that e-mail message. The Web page allowed visitors to make a donation and asked them to provide their e-mail addresses, which would be used to send updates on the Sudan project. A second e-mail was sent two weeks later to addresses on the list that had not yet responded. This second e-mail included a video file that played automatically when the e-mail was opened. The video conveyed the message that Oxfam had delivered $300,000 in aid to the camps but that more help was urgently needed in the region. This second e-mail included three links that led to the Web page created for the first e-mail. Two weeks later, a final e-mail was sent to addresses on the list that had not responded to either of the first two e-mails. This third e-mail included an audio recording in which Oxfam's executive director made a plea for the cause. The e-mail also included text that provided examples of which aid items could be provided for specific donation amounts.

Oxfam's three-part e-mail campaign was considered a success by direct marketing standards. The first e-mail was opened by 32 percent of recipients and had a click-through rate of 8 percent. The second e-mail had similar, but somewhat higher, results (33 percent opened, 10 percent clicked-through). 90 percent of those who opened the e-mail watched the video. The third e-mail continued the slightly increasing trends for opening and attention (34 percent opened, and 94 percent listened to the audio), but the click-through rate was much higher than the previous two e-mails (14 percent). Also, the dollar amount of donations increased with each subsequent e-mailing. The e-mail campaign raised more than $450,000 in its six-week period.

Oxfam coordinated this e-mail effort with other awareness activities it was conducting in the same time period. The organization sent letters to supporters who had not provided e-mail addresses and ran ads in two newspapers (*The Independent* and *The Guardian*) that carried messages similar to those in the e-mails.

Required:

1. Oxfam chose not to use online banner ads this campaign. In about 100 words, explain the advantages and disadvantages that Oxfam would have experienced by using banner ads to achieve the objectives of this campaign.

2. Oxfam used only its existing e-mail list for this campaign, it did not purchase (or borrow from other charitable organizations) any additional e-mail addresses. Evaluate this decision. In about 300 words explain the advantages and disadvantages of acquiring other e-mail addresses for a campaign of this nature.

3. For this campaign, Oxfam chose to use e-mails that contained HTML, audio, and video elements rather than using plain-text e-mails. In about 100 words, describe the advantages and disadvantages of using formats other than plain-text in this type of e-mail campaign. Be sure to identify any specific trade-offs that Oxfam faced in deciding not to use plain-text e-mail.

4. Oxfam used HTML in the its first e-mail, video in the second, and audio in the third. Evaluate the use of different e-mail formats for this type of message and consider the sequencing of the formats that Oxfam used in this campaign. In about 300 words, summarize the considerations that would affect a decision to use a particular sequence of e-mail formats in a campaign such as this and evaluate the sequence that Oxfam used.

5. A manager at Oxfam might be tempted to conclude that the sequence of formats used in the e-mail messages was related to the increase in donations over the six weeks of the campaign. In about 100 words, present at least two reasons why this would be an incorrect conclusion.

Note: Your instructor might assign you to a group to complete this case, and might ask you to prepare a formal presentation of your results to your class.

C2. Montana Mountain Biking

Jerry Singleton founded Montana Mountain Biking (MMB) 16 years ago. MMB offers one-week guided mountain biking expeditions based in four Montana locations. Most of MMB's new customers hear about the company and its tours from existing customers. Many of MMB's customers come back every year for a mountain biking expedition; about 80 percent of the riders on any given expedition are repeat customers.

Jerry is happy with this high repeat percentage, but he is worried that MMB is missing a large potential market. He has been reluctant to spend a lot of money on advertising. About 10 years ago, he spent $80,000 on a print advertising campaign that included ads in several outdoor interest and sports magazines, but the ads did not generate enough additional customers to cover the cost of the advertising. Five years ago, a marketing consultant advised Jerry that the ads had not been placed well. The magazines did not reach the serious mountain bike enthusiast, which is MMB's true target market. After all, a casual mountain bike rider would probably not be drawn to a week-long expedition.

Another concern of Jerry's is that more than 90 percent of MMB's customers come from neighboring states. Jerry has always thought that MMB was not reaching the sizable market of serious mountain bike enthusiasts in California. He talked to the marketing consultant about buying an address list and sending out a promotional mailing, but producing and mailing the letters seemed too expensive. The cost of renting the list was $0.10 per name, but the printing and mailing were $4 per letter. There were 60,000 addresses on the list, and the consultant told him to expect a conversion rate of between 1 percent and 3 percent. At best, the mailing would yield 1800 new customers and MMB's profit on the one-week expedition was only about $100 per customer. It looked like the conversion cost would be about $246,000 (60,000 x $4.10) to obtain a profit of $180,000 (1800 x $100). The consultant explained that it was an investment; because MMB had such a high customer retention rate, the profit from the new customers in the second or third years would exceed the one-time cost of the mailing in the first year. Jerry was not convinced.

Six years ago, MMB launched its first Web site. It included information about the company and its tours, but Jerry did not see any need to include an expedition-booking function on the site. He did think about selling caps and jackets with the MMB logo, but that idea never was implemented. The MMB logo is well known in the mountain biking community in the upper Midwest.

The MMB Web site includes an e-mail address so that visitors to the site can send an e-mail requesting more information about the expeditions. Robin Davis, one of MMB's expedition leaders, is an amateur photographer who has taken many photos while on the trails over the years. Last year, she had those photos digitized and put them on the MMB Web site. The number of e-mail inquiries increased dramatically within a month. Many of the inquiries were about MMB's expeditions, but a surprising number asked for permission to use the photos, or asked if MMB had more photos like those for sale. Jerry is not quite sure what to make of the popularity of those photos. He is, after all, in the mountain bike expedition business.

Required:

1. Review the five stages of customer loyalty shown in Figure 4-10 and prepare a report in which you classify MMB's customers. Estimate the percentage of MMB customers who fall into each of the five categories. Support your classification with logic and evidence from the case narrative.

2. Recommend an e-mail marketing strategy for MMB. In your recommendation, consider the results of MMB's earlier print mail advertising campaign, your answer to the first requirement, and the potential offered by permission marketing.

3. Explain how MMB could use viral marketing to gain new customers and cement its relationships with existing customers. In your answer, be sure to discuss features that MMB should include on its Web site to support the viral marketing initiative.

4. Outline an affiliate marketing strategy for MMB. Include a description of the types of Web sites that MMB should attempt to recruit as affiliates, and present at least five examples of specific sites that would be good referral sources.

Note: Your instructor might assign you to a group to complete this case, and might ask you to prepare a formal presentation of your results to your class.

For Further Study and Research

Agarwal, A., D. Harding, and J. Schumacher. 2004. "Organizing for CRM," *The McKinsey Quarterly*, June, 80–91.

Andrews, R. and I. Currim. 2004. "Behavioral Differences Between Consumers Attracted to Shopping Online Vs. Traditional Supermarkets: Implications for Enterprise Design and Marketing," *International Journal of Internet Marketing and Advertising*, 1(1), January–March, 38–61.

Armitt, C. 2004. "Case Study: Crisis in Sudan Email Campaign," *New Media Age*, September 2, 22.

Bergert, S. and K. Kazimer-Shockley. 2001. "The Customer Rules," *Intelligent Enterprise*, 4(11), July 23, 31–34.

Blair, J. 2001. "Behind Kozmo's Demise: Thin Profit Margins," *The New York Times*, April 13. (http://www.nytimes.com/2001/04/13/technology/13KOZM.html)

Bond, P. 2004. "Net Ads Pop Up To $7.2 Billion per Year," *Shoot*, 45(22), June 4, 20–21.

Chan, A., Dodd, J. and R. Stevens. 2004. *The Efficacy of Pop-ups and the Resulting Effect on Brands.* Oxfordshire, UK: Bunnyfoot Universality.

Dahlen, M. and J. Bergendahl. 2001. "Informing and Transforming on the Web: An Empirical Study of Response to Banner Ads for Functional and Expressive Products," *International Journal of Advertising*, 20(2), 189–206.

Del Franco, M. and P. Miller. 2003. "Reevaluating Affiliate Marketing," *Catalog Age*, May 1, 6.

Delio, M. 2001. "Kozmo Kills the Messenger," *Wired News*, April 13. (http://www.wired.com/news/business/0,1367,43025,00.html)

Dysart, J. 2004. "Search Engine Strategies 2004," *Information Today*, 21(4), September, 27–28.

Fallows, J. 2004. "How Google Took the Work Out of Selling Advertising," *The New York Times*, June 13, C5.

Forsyth, J., J. Lavoie, and T. McGuire. 2000. "Segmenting the E-market," *The McKinsey Quarterly*, October. (http://mckinseyquarterly.com/article_page.asp?tk=302016:939:24&ar=939&L2=24&L3=44)

Gardner, E. 1999. "Art.com," *Internet World*, March 15, 13. (http://www.iw.com/print/1999/03/15/)

Geyskens, I., K. Gielens, and M. Dekimpe. 2002. "The Market Valuation of Internet Channel Additions," *Journal of Marketing*, 66(2), April, 102–116.

Godin, S. and D. Peppers. 1999. *Permission Marketing: Turning Strangers into Friends, and Friends into Customers.* New York: Simon & Schuster.

Green, H. and P. Gogoi. 2003. "Online Ads Take Off—Again: Mainstream Marketers Are Hiking Their Internet Ad Spending," *Business Week*, May 5, 75.

Halliday, J. 2001. "Edmunds.com Overhauls Auto Site: Ad Profits Needed to Cover Massive Expansion," *Advertising Age*, April 9, 28.

Halliday, J. 2003. "Study: TV Ads Don't Sell Cars: Consumers Cite Internet and Direct Mail as More Influential," *Ad Age*, October 13. (http://www.adage.com/news.cms?newsId=38933)

Harvard Business Review. 2003. "How to Measure the Profitability of Your Customers," 81(6), June, 74.

Harwood, S. 2004. "Online Ad Industry Divided Over Single Standard Plans," *New Media Age*, June 17, 14.

Hilzenrath, D. 2001. "Saylor Firm Spent Millions Investing in Web Addresses," *Washington Post*, April 10, E1.

Hoffman, D. and T. Novak. 2000. "How to Acquire Customers on the Web," *Harvard Business Review*, 78(3), May–June, 179–188.

Jarvis, S. 2002. "Bright Spots in Marketing," *Marketing News*, 36(8), April 15, 1–4.

Kaihla, P. 2001. "Five Battle-Tested Rules of Online Retail," *eCompany*, April. (http://www.ecompany.com/articles/mag/0,1640,9599,00.html)

Kessler, S. 2002. "Online Advertising is Ready to Click," *Business Week*, January 27. (http://www.businessweek.com/investor/content/jan2002/pi20020127_4576.htm)

Koprowski, G. 1998. "The (New) Hidden Persuaders: What Marketers Have Learned About How Consumers Buy on the Web," *The Wall Street Journal*, December 7, R10.

Lefton, T. 2001. "The Great Flameout," *The Industry Standard*, March 19. (http://www.thestandard.com/article/0,1902,22685,00.html)

Livingston, B. 2002. "How an Ad-Based Online Company Grew in 2001," *InfoWorld*, April 2. (http://www.infoworld.com/articles/op/xml/02/04/04/020404opsecrets.xml)

Lytel, J. 2000. "Domain-Name Disputes Get Personal," *BizReport*, September 22. (http://www.bizreport.com/marketing/2000/09/2000922-1.htm)

216

MacPherson, K. 2001. *Permission-Based E-Mail Marketing That Works!* Chicago: Dearborn Trade Press.

Maddox, K. 2004. "The Return of the Boom," *B to B*, 89(7), 23.

Marckini, F. 2001. *Search Engine Positioning*. San Antonio, TX: Republic of Texas Press.

McWilliams, B. 2002. "Dot-Com Noir: When Internet Marketing Goes Sour," *Salon.com*, July 1. (http://www.salon.com/tech/feature/2002/07/01/spyware_inc/index.html)

New Media Age. 2004. "Has Branding Got Lost Amid Search?" September 2, 21–22.

Orenstein, S. 2000. "Boo.com: A Cautionary Tale," *The Industry Standard*, June 5, 106–113.

Overholt, A. 2004. "Search for Tomorrow," *Fast Company*, August, 69–71.

Pastore, M. and C. Saunders. 2001. "Banners Can Brand, Honestly They Can," *Internet News*, July 19. (http://www.internetnews.com/IAR/article/0,,12_804771,00.html)

Plosker, G. 2004. "What Does Paid Search Mean to You?" *Online*, 28(5), September-October, 2004, 49–51.

Rapoza, J. 2004. "Annoying Web Ads Redux," *eWeek*, 21(15), April 12, 70.

Rayport, J. and J. Sviokla. 1994. "Managing in the Marketspace," *Harvard Business Review*, 72(6), November–December, 141–150.

Rayport, J. and J. Sviokla. 1995. "Exploiting the Virtual Value Chain," *Harvard Business Review*, 73(6), November–December, 75–85.

Rozanski, H., G. Bollman, and M. Lipman. 2001. "Seize the Occasion: Usage-Based Segmentation for Internet Marketers," *eInsights*, March 20, 1–13. (http://www.strategy-business.com/enews/032001/032001.html)

Sandoval, G. 2001. "Kozmo to Shut Down, Lay Off 1,100," *News.com*, April 11. (http://www.zdnet.com/ecommerce/stories/main/0,10475,5081050,00.html)

Saunders, C. 2001. "Nader Group Criticizes Pay for Placement Search Engines," *InternetNews*, July 17. (http://www.internetnews.com/IAR/article/0,,12_803451,00.html)

Shelat, B. 2002. "From Usability to Credibility: On-line Trust and How to Build It," *System Concepts*, July 3. (http://www.system-concepts.com/articles/trust.html)

Shelton, J. 2002. "Shopping Cart Abandonment," White Paper. Global Millenia Marketing, Kirkland, Quebec. (http://www.globalmillenniamarketing.com/reports.htm)

Sullivan, D. 2003. "Contextual Advertising in Context: Part 1," *ClickZ*, March 19. (www.clickz.com/search/opt/article.php/2114501)

Tedeschi, B. 2003. "If You Liked the Web Page, You'll Love the Ad," *The New York Times*, August 4, C1–C2.

Time, 2001. "Converting Web Surfers to Buyers," July 23, 46–47.

Vickrey, L. 2001. "Keeping the Cachet," *The Wall Street Journal*, April 23, R28.

Weber, T. 2001. "Can You Say 'Cheese'? Intrusive Web Ads Could Drive Us Nuts," *The Wall Street Journal*, May 21, B1.

Wind, Y. and V. Mahajan. 2002. "Convergence Marketing," *Journal of Interactive Marketing*, 16(2), Spring, 64–80.

CHAPTER **5**

BUSINESS-TO-BUSINESS STRATEGIES: FROM ELECTRONIC DATA INTERCHANGE TO ELECTRONIC COMMERCE

LEARNING OBJECTIVES

In this chapter, you will learn about:

- Strategies that businesses use to improve purchasing, logistics, and other support activities
- Electronic data interchange and how it works
- How businesses are moving electronic data interchange operations to the Internet
- Supply chain management and how businesses are using Internet technologies to improve it
- Electronic marketplaces and portals that make purchase-sale negotiations easier and more efficient

INTRODUCTION

General Electric (GE) is one of the largest and most successful companies in the world. It engages in a wide range of businesses around the world, including the production of appliances and electrical and electronic products, broadcasting, and a variety of financial and insurance activities. One of its oldest lines of business is **GE Lighting**, which produces over 30,000 different kinds of lightbulbs in its 28 North American plants and other locations around the world. The raw materials used in making lightbulbs are fairly standard items: glass, aluminum, various insulating plastics, and filament materials. However, a major portion of each lightbulb's cost is the money that GE Lighting spends on indirect materials and parts for the machines used to fabricate and assemble the bulbs. These indirect materials and parts must conform to detailed specifications that GE stores on more than 3 million blueprints and other design drawings.

Because the technologies for making lightbulbs are mature and well-known, GE Lighting can solicit bids from a variety of suppliers for indirect materials and machinery replacement parts without worrying about the possible disclosure of trade secrets. Unfortunately, the bidding process at GE Lighting had become very slow and inefficient. Each transaction required the Purchasing Department to request the relevant blueprints, photocopy them, attach them to other material specification documents, and mail the whole package to suppliers that might be interested in bidding on the item. It would often take Purchasing personnel more than four weeks to gather the information, send it to potential suppliers, obtain and evaluate suppliers' bids, negotiate with the chosen suppliers, and place an order. These long delays were limiting GE Lighting's flexibility and ability to respond to requests from its customers.

By applying the tools of electronic commerce to these purchase transactions, GE Lighting was able to make major improvements to the entire parts acquisition process. Today, Purchasing personnel have access to a procurement system through their desktop computers. When they need to buy replacement parts for a machine, they create a new purchase file that includes basic quantity, delivery date, and delivery location information. Then, from a list generated by a continually updated supplier database, they select suppliers from which they request quotes. Finally, they attach electronic copies of all necessary blueprints and engineering drawings, which are now digitized and stored in another database; with a mouse click, they send the entire bid package off in an encrypted format to all the selected suppliers. Assembling the bid package now takes hours instead of a week or more. Suppliers are asked to respond within a short time period—usually a week—through the Internet. The Purchasing staff member can evaluate the returned bids and award a contract online, completing the entire process in about 10 days.

The most significant savings for GE Lighting were in process-time reduction—from four weeks or more to 10 days—and in the elimination of paper and the costs of handling paper. However, the company also realized other benefits. Because the online system made it easier to send out bid packages, the Purchasing Department could send out more bids to a wider range of suppliers. In particular, many foreign suppliers that had been difficult to reach with mailed bid packages could be included in the solicitation for quotes. The increased competition drove down prices; GE Lighting has saved up to 20 percent on many of these items since moving the bid process online. Suppliers welcome the reduced time lag between submitting the bid and learning whether GE Lighting will award them the contract; this makes their production planning easier.

PURCHASING, LOGISTICS, AND SUPPORT ACTIVITIES

In the previous two chapters, you learned about strategy issues that arise when businesses and other organizations provide information to potential customers. In terms of the value chain model described in Chapter 1, you learned about the primary activities: identify customers, market and sell, and deliver. You also became familiar with a number of business models for selling on the Web. Although many of these business models are used in business-to-business electronic commerce, the emphasis in Chapters 3 and 4 was on business-to-consumer advertising, promotion, and sales activities.

In this chapter, you will learn how companies use electronic commerce to improve their purchasing and logistics primary activities, and all of their support activities (which include finance and administration, human resources, and technology development). You can refer to Figure 1-9 in Chapter 1 for a review of primary activities and support activities. Although the work might not be as glamorous as designing a Web site or creating an advertising campaign, the potential for cost reductions and business process improvements in purchasing, logistics, and support activities is tremendous. Although governments seldom sell products or services to customers, they perform many functions for the individual citizens, businesses, and other organizations that they serve. Governments increasingly are using electronic commerce to improve the efficiency with which they undertake their own support activities and serve their stakeholders better. These electronic commerce activities are collectively referred to as **e-government**.

As Internet technologies become commonplace in businesses, the potential for synergies increases. Many of these synergies are forming the basis for second-wave electronic commerce opportunities. You will learn about a number of these second-wave opportunities in this chapter.

An emerging characteristic of purchasing, logistics, and support activities is that they need to be flexible. A purchasing or logistics strategy that works this year may not work next year. Fortunately, economic organizations are evolving from the hierarchical structures used since the Industrial Revolution to new, more flexible network structures. These network structures are, in many cases, made possible by the transaction cost reductions that companies realize when they use Internet and Web technologies to carry out business processes.

Purchasing Activities

Purchasing activities include identifying vendors, evaluating vendors, selecting specific products, placing orders, and resolving any issues that arise after receiving the ordered goods or services. These issues might include late deliveries, incorrect quantities, incorrect items, and defective items. By monitoring all relevant elements of purchase transactions, purchasing managers can play an important role in maintaining and improving product quality and reducing cost. In Chapter 1, you learned how companies can organize their strategic business unit activities using an industry value chain. The part of an industry value chain that precedes a particular strategic business unit is often called a **supply chain**. A company's supply chain for a particular product or service includes all the activities undertaken by every predecessor in the value chain to design, produce, promote, market, deliver, and support each individual component of that product or service. For example, the supply chain of an automobile manufacturer includes every activity undertaken by each individual component supplier, including engine manufacturers, steel fabricators, glass manufacturers, wiring harness assemblers, and thousands of others.

The Purchasing Department within most companies traditionally has been charged with buying all of these components at the lowest price possible. Usually, Purchasing staff did this by identifying qualified vendors and asking them to prepare bids that described what they would supply and how much they would charge. The Purchasing staff would then select the lowest bid that still met the quality standards for the component. This bidding process led to a very competitive environment with a large number of suppliers; this process focused excessively on the cost of individual components and ignored the total supply chain costs, including the cost to the manufacturing organization of dealing with such a large number of suppliers. As you learned in Chapter 1, many managers call this function "procurement" instead of "purchasing" to distinguish the broader range of responsibilities. Procurement generally includes all purchasing activities, plus the monitoring of all elements of purchase transactions. It also includes managing and developing relationships with key suppliers. Another term that is used to describe procurement activities is supply management. In many companies, procurement staff must have high levels of product knowledge to identify and evaluate appropriate suppliers. The part of procurement activity devoted to identifying suppliers and determining the qualifications of those suppliers is called **sourcing**. In Chapter 1, you learned that the use of Internet technologies in procurement activities is called e-procurement. Similarly, the use of Internet technologies in sourcing activities is called **e-sourcing**. Specialized Web purchasing sites can be particularly useful to procurement professionals responsible for sourcing. The business purchasing process is usually much more complex than most consumer purchasing processes. Figure 5-1 (on the next page) shows the steps in a typical business purchasing process.

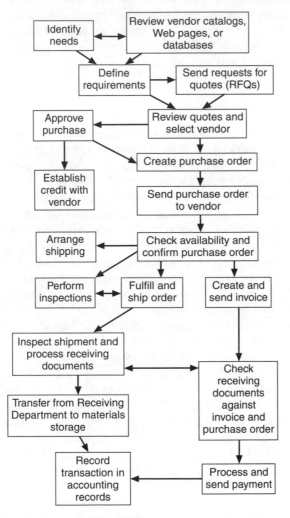

FIGURE 5-1 Steps in a typical business purchase process

As you can see, the business purchasing process includes many steps. The business purchasing process also requires a number of people to coordinate their individual activities as part of the process. In large companies, the Procurement Department that supervises the purchasing process may include hundreds of employees who supervise the purchasing of materials, inventory for resale, supplies, and all of the other items that the company needs to buy. The total dollar amount of the goods and services that a company buys during a year is called its **spend**. In large companies, the spend can be many billions of dollars. Managing the spend in those companies is an important function and can be a key element in a company's overall profitability.

In 2002, Motorola implemented a set of Internet technologies in its procurement operation. Motorola's spend is about $48 billion; it involves one million purchase orders and six million inventory receipts. Motorola estimates that it saved $2.5 billion by using the new Internet technologies to accomplish a variety of procurement tasks more efficiently and at lower cost.

For many years, the National Association of Purchasing Management has been the main organization for procurement professionals. In 2002, the association changed its name to the **Institute for Supply Management (ISM)**. ISM runs conferences, publishes a monthly journal (***Inside Supply Management***), and offers helpful information on its Web site. Many of the articles in recent issues of the journal have dealt with electronic commerce. Full-time students who want to learn more about supply management can join ISM at no cost.

Direct vs. Indirect Materials Purchasing

Businesses make a distinction between direct and indirect materials. **Direct materials** are those materials that become part of the finished product in a manufacturing process. Steel manufacturers, for example, consider the iron ore that they buy to be a direct material. The procurement process for direct materials is an important part of any manufacturing business because the cost of direct materials is usually a very large part of the cost of the finished product. Large manufacturing companies, such as auto manufacturers, engage in two types of direct materials purchasing. In the first type, called **replenishment purchasing** (or **contract purchasing**), the company negotiates long-term contracts for most of the materials that it will need. For example, an auto manufacturer estimates how many cars it will make during a year and contracts with two or three steel mills to supply most of the steel it will need to build those cars. By negotiating the contracts in advance and guaranteeing the purchase, the auto manufacturer obtains low prices and good delivery terms. Of course, actual demand never matches expected demand perfectly. If demand is higher than the auto company's estimate, it must buy additional steel during the year. These purchases are made in a loosely organized market that includes steel mills, warehouses, speculators (who buy and sell contracts for future delivery of steel), and companies that have excess steel that they purchased on contract (demand for their products was lower than they had anticipated). This market is called a **spot market**, and buying in this market, the second type of direct materials purchasing, is called **spot purchasing**. **Indirect materials** are all other materials that the company purchases, including factory supplies such as sandpaper, hand tools, and replacement parts for manufacturing machinery.

Large companies usually assign responsibility for purchasing direct and indirect materials to separate departments. Most companies include the purchase of nonmanufacturing goods and services—such as office supplies, computer hardware and software, and travel expenses—in the responsibilities of the indirect materials Procurement Department. Many vendors that manufacture general industrial merchandise and standard machine tools for a variety of industries have created Web sites through which their customers can purchase materials. A number of customers buy these indirect materials products on a recurring basis, and many of them are commodities, that is, standard items that buyers usually select using price as their main criterion. These indirect materials items are often called **maintenance, repair, and operating (MRO)** supplies. Increasingly, procurement professionals are using the terms "indirect materials" and "MRO supplies" interchangeably. Most companies have a difficult time controlling MRO spending from a centralized procurement office because many MRO purchases are numerous and small in dollar value. One way that Procurement Departments control MRO spending is by issuing **purchasing cards** (usually called **p-cards**). These cards, which resemble credit cards, give individual managers the ability to make multiple small purchases at their discretion while providing cost-tracking information to the procurement office.

By using a Web site to process orders, the vendors in this market can save the costs of printing and shipping catalogs and handling telephone orders. They can also keep price and quantity information continually updated, which would be impossible to do in a printed catalog. Some industry analysts estimate that the cost to process an MRO order through a Web site can be less than one-tenth of the cost of handling the same order by telephone.

Two of the largest MRO suppliers in the world are **McMaster-Carr** and **W.W. Grainger**. The W.W. Grainger Web site offers more than 220,000 different products for sale. Grainger's Web store, which appears in Figure 5-2, offers visitors a variety of ways to access information about and order Grainger products.

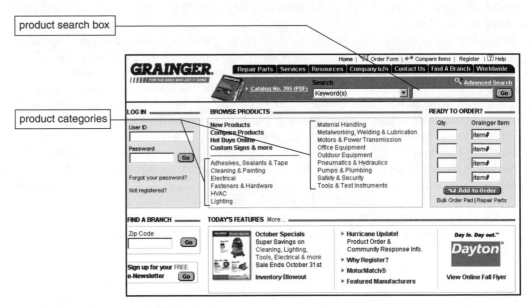

FIGURE 5-2   Grainger.com Web store

A visitor can enter the online catalog, use the product search box at the top of the page, or search by clicking a hyperlink to one of the categories listed in the middle of the page.

Office equipment and supplies are also items that are used by a wide variety of businesses. Market leaders **Office Depot** and **Staples** each have well-designed Web sites devoted to helping business Purchasing Departments buy these routine items as easily as possible. On their business-to-business Web sites, **Digi-Key** and **Newark Electronics** sell electronic parts, and **Global Computer Supplies** sells computers and related items.

Logistics Activities

The classic objective of logistics has always been to provide the right goods in the right quantities in the right place at the right time. Logistics management is an important support activity for both the sales and the purchasing activities in a company. Businesses need to ensure that the products they sell to customers are delivered on time and that the raw materials they buy from vendors and use to create their products arrive when needed. The management of materials as they go from the raw materials storage area through production processes to become finished goods is also an important part of logistics.

Logistics activities include managing the inbound movements of materials and supplies and the outbound movements of finished goods and services. Thus, receiving, warehousing, controlling inventory, scheduling and controlling vehicles, and distributing finished goods are all logistics activities. The Web and the Internet are providing an increasing number of opportunities to manage these activities better as they lower transaction costs and provide constant connectivity between firms engaged in logistics management. Web-enabled automated warehousing operations are saving companies millions of dollars each year, and major transportation companies such as **Schneider Logistics**, **Ryder System**, and **J.B. Hunt** now want to be seen by their customers as information management firms as well as freight carriers.

For example, the Schneider Track and Trace system delivers real-time shipment information to Web browsers on its customers' computers. This system shows the customer which freight carrier is transporting a shipment, where the shipment is, and when it should arrive at its destination. J.B. Hunt, which operates more than 70,000 trucks, trailers, and containers, implemented a Web site that lets its customers track their shipments themselves. With customers doing their own tracking, J.B. Hunt needs far fewer customer service representatives. Also, J.B. Hunt found that its customers could monitor their own shipments more effectively than the company. Thus, this Web site saves more than $12,000 per week in labor and lost shipment costs. When transportation and freight companies engage in the business of operating all or a large portion of a customer's materials movement activities, the company is called a **third-party logistics (3PL) provider**. For example, Ryder has a multiyear contract to design, manage, and operate all of Whirlpool's inbound freight activities and is considered a 3PL provider to Whirlpool.

FedEx has freight-tracking Web pages available to its customers, as does UPS. Firms that run their own trucking operations have implemented tracking systems that use global positioning satellite (GPS) technology to monitor vehicle movements. These freight-handling companies are also moving into the 3PL provider business as a way to generate additional revenue from the investment they made in information technology to support their core businesses. The marriage of GPS and portable computing technologies with the Internet is an excellent example of second-wave electronic commerce.

Support Activities

Support activities include the general categories of finance and administration, human resources, and technology development. Finance and administration includes activities such as making payments, processing payments received from customers, planning capital expenditures, and budgeting and planning to ensure that sufficient funds will be available to meet the organization's obligations as they come due. The operation of the computing infrastructure of the organization is also an administration activity. Human resources activities include hiring, training, and evaluating employees; administering benefits; and complying with government record-keeping regulations. Technology development can include a wide variety of activities, depending on the nature of the business or organization. It can include networking research scientists into virtual collaborative workgroups, posting research results, publishing research papers online, and providing connections to outside sources of research and development services.

A 1999 article in *Inc* magazine described the challenges facing Allegiance Telecom. The company was growing very rapidly and hiring over 100 people each month to staff its sales offices throughout the United States. Each new hire had to receive a full briefing on medical, dental, and retirement benefits plans, and then he or she had to select from among several options for each. Because Allegiance was growing so rapidly, its human resources staff was spread thin and could not be in every sales office for every hire. The company turned to **Online Benefits**, a firm that duplicates its clients' human resources functions on a password-protected Web site that is accessible to clients' employees. The employees can then access their employers' benefits information, find the answers to frequently asked questions, and even perform complex benefit option calculations. Other firms that offer support activities services include **CyLex Systems**, which offers document storage services, and **PayMaxx**, which offers payroll processing. Larger firms are building these types of functions into their intranets. These larger firms are also including Web-enabled sales support and sales force automation functions in their extranets.

One common support activity that underlies multiple primary activities is training. In many companies, the Human Resources Department handles training. Other companies may decentralize this function and have individual departments administer it. For example, insurance firms expend large amounts of resources on sales training. In most insurance companies, the Sales and Marketing Department administers this training. By putting training materials on the company intranet, insurance companies can distribute the training materials to many different sales offices, yet coordinate the use of those materials in the corporate headquarters sales office.

In 1999, the Swedish telecommunications giant Ericsson launched an extranet for current and former employees, families of those employees, and employees of approved business partners. Ericsson has more than 100,000 employees scattered across the globe. One part of this extranet included a Web site that enabled current employees, retirees, and other recipients of payments from the company's medical and retirement plans to efficiently track their benefits. Another part of the extranet included a Web site that was designed to facilitate knowledge management. **Knowledge management** is the intentional collection, classification, and dissemination of information about a company, its products, and its processes. This type of knowledge is developed over time by individuals working for or with a company and is often difficult to gather and distill.

Ericsson managers hope that their knowledge network will generate new ideas, help solve problems, and improve business processes throughout the international organization. Designers of the system have identified their biggest challenge: to direct the information they collect in the extranet to projects and product development activities that will benefit from that information.

BroadVision, a software development and consulting firm, installed an internal system called K-Net, or Knowledge Network, that organizes all information sources used regularly by its employees. It found that many of its employees were visiting between 10 and 20 Web sites each day in the course of doing their jobs. K-Net brings together all of the information that each employee needs and combines it into one dashboard-style interface presented on a Web browser. Much of the interface is customized for individual employees, although some parts of the interface—such as health insurance, vacation days, and other human resources information—are standardized for all employees. BroadVision has found the K-Net system to be so useful that it is partnering with Bank of America, Hewlett-Packard, and Amadeus (a European travel services company) to develop a version of K-Net to sell to other companies. You can learn more about knowledge management in general at the **KMWorld** Web site. In Chapter 9, you will learn about software that companies can use to build knowledge management systems.

E-Government

Although governments do not typically sell products or services to customers, they perform many functions for their stakeholders. Many of these functions can be enhanced by the use of electronic commerce. Governments also operate businesslike activities; for example, they employ people, buy supplies from vendors, and distribute benefit payments of many kinds. They also collect a variety of taxes and fees from their constituents (you will learn more about how governments use the Web in administering their tax laws in Chapter 7). The use of electronic commerce by governments and government agencies to perform these functions is often called e-government.

In 2000, the U.S. government's Financial Management Service (FMS) opened its **Pay.gov** Web site. The FMS is the agency responsible for receiving the government's tax, license, and other fee revenue (more than $2 trillion per year). It is also responsible for paying out more than $1.2 trillion per year in Social Security benefits, veterans benefits, tax refunds, and other disbursements. Federal agencies can link their Web sites to Pay.gov, which lets site visitors pay taxes and fees they owe to these agencies using their credit cards, debit cards, or various forms of electronic funds transfer. The U.S. government's Bureau of Public Debt operates the **TreasuryDirect** site, which allows individuals to buy savings bonds and financial institutions to buy treasury bills, bonds, and notes.

Following the terrorist attacks of September 11, 2001, the U.S. government became aware of a lack of activity coordination and information sharing among several of its agencies including the Federal Bureau of Investigation (FBI), the Central Intelligence Agency (CIA), and the Bureau of Customs and Border Protection. A number of initiatives that use Internet technologies are under way to increase the availability of information within and among these agencies under the auspices of the new **Department of Homeland Security (DHS)**.

Other countries' national governments are finding that e-government can reduce administrative costs and provide better service to stakeholders. In the United Kingdom, the

Department for Work and Pensions Web site provides information on unemployment, pension, and social security benefits. Smaller countries are also launching Web sites, such as Singapore's **SINGOV** site, that provide information to stakeholders and ways for citizens to interact with their governments online.

State governments are also creating Web sites for conducting business and interacting with their stakeholders. In 2001, the State of California opened its one-stop portal site, **my.ca.gov**, which appears in Figure 5-3.

FIGURE 5-3   State of California portal site my.ca.gov

This site gives visitors access to virtually every California government agency and state operation. Site visitors can transact a wide array of business with the state, from renewing a driver's license to reserving a camp site. The goal of the site is to give constituents one site through which they can conduct all of their business with the State of California. For businesses, the site offers the full text of all California business laws and regulations. It also provides information about how to sell to and buy from the state and its agencies.

Many other U.S. state governments (and, in other countries, provincial or regional governments) have, or are currently developing, similar Web sites. States can reduce the cost of providing services while providing those services more efficiently by using Web technologies to serve their stakeholders. The most common services offered by states and similar regional governments are: access to the text of state laws and regulations, renewal of licenses, promotion of the state to businesses considering new locations, job listings, promotion of tourism in the state, tax forms and filing information, and information for companies that want to do business with the state. The State of New York has separate Web sites for individuals living or working in the state (**New York State Citizen Guide**) and for companies that do business with the state (**e-bizNYS**).

Many local governments now have Web sites that offer residents a variety of information. The Web sites of larger cities (such as **Minneapolis** or **New Orleans**) include transcripts of city council meetings, local laws and regulations, business license and tax administration functions, and promotional information about the city for new residents or businesses seeking new locations. Smaller cities, towns, and villages are also using the Web to communicate with residents (see the **Cheviot, Ohio** Web site for one example).

New York City (**MyNYC.gov**) hired a management consulting firm to help review the city's ability to respond to terrorist attacks such as those of September 11, 2001. The review noted a number of weaknesses in the coordination of communications and data access among the New York Police Department, the Fire Department of New York, and other city departments. In response to the review, New York City revised its information systems to allow better coordination of activities during large-scale emergencies. Many New York City departments can now access each others' databases through Web interfaces that use electronic commerce technologies. You can learn more about applications of Internet technologies in state and local governments by reviewing articles on e.Republic's **Government Technology** Web site.

Network Model of Economic Organization

In Chapter 1, you learned about the three different forms of economic organization: markets, hierarchies, and networks. One trend that is becoming clear in purchasing, logistics, and support activities is the shift away from hierarchical structures toward network structures. The traditional purchasing model had one hierarchically structured firm negotiating purchase terms with several similarly structured supplier firms, playing each supplier against the others. As is typical in a network organization, more businesses are now giving their Procurement Departments new tools to negotiate with suppliers, including the possibility of forming strategic alliances. For example, a buying firm might enter into an alliance with a supplier to develop a new technology that will reduce overall product costs. The technology development might be done by a third firm using research conducted by a fourth firm. Such alliances and outsourcing contracts are examples of the move toward network economic structures that you learned about in Chapter 1.

While reading the previous sections in this chapter, you might have noticed that companies can have other firms perform various support activities for them. Again, these are examples of firms moving to a network model of economic organization. Imagine a business that uses one supplier to manage its payroll, another to administer its employee benefits plans, and a third to handle its document storage needs. The document storage service supplier might store the documents of the payroll service supplier and the benefits administration firm. The payroll service supplier might handle the payroll for the benefits administration firm. A fourth firm might provide online backup storage for the files of the other three companies. Of course, the payroll firm and the employee benefits firm might form a marketing partnership to sell both of their services to particular market segments. The document storage firm and the online backup storage firm might form a similar strategic alliance. Some researchers who study the interaction of firms within an industry value chain are beginning to use the term "**supply web**" instead of "supply chain" because many industry value chains no longer consist of a single sequence of companies linked in a single line, but include many parallel lines that are interconnected in a web or network configuration.

Highly specialized firms can now exist and trade services very efficiently on the Web. The Web is enabling this shift from hierarchical to network forms of economic organization. These emerging networks of firms are more flexible and can respond to changes in the economic environment much more quickly than hierarchically structured businesses. You can learn more about the economics of networked organizations at the **Network Economics** Web site maintained by the University of California, Berkeley. The roots of Web technology for business-to-business transactions, however, lie in a very hierarchically structured approach to interfirm information transfer: electronic data interchange.

ELECTRONIC DATA INTERCHANGE

You learned in Chapter 1 that electronic data interchange (EDI) is a computer-to-computer transfer of business information between two businesses that uses a standard format of some kind. The two businesses that are exchanging information are trading partners. Firms that exchange data in specific standard formats are said to be **EDI compatible**. The business information exchanged is often transaction data; however, it can also include other information related to transactions, such as price quotes and order status inquiries. Transaction data in business-to-business transactions includes the information traditionally included on paper invoices, purchase orders, requests for quotations, bills of lading, and receiving reports. The data on these five types of forms accounts for over 75 percent of all information exchanged by trading partners in the United States. Thus, EDI was the first form of electronic commerce to be widely used in business—some 20 years before anyone used the term "electronic commerce" to describe anything!

It is very important that you understand what EDI is designed to accomplish and how it came to be the preferred way for businesses to exchange information, because most B2B electronic commerce is an adaptation of EDI or is based on EDI principles. Another important reason for being familiar with EDI is that EDI is still the method used for most electronic B2B transactions. According to one study (see the article by Richard Villars cited in For Further Study and Research at the end of this chapter), the dollar amount of EDI transactions in 2002 was three times the total amount of all other B2B electronic transactions.

This section provides you with a brief history of EDI and explains how it works. It also explains why conducting EDI is better than processing mountains of paper transactions.

Early Business Information Interchange Efforts

The emergence of large business organizations in the late 1800s and early 1900s brought with it the need to create formal records of business transactions. In the 1950s, companies began to use computers to store and process internal transaction records, but the information flows between businesses continued to be printed on paper; purchase orders, invoices, bills of lading, checks, remittance advices, and other standard forms were used to document transactions.

The process of using a person or computer to generate a paper form, mailing that form, and then having another person enter the data into the trading partner's computer was slow, inefficient, expensive, redundant, and unreliable. By the 1960s, businesses that engaged in large volumes of transactions with each other had begun exchanging transaction information on punched cards or magnetic tape. Advances in data communications technology during the 1960s and 1970s allowed trading partners to transfer data over telephone lines instead of shipping punched cards or magnetic tapes to each other.

Although these information transfer agreements between trading partners increased efficiency and reduced errors, they were not an ideal solution. Because the data translation programs that one trading partner wrote usually would not work for other trading partners, each company participating in this information exchange had to make a substantial investment in computing infrastructure. Only large trading partners could afford this investment, and even those companies had to perform a significant number of transactions to justify the cost. Smaller or lower-volume trading partners could not afford to participate in the benefits of these paper-free exchanges.

In 1968, a number of freight and shipping companies joined together to form the Transportation Data Coordinating Committee (TDCC), which was charged with exploring ways to reduce the paperwork burden that shippers and carriers faced. The TDCC created a standardized information set that included all the data elements that shippers commonly included on bills of lading, freight invoices, shipping manifests, and other paper forms. Instead of printing a paper form, shippers could convert information about shipments into a computer file that conformed to the TDCC standard format. The shipper could electronically transmit that computer file to any freight company that had adopted the TDCC format. The freight company translated the TDCC format into data it could use in its own information systems. The savings from not printing and handling forms, not entering the data twice, and not having to worry about error-correction procedures were significant for most shippers and freight carriers.

Although these early industry-specific data interchange efforts were very helpful, their benefits were limited to members of the industries that created standard-setting groups. In addition, most businesses that are in a particular industry buy goods and services from businesses that are in other industries. For example, a machinery manufacturer might buy materials from steel mills, paint distributors, electrical assembly contractors, and container manufacturers. Also, almost every business needs to buy office supplies and the services of freight and transportation companies. Thus, full realization of EDI's economies and efficiencies required standards that could be used by companies in all industries.

Emergence of Broader EDI Standards

After a decade of fragmented attempts at setting broader EDI standards, a number of industry groups and several large companies decided to mount a major effort to create a set of cross-industry standards for electronic components, mechanical equipment, and other widely used items. The **American National Standards Institute (ANSI)** has been the coordinating body for standards in the United States since 1918. **ANSI** does not set standards itself, but it has created a set of procedures for the development of national standards and it accredits committees that follow those procedures.

In 1979, ANSI chartered a new committee to develop uniform EDI standards. This committee is called the **Accredited Standards Committee X12 (ASC X12)**. The **ASC X12** committee meets three times each year to develop and maintain EDI standards. The committee and its subcommittees include information systems professionals from over 800 businesses and other organizations. Membership is open to organizations and individuals who have an interest in the standards. The administrative body that coordinates ASC X12 activities is the **Data Interchange Standards Association (DISA)**.

The ASC X12 standard has benefited from the participation of members from a wide variety of industries. The standard currently includes specifications for several hundred **transaction sets**, which are the names of the formats for specific business data interchanges. Figure 5-4 lists some of the more commonly used ASC X12 transaction sets.

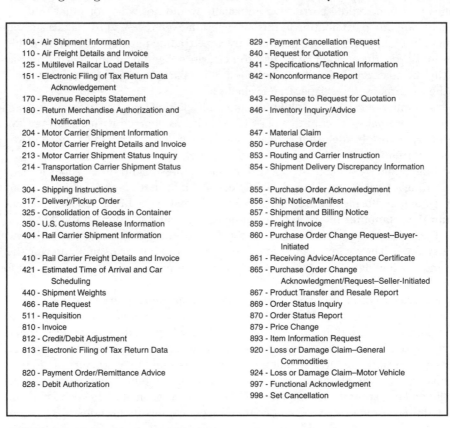

104 - Air Shipment Information	829 - Payment Cancellation Request
110 - Air Freight Details and Invoice	840 - Request for Quotation
125 - Multilevel Railcar Load Details	841 - Specifications/Technical Information
151 - Electronic Filing of Tax Return Data Acknowledgement	842 - Nonconformance Report
170 - Revenue Receipts Statement	843 - Response to Request for Quotation
180 - Return Merchandise Authorization and Notification	846 - Inventory Inquiry/Advice
204 - Motor Carrier Shipment Information	847 - Material Claim
210 - Motor Carrier Freight Details and Invoice	850 - Purchase Order
213 - Motor Carrier Shipment Status Inquiry	853 - Routing and Carrier Instruction
214 - Transportation Carrier Shipment Status Message	854 - Shipment Delivery Discrepancy Information
304 - Shipping Instructions	855 - Purchase Order Acknowledgment
317 - Delivery/Pickup Order	856 - Ship Notice/Manifest
325 - Consolidation of Goods in Container	857 - Shipment and Billing Notice
350 - U.S. Customs Release Information	859 - Freight Invoice
404 - Rail Carrier Shipment Information	860 - Purchase Order Change Request–Buyer-Initiated
410 - Rail Carrier Freight Details and Invoice	861 - Receiving Advice/Acceptance Certificate
421 - Estimated Time of Arrival and Car Scheduling	865 - Purchase Order Change Acknowledgment/Request–Seller-Initiated
440 - Shipment Weights	867 - Product Transfer and Resale Report
466 - Rate Request	869 - Order Status Inquiry
511 - Requisition	870 - Order Status Report
810 - Invoice	879 - Price Change
812 - Credit/Debit Adjustment	893 - Item Information Request
813 - Electronic Filing of Tax Return Data	920 - Loss or Damage Claim–General Commodities
820 - Payment Order/Remittance Advice	924 - Loss or Damage Claim–Motor Vehicle
828 - Debit Authorization	997 - Functional Acknowledgment
	998 - Set Cancellation

FIGURE 5-4   Commonly used ASC X12 transaction sets

Although the X12 standards were quickly adopted by major firms in the United States, in many cases, businesses in other countries continued to use their own national standards. In the mid-1980s, the United Nations Economic Commission for Europe invited both North American and European EDI experts to work together on designing a common set of EDI standards based on the successful experiences of U.S. firms in using the ASC X12 standards. In 1987, the United Nations published its first standards under the title **EDI for Administration, Commerce, and Transport (EDIFACT, or UN/EDIFACT)**. As you can see from Figure 5-5, a number of the commonly used UN/EDIFACT standard transaction sets are similar to those in the ASC X12 standard.

AUTHOR	-	Authorization	IFTCCA	-	Forwarding/Transport Shipment Charge Calculation
BOPCUS	-	Balance of Payment Customer Transaction Report	IFTDGN	-	Dangerous Goods Notification
BOPDIR	-	Direct Balance of Payment Declaration	IFTFCC	-	International Transport Freight Costs/Other Charges
BOPINF	-	Balance of Payment Information from Customer	IFTMAN	-	Arrival Notice
COARRI	-	Container Discharge/Loading Report	INVOIC	-	Invoice
COHAOR	-	Container Special Handling Order	INVRPT	-	Inventory Report
CONAPW	-	Advice on Pending Works	ORDCHG	-	Purchase Order Change Request
CONDPV	-	Direct Payment Valuation	ORDERS	-	Purchase Order
CONITT	-	Invitation to Tender	ORDRSP	-	Purchase Order Response
CONPVA	-	Payment Valuation	PAXLST	-	Passenger List
CONQVA	-	Quantity Valuation	PAYMUL	-	Multiple Payment Order
COPRAR	-	Container Discharge/Loading Order	PAYORD	-	Payment Order
COREOR	-	Container Release Order	PRODEX	-	Product Exchange Reconciliation
COSTCO	-	Container Stuffing/Stripping Confirmation	QALITY	-	Quality Data
COSTOR	-	Container Stuffing/Stripping Order	QUOTES	-	Quote
CREADV	-	Credit Advice	RECADV	-	Receiving Advice
CUSDEC	-	Customs Declaration	REMADV	-	Remittance Advice
CUSRES	-	Customs Response	REQDOC	-	Request for Document
DEBADV	-	Debit Advice	REQOTE	-	Request for Quote
DELFOR	-	Delivery Schedule	SSREGW	-	Notification of Registration of a Worker
HANMOV	-	Cargo/Goods Handling and Movement	STATAC	-	Statement of Account
IFCSUM	-	Forwarding and Consolidation Summary	SUPRES	-	Supplier Response

FIGURE 5-5 Commonly used UN/EDIFACT transaction sets

The ASC X12 organization and the UN/EDIFACT group agreed in late 2000 to develop one common set of international standards; however, no date for implementation of the common standards has been set. Both organizations created their transaction sets by extracting the information items from the paper forms used to document business transactions. Some critics of the current EDI standards argue that this reliance on forms has made it difficult for businesses to integrate EDI data flows into their business process-oriented information systems. Unfortunately, changing EDI transaction sets to follow business processes instead of paper transaction forms would require a complete redesign of standards that have become part of many organizations' computing infrastructures over the past 30 years.

How EDI Works

Although the basic idea behind EDI is straightforward, its implementation can be complicated, even in fairly simple business situations. For example, consider a company that needs a replacement for one of its metal-cutting machines. This section describes the steps involved in making this purchase using a paper-based system, and then explains how the process would change using EDI. In both of these examples, assume that the vendor uses its own vehicles instead of a common carrier to deliver the purchased machine.

Paper-Based Purchasing Process

The buyer and the vendor in this example are not using any integrated software for business processes internally; thus, each information processing step results in the production of a paper document that must be delivered to the department handling the next step. Information transfer between the buyer and vendor is also paper based and can be delivered by mail, courier, or fax. The information flows that occur in the paper-based version of the purchasing process example are shown in Figure 5-6.

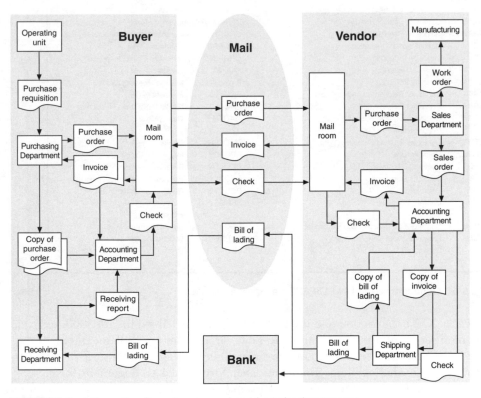

FIGURE 5-6   Information flows in a paper-based purchasing process

Once the production manager in the operating unit decides that the metal-cutting machine needs to be replaced, the following process begins:

- The production manager completes a purchase requisition form and sends it to Purchasing. This requisition describes the machine that is needed to perform the metal-cutting operation.
- Purchasing contacts vendors to negotiate price and terms of delivery. When Purchasing has selected a vendor, it prepares a purchase order and forwards it to the mail room.
- Purchasing also sends one copy of the purchase order to the Receiving Department so that Receiving can plan to accept delivery when scheduled; Purchasing sends another copy to Accounting to advise it of the financial implications of the order.
- The mail room sends the purchase order it received from Purchasing to the selected vendor by mail or courier.
- The vendor's mail room receives the purchase order and forwards it to its Sales Department.
- The vendor's Sales Department prepares a sales order that it sends to its Accounting Department and a work order that it sends to Manufacturing. The work order describes the machine's specifications and authorizes Manufacturing to begin work on it.
- When the machine is completed, Manufacturing notifies Accounting and sends the machine to shipping.
- The Accounting Department sends the original invoice to the mail room and a copy of the invoice to the Shipping Department.
- The mail room sends the invoice to the buyer by mail or courier.
- The vendor's Shipping Department uses its copy of the invoice to create a bill of lading and sends it with the machine to the buyer.
- The buyer's mail room receives the invoice at about the same time as its Receiving Department receives the machine with its bill of lading.
- The buyer's mail room sends one copy of the invoice to Purchasing so the Purchasing Department knows that the machine was received, and sends the original invoice to Accounting.
- The buyer's Receiving Department checks the machine against the bill of lading and its copy of the purchase order. If the machine is in good condition and matches the specifications on the bill of lading and the purchase order, Receiving completes a receiving report and delivers the machine to the operating unit.
- Receiving sends a completed receiving report to Accounting.
- Accounting makes sure that all details on its copy of the purchase order, the receiving report, and the original invoice match. If they do, Accounting issues a check and forwards it to the mail room.
- The buyer's mail room sends the check by mail or courier to the vendor.
- The vendor's mail room receives the check and sends it to Accounting.
- Accounting compares the check to its copies of the invoice, bill of lading, and sales order. If all details match, Accounting deposits the check in the vendor's bank and records the payment received.

EDI Purchasing Process

The information flows that occur in the EDI version of this sample purchasing process are shown in Figure 5-7. The mail service has been replaced with the data communications of an EDI network, and the flows of paper within the buyer's and vendor's organizations have been replaced with computers running EDI translation software.

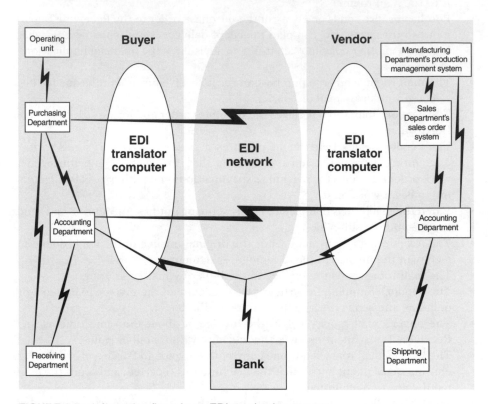

FIGURE 5-7   Information flows in an EDI purchasing process

In the EDI purchasing process, when the operating unit manager decides that the metal-cutting machine needs to be replaced, the following process begins:

- The operating unit manager sends an electronic message to its Purchasing Department. This message describes the machine that is needed to perform the metal-cutting operation.
- Purchasing contacts vendors by telephone, e-mail, or through their Web sites to negotiate price and terms of delivery. After selecting a vendor, Purchasing sends a message to the Sales Department announcing the selection.
- The buyer's EDI translator computer converts this message to a standard format purchase order transaction set, and then forwards the message through an EDI network to the vendor.

- Purchasing also sends one electronic message to the buyer's Receiving Department so it can plan to accept delivery when it is scheduled; Purchasing sends another electronic message to the buyer's Accounting Department that includes details such as the agreed purchase price.

- The vendor's EDI translator computer receives the purchase order transaction set message and converts it to the file format used by the vendor's information systems.

- The converted purchase order details appear in the Sales Department's sales order system and are automatically forwarded to the production management system in Manufacturing and to the accounting system.

- The information that was automatically forwarded to Manufacturing describes the machine's specifications and authorizes Manufacturing to begin work on it.

- When the machine is completed, Manufacturing notifies Accounting and sends the machine to the vendor's Shipping Department.

- The vendor's Shipping Department sends an electronic message to its Accounting Department indicating that the machine is ready to ship.

- The vendor's Accounting Department sends a message to its EDI translator computer, which converts the message to the standard invoice transaction set and forwards it through the EDI network to the buyer.

- The buyer's EDI translator computer receives the invoice transaction set before its Receiving Department receives the machine. The computer then converts the invoice data to a format that the buyer's information systems can use. The invoice data becomes immediately available to both the buyer's Accounting and Receiving Departments.

- When the machine arrives, the buyer's Receiving Department checks the machine against the invoice information on its computer system. If the machine is in good condition and matches the specifications shown in the buyer's system, Receiving sends a message to Accounting confirming that the machine has been received in good order. It then delivers the machine to the operating unit.

- The buyer's Accounting Department system compares all details in the purchase order data, receiving data, and decoded invoice transaction set from the vendor. If all the details match, the accounting system notifies its bank to reduce the buyer's account and increase the vendor's account by the amount of the invoice. The EDI network may provide services that perform this task.

Value-Added Networks

As you can see by comparing the paper-based purchasing process in Figure 5-6 to the EDI purchasing process in Figure 5-7, the departments are exchanging the same messages among themselves, but EDI reduces paper flow and streamlines the interchange of information among departments within a company and between companies. These efficiencies were responsible for the benefits described in the GE Lighting example presented in the introduction to this chapter. The three key elements shown in Figure 5-7 that alter the process so dramatically are the EDI network (instead of the mail service) that connects the two companies and the two EDI translator computers that handle the conversion of data from the formats used internally by the buyer and the vendor to standard EDI transaction sets. Trading partners can implement the EDI network and EDI translation processes in several ways. Each of these ways uses one of two basic approaches: direct connection or indirect connection.

The first approach, called **direct connection EDI**, requires each business in the network to operate its own on-site EDI translator computer (as shown in Figure 5-7). These EDI translator computers are then connected directly to each other using modems and dial-up telephone lines or dedicated leased lines. The dial-up option becomes troublesome when customers or vendors are located in different time zones, and when transactions are time sensitive or high in volume. The dedicated leased-line option can become very expensive for businesses that must maintain many connections with customers or vendors. Trading partners that use different communications protocols can make either of the direct connection methods difficult to implement.

Instead of connecting directly to each of its trading partners, a company might decide to use the services of a value-added network. As you learned in Chapter 1, a value-added network (VAN) is a company that provides communications equipment, software, and skills needed to receive, store, and forward electronic messages that contain EDI transaction sets. To use the services of a VAN, a company must install EDI translator software that is compatible with the VAN. Often, the VAN supplies this software as part of its operating agreement.

To send an EDI transaction set to a trading partner, the VAN customer connects to the VAN using a dedicated or dial-up telephone line and then forwards the EDI-formatted message to the VAN. The VAN logs the message and delivers it to the trading partner's mailbox on the VAN computer. The trading partner then dials in to the VAN and retrieves its EDI-formatted messages from that mailbox. This approach is called **indirect connection EDI** because the trading partners pass messages through the VAN instead of connecting their computers directly to each other. Figures 5-8 and 5-9 show the differences between direct connection EDI and indirect connection EDI using a VAN.

Companies that provide VAN services include **Computer Associates**, **Descartes VAN Services**, **EC/EDI**, **GPAS**, **IBM Global Services**, **Kleinschmidt**, and the **Sterling Electronic Trading Network**. Advantages of using a VAN are as follows:

- Users need to support only the VAN's one communications protocol instead of many possible protocols used by trading partners.
- The VAN records message activity in an audit log. This VAN audit log becomes an independent record of transactions, and this record can be helpful in resolving disputes between trading partners.
- The VAN can provide translation between different transaction sets used by trading partners (for example, the VAN can translate an ASC X12 set into a UN/EDIFACT set).
- The VAN can perform automatic compliance checking to ensure that the transaction set is in the specified EDI format.

VANs do have some disadvantages, however. One major issue is cost. Most VANs require an enrollment fee, a monthly maintenance fee, and a transaction fee. The transaction fee can be based on transaction volume, transaction length, or both. Trading partners with

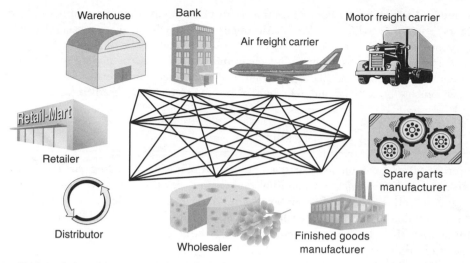

FIGURE 5-8 Direct connection EDI

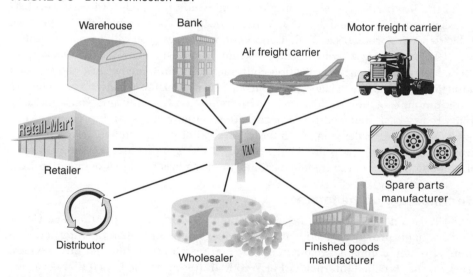

FIGURE 5-9 Indirect connection EDI through a VAN

few transactions often find it difficult to justify the high fixed costs of the enrollment and monthly maintenance fees. For example, the up-front cost of implementing indirect connection EDI, including software, VAN enrollment fee, and hardware, can exceed $50,000. Other trading partners with high transaction volumes find the VAN's ongoing transaction-based fees prohibitive.

In the past, many vendors were forced into bearing the high costs of participating in EDI to satisfy the needs of one or two large customers. This happened frequently to suppliers of the auto industry and the retail merchandising industry. Using VANs can become cumbersome and expensive for companies that want to do business with a number of trading partners, each using different VANs. Although some VANs do offer the service of exchanging messages with other VANs, the cost of this service can be unpredictable. Also, inter-VAN transfers do not always provide a clear audit trail for use in dispute resolution. Firms precluded from adopting EDI by its high cost welcomed the Internet as a low-cost communications medium that could help them overcome some of the disadvantages of traditional EDI.

EDI ON THE INTERNET

As the Internet gained prominence as a tool for conducting business, trading partners using EDI began to view the Internet as a potential replacement for the expensive leased lines and dial-up connections required to support both direct and VAN-aided EDI. Companies that had been unable to afford EDI began to look at the Internet as an enabling technology that might get them back in the game of selling to large customers that demanded EDI capabilities from their suppliers.

The major roadblocks to conducting EDI over the Internet initially were concerns about security and the Internet's general inability to provide audit logs and third-party verification of message transmission and delivery. As the basic TCP/IP structure of the Internet was enhanced with secure protocols and other encryption schemes (you will learn about these in Chapter 10), businesses worried less about security issues; however, concerns still existed. The lack of third-party verification continues to be an issue because the Internet has no built-in facility for it. Because EDI transactions are business contracts and often involve large amounts of money, the issue of nonrepudiation is significant. **Nonrepudiation** is the ability to establish that a particular transaction actually occurred. It prevents either party from repudiating, or denying, the transaction's validity or existence. In the past, the nonrepudiation function was provided either by a VAN's audit logs for indirect connection EDI or a comparison of the trading partners' message logs for direct connection EDI.

Open Architecture of the Internet

In the mid-1990s, a number of firms began providing EDI services on the Internet. Companies that originally provided traditional VAN services now offer EDI on the Internet, along with a number of new companies that entered the market with their Internet EDI services. EDI on the Internet is called **Internet EDI** or **Web EDI**. It is also called **open EDI** because the Internet is an open architecture network, as you learned in Chapter 2. Many of the new EDI offerings go beyond traditional EDI and help trading partners accomplish information interchanges that are more complex than the EDI standard transaction sets. Internet EDI has grown rapidly, but has not replaced traditional EDI because so many large companies have significant investments in the computing infrastructure they use for traditional EDI. Most VANs today offer Internet EDI services, but they continue to provide traditional EDI services.

The open architecture of the Internet allows trading partners virtually unlimited opportunities for customizing their information interchanges. New tools such as XML are helping trading partners be even more flexible in exchanging detailed information. Several groups, including an ASC X12 task group, have attempted to convert the ASC X12 EDI data elements and transaction set structures to XML in a way that retains a one-to-one mapping between the existing ASC X12 and the new XML data elements. These efforts have been stalled while an XML specification for electronic business called ebXML is refined. Meanwhile, EDI and XML are both used in many businesses to handle the electronic exchange of transaction information. A number of companies use both approaches simultaneously. You can learn more about the ebXML development effort by following the link to **ebXML.org** in the Online Companion.

Other firms are extending their internal networks (intranets) to their trading partners, which turns the intranets into extranets. Technologies such as virtual private networks (VPNs), which you learned about in Chapter 2, are providing the security that makes such extranets increasingly attractive. For example, Nintendo USA uses an EDI-based product registration system to prevent fraudulent returns. The system allows retailers to send the serial numbers of Nintendo products that they have sold directly to Nintendo USA. This system works well for large retailers, but the benefits do not offset the costs for smaller toy stores. Therefore, in 1998, Nintendo expanded the registration system to include non-EDI adopters. Nintendo bought a software package from IPNet that captures serial number and other warranty information at the cash register and then sends it over the Internet to Nintendo; this allows smaller retailers to now enjoy all the benefits of EDI at a much lower cost than traditional EDI.

Financial EDI

Although Internet EDI is growing and offering new, flexible information interchange solutions for many trading partners, some elements of EDI remain difficult to transfer to the Internet. The EDI transaction sets that provide instructions to a trading partner's bank are called **financial EDI (FEDI)**. All banks have the ability to perform electronic funds transfers (EFTs), which are the movement of money from one bank account to another. You learned about EFTs in Chapter 1. The bank accounts involved in EFTs may be customer accounts or the accounts that banks keep on their own behalf with each other. When EFTs involve two banks, they are executed using an **automated clearing house (ACH)** system, which is a service that banks use to manage their accounts with each other. In the United States, banks can use the ACH operated by the U.S. Federal Reserve Banks or one of the private ACHs operated by a group of banks or a separate company. **EDI-capable banks** are banks that are equipped to exchange payment and remittance data through VANs. Some banks also offer VAN services for nonfinancial transactions. These banks are called **value-added banks (VABs)**. Nonbank VANs that can translate financial transaction sets into ACH formats and transmit them to banks that are not EDI capable are sometimes called **financial VANs (FVANs)**.

Many companies are reluctant to use the Internet to transmit FEDI transaction sets that contain transfer instructions for large amounts of money—in some cases, millions of dollars—because of the perceived low level of security on the Internet. FEDI transaction sets are negotiable instruments—the electronic equivalent of checks. The reliability of FEDI itself is an issue, too. Because FEDI uses the Internet, it can be exposed to problems that are less likely to occur on the dedicated leased telephone lines used for connecting to a VAN. For example, if an Internet router outage delays an instruction to transfer $10 million, a trading partner could easily lose a day's interest on the funds. Thus, companies that have established indirect connection EDI through a VAN are likely to continue doing so to ensure added security for FEDI transaction sets, even though the cost of using a VAN is much higher than the cost of using the Internet for FEDI. You will learn more about payment processing for all types of transactions in Chapter 11.

EDI was the original form of electronic commerce, and it appears that it will continue to evolve and be a part of the electronic commerce boom on the Internet. Many articles in the business and information technology press over the past six years have announced the impending death of EDI. However, large companies have a huge investment in EDI systems and trained personnel. They are reluctant to change their business processes and move to Internet EDI or other proposed intercompany transaction processing approaches based on XML technologies. As these systems grow older, however, the move away from EDI will gradually occur. Total EDI transaction volume is growing steadily at a rate of about 5 percent each year and is expected to exceed $2.6 trillion by 2007. The share of EDI carried by traditional VANs has been declining. In 1997, traditional VANs carried more than 95 percent of all EDI traffic. Experts predict that VANs' share of the transaction volume will decline to about 10 percent by 2007. This shift reflects an expected increase in the number of smaller businesses that can afford to engage in Internet EDI now that an expensive VAN and dedicated leased telephone lines are no longer required.

SUPPLY CHAIN MANAGEMENT USING INTERNET TECHNOLOGIES

You learned earlier in this chapter that the part of an industry value chain that precedes a particular strategic business unit is called a supply chain. Many companies use strategic alliances, partnerships, and long-term contracts to create relationships with other companies in the supply chains for the products that they manufacture or sell. In many cases, companies are able to reduce costs by developing close relationships with a few suppliers rather than negotiating with a large number of suppliers each time they need to buy materials or supplies. When companies integrate their supply management and logistics activities across multiple participants in a particular product's supply chain, the job of managing that integration is called **supply chain management**. The ultimate goal of supply chain management is to achieve a higher-quality or lower-cost product at the end of the chain.

Value Creation in the Supply Chain

In recent years, businesses have realized that they can save money and increase product quality by taking a more active role in negotiations with suppliers. By engaging suppliers in cooperative, long-term relationships, companies have found that they can work together with these suppliers to identify new ways to provide their own customers with faster, cheaper, and better service. By coordinating the efforts of supply chain participants, firms that engage in supply chain management are reaching beyond the limits of their own organization's hierarchical structure and creating a new network form of organization among the members of the supply chain.

Supply chain management was originally developed as a way to reduce costs. It focused on very specific elements in the supply chain and tried to identify opportunities for process efficiency. Today, supply chain management is used to add value in the form of benefits to the ultimate consumer at the end of the supply chain. This requires a more holistic view of the entire supply chain than had been common in the early days of supply chain management.

Businesses that engage in supply chain management work to establish long-term relationships with a small number of very capable suppliers. These suppliers, called **tier one suppliers**, in turn develop long-term relationships with a larger number of suppliers that provide components and raw materials to them. These **tier two suppliers** manage relationships with the next level of suppliers, called **tier three suppliers**, that provide them with components and raw materials. A key element of these relationships is trust between the parties. The long-term relationships created among participants in the supply chain are called **supply alliances**. The level of information sharing that must take place among the supply chain participants can be a major barrier to entering into these alliances. Firms are not accustomed to disclosing detailed operating information and often perceive that information disclosure might hurt the firm by placing it at a competitive disadvantage.

Dell Computer is one company that has been able to reduce supply chain costs by sharing information with its suppliers. The moment Dell receives an order from a customer, it makes that information available to its tier one suppliers, who can then better plan their production based on Dell's exact demand trends. For example, a supplier of disk drives can change its production plans immediately when it sees a shift in Dell's customer orders from computers with one size disk drive to another, usually larger, size disk drive. This prevents the supplier from overproducing the smaller drive, which reduces the supplier's costs (for unsold drives) and costs in the supply chain overall (the supplier does not need to charge more for the disk drives it does sell to Dell to recover the cost of the unsold drives).

In exchange for the stability of the closer, long-term relationships, buyers expect annual price reductions and quality improvements from suppliers at each stage of the supply chain. However, all supply chain participants share information and work together to create value. Ideally, the supply chain coordination creates enough value that each level of supplier can share the benefits of reduced cost and more efficient operations. Supply chain management has been gaining momentum during the past decade and is supported by major purchasing groups such as the **Supply Chain Council**. By working together, supply chain members can reduce costs and increase the value of the product or service to the ultimate consumer.

One area in which differences in organizational goals often arise is described by Marshall Fisher in his 1997 *Harvard Business Review* article. He explains that firms often organize themselves to achieve either efficient process goals or market-responsive flexibility goals. Some companies structure themselves to be efficient producers, whereas others structure themselves to be flexible producers. The kinds of things that allow a firm to be an efficient, low-cost producer are exactly the things that prevent a firm from being flexible enough to respond to market changes. For example, the efficient producer invests in expensive machines that can stamp out large numbers of low-cost items. This investment drives down the cost of production, but makes it difficult for the producer to be flexible. A large investment in specialized machinery prevents that producer from reconfiguring the plant layout. If even one member of the supply chain for a product that requires flexible production operates as an efficient producer (instead of as a flexible producer), every other firm in the supply chain suffers. The efficient producer creates bottlenecks that hamper the best efforts of all other supply chain members. Clear communication up and down the supply chain can keep each participant informed of what the ultimate consumer demands. The participants can then plot a strategy to meet those demands.

Clear communications, and quick responses to those communications, are key elements of successful supply chain management. Technologies, and especially the technologies of the Internet and the Web, can be very effective communications enhancers. For the first time, firms can effectively manage the details of their own internal processes and the processes of other members of their supply chains. Software that uses the Internet can help all members of the supply chain review past performance, monitor current performance, and predict when and how much of certain products need to be produced. Figure 5-10 lists the advantages of using Internet technologies in supply chain management. The only major disadvantage of using Internet technologies in supply chain management is the cost of the technologies. In most cases, however, the advantages provide value that greatly exceeds the cost of implementing and maintaining the technologies.

Suppliers can:

- Share information about customer demand fluctuations
- Receive rapid notification of product design changes and adjustments
- Provide specifications and drawings more efficiently
- Increase the speed of processing transactions
- Reduce the cost of handling transactions
- Reduce errors in entering transaction data
- Share information about defect rates and types

FIGURE 5-10   Advantages of using Internet technologies in supply chain management

Increasing Supply Chain Efficiencies

Many companies are using Internet and Web technologies to manage supply chains in ways that yield increases in efficiency throughout the chain. These companies have found ways to increase process speed, reduce costs, and increase manufacturing flexibility so that they can respond to changes in the quantity and nature of ultimate consumer demand.

For example, **Boeing**, the largest producer of commercial aircraft in the world, faces a huge task in keeping its production on schedule. Each airplane requires more than 1 million individual parts and assemblies, and each airplane is custom configured to meet the purchasing airline's exact specifications. These parts and assemblies must be completed and delivered on schedule or the production process comes to a halt.

In 1997, production and scheduling errors required Boeing to shut down two entire assembly operations for several weeks, costing the company over $1.5 billion. To prevent this from ever happening again, Boeing invested in a number of new information systems that increase production efficiency by providing planning and control over logistics in every element of its supply chain. Using EDI and Internet links, Boeing is working with suppliers so that they can provide exactly the right part or assembly at exactly the right time. Even before starting an airplane into production, Boeing makes the engineering specifications and drawings available to its suppliers through secure Internet connections. As work on the airplane progresses, Boeing keeps every member of the supply chain continually informed of completion milestones achieved and necessary schedule changes.

By its second year of using these new systems, Boeing had cut in half the time needed to complete individual assembly processes. It has realized similar reductions in part defect costs. The combined effects of these increased efficiencies are helping Boeing do a much better job of meeting its customers' needs. Instead of waiting 36 months for delivery, customers can now have their new airplanes in 10 to 12 months.

To further benefit its customers, Boeing launched a spare parts Web site, **Boeing PART** (part analysis and requirements tracking). Over 500 airlines that are Boeing customers do not use EDI to order replacement parts. Boeing PART lets these customers register and then order parts using their Web browsers. The site is processing over 5000 transactions per day at a significantly lower cost to Boeing than if it were handling faxes, telephone calls, and mailed purchase orders. Boeing can deliver most parts ordered through Boeing PART on the same or next day.

Although **Dell Computer** has become famous for its use of the Web to sell custom-configured computers to individuals and businesses, it has also used technology-enabled supply chain management to give customers exactly what they want. Dell reduced the amount of inventory it keeps on hand from three weeks' sales to six days' sales. Ultimately, Dell wants to see inventory levels measured in minutes. By increasing the amount of information it has about its customers, Dell has been able to dramatically reduce the amount of inventory it must hold. Dell has also shared this information with members of its supply chain.

Dell's top suppliers have access to a secure Web site that shows them Dell's latest sales forecasts, along with other information about planned product changes, defect rates, and warranty claims. In addition, the Web site tells suppliers who Dell's customers are and what they are buying. All of this information helps these tier one suppliers plan their production much better than they could otherwise. The information sharing goes in both directions in Dell's supply chain: Tier one suppliers are required to provide Dell with current information on their defect rates and production problems. As a result, all members of the supply chain work together to reduce inventories, increase quality, and provide high value to the ultimate consumer. Much of this cooperative work requires a high level of trust. To enhance this trust and develop a sense of community, Dell maintains bulletin boards as an open forum in which its supply chain members can share their experiences in dealing with Dell and with each other.

For Boeing, Dell, and other firms, the use of Internet and Web technologies in managing supply chains has yielded significantly increased process speed, reduced costs, and increased flexibility. All of these attributes combine to allow a coordinated supply chain to produce products and services that better meet the needs of the ultimate consumer.

Using Materials-Tracking Technologies with EDI and Electronic Commerce

Tracking materials as they move from one company to another and as they move within the company has always been a troublesome task. Companies have been using optical scanners and bar codes for many years to help track the movement of materials. In many industries, the integration of bar coding and EDI has become prevalent. Figure 5-11 shows a typical bar coded shipping label that is used in the auto industry. Each bar coded element is a representation of a segment in the ASC X12 transaction set number 856, Ship Notice/Manifest. If you examine the figure carefully, you can see that five of the 856 transaction set's segments have been bar-coded (including Part Number, Quantity Shipped, Purchase Order Number, Serial Number, and Packing List Number).

FIGURE 5-11   Shipping label with bar-coded segments from ASC X12 transaction set 856, Ship Notice/Manifest

These bar codes allow companies to scan materials as they are received and to track them as they move from the materials warehouse into production. Companies can use this bar-coded information along with information from their EDI systems to manage inventory flows and forecast materials needs across their supply chains.

In the second wave of electronic commerce, companies are integrating new types of tracking into their Internet-based materials-tracking systems. The most promising technology now being used is **radio frequency identification devices (RFIDs)**, which are small chips that use radio transmissions to track inventory. RFID technology has existed for many years, but until recently, it required each RFID to have its own power supply (usually a battery).

The new development in RFID technology is the passive RFID tag, which can be made cheaply and in very small sizes. A passive RFID tag does not need a power source. It receives a radio signal from a nearby transmitter and extracts a tiny amount of power from that signal. It uses the power it extracts to send a signal back to the transmitter. That signal includes information about the inventory item to which the RFID tag has been affixed. RFID tags are small enough to be installed on the face of credit cards or sewn into clothing items. In 2003, Wal-Mart began testing the use of RFID tags on its merchandise for inventory tracking and control. You can learn more about current developments in this technology by visiting the **RFID Journal** online.

Creating an Ultimate Consumer Orientation in the Supply Chain

One of the main goals of supply chain management is to help each company in the chain focus on meeting the needs of the consumer at the end of the supply chain. Companies in industries with long supply chains have, in the past, often found it difficult to maintain this customer focus, which is often called an **ultimate consumer orientation**. Instead, companies have directed their efforts toward meeting the needs of the next member in the supply chain. This short-sighted approach can cause companies to miss opportunities to add value in subsequent steps of the chain.

One company that pioneered the use of Internet technology to go beyond the next step in its value chain is Michelin North America. Michelin has a highly respected brand name and reputation in the tire business. However, most consumers rely on local tire dealers to make specific recommendations when they need replacement tires for their vehicles. Michelin spends a great deal of money on direct advertising to its ultimate consumers. This advertising is directed at maintaining Michelin's powerful brand and convincing the consumer of the value of Michelin tires. The advertising and brand building effort can be wasted, however, if the consumer goes to a local tire dealer who recommends another brand.

Michelin launched an electronic commerce initiative in 1995 called BIB NET (after the company's famous "Michelin Man" mascot, whose name is Bibendum). The goal of this initiative was to sell more Michelin tires to consumers, but the initiative was directed at Michelin's tire dealers, not the ultimate consumers. BIB NET was an extranet that allowed tire dealers to access tire specifications, inventory status, and promotional information about Michelin products through a simple-to-use Web browser interface. Before BIB NET, dealers calling Michelin for product information were sometimes placed on hold. A dealer who is talking to a customer cannot afford to wait on hold. By giving dealers the power to access Michelin product information directly and immediately, Michelin saved money (maintaining a Web page is much less expensive than answering thousands of phone calls) and gave dealers better service. Dealers using BIB NET are much less likely to recommend a competitor's tires to their customers.

Because Internet technologies are tools that improve communications at a very low cost, they are ideal aids for enhancing the creation of a highly coordinated and effective supply chain. A number of polls and studies confirm that most information technology and purchasing managers believe that information technology is helping to improve their firms' relationships with suppliers and supply chain management initiatives.

Building and Maintaining Trust in the Supply Chain

The major issue that most companies must deal with in forming supply chain alliances is developing trust. Continual communication and information sharing are key elements in building trust. Because the Internet and the Web provide excellent ways to communicate and share information, they offer new avenues for building trust. Most procurement professionals have built trust on years of doing business with the same vendors. In many industries, vendors send sales representatives to call on buyers regularly. Vendors also participate actively in trade shows and conferences. By giving buyers frequent opportunities to interact with vendor representatives, vendors help build trust.

Vendors are finding that the Web gives them an opportunity to stay in contact with their customers more easily and less expensively. Although most buyers still see sales representatives regularly, e-mail and the Web give them nearly instant access to their sales representative and other vendor personnel. By providing comprehensive information at a moment's notice, vendors can build buyers' trust in the vendor's ability to deliver products and provide the personalized service that buyers need. Many supply chain management researchers are working on new ways to accumulate information about supplier performance and report that information to supply chain partners. This type of monitoring and reporting could help companies establish trust more quickly. Many issues, such as the objectivity and validity of performance measurements, must still be resolved before these information networks become generally accepted and used by the supply chain community. The task of developing information exchange resources that can provide supplier performance summaries is one of the great challenges that B2B electronic commerce faces as it moves into its second wave.

ELECTRONIC MARKETPLACES AND PORTALS

As the Web emerged in the mid-1990s, many business researchers and consultants believed that it would provide an opportunity for companies to establish information hubs for each major industry. These industry hubs would offer news, research reports, analyses of trends, and in-depth reports on companies in the industry—much as specialized industry trade magazines had provided in print format for years. In addition to information, these hubs would offer marketplaces and auctions in which companies in the industry could contact each other and transact business. Because these hubs would offer a doorway (or portal) to the Internet for industry members, and because these hubs would be vertically integrated (that is, each hub would offer services to just one industry), these planned enterprises were called **vertical portals**, or **vortals**. These are types of portals, which you learned about in Chapter 3.

As with many electronic commerce predictions, the prediction that vertical portals would change business forever did not turn out to be exactly correct. In this section, you will learn how B2B electronic marketplaces were conceived, developed, and operated as this sector of electronic commerce matured from 1997 through the present.

Independent Industry Marketplaces

The first companies to launch industry hubs that followed the vertical portal model created trading exchanges that were focused on a particular industry. These vertical portals became known by various names that highlighted different elements of their collective nature, including **industry marketplaces** (focused on a single industry), **independent exchanges** (not controlled by a company that was an established buyer or seller in the industry), or **public marketplaces** (open to new buyers and sellers just entering the industry). These portals are also known collectively as **independent industry marketplaces**. Ventro opened its first industry marketplace, Chemdex, in early 1997 to trade in bulk chemicals. To leverage the high investment it had made in trading exchange technology, Ventro followed Chemdex with other Web marketplaces, including Promedix in specialty medical supplies, Amphire Solutions in food service, MarketMile in general business products and services, and a number of others. Other companies were quick to follow in Ventro's chosen markets and many others. **SciQuest** founded an industry marketplace in life science chemicals. The home page of **ChemConnect.com**, which is a surviving industry marketplace in the bulk chemicals market, appears in Figure 5-12.

FIGURE 5-12 ChemConnect home page

The number of new entrants into these businesses grew rapidly during the next two years. By mid-2000, there were more than 2200 independent exchanges in a wide variety of industries. For example, there were 200 exchanges operating in the metals industry alone (see Learning from Failures: MetalSite). As venture capital funding became scarce for companies that were not earning profits—and virtually all of these marketplaces were not earning profits—many of them closed. By 2002, there were fewer than 100 industry marketplaces still operating. Ventro, for example, has closed all of the dozens of marketplaces it had opened during the boom years. It simply did not make economic sense to have more

than one or two independent marketplaces in any particular industry. Some of the industry pioneers who closed their industry marketplace operations, such as Ventro, began selling the software and technology that they developed to run their marketplaces. Their new customers were operators of other B2B marketplace models that arose to take business away from the independent marketplaces. You will learn about four of these models—private stores, customer portals, private company marketplaces, and industry consortia-sponsored marketplaces—in the remainder of this section.

LEARNING FROM FAILURES

METALSITE

Although a number of small steel manufacturing plants (called minimills) have opened in the past 20 years, most of the world's steel is still produced in very large steel mills. In these steel mills, it is economical to produce steel only in large batches. Because of the high cost of reconfiguring machinery, a steel mill set up to create one type of steel (for example, rolled sheets) requires significant time and money to change over to produce another type of steel (for example, bar steel). To minimize these changeover costs, steel mills produce steel products in large batches to meet estimated demand rather than actual orders. Because production quantities are designed to meet estimated demand instead of actual demand, steel mills often have overproduction of some items.

Companies such as Bethlehem Steel, with annual revenues of over $4 billion and 14,000 employees, solved this problem in the past by sending faxes to potential buyers of their excess production. Buyers would respond with a bid on the product in which they were interested, and Bethlehem would negotiate with them to determine price and delivery terms.

In 1998, MetalSite was one of the first metal trading exchanges to begin doing business on the Web. These exchanges offered manufacturers such as Bethlehem an efficient way to reach a larger market for their excess production. By mid-2000, there were more than 200 metal exchanges operating on the Web. These exchanges were following a reintermediation strategy; that is, they were entering the supply chain of the steel industry to provide some added value that had not existed in the supply chain before. However, most industry analysts agreed that there was no need for more than one or two exchanges in the steel industry. In 2001, metal trading exchange sites began to fail.

MetalSite had grown rapidly. With over $35 million of investors' money, MetalSite was able to sign up 24,000 registered users and by mid-2001, was trading about $30 million worth of steel each month. However, its commissions of between 1 percent and 2 percent on each trade did not yield enough money to cover operating costs. The steel business was in a downturn along with the rest of the U.S. economy, and the downward pressure on commissions from competing exchanges was increasing rapidly. The major steel companies were discussing ways to form alliances to operate their own exchanges. After three years of operation and a desperate last-minute search for new investors, MetalSite closed in August 2001.

continued

MetalSite had entered a business that could not support more than a few companies, and it was unable to become one of the survivors. The lesson from MetalSite's experience is that a reintermediation strategy must add significant value to the supply chain, and the company pursuing that strategy must be able to construct significant barriers that competitors must overcome to enter the business. MetalSite was unable to do either and thus failed. Many other B2B exchange sites that found themselves in similar competitive situations have also failed.

Private Stores and Customer Portals

As established companies in various industries watched new businesses open marketplaces, they became concerned that these independent operators would take control of transactions from them in supply chains—control that the established companies had spent years developing. Large companies that sell to many relatively small customers can exert great power in negotiating price, quality, and delivery terms with those customers. These sellers feared that industry marketplaces would dilute that power.

Many of these large sellers had already invested heavily in Web sites that they believed would meet the needs of their customers better than any industry marketplace would. For example, Cisco and Dell offer private stores for each of their major customers within their selling Web sites. A **private store** has a password-protected entrance and offers negotiated price reductions on a limited selection of products—usually those that the customer has agreed to purchase in certain minimum quantities. Other companies, such as Grainger, provide additional services for customers on their selling Web sites. These **customer portal** sites offer private stores along with services such as part number cross-referencing, product usage guidelines, safety information, and other services that would be needlessly duplicated if the sellers were to participate in an industry marketplace.

Private Company Marketplaces

Similarly, large companies that purchase from relatively small vendors can exert similar power over those vendors in purchasing negotiations. The Procurement Departments of these companies can invest in procurement software from companies such as Ariba and CommerceOne (you will learn more about all types of electronic commerce software in Chapter 9). This software, generally referred to as **e-procurement software**, allows a company to manage its purchasing function through a Web interface. It automates many of the authorizations and other steps, described in Figure 5-1, that are part of business procurement operations. Although e-procurement software was originally designed to help manage the MRO procurement process, recent releases of this software have begun to include other marketplace functions, such as request for quote posting areas, auctions, and integrated support for purchasing direct materials. Most industry observers expect these features to be improved and expanded in future versions.

Companies that implement e-procurement software usually require their suppliers to bid on their business. For example, an office supplies provider would create a schedule of prices at which it would sell to the company. The company would then compare that pricing to bids from other suppliers. The selected supplier would provide product price and description information to the company, which would insert that information into its e-procurement software. This permits authorized employees to order office supplies at the negotiated prices through a Web interface.

When industry marketplaces opened for business, these larger companies were reluctant to abandon their investments in e-procurement software or to make the software work with industry marketplaces' software—especially in the early years of industry marketplaces when there were many of them in each industry. These companies use their power in the supply chain to force suppliers to deal with them on their own terms rather than negotiate with suppliers in an industry marketplace.

As marketplace software became more reliable, many of these companies purchased software and technology consulting services from companies, such as Ventro and e-Steel, that had abandoned their industry marketplace businesses and were offering the software they had developed to companies that wanted to develop private marketplaces. A **private company marketplace** is a marketplace that provides auctions, request for quote postings, and other features (many of which are similar to those of e-procurement software) to companies that want to operate their own marketplaces. United Technologies, which annually sells more than $35 billion of high-technology products and services to the aerospace and building systems industries, was one of the first major companies to open a private company marketplace, launching its site in 1996. Since then, United Technologies has purchased more than $10 billion in goods through its private marketplace and estimates that it has saved more than $2 billion through lower prices and transaction cost savings on those purchases.

Industry Consortia-Sponsored Marketplaces

Some companies had relatively strong negotiating positions in their industry supply chains, but did not have enough power to force suppliers to deal with them through a private company marketplace. These companies began to form consortia to sponsor marketplaces. An **industry consortia-sponsored marketplace** is a marketplace formed by several large buyers in a particular industry.

One of the first such marketplaces was Covisint, which was created in 2000 by a consortium of DaimlerChrysler, Ford, and General Motors. Several thousand auto industry suppliers now belong to **Covisint**. In the hotel industry, Marriott, Hyatt, and three other major hotel chains formed a consortium to create the **Avendra** marketplace. Boeing led a group of companies in the aerospace industry to create the **Exostar** marketplace. In the consumer packaged goods industry, Procter & Gamble joined with Sara Lee, Coca Cola, and several other companies to launch the **Transora** marketplace.

These consortia-based marketplaces—along with private company marketplaces, private Web stores, and customer portals—have taken a large part of the market from the industry marketplaces that appeared to be so promising in the early days of B2B electronic commerce. One concern that suppliers have when using an industry marketplace is its ownership structure. For example, Covisint was created by a consortium of buyers in the auto industry. In 2004, the consortium decided to sell Covisint to an independent operator, Compuware, that had no ties to the founding companies. Covisint was sold, at least in part, to convince suppliers that the marketplace would not be operated to keep them at a bargaining disadvantage with the large buyers in that industry.

Figure 5-13 summarizes the characteristics of five general forms of marketplaces that exist in B2B electronic commerce today. The information in the figure comes from several sources, but the structure of the figure is adapted from one presented by Warren Raisch, a Web marketplace consultant, in his book *The eMarketplace*.

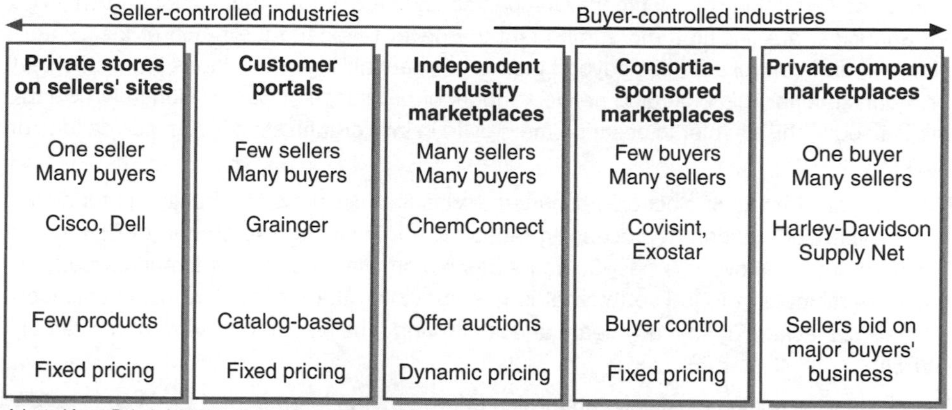

Seller-controlled industries			Buyer-controlled industries	
Private stores on sellers' sites	**Customer portals**	**Independent Industry marketplaces**	**Consortia-sponsored marketplaces**	**Private company marketplaces**
One seller Many buyers	Few sellers Many buyers	Many sellers Many buyers	Few buyers Many sellers	One buyer Many sellers
Cisco, Dell	Grainger	ChemConnect	Covisint, Exostar	Harley-Davidson Supply Net
Few products	Catalog-based	Offer auctions	Buyer control	Sellers bid on major buyers' business
Fixed pricing	Fixed pricing	Dynamic pricing	Fixed pricing	

Adapted from: Raisch, W. 2001. *The eMarketplace, p. 225.*

FIGURE 5-13 Characteristics of B2B marketplaces

Summary

In this chapter, you learned that companies are using Internet and Web technologies in a variety of ways to improve their purchasing and logistics primary activities. Businesses are also making similar improvements in a wide range of support activities such as human resources, accounting, and technology development. Companies and other large organizations, such as government agencies, are finding it more important than ever to extend the reach of their enterprise planning and control activities beyond their organizations' legal definitions to include parts of other organizations. This emerging network model of organization was introduced in Chapter 1 and is used in this chapter to describe the growth in interorganizational communications and coordination.

EDI, the first example of electronic commerce, was first developed by freight companies to reduce the paperwork burden of processing repetitive transactions. The spread of EDI to virtually all large companies over the past 30 years has led smaller businesses to seek an affordable way to participate in EDI. The Internet is now providing the inexpensive communications channel that EDI lacked for so many years and is allowing smaller companies to participate in Internet EDI.

The increase in communications capabilities offered by the Internet and the Web is, and will continue to be, an important force driving the adoption of supply chain management techniques in a variety of industries. Supply chain management incorporates several elements that can be implemented and enhanced through the use of the Internet and the Web. Increasingly, firms are connecting with their supply chain alliance partners and other companies, such as 3PL providers, to become more efficient and provide more value to the ultimate consumer of their value chains' products and services.

The emergence of industry electronic marketplaces in the mid-1990s gave way to the development of several different models for B2B electronic commerce, including private stores, customer portals, private marketplaces, and industry consortia-sponsored marketplaces. Today, all four of these models continue to coexist with the original industry marketplace model. It is not clear which of these models will dominate B2B electronic commerce in the future; however, industry consortia-sponsored marketplaces appear to be the most successful at this point.

Key Terms

Accredited Standards Committee X12
  (ASC X12)

American National Standards Institute (ANSI)

Automated clearing house (ACH)

Contract purchasing

Customer portal

Direct connection EDI

Direct materials

EDI-capable banks

EDI compatible

EDI for Administration, Commerce, and
  Transport (EDIFACT or UN/EDIFACT)

E-government

E-procurement software

E-sourcing

Financial EDI (FEDI)

Financial VANs (FVANs)

Independent exchange

Independent industry marketplace

Indirect connection EDI

Indirect materials

Industry consortia-sponsored marketplace	Spot market
Industry marketplace	Spot purchasing
Internet EDI	Supply alliances
Knowledge management	Supply chain
Maintenance, repair, and operating (MRO)	Supply chain management
Nonrepudiation	Supply web
Open EDI	Third-party logistics (3PL) provider
Private company marketplace	Tier one suppliers
Private store	Tier three suppliers
Public marketplace	Tier two suppliers
Purchasing card (p-card)	Transaction sets
Radio frequency identification device (RFID)	Ultimate consumer orientation
Replenishment purchasing	Value-added banks (VABs)
Sourcing	Vertical portal (vortal)
Spend	Web EDI

255

Review Questions

RQ1. Define "direct materials" and "indirect materials." List reasons for a large company having two separate departments to manage the purchasing of each.

RQ2. Which industries were the first to establish standard EDI transaction sets? In about 100 words, state why, in your opinion, these industries were more interested in setting standards than other industries.

RQ3. Define "knowledge management." In one paragraph, describe three advantages that a management consulting firm could gain over its competitors by creating an internal knowledge management system.

RQ4. Companies in a particular supply chain can work together to eliminate costs from the supply chain. In many cases, these cost savings are not shared evenly among the companies in the supply chain. Using research resources on the Web or in your library, identify an industry in which savings are not shared equally. In two or three paragraphs, explain why some supply chain participants in your chosen industry can obtain more benefit than others from cost reductions in the supply chain.

RQ5. In about 300 words, describe the reasons a buyer might have for wanting to participate in an industry consortium marketplace instead of setting up its own private company marketplace.

Exercises

E1. Use the **Thomas Register of American Manufacturers** Web site (note: free registration is required to access this site) to locate an industrial product with which you are completely unfamiliar. Note how many companies offer that product, have catalogs or Web sites, offer online ordering, and offer literature by fax. Summarize what you learn about the product and its availability on the Web in a report of approximately 400 words.

Note: Your instructor might ask you to prepare a formal presentation of your findings in class.

E2. You work for Andrew Wheeler, who is the president of Fabro-Max, a small plastic parts fabricator. He wants you to look into improving customer interactions through Internet EDI. You know that a number of companies that provide VAN services for companies engaging in EDI, such as **Descartes VAN Services**, **EC/EDI**, **GPAS**, **IBM Global Services**, and **Kleinschmidt**, also offer Internet EDI services. Some of these VAN operators are targeting smaller businesses that, in the past, would not have been able to afford to implement EDI. Choose two VAN providers from this list, or that you find on the Web, and examine their Web sites. For the two VAN providers that you choose, determine whether they are offering Internet EDI services. Also decide whether, in your opinion, their Web sites are targeting smaller businesses such as Andrew's. In a memo to Andrew of approximately 150 words, summarize your findings for the two VAN providers you chose.

E3. A number of standard-setting organizations offer memberships to business firms. You are working for Grace Henry, chief information officer (CIO) of Flex-Electric, a medium-size company that manufactures components for electronic medical and laboratory instruments. Grace asks you to investigate the benefits of joining an industry standard-setting organization, **RosettaNet**. Prepare a memo to Grace in which you outline the purposes of the organization and the costs and benefits of becoming a member. Close your memo with a recommendation regarding whether your company should join the organization.

Cases

C1. Harley-Davidson

Harley-Davidson manufactures high-end motorcycles and sells them worldwide. The company sells more than $4 billion in motorcycles and related products each year, and has one of the most recognized brands in the world. However, business was not always so good for the company. In the 1980s, the company was on the brink of bankruptcy. Facing increasing competition from Japanese and German manufacturers, Harley-Davidson had allowed its quality standards and cost controls to slip. In a legendary business turnaround, the company rebuilt itself. Harley-Davidson completely changed its supply chain to fulfill the expectations of its brand-aware customers.

Over a period of several years, Harley-Davidson reduced its number of suppliers from 4000 to fewer than 350. More importantly, it began to work with those suppliers to reduce costs throughout the supply chain. Each supplier is expected to find ways (with the help and cooperation of Harley-Davidson) to reduce manufacturing costs and improve quality every year. This was the only way Harley-Davidson believed it could avoid moving its factories to lower cost locations in other countries. The efforts paid off and the company still manufactures its motorcycles only in the United States.

In 2000, the company decided to focus its cost reduction and quality improvement efforts on its information technology infrastructure. Since it had been so successful in working with its suppliers to reduce manufacturing costs and improve quality, Harley-Davidson wanted to do the same thing with information technology. By using Internet technologies to share information throughout the supply chain, the company hoped to find opportunities for efficiencies and cost reductions at all stages of the process of creating motorcycles.

When the company first talked with its suppliers about its information technology initiative, those suppliers noted that each of Harley-Davidson's main factories used different invoices, production schedules, and purchasing procedures. The suppliers explained that this created difficulties for them when they dealt with more than one factory and increased their cost of doing business with Harley-Davidson. Thus, one of the first things the company did was to standardize forms and procedures. Then it moved to require all suppliers to use EDI. For smaller suppliers, the company set up a Web site that had Internet EDI capabilities. The smaller suppliers could simply log in to the Web site and conduct EDI transactions through their Web browsers.

This Web browser interface grew to become a complete extranet portal called **Harley-Davidson Supply Net**. All suppliers now use the portal to consolidate orders, track production schedule changes, obtain inventory forecasts in real time, and obtain payments for materials shipped. The portal also allows suppliers to obtain product testing information, part specifications, and product design drawings.

Key elements in both EDI and the Web portal systems have been bar codes and scanners. Most individual parts and all shipments are bar-coded. The bar code information is integrated with the materials tracking, invoicing, and payment information in the systems and is made available, as appropriate, to suppliers. Harley-Davidson uses bar code standards developed by the **Automotive Industry Action Group**.

Required:

1. Become familiar with RFID technology and its potential uses in Harley-Davidson's supply chain using the information presented in this chapter and information you obtain through the Online Companion links, your favorite search engine, and your library. In about 400 words, evaluate the advantages and disadvantages for Harley-Davidson of replacing its bar codes and scanners technology with RFID.

2. Develop and present a timetable for adoption of RFID technology with specific recommendations on where it should be implemented first. Justify the time delays you propose in the adoption of RFID at each stage of the supply chain.

Note: Your instructor might assign you to a group to complete this case, and might ask you to prepare a formal presentation of your results to your class.

C2. American Packaging Machinery

American Packaging Machinery (APM) is a company that provides repair and maintenance services to companies that operate large packaging systems. Packaging systems are arrangements of machinery that place items in containers such as boxes or bags and apply plastic shrink wrap to the containers. These machines must be adjusted regularly, and they have hundreds of parts that can wear out or fail. APM offers service contracts on most major packaging systems. A typical service contract provides for an APM technician to make regular visits to the customer site to perform preventive maintenance. The service contract also includes a certain number of emergency repair visits per year. APM also sends technicians to perform repairs for companies that do not have service contracts.

APM technicians are paid by the hour, with additional pay for overtime hours and time they work outside of standard working hours, such as weekends and holidays. APM technicians are members of a labor union, the International Brotherhood of Electrical Workers (IBEW), which negotiates pay rates and working conditions for the technicians. APM subtracts union dues from each technician's weekly paycheck and submits the total dues collected each week to the IBEW regional office. The union contract currently provides that APM technicians are covered by a medical insurance plan underwritten by the Prudential Trust Insurance Company. Although APM pays most of the insurance premium, technicians do pay a part of the premium cost. This contribution to the premium is withheld from their paychecks each week.

You are the director of electronic commerce for APM and you report to Laura Adams, APM's chief information officer. Laura asks for your help in outlining a new automated system she wants to install, which would use EDI and EFTs to handle APM's technician payroll and related transactions. She has provided the following narrative that describes how the system will work:

1. Technicians will record their time worked by entering the start and stop times for each job into a program that runs on their handheld computers (the technicians already use these handheld computers to look up wiring and mechanical diagrams for the machinery on which they work and to receive their job assignments). The time worked information will be transmitted from the handheld computer to APM's Payroll Department.

2. The Payroll Department will summarize the time worked information and send it to supervisors' desktop computers. Each supervisor will indicate an authorization for each technician's time worked, overtime, and holiday/weekend hours. That authorization will be returned by the system each day to the Payroll Department.

3. The Payroll Department will summarize the time worked information each week and calculate gross pay, deductions, and net pay for each employee. The deductions include the federal and state taxes that must be withheld by law, the contribution to the medical insurance premium, and the union dues that are withheld under the IBEW union contract.

4. The Payroll Department will send an electronic summary of the payroll information, including deductions, to the Accounting Department, which will prepare payroll tax returns and make the necessary entries in the APM accounting system to record payroll and the related tax expenses.

5. The Payroll Department will send electronic authorizations to APM's bank to make the necessary EFTs to deposit: the amount of each technician's net pay to that technician's bank account; the amount of each tax withheld to the account of the appropriate government agency; the amount of the total contributions to the medical insurance premium to the insurance company's account; and the amount of the union dues withheld to the IBEW's account. Most of these accounts are at other banks.

6. The Payroll Department will send electronic notifications to Prudential Trust and the IBEW regional office, notifying them of the transferred amounts each week.

7. The Payroll Department will send an electronic summary of the hours worked by each technician and the amount of gross pay, including overtime and holiday/weekend pay, to the APM union steward's desktop computer. The union steward is an APM technician who is elected by the technicians to monitor the terms of the union contract and handle any grievances that arise between the technicians and APM management.

Required:

1. Draw a diagram of the proposed payroll EDI and EFT system (you can use Figure 5-7 as a guide).

2. List and briefly describe any problems or issues that you think might arise in the implementation of this system.

3. Provide a rationale and recommendation as to which, if any, elements of this system you think APM should hire an outside company to implement.

Note: Your instructor might assign you to a group to complete this case, and might ask you to prepare a formal presentation of your results to your class.

For Further Study and Research

Al-Kibsi, G., K. de Boer, M. Mourshed, and N. Rea. 2001. "Putting Citizens On-Line, not In Line," *The McKinsey Quarterly*, April, 64–73.

Ayers, J. 1999. "Supply Chain Strategies," *Information Systems Management*, 16(2), Spring, 72–80.

Bacheldor, B., L. Sullivan, C. Murphy, and R. Whiting. 2004. "RFID Kick-start," *Information Week*, May 24, 20–22.

Barlas, D. 2003. "Motorola's E-Business Intelligence," *Line56: The E-Business Executive Daily*, October 24. (http://www.line56.com/articles/default.asp?ArticleID=5104)

Binns, S. 2004. "Businesses Miss Benefits of High-Tech Radio Tagging," *Supply Management*, 9(2), January 22, 13.

Black, J. 2001. "Build Lasting Partnerships: Collaboration Is the Name of the Game at General Motors," *Internet World*, August 1, 24.

Bovel, D. and M. Joseph. 2000. "From Supply Chain to Value Net," *Journal of Business Strategy*, 21(4), July–August, 24–28.

Brockmann, P. 2003. "EDI and XML," *World Trade*, 16(9), September, 68–69.

Clark, L. 2003. "Covisint Leads Drive to Move Motor Industry From EDI to XML," *Computer Weekly*, May 20, 16.

Clark, P. 2001. "MetalSite Kills Exchange, Seeks Funding," *B to B*, 86(13), June 25, 3.

Cleary, M. 2001. "Metal Meltdown Doesn't Deter New Ventures," *Interactive Week*, 8(27), July 9, 29.

Colberg, T., N. Gardner, K. Horan, D. McGinnis, P. McLauchlin, and Y-H. So. 1995. *The Price Waterhouse EDI Handbook*. New York: John Wiley & Sons.

Computerworld. 2001. "Autopsy of a Dot Com: Chemdex," February 15. (http://computerworld.com/cwi/story/0,1199,NAV65-665_STO57746,00.html)

Cooke, J. 2002. "Is XML the Next Big Thing? Maybe Not for a While," *Logistics Management & Distribution Report*, 41(5), May, 53–55.

Delfmann, W., S. Albers, and M. Gehring. 2002. "The Impact of Electronic Commerce on Logistics Service Providers," *International Journal of Physical Distribution & Logistics Management*, 32(3), 203–222.

Dobbs, J. 1999. *Competition's New Battleground: The Integrated Value Chain*. Cambridge, MA: Cambridge Technology Partners.

Drickhamer, D. 2003. "EDI is Dead! Long Live EDI!" *Industry Week/IW*, 252(4), April, 31–35.

Eid, R., M. Trueman, and A. Ahmed. 2002. "A Cross-Industry Review of B2B Critical Success Factors," *Internet Research: Electronic Networking Applications and Policy*, 12(2), 110–123.

Ekoniak, J. 2001. "Economic Analysis: Automate Your Supply Chain," *Upside*, 13(5), May, 90–91.

Fickel, L. 1998. "MicroAge's Internet-Based Training Program," *CIO Magazine*, July 15, 72.

Fisher, M. 1997. "What Is the Right Supply Chain for Your Product?" *Harvard Business Review*, 75(2), March–April, 105–116.

Fox, P. 2001. "Boeing Shows How XML Can Help Business," *Computerworld*, 35(11), March 12, 28–29.

Garcia-Dastugue, S. and D. Lambert. 2003. "Internet-Enabled Coordination in the Supply Chain," *Industrial Marketing Management*, 32(2), June, 251–263.

Graham, D. 2002. "Why Translating EDI to XML Is So Difficult," *eBizQ*, February 11. (http://b2b.ebizq.net/ std/graham_1.html)

Hoffman, C. 2000. "Run XBRL Right Now," *Journal of Accountancy*, 190(2), August, 28–29.

Jap, S. and J. Mohr. 2002. "Leveraging Internet Technologies in B2B Relationships," *California Management Review*, 44(4), Summer, 24–39.

Johnston, M. 2000. "Government Adopts Electronic Transactions," *InfoWorld*, 22(31), July 31, 5.

Kanakamedala, K., J. King, and G. Ramsdell. 2003. "The Truth about XML," *McKinsey Quarterly*, July, 9–12.

Kaplan, S. and M. Sawhney. 2000. "E-Hubs: The New B2B Marketplaces," *Harvard Business Review*, 78(3), May–June, 97–103.

Karpinski, R. 2001. "EDI-to-Web Move Promises ROI," *InternetWeek*, August 20, 9.

Karpinski, R. 2002. "Wal-Mart Mandates Secure, Internet-Based EDI for Suppliers," *InternetWeek*, September 12. (http://www.internetwk.com/security02/INW20020912S0011)

Kay, E. 2000. "From EDI to XML," *Computerworld*, 34(25), June 19, 84–85.

Kemp, T., D. Drucker, and C. Moozakis. 2001. "Reload: Slower Buy-In," *InternetWeek*, June 23, 10.

Levinson, M. 2002. "How to Grow Your Own B2B Network," *CIO Magazine*, June 15, 60–64.

Lewis, W. 2000. "Pillar of the Community: XML Is Becoming the Standard Platform," *Intelligent Enterprise*, 3(13), August 18, 32–38.

Lykins, D. 2002. "Optimize Your Supply Chain," *E-Business Advisor*, June, 14–19.

Marcella, A. and S. Chan. 1993. *EDI Security, Control, and Audit*. Norwood, MA: Artech House.

Massetti, B. and R. Zmud. 1996. "Measuring the Extent of EDI Usage in Complex Organizations: Strategies and Illustrative Examples," *MIS Quarterly*, 20(3), September, 331–345.

Meehan, M. 2001. "EDI, ebXML Groups Agree to Cooperate: B2B Standards Inch Forward," *Computerworld*, 35(27), July 20, 1–2.

Meehan, M. 2001. "Michelin Sees Long Road to B2B Adoption," *Computerworld*, 35(32), August 6, 10.

Moozakis, C. and D. Joachim. 2001. "Auto Hub Revamps," *InternetWeek*, August 20, 9.

Morgan, J. and R. Monczka. 2003. "Why Supply Chains Must Be Strategic," *Purchasing*, April 17, 42–45.

Osmonbekov, T., D. Bello, and D. Gilliland. 2002. "Adoption of Electronic Commerce Tools in Business Procurement: Enhanced Buying Center Structure and Processes," *Journal of Business & Industrial Marketing*, 17(2/3), 151–166.

Ovans, A. 2000. "E-Procurement at Schlumberger," *Harvard Business Review*, 78(3), 21–22.

Papazoglou, M. 2001. "Agent-Oriented Technology in Support of E-Business," *Communications of the ACM*, 44(4), April, 71–77.

Poirier, C. and M. Bauer. 2001. *E-Supply Chain: Using the Internet to Revolutionize your Business.* San Francisco: Barrett-Koehler.

Power, C. 1999. "Internet Systems Imperil EDI for Corporate Buying," *American Banker*, 164(73), April 19, 15.

Power, D. and A. Sohal. 2002. "Implementation and Usage of Electronic Commerce in Managing the Supply Chain: A Comparative Study of Ten Australian Companies," *Benchmarking: An International Journal*, 9(2), 190–208.

Premkumar, G. 2000. "Interorganization Systems and Supply Chain Management: An Information Processing Perspective," *Information Systems Management*, 17(3), Summer, 56–69.

Purchasing. 2001. "MetalSite Shuts Operations While Seeking New Owner," July 5, 32.

Purchasing. 2004. "Easing into E-procurement with Indirect Spend," February 19, 35–36.

Pushkin, A. and B. Morris. 1997. "Understanding Financial EDI," *Management Accounting*, 70(5), November, 42–46.

Raisch, W. 2001. *The eMarketplace: Strategies for Success in B2B Ecommerce.* New York: McGraw-Hill.

Rinat, Z. 2001. "Beyond Private Exchanges: The Private Business Network," *E-Business Advisor*, July–August, 20.

Roberts, B. 1998. "Portals, You Say? This One's Private: Ericsson's Intranet Is a Give-and-Take Affair with Employees," *Intranet Design Magazine*, December 14. (http://idm.internet.com/articles/200003/ pt_03_15_00f.html)

Roberts-Witt, S. 2001. "Steel Gets Wired," *PC Magazine*, 20(11), June 12, 14.

Rosencrance, L. 2000. "Transporters Move to Deliver on E-Commerce," *Computerworld*, 34(27), July 3, 22.

Senn, J. 1992. "Electronic Data Interchange," *Information Systems Management*, 9(1), Winter, 45–53.

Smith, T. 2002. "Winn-Dixie's Supply Chain Success Story," *InternetWeek*, August 2. (http://www.internetwk.com/story/INW20020802S0001)

Songini, M. 2001. "Group Maps RosettaNet to Supply-Chain Process," *Computerworld*, 35(16), April 16, 8.

Songini, M. 2004. "Supply Chain System Failures Hampered Army Units in Iraq," *Computerworld*, 38(30), July 26, 1–2.

Stockdale, R. and C. Standing. 2002. "A Framework for the Selection of Electronic Marketplaces: A Content Analysis Approach," *Internet Research: Electronic Networking Applications and Policy*, 12(3), 221–234.

Sullivan, L. 2004. "Ready to Roll," *Information Week*, March 8, 45–47.

Sullivan, M. 2001. "High-Octane Hog," *Forbes*, 168(6), September 10, 8–10.

Supplier Selection & Management Report. 2003. "How Harley-Davidson Teamed With 16 Major Suppliers To Cut Costs," 3(1), January, 1–3.

Taylor, D. 2004. "No Time to Spare: A Guide to Supply Chain Performance Management," *Intelligent Enterprise*, 7(10), June 12, 20–24.

Taylor, D. and A. Terhune. 2001. *Doing E-Business: Strategies for Thriving in an Electronic Marketplace.* New York: John Wiley & Sons.

Teschler, L. 2000. "New Role for B-to-B Exchanges: Helping Developers Collaborate," *Machine Design*, 72(19), October 5, 52–58.

Thaler, M. 2001. "Private Exchanges: Are They All They're Cracked Up to Be?" *E-Business Advisor*, July–August, 16.

Tie, R. 2000. "Comments Encouraged on Newly Named XBRL," *Journal of Accountancy*, 189(6), June, 14–15.

Tillett, S. 2001. "Medical Companies Track E-Learning," *InternetWeek*, August 20, 13.

Vilar, A. 2001. "B2B: The Best Is Yet to Be," *Forbes*, September 10. (http://www.forbes.com/global/ 2001/0917/070.html)

Villars, R. 2002. "EDI Service Providers: A Ray of Hope in Difficult Seas," *xSP Advisor*, January 2. (http://www.idc.com/getdoc.jhtml?containerId=xa20020102)

Waugh, R. and S. Elliff. 1998. "Using the Internet to Achieve Purchasing Improvements at General Electric," *Hospital Material Management Quarterly*, 20(2), November, 81–83.

Weinberg, N. 2001. "B2B Grows Up," *Forbes*, 168(6), September 10, 18–20.

Willis, D. 2002. "EDI: What Death? The Convergence of EDI, XML, and Internet Technology," *Line56: The E-Business Executive Daily*, February 26. (http://www.line56.com/articles/default. asp?articleid=3338)

Young, E. 2002. "Web Marketplaces That Really Work," *Fortune/CNET Tech Review*, Winter, 78–86.

CHAPTER **6**

ONLINE AUCTIONS, VIRTUAL COMMUNITIES, AND WEB PORTALS

LEARNING OBJECTIVES

In this chapter, you will learn about:

- Origins and key characteristics of the seven major auction types
- Strategies for Web auction sites and auction-related businesses
- Virtual communities and Web portals

INTRODUCTION

In 1995, Pierre Omidyar was working as a Web programmer for General Magic, Inc. and, in his spare time, operating a small personal Web site that provided, among other things, updates on the Ebola virus. Omidyar decided that the Web provided a good environment for bringing auction buyers and sellers together, so he built a small auction function into his site and called it AuctionWeb. Interest in the site's auctions grew so rapidly that within a year, Omidyar had quit his job to devote his full energies to the Web auction business he had created. By the end of its second year in operation, Omidyar's Web site, which he had renamed **eBay**, had auctioned over $95 million worth of goods and was showing a small profit.

Inspired by this success, Omidyar obtained $5 million in funding from Benchmark Capital in 1997. Benchmark also helped him recruit a top-notch management team for the business, including current CEO Meg Whitman. In September 1998, eBay offered its stock to the public and raised $63 million. Like many Internet companies, eBay experienced extremely rapid growth. Unlike many of the new dot-com companies of the time, eBay was profitable from its inception; its net income in 1998 was over $2 million. Because of its high growth rate and solid profitability, eBay was able to return to the stock market and raise an additional $765 million in April 1999. In just three years, eBay established itself as the dominant Web site for general consumer auctions with over 2 million registered buyers and sellers.

Today, eBay has more than 100 million registered users and hosts auctions for goods valued at more than $30 billion each year. The company earns an annual net income of several hundred million dollars from its auctions and related businesses, and net income is expected to continue to grow rapidly during the next few years.

Because eBay was one of the first auction Web sites and because it pursued an aggressive promotion strategy, it has become the first-choice site for many people who want to participate in auctions. Both buyers and sellers benefit from a large marketplace such as the one eBay created. eBay's early advantage in the online auction business will be very difficult for competitors to overcome.

AUCTION OVERVIEW

In Chapters 3 and 4, you learned how businesses are using the Web to create online identities, reach customers, and sell to them. In Chapter 5, you learned how businesses are using the Web to purchase goods and work with their suppliers more effectively. In all three of these chapters, the focus was on how companies can use the Web to improve the things that they have been doing for years: buying and selling. In this chapter, you will learn how companies are using the Web to do things that they have never done before. These new things include running auctions, creating virtual communities, and operating Web portals.

In many ways, online auctions provide a business opportunity that is perfect for the Web. An auction site can charge both buyers and sellers to participate, and it can sell advertising on its pages. People interested in trading specific items can form a market segment that advertisers will pay extra to reach. Thus, the same kind of targeted advertising opportunities that search engine sites generate with their results pages are available to advertisers on auction sites. This combination of revenue-generating characteristics makes it relatively easy to develop online auctions that yield profits early in the life of the project.

One of the Internet's strengths is that it can bring together people who share narrow interests but are geographically dispersed. Online auctions can capitalize on that ability by either catering to a narrow interest or providing a general auction site that has sections devoted to specific interests.

Origins of Auctions

The earliest written records of auctions are from Babylon and date from 500 BC. In those auctions, men bid against each other for the women they wished to marry. Roman soldiers used auctions to liquidate the property they took from their vanquished foes. In AD 193, the Praetorian Guard auctioned off the entire Roman Empire after killing the Emperor Pertinax. In later years, Buddhist temples held auctions to sell off the possessions of deceased monks.

Auctions became common activities in 17th-century England, where taverns held regular auctions of art and furniture. The 18th century saw the birth of two British auction houses—**Sotheby's** in 1744 and **Christie's** in 1766—that continue to be major auction firms today. The British settlers of the colonies that would become the United States brought auctions with them. Colonial auctions were used to sell farm equipment, animals, tobacco, and, sad to say, human beings.

In an auction, a seller offers an item or items for sale, but does not establish a price. This is called "putting an item up for bid" or "putting an item on the (auction) block." Potential buyers are given information about the item or some opportunity to examine it; they then offer **bids**, which are the prices they are willing to pay for the item. The potential buyers, or **bidders**, each have developed **private valuations**, or amounts they are willing to pay for the item. The whole auction process is managed by an **auctioneer**. In some auctions, people employed by the seller or the auctioneer can make bids on behalf of the seller. These people are called **shill bidders**. Shill bidders can artificially inflate the price of an item and may be prohibited from bidding by the rules of a particular auction.

English Auctions

Many different kinds of auctions exist. Most people who have attended or seen an auction on television have experienced only one type of auction, the **English auction**, in which bidders publicly announce their successive higher bids until no higher bid is forthcoming. At that point, the auctioneer pronounces the item sold to the highest bidder at that bidder's price. This type of auction is also called an **ascending-price auction**. An English auction is sometimes called an **open auction** (or **open-outcry auction**) because the bids are publicly announced; however, there are other types of auctions that use publicly announced bids that are also called open auctions.

In some cases, an English auction has a minimum bid, or reserve price. A **minimum bid** is the price at which an auction begins. If no bidders are willing to pay that price, the item is removed from the auction and not sold. In some auctions, a minimum bid is not announced, but sellers can establish a minimum acceptable price, called a **reserve price**, or simply **reserve**. If the reserve price is not exceeded, the item is withdrawn from the auction and not sold.

English auctions that offer multiple units of an item for sale and allow bidders to specify the quantity they want to buy are called **Yankee auctions**. When the bidding concludes in a Yankee auction, the highest bidder is allotted the quantity he or she bid. If items remain after satisfying the highest bidder, those remaining items are allocated to successive lower (next-highest) bidders until all items are distributed. Although all successful bidders (except possibly the lowest successful bidder) receive the quantity of items on which they bid, they only pay the price bid by the *lowest* successful bidder.

To understand Yankee auctions better, consider this example. A seller places nine items up for bid. When the bidders stop increasing their bids, the successful bidders include: the highest bidder, who bid $85, quantity five; the second-highest bidder, who bid $83, quantity three; and the third-highest bidder, who bid $81, quantity four. All three of the successful bidders pay $81 per item, but the highest bidder receives five items, the second-highest bidder receives three items, and the third-highest bidder receives the one remaining item, despite having bid for a quantity of four, because only one is left after satisfying the quantity bids of the higher bidders.

English auctions have drawbacks for both sellers and bidders. Because the winning bidder is only required to bid a small amount more than the next-highest bidder, winning bidders tend not to bid their full private valuations, which prevents sellers from obtaining the maximum possible price. Bidders risk becoming caught up in the excitement of competitive bidding and then bidding more than their private valuations. This psychological phenomenon, called the **winner's curse**, has been extensively documented by William Thaler and other behavioral economists.

Dutch Auctions

The **Dutch auction** is a form of open auction in which bidding starts at a high price and drops until a bidder accepts the price. Because the price drops until a bidder claims the item, Dutch auctions are also called **descending-price auctions**. Farmers' cooperatives in the Netherlands use this type of auction to sell perishable goods such as produce and flowers, which is how it came to be known as a "Dutch" auction.

In most Dutch auctions, the seller offers a number of similar items for sale. One common implementation of a Dutch auction uses a clock that drops the price with each tick. The first bidder to call out "stop," which stops the clock, becomes the winning bidder. The winning bidder can take all or any part of the auctioned items at that price. If any items remain, the clock is restarted and continues to run until all the items are taken by successive lower bidders. A Dutch auction is often better for the seller because the bidder with the highest private valuation will not let the bid drop much below that valuation for fear of losing the item to another bidder. Dutch auctions are particularly good for moving large numbers of commodity items quickly. A few online stores have offered Dutch auctions from time to time. For several years, **Coldwater Creek** used Dutch auctions to sell closeout items on its site, shown in Figure 6-1.

Most online retailers who have tried Dutch auctions have found that they do not increase sales or generate interest in the products well enough to justify the costs of operating the auction. They also have found that their customers are confused by sites that include a Dutch auction as an alternative to regular sales of closeout or marked-down items. This does not mean that Dutch auctions are never useful. In 2004, Google used a Dutch auction to sell its stock to investors in its initial public offering. The financial community considered this use of a Dutch auction to be highly innovative and very successful.

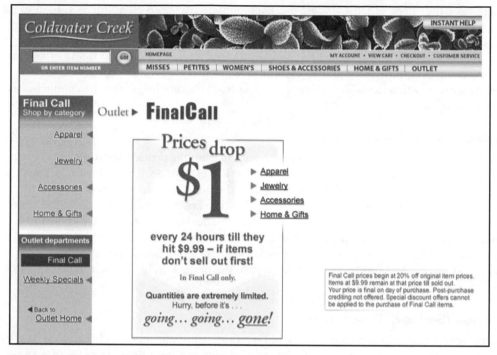

FIGURE 6-1   Coldwater Creek Dutch auction of closeout merchandise

First-Price Sealed-Bid Auctions

In **sealed-bid auctions**, bidders submit their bids independently and are usually prohibited from sharing information with each other. In a **first-price sealed-bid auction**, the highest bidder wins. If multiple items are auctioned, successive lower (next-highest) bidders are awarded the remaining items at the prices they bid.

Second-Price Sealed-Bid Auctions

The **second-price sealed-bid auction** is the same as the first-price sealed-bid auction except that the highest bidder is awarded the item at the price bid by the *second*-highest bidder. At first glance, one might wonder why a seller would even consider such an auction because it gives the item to the winning bidder at a lower price. William Vickrey won the **1996 Nobel Prize in Economics** for his studies of the properties of this auction type. He concluded that it yields higher returns for the seller, encourages all bidders to bid the amounts of their private valuations, and reduces the tendency for bidders to collude. Because the winning bidder is protected from an erroneously high bid, all bidders tend to bid higher than they would in a first-price sealed-bid auction. Second-price sealed-bid auctions are commonly called **Vickrey auctions**.

Open-Outcry Double Auctions

The **Chicago Board of Trade** conducts **open-outcry double auctions** of commodity futures and stock options. The buy and sell offers are shouted by traders standing in a small area on the exchange floor called a trading pit. Each commodity or stock option is traded in its own pit. The action in a trading pit can become quite frenzied as 20 or 30 traders shout offers aloud.

Double auctions, either sealed-bid or open-outcry, work well only for items of known quality, such as securities or graded agricultural products, that are regularly traded in large quantities. Such items can be auctioned without bidders inspecting the items before placing their bids.

Sealed-Bid Double Auctions

In a **double auction**, buyers and sellers each submit combined price-quantity bids to an auctioneer. The auctioneer matches the sellers' offers (starting with the lowest price and then going up) to the buyers' offers (starting with the highest price and then going down) until all the quantities offered for sale are sold to buyers. Double auctions can be operated in either sealed-bid or open-outcry formats. The **New York Stock Exchange** conducts sealed-bid double auctions of stocks and bonds in which the auctioneer, called a specialist, manages the market for a particular stock or bond issue. The specialist company must use its own funds, when necessary, to maintain a stable market in the specific security it manages.

Reverse (Seller-Bid) Auctions

In a **reverse auction** (also called a **seller-bid** auction), multiple sellers submit price bids to an auctioneer who represents a single buyer. The bids are for a given amount of a specific item that the buyer wants to purchase. The prices go down as the bidding continues until no seller is willing to bid lower. Reverse auctions are used by consumers, but the vast majority of these auctions (and by far the largest portion of the dollar volume of these auctions) involves businesses that are both buyers and sellers. In many business reverse auctions, the buyer acts as auctioneer and screens sellers before they can participate. You will learn more about specific implementations of reverse auctions, both consumer and business, later in this chapter.

The seven auction types described in this section are the most commonly used in business today. Figure 6-2 summarizes the key characteristics of each of these seven major auction types.

Auction type	Key characteristics
English auction	Starting from a low price, bidding increases until no bidder is willing to bid higher.
Dutch auction	Starting from a high price, bidding automatically decreases until the bidder accepts the price.
First-price sealed-bid auction	Secret bidding process; the highest bidder pays the amount of the highest bid.
Second-price sealed-bid auction (Vickrey auction)	Secret bidding process; the highest bidder pays the amount of the *second*-highest bid.
Double auction (open-outcry)	Buyers and sellers declare combined price-quantity bids. The auctioneer matches seller offers (lowest to highest) with buyer offers (highest to lowest). Buyers and sellers can modify bids based on knowledge gained from other bids.
Double auction (sealed-bid)	Buyers and sellers declare combined price-quantity bids. The auctioneer (specialist) matches seller offers (lowest to highest) with buyer offers (highest to lowest). Buyers and sellers cannot modify their bids.
Reverse auction (seller-bid)	Multiple sellers submit price bids to an auctioneer that represents a single buyer. The bids are for a given amount of a specific item that the buyer wants to purchase. Prices go down as the bidding continues until no seller is willing to bid lower.

FIGURE 6-2 Key characteristics of seven major auction types

ONLINE AUCTIONS AND RELATED BUSINESSES

Online auctions are one of the fastest growing segments of online business today. Millions of people buy and sell all types of goods on consumer auction sites each year. Industry analysts predict that online auctions will account for 50 percent of all electronic commerce dollar volume by 2007. Although the online auction business is changing rapidly as it grows, three broad categories of auction Web sites have emerged: general consumer auctions, specialty consumer auctions, and business-to-business auctions. Some industry analysts consider the two types of consumer auctions to be business-to-consumer electronic commerce. Other analysts believe that a more appropriate term for the electronic commerce that occurs in general consumer auctions is consumer-to-consumer or even consumer-to-business (because the bidders at a general consumer auction might be businesses). Their argument is that many sellers who participate in general consumer auctions are not really businesses; they are ordinary people who use these auctions to sell personal items instead of holding a garage sale, for example. Whether you prefer to think of online auctions as business-to-consumer, consumer-to-consumer, or consumer-to-business, the largest number of transactions occurs on general consumer auction sites.

General Consumer Auctions

The most successful consumer auction Web site today (by far) is **eBay**, the company described in the introduction to this chapter. The eBay home page, which appears in Figure 6-3, includes links to categories of items. Alternatively, a potential bidder can use the Smart Search feature to find a specific item by entering descriptive terms. The bottom of the page includes a link to the **third-party assurance provider TRUSTe**. Organizations such as TRUSTe provide assurance that the privacy policies of the Web sites meet certain standards. You will learn more about TRUSTe and other assurance providers in Chapter 10.

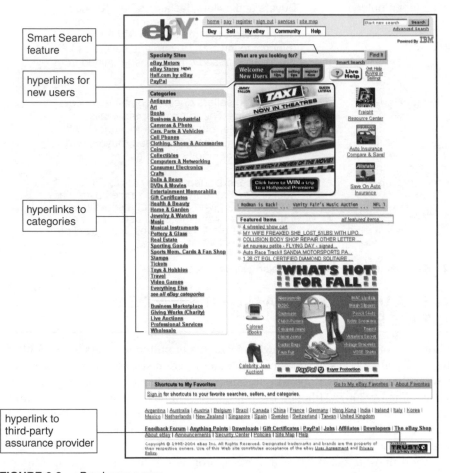

FIGURE 6-3 eBay home page

Sellers and buyers must register with eBay and agree to the site's basic terms of doing business. Sellers pay eBay a listing fee and a sliding percentage of the final selling price.

Buyers pay nothing to eBay. In addition to paying the basic fees, sellers can choose from a variety of enhanced and extra-cost services, including having their auctions listed in bold-face type and featured in lists of preferred auctions.

In an attempt to address buyer concerns about seller reliability, eBay instituted a rating system. Buyers can submit ratings of sellers after doing business with them. These ratings are converted into graphics that appear with the seller's nickname in each auction in which that seller participates. Although this system is not without flaws, many eBay bidders feel that it affords them some level of protection from unscrupulous sellers. The converse is true also; sellers rate buyers, which provides sellers some protection from unscrupulous buyers.

Although eBay does not release any statistics about buyer and seller frauds, most industry observers agree that sellers face larger potential losses than buyers. Sellers' greatest risks are from buyers who use stolen credit card numbers or who place the winning bid but never contact the seller to conclude the transaction. Buyers' risks include sellers who never deliver or who misrepresent their merchandise. You will learn about ways that sellers and buyers can protect themselves later in this chapter.

The most common format used on eBay is a computerized version of the English auction. The eBay English auction allows the seller to set a reserve price. In eBay English auctions, the bidders are listed, but the bid amounts are not disclosed until after the auction is over. This is a slight variation on the in-person English auction, but because eBay always shows a continually updated high bid amount, a bidder who monitors the auction can see the bidding pattern as it occurs. The main difference between eBay and a live English auction is that bidders do not know who placed which bid until the auction is over. The eBay English auction also allows sellers to specify that an auction be made private. In an eBay private auction, the site never discloses bidders' identities and the prices they bid. At the conclusion of the auction, eBay notifies only the seller and the highest bidder. Another auction type offered by eBay is an increasing-price format for multiple-item auctions that eBay calls a Dutch auction. However, eBay auctions in this format are not true Dutch auctions; they are considered Yankee auctions.

In either type of eBay auction, bidders must constantly monitor the bidding activity. All eBay auctions have a **minimum bid increment**, the amount by which one bid must exceed the previous bid, which is about 3 percent of the bid amount. To make bidding easier, eBay allows bidders to make a proxy bid. In a **proxy bid**, the bidder specifies a maximum bid. If that maximum bid exceeds the current bid, the eBay site automatically enters a bid that is one minimum bid increment higher than the current bid. As new bidders enter the auction, the eBay site software continually enters higher bids for all bidders who placed proxy bids. Although this feature is designed to make bidding require less bidder attention, if a number of bidders enter proxy bids on one item, the bidding rises rapidly to the highest proxy bid offered. This rapid rise in the current bid often occurs in the closing hours of an eBay auction.

eBay has been so successful because it was the first major Web auction site for consumers that did not cater to a specific audience and because it advertises widely. eBay spends more than $100 million each year to market and promote its Web site. A significant portion of this promotional budget is devoted to traditional mass media outlets, such as television advertising. For eBay, such advertising has proven to be the best way to reach its main market: people who have a hobby or a very specific interest in items that are not

271

Online Auctions, Virtual Communities, and Web Portals

locally available. Whether those items are jewelry, antique furniture, coins, first-edition books, or stuffed animals, eBay has created a place where people can become collectors, dispose of their collections, or trade out of their collections.

LEARNING FROM FAILURES

Auction Universe

One of the most promising new entrants into the general consumer auction business was Auction Universe. Times Mirror, the parent company of the *Los Angeles Times* newspaper, started Auction Universe in 1997 and then sold it in 1998 to a partnership of eight major newspaper companies (including Times Mirror itself) called **Classified Ventures**. These companies were concerned that classified advertising on the Web posed a threat to their newspapers' classified advertising, which is one of the most profitable elements in the newspaper business. Through their Classified Ventures partnership, these newspaper companies started their own Web sites for classified ads such as Apartments.com, Cars.com, and NewHomeNetwork.com. These sites earn revenue by charging for running ads, by selling advertising on their pages, or both. Classified Ventures believed that the Auction Universe site could become an important and profitable part of its Web presence.

Auction Universe closed in August 2000. Classified Ventures' classified ad sites continue to operate. The Auction Universe site was modeled on eBay and offered similar types of auctions and services for buyers and sellers. Some critics believed that the Auction Universe interface was more intuitive than eBay's and included a better search engine; however, the site failed to mount a sustained challenge to eBay's dominance. Even with major corporate sponsorship and a $10 million advertising campaign behind it, Auction Universe was unable to displace the advantage eBay obtained as the first Web auction site for general consumers.

Because one of the major determinants of Web auction site success is attracting enough buyers and sellers to create markets in many different items, some Web sites that already have a large number of visitors have entered this business. Portal sites such as **Yahoo!** have created general consumer auctions patterned after eBay. The Yahoo! Shopping and Auctions home page appears in Figure 6-4.

As you examine the home page in Figure 6-4, notice that it includes many of the same features as the eBay home page. For example, it has links to categories of auction items and a search function.

To attract sellers who frequently offer items or who continually offer large numbers of items, eBay offers a platform called eBay Stores within its auction site. At a very low cost, sellers can establish eBay stores that show items for sale as well as items being auctioned. This can help sellers generate additional profits from sales of items related to those offered in their auctions. These eBay stores are integrated into the auction site; that is, when a bidder searches for an item, the results page includes auctions and listings from sellers' eBay stores.

Yahoo! had some early success in attracting large numbers of auction participants, in part because it offered its auction service to sellers at no charge. Yahoo! was less successful in attracting buyers, resulting in less bidding action in each auction than generally occurs

hyperlinks for new users

hyperlinks to categories

FIGURE 6-4 Yahoo! Auctions and Shopping Web page

on eBay. In January 2001, Yahoo! began charging sellers in the face of dropping ad revenues in its other Web operations. Within one month, Yahoo! lost about 80 percent of its auction listings; however, the percentage of listed items that ended in a sale increased six fold, and the dollar amount of completed auctions remained constant. Because Yahoo! draws a large number of visitors every month, it is possible that the site will be able to further increase participation in its auctions and attract some of the sellers who left in reaction to the fees.

Amazon.com, the pioneering Web bookseller, also added auctions to its list of products and services. Unlike eBay, which was profitable from the start, Amazon took seven years to earn its first small profits. Some industry observers note that Amazon might earn more by charging a commission on the auction of a used book than it could earn by selling the same title as a new book. In the auction of a used book, Amazon does not incur the costs of buying, handling, and shipping inventory; it simply collects a commission on the sale.

One of the aggressive marketing tactics that Amazon used to promote its auction business was its "Auctions Guarantee." This guarantee directly addressed concerns raised in the media by eBay customers about being cheated by sellers. When Amazon opened its Auctions site, it agreed to reimburse any buyer for merchandise purchased in an auction that was not delivered or that was "materially different" from the seller's representations. Amazon limited its guarantee to items costing $250 or less; however, buyers of more expensive items can protect themselves by using a third-party **escrow service**, which holds the buyer's payment until he or she receives and is satisfied with the purchased item. You will learn more about escrow services later in this chapter.

Online Auctions, Virtual Communities, and Web Portals

In response to Amazon's guarantee, eBay immediately offered its customers a similar guarantee, but not before Amazon gained free advertising from the media coverage of its guarantee. In 2003, eBay increased its guarantee to $500 in the hopes that it would induce new customers to buy at eBay auctions. The experiment worked well; in fact, eBay increased its guarantee again in 2004 to $1000. Some eBay users have complained that the company does not act quickly on claims under the guarantee and does its best to avoid paying claims; however, the guarantee remains a powerful marketing tool.

Amazon also used other strategies to compete with eBay. For example, Amazon established an online joint venture with Sotheby's, the famous British auction house, to hold online auctions of fine art, antiques, jewelry, and other high-value collectibles. In general, it is difficult to sell these types of items on the Web because of the importance of direct, in-person inspection. Such inspections help establish the item's authenticity and condition. Sotheby's and its international network of dealers obtain the items for their online auctions and guarantee the authenticity and condition of items, just as at a Sotheby's in-person auction. Again, Amazon is addressing a serious concern of some of eBay's most prized customers, those who participate in auctions of high-value items. The Sotheby's joint venture suggests that Amazon is trying hard to differentiate its auction site from eBay's as a more attractive home for the upper end of the auction market.

At the lower end of the market, Amazon integrated its zShops platform with its auction operation. This gives small sellers the same kind of combined selling space that eBay does with its combination of auction listings with eBay Store listings. Amazon further increases the value of this marketplace by including zShops and Amazon Auction listings on search results pages for customers who are shopping in any part of the Amazon site. For example, a site visitor searching for a DVD at Amazon will see Amazon Auction listings and offers to sell new or used copies of that DVD from zShops sellers in the page that provides the link to purchase a new copy of the DVD from Amazon. Many industry analysts agree that Amazon's zShops has taken a significant amount of business away from eBay.

Despite the innovations and large customer bases of Yahoo! and Amazon, the premier general consumer U.S. Web auction site today is still eBay. Any competitors, even large and well-financed companies, must overcome the strong advantage built by eBay. Any challenger to eBay will find that the economic structure of markets is biased against new entrants. Because markets become more efficient (yielding fairer prices to both buyers and sellers) as the number of buyers and sellers increases, new auction participants are inclined to patronize established marketplaces. Thus, existing auction sites, such as eBay, are inherently more valuable to customers than new auction sites. This basic economic fact, which economists call a **lock-in effect**, will make the task of creating other successful general consumer Web auction sites even more difficult in the future.

A somewhat ironic example of the lock-in effect exists in the Japanese general consumer auction market. In this market, unlike in the United States, Yahoo! was the first major company to offer online auctions. At the time (early 1999), Yahoo! did not charge fees to sellers. When eBay entered the Japanese market five months later, it charged fees and found few people interested in its services. Although Yahoo! has since instituted fees for its auctions, the lock-in effect has preserved its strong lead in Japan. Yahoo! Auctions holds more than 90 percent of the $3 billion market in Japan, while eBay's market share is less than 3 percent.

Specialty Consumer Auctions

Rather than struggle to compete with a well-established rival such as eBay in the general consumer auction market, a number of firms have decided to identify special-interest market targets and create specialized Web auction sites that meet the needs of those market segments. Several early Web auction sites started by featuring technology items such as computers, computer parts, photographic equipment, and consumer electronics.

Doug Salot was buying and selling computer equipment on the Internet's Usenet newsgroups before the Web existed. He saw the potential for the Web's graphical user interface in creating auctions, and, in September 1996, started an auction site, Haggle Online, for computer equipment. Haggle was bought and sold several times between 1999 and 2002. Today, the Haggle Online auction business is operated under the brand name **uBid**. Unlike most online auction sites, uBid sells its own inventory of closeouts, refurbished computers, and computer-related items.

Although computers and technology were obvious early market segments that would find online auctions appealing in the first wave of electronic commerce, a number of other specialized Web auction sites emerged as the Web matured. Although their operations are much smaller than those of general-consumer auction sites, some companies that operate specialty consumer auctions have succeeded in building loyal followings. **PotteryAuction.com** and **JustBeads.com** are two examples of auction sites that cater to buyers and sellers who are geographically dispersed but share highly focused interests. **StubHub** operates an auction site for event tickets. The site includes tickets offered for sale by ticket brokers and also by individuals for fixed prices. The StubHub home page appears in Figure 6-5.

Other specialty consumer auction sites include **Cigarbid.com**, **Golf Club Exchange**, and **Winebid**. These sites gain an advantage by identifying a strong market segment with readily identifiable products that are desired by people with relatively high levels of disposable income. Golf clubs, cigars, wine, and technology products all meet these requirements. As other Web auction site developers identify similar market segments, these specialized consumer auctions might become profitable niches that can successfully coexist with large general consumer sites, such as eBay.

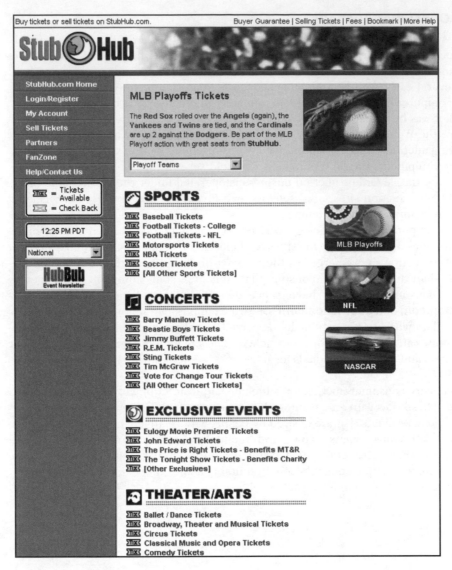

FIGURE 6-5 StubHub home page

Consumer Reverse Auctions and Group Purchasing Sites

Sites such as **Respond.com** offer reverse auctions directed at a consumer market. For example, at the Respond.com site, a site visitor fills out a form that describes the item or service in which he or she is interested. The site then routes the visitor's request to a group of participating merchants who reply to the visitor by e-mail with offers to supply the item at a particular price. This type of offer is often called a **reverse bid**. The buyer can then accept the lowest offer or the offer that best matches the buyer's criteria.

Many people think of **Priceline.com** as a seller-bid auction site. Priceline.com allows site visitors to state a price they are willing to pay for airline tickets, car rentals, hotel rooms, and a few other services. If the price is sufficiently high, the transaction is completed. However, Priceline.com completes many of its transactions from an inventory that it has purchased from airlines, car rental agencies, and hotels. To the extent that Priceline sells out of its inventory, it operates more as a liquidation broker (you will learn more about liquidation brokers in the next section) than as a true reverse auction site. The Priceline.com home page appears in Figure 6-6.

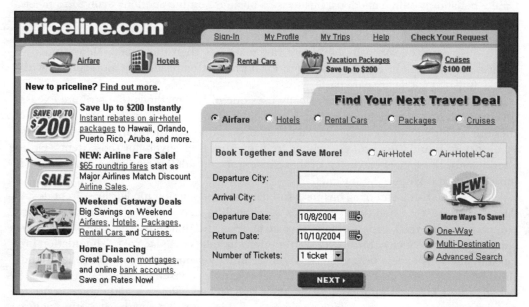

FIGURE 6-6   Priceline.com home page

Another new type of business made possible by the Internet is the group purchasing site, which is similar to a consumer reverse auction. On a **group purchasing site**, the seller posts an item with a price. As individual buyers enter bids on an item (these bids are agreements to buy one unit of that item, but no price is specified), the site can negotiate a better price with the item's provider. The posted price ultimately decreases as the number of bids increases, but only if the number of bids increases. Thus, a group purchasing site builds up a number of buyers who will force the seller to reduce its price. The effect is very much like the one achieved by a consumer reverse auction.

The types of products that are ideal for group purchasing sites are branded products with well-established reputations. This allows buyers to feel confident that they are getting a good bargain and are not trading off price for reduced quality. The products should also have a high value-to-size ratio and should not be perishable.

Mercata was the first major group purchasing site, but it closed its doors in January 2001. Some sites still offer group purchasing, such as the European site **LetsBuyIt.com**, but consumer group purchasing sites have had difficulty attracting sellers' interest. Few companies selling products that are well suited to group purchasing efforts—such as computers, consumer electronics, and small appliances—have been willing to work with the group purchasing sites. These sellers have not found any compelling advantage in offering reduced prices on their merchandise to the group purchasing sites. Most of these sellers believe that these sites cannibalize product sales in their existing sales channels and are reluctant to offend the current distributors of their products by selling through group purchasing sites.

Business-to-Business Auctions

Unlike consumer online auctions, business-to-business online auctions evolved to meet a specific existing need. Many manufacturing companies periodically need to dispose of unusable or excess inventory. Despite the best efforts of procurement and production management, businesses occasionally buy more raw materials than they need. Many times, unforeseen changes in customer demand for a product can saddle manufacturers with excess finished goods or spare parts.

Depending on its size, a firm typically uses one of two methods to distribute excess inventory. Large companies sometimes have liquidation specialists who find buyers for these unusable inventory items. Smaller businesses often sell their unusable and excess inventory to **liquidation brokers**, which are firms that find buyers for these items. Online auctions are the logical extension of these inventory liquidation activities to a new and more efficient channel, the Internet.

Two of the three emerging business-to-business Web auction models are direct descendants of these two traditional methods for handling excess inventory. In the large-company model, the business creates its own auction site that sells excess inventory. In the small-company model, a third-party Web auction site takes the place of the liquidation broker and auctions excess inventory listed on the site by a number of smaller sellers. The third business-to-business Web auction model resembles consumer online auctions. In this model, a new business entity enters a market that lacked efficiency and creates a site at which buyers and sellers who have not historically done business with each other can participate in auctions. An alternative implementation of this model occurs when a Web auction replaces an existing sales channel.

One of the earliest examples of the large-company model is Ingram Micro's Auction Block site, which Ingram Micro started in 1997. **Ingram Micro** is a major distributor of computers and related equipment to value-added resellers (VARs), which are companies that configure computer hardware and software, such as network servers, for business users. Because computer technology changes rapidly, Ingram Micro often finds itself with outdated disk drives, computer chips, and other items that it formerly turned over to liquidation brokers.

Ingram Micro now auctions those items to its established customers through its internally operated Auction Block site. Auction volume is more than $6 million per year and the VARs that are Ingram Micro's main customers now have the option of putting the Auction Block program on their own sites, which allows their customers to participate in the bidding. The software used by Ingram Micro and its customers was developed by **Moai Technologies**, which now sells the software to other companies that want to follow Ingram Micro's strategy.

Ingram Micro estimates that the auction prices it receives on the site average about 60 percent of the items' costs. This percentage compares favorably to the average of 10 percent to 25 percent of cost that Ingram Micro was obtaining from liquidation brokers. In effect, large companies such as Ingram Micro are removing the liquidation brokers from the value chain and claiming the brokers' intermediary profits. Recall that this process is called disintermediation.

Another large computer technology company that decided to build its own auction site to dispose of obsolete inventory is CompUSA. Although CompUSA sells to individuals, it makes a significant portion of its sales to corporate customers. Instead of selling through liquidation brokers, CompUSA decided to let midsized and smaller businesses bid directly on its technology inventory. Its Web auction site, **CompUSA Auctions**, appears in Figure 6-7.

FIGURE 6-7   CompUSA auctions home page

In the second business-to-business auction model, smaller firms sell their obsolete inventory through an independent third-party auction site. In some cases, these online auctions are conducted by the same liquidation brokers that have always handled the disposition of obsolete inventory. These brokers adapted to the changed environment and implemented electronic commerce to stay in business. One example is the **Dove Bid** site established by the Ross-Dove Company, a traditional liquidation broker for many years.

Gordon Brothers Group, another liquidation broker, has been selling the inventory of failed retailers since 1903. The company has used its expertise to launch or help others launch Web sites that liquidate retailer inventories, including **GB RetailExchange** and

SmartBargains.com. As many dot-com companies began to fail, the savvy liquidation company identified yet another business opportunity. Gordon Brothers created a separate subsidiary that sells entire Web sites, software, hardware, and even the intellectual property left in the wake of failed Web ventures.

Other third-party auction sites have been started by newcomers or companies that want to liquidate their inventory and are willing to do the same for other companies in their industry. In some industries, new auction markets on the Web are replacing older ways of doing business. For example, telecommunications companies can buy or sell time on their networks to each other through the **Band-X** Web auction site. Sellers list the number of minutes they have available, and the price of airtime minutes fluctuates in response to buyers' bids on those minutes.

Established securities trading organizations such as the New York Stock Exchange (NYSE) and the Chicago Board of Trade (CBOT) are facing an electronic challenge to their time-honored ways of doing business. In 1998, a new venture called the **International Securities Exchange (ISE)** was funded by electronic brokers E*TRADE and Ameritrade Holdings, with contributions from several other brokerage firms. This new securities exchange was the first to be registered in the United States since 1973. In May 2000, the ISE began its operations with trading in 82 of the most actively traded stock options contracts. By 2003, the ISE had completed more than 300 million trades and had become the largest equity options trading company in the United States. In 2000, the **Pacific Exchange,** a traditional stock exchange that has been in business since 1862, joined with Archipelago Holdings to develop an electronic exchange, **ArcaEx**, which replaced the Pacific Exchange's physical trading floor in March, 2002. ArcaEx trades securities listed on the NYSE, the American Stock Exchange, the Pacific Exchange, and NASDAQ Stock Market.

Electronic securities exchanges pose a threat to all existing physical securities exchanges because their lower fees might attract the most lucrative large trades of active issues from existing exchanges. Industry analysts question whether traditional exchanges such as the NYSE and the CBOT can continue to exist once electronic exchanges become better established.

Another online auction innovation is the new approach to bidding pioneered by FreeMarkets, now a part of **Ariba**. Instead of using a public online auction site, the FreeMarkets approach provides software and hardware tools to coordinate private online auctions that allow businesses to solicit bids from suppliers. Instead of sending out request for proposal packages to many suppliers, a business can list its request for proposals with Ariba. Companies that have used this approach report savings of 10 percent to 20 percent in their procurement costs. In effect, Ariba has moved the traditional first-price sealed-bid auction form onto the Internet.

A growing number of hospitals and other organizations are using online auctions to fill temporary employment openings. Health care workers, such as nurses, perform similar duties in specific health care settings in most hospitals. For example, the duties performed by an intensive care unit nurse are almost identical across hospitals. State regulations on nurse licensing require that nurses have similar levels of knowledge, skills, and abilities. Having similar job functions in workplaces and having similarly qualified persons working in those jobs allows both nurses and employers to treat the nursing function as a commodity. Therefore, nurses can easily work for a variety of employers and do not require long periods of training or learning procedures specific to a particular hospital. In the past, nurse agencies would coordinate placement, matching nurses who wanted to work particular days or shifts with hospitals and other health care organizations who had shifts to fill. The agency would earn a commission on each placement. Today, companies such as **BidShift** sell software to employers that lets them operate their own shift auctions. Nurses bid on the shifts they would prefer to work and the software manages the auctions. In an efficient matching of supply and demand, employers meet their staffing needs efficiently, nurses get to work when they want, and the agency fee is avoided.

Business-to-Business Reverse Auctions

In Chapter 5, you learned how businesses are creating various types of electronic marketplaces to conduct B2B transactions. Many of these marketplaces include auctions and reverse auctions. In 2001, glass and building materials producer Owens Corning held more than 200 reverse auctions for a variety of items including chemicals (direct materials), conveyors (fixed assets), and pipe fittings (MRO). Owens Corning even held a reverse auction to buy bottled water. Asking its suppliers to bid has reduced the cost of those items by an average of 10 percent. Because Owens Corning buys about $3.4 billion of materials, fixed assets, and MRO items each year, the potential for future cost savings is significant. Both the U.S. Navy and the federal government's General Services Administration are experimenting with reverse auctions to acquire a small part of the billions of dollars worth of materials and supplies they purchase each year. Companies that use reverse auctions include Agilent, BankOne, Bechtel, Boeing, Raytheon, and Sony.

Not all companies are enthusiastic about reverse auctions. Some purchasing executives argue that reverse auctions cause suppliers to compete on price alone, which can lead suppliers to cut corners on quality or miss scheduled delivery dates. Others argue that reverse auctions can be useful for nonstrategic commodity items with established quality standards. However, as R. Gene Richter (a supply management pioneer at IBM) noted in a 2001 interview published in *Purchasing*, "Everything is strategic to somebody. Talk about ballpoint pens. A secretary has spots all over her brand new blouse because the pen you bought for a cent and a half is leaking." Companies that have considered reverse auctions and decided not to use them include Cisco, Cubic, IBM, and Solar Turbines.

With compelling arguments on both sides, the extent to which reverse auctions will be used in the B2B sector is not yet clear; however, some guidelines for deciding whether to use reverse auctions are beginning to emerge. In some industry supply chains, the need for trust and long-term strategic relationships with suppliers makes reverse auctions less attractive. In fact, the trend in purchasing management over the last 20 years has been to increase trust-based relationships that endure for many years. Using reverse auctions replaces trusting relationships with a bidding activity that pits suppliers against each other and is seen by many purchasing managers as a step backward.

In some industries, suppliers are larger and more powerful than the buyers. In those industries, suppliers simply do not agree to participate in reverse auctions. If enough important suppliers refuse to participate, it is impossible to conduct reverse auctions. In industries where there is a high degree of competition among suppliers, however, reverse auctions can be an efficient way to conduct and manage the price bidding that would naturally occur in that market. Figure 6-8 lists the supply chain characteristics that support or discourage reverse auctions identified in ongoing research being conducted by Dima Ghawi and the author.

Supply Chain Characteristics that Support Reverse Auctions:

- Suppliers are highly competitive.
- Product features can be clearly specified.
- Suppliers are willing to reduce the margin they earn on this product.
- Suppliers are willing to participate in reverse auctions.

Supply Chain Characteristics that Discourage Reverse Auctions:

- Product is highly complex or requires regular changes in design.
- Product has customized features.
- Long-term strategic relationships are important to buyers and suppliers.
- Switching costs are high.

FIGURE 6-8   Supply chain characteristics and reverse auctions

Auction-Related Services

The growth of eBay and other auction sites has encouraged entrepreneurs to create businesses that provide auction-related services of various kinds. These include escrow services, auction directory and information services, auction software (for both sellers and buyers), and auction consignment services. This section describes each of these new industries that have arisen to meet the needs of auction participants. You will learn about yet another auction-related business, payment-processing services, in Chapter 11.

Auction Escrow Services

A common concern among people bidding in online auctions is the reliability of the sellers. Surveys indicate that as many as 15 percent of all Web auction buyers either do not receive the items they purchased, or find the items to be different from the seller's representation in some significant way. About half of those buyers are unable to resolve their disputes to their satisfaction. When purchasing high-value items, buyers can use an escrow service to protect their interests.

You learned earlier in this chapter that an escrow service is an independent party that holds a buyer's payment until the buyer receives the purchased item and is satisfied that the item is what the seller represented it to be. Some escrow services take delivery of the item from the seller and perform the inspection for the buyer. In such situations, buyers give the escrow service authority to examine. Usually, escrow agents that perform this service are art appraisers, antique appraisers, and the like who are qualified to judge quality, usually with better judgment than the buyer. Escrow services do, however, charge fees ranging from 1 percent to 10 percent of the item's cost, subject to a minimum fee, typically between $5 and $50. The minimum fee provision can make escrow services too expensive for small purchases. Escrow services that handle Web auction transactions include **Escrow.com** and **SafeBuyer.com**. Some of these escrow firms also sell auction buyer's insurance, which can protect buyers from nondelivery and some quality risks.

Wary bidders in low-price auctions (for which the minimum escrow charges would be excessive) do have some other ways to protect themselves. One way is to consult the seller's record on the auction site to see how the seller is rated. Also, some Web sites offer lists of auction sellers who have failed to deliver merchandise or who have otherwise cheated bidders in the past. These sites are operated as free services (often by bidders who have been cheated), so they sometimes contain unreliable information and they open and close periodically, but you can use your favorite search engine to locate sites that currently carry such lists.

Auction Directory and Information Services

Another service offered by some firms on the Web is a directory of auctions. Sites such as **Auctionguide.com** offer guidance for new auction participants and helpful hints and tips for more experienced buyers and sellers along with directories of online auction sites. **AuctionBytes** is an auction information site that publishes an e-mail newsletter with articles about developments in the online auction industry. **BidXS** is a search engine site that specializes in searches of auction sites. It allows users to monitor bidding action on specific items or categories of items as it occurs on many different online auction sites.

The **StrongNumbers** site offers information about fair market value for a wide variety of products and collectibles. This information can be useful to sellers who are trying to set a reserve price or to buyers deciding whether to bid or how much to bid. **Price Watch** is an advertiser-supported site on which those advertisers post their current selling prices for computer hardware, software, and consumer electronics items. Although this monitoring is a retail pricing service designed to help shoppers find the best price on new items, Web auction participants find it can help them with their bidding strategies. **PriceSCAN** is a similar price-monitoring service that also includes prices on books, movies, music, and sporting goods, in addition to the types of items monitored by Price Watch.

Auction Software

Both auction buyers and sellers purchase software to help them manage their online auctions. Sellers often run many auctions at the same time. Companies such as **Andale**, **AuctionHawk**, and **Vendio** sell auction management software and services for both buyers and sellers. For sellers, these companies offer software and services that can help with or automate tasks such as image hosting, advertising, page design, bulk repeatable listings, feedback tracking and management, report tracking, and e-mail management. Using these tools, sellers can create attractive layouts for their pages and manage hundreds of auctions. For example, one of Andale's offerings is a combination of software and services that manages about 100 auctions for a monthly fee of under $20. Figure 6-9 includes a portion of the Andale products page that lists some of the seller management software and services that the company offers.

For buyers, a number of companies sell auction sniping software. **Sniping software** observes auction progress until the last second or two of the auction clock. Just as the auction is about to expire, the sniping software places a bid high enough to win the auction (unless that bid exceeds a limit set by the sniping software's owner). The act of placing a winning bid at the last second is called a **snipe**. Because sniping software synchronizes its internal clock to the auction site clock and executes its bid with a computer's precision, the

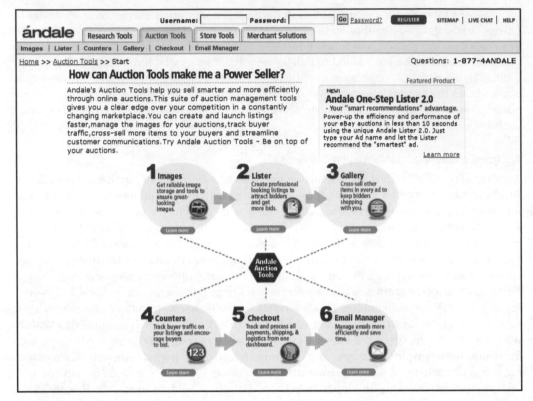

FIGURE 6-9   Andale auction tools products

software almost always wins out over a human bidder. The first sniping software, named Cricket Jr., was written by David Eccles in 1997. He sells the software on his **Cricket Sniping Software** site. A number of other sniping software sellers have entered the market—each claiming that its software will outbid other sniping software. Some sites offer sniping services; that is, the sniping software runs on their Web site and customers enter their sniping instructions on that site. Some of these companies offer subscriptions; others use a mixed revenue model in which they offer some free snipes supported by advertising, but require payment for additional snipes. A good source for current information about the sniping software and services business is the **AuctionBytes** Web site. The home page of AuctionBytes is shown in Figure 6-10.

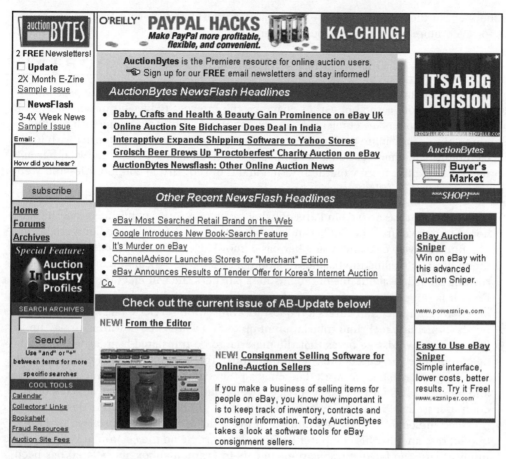

FIGURE 6-10   AuctionBytes home page

Auction Consignment Services

Several entrepreneurs have identified yet another auction-related business that meets the needs of people and small businesses who want to use an online auction, but do not have the skills or the time to become a seller. These companies, called **auction consignment services**, take an item and create an online auction for that item, handle the transaction, and remit the balance of the proceeds after deducting a fee that ranges from 10 percent to 40 percent of the selling price obtained. Items that do not sell are returned or donated to charity.

The main auction consignment businesses include **AuctionDrop**, **QuickDrop**, and **Picture it SOLD**. Because one key to success in this business is having convenient locations at which customers can drop off their items, all of these companies are planning to open their own stores and franchise stores as rapidly as possible.

All four of these auction-related businesses are excellent examples of the second wave of electronic commerce. In the first wave, the online auction business was made possible by the Web. In the second wave, the online auction business has itself created opportunities for even more entirely new types of business.

VIRTUAL COMMUNITIES AND WEB PORTALS

Online auctions and related activities are not the only new businesses made possible by the Internet. As you learned in earlier chapters, the Internet reduces transaction costs in value chains and offers an efficient means of communication to anyone with an Internet connection. Combining the Internet's transaction cost reduction potential with its role as a facilitator of communication among people, companies have developed two other new approaches to making money on the Internet and the Web: virtual communities and Web portals. Consider the following scenario:

> Fran Dennison has arrived in Paris one day early for a series of business meetings. She hopes to recover from her jet lag and enjoy a little French food before her work begins. She finds a lovely café and, using her basic knowledge of French, successfully orders lunch. Fran is reading the business section of *Le Monde*, a local newspaper. She begins reading an article about one of the business partners she will meet tomorrow, but her French is not good enough to completely understand the article. Fran opens her notebook computer and enters a request for translation services. She specifies that she needs immediate real-time translation of up to 500 words and is willing to pay up to 20 cents per word. She notes that the material to be translated is an article in today's *Le Monde*; she also enters the title of the article. Her computer, which contains a cellular link to her office network, launches an immediate search of online communities and marketplaces for this exact service. Two minutes later, a message appears on her computer from a French graduate student in the United States, Philippe Desmarest. His message indicates that he is willing to provide an immediate translation at Fran's quoted rate and that his computer has found the article on the *Le Monde* Web site. Five minutes later, an English translation appears in Fran's mailbox and $94.20 has been moved from her checking account to Philippe's. Fran has time to read the article and think about how she will adjust her presentation at tomorrow's meeting before her salad arrives.

This scenario is very close to becoming possible today. Three key elements are required to make things such as Fran's on-demand translation a reality: cellular-satellite (mobile) communications technology, electronic marketplaces, and software agents. All three of these elements exist today, but they have yet to be completely integrated. You will learn about each of these elements in the following sections.

Mobile Communications Technology

Cellular-satellite communications technology capable of linking Fran to the Internet can be packaged with notebook computers, personal digital assistants (PDAs), and mobile phones. A PDA displaying a Web page appears in Figure 6-11.

The PDA shown in the figure displays a Web page sent using the **Wireless Application Protocol (WAP)**. WAP allows Web pages formatted in HTML to be displayed on devices with small screens, such as PDAs and mobile phones. As mobile technology improves, more and

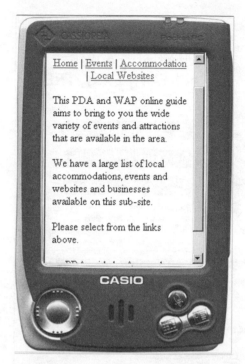

FIGURE 6-11   Web page displayed on a PDA

more devices will become Web-enabled and will include other features that will open doors to a second wave of electronic commerce. For example, **Garmin** has introduced a PDA that includes a Global Positioning Service (GPS) receiver. The user enters a destination address and the PDA displays a map. As the user drives toward the destination, the GPS receiver tracks the PDA's current location (using signals from GPS satellites) and causes the PDA to announce driving directions, such as "turn right 300 feet ahead." The PDA makes the announcements based on information contained in the GPS signals it receives and a map of the area that is stored in its memory.

Electronic Marketplaces

In Chapter 5, you learned that electronic marketplaces have grown in the B2B sector. As wireless and satellite data transmission technologies become integrated with marketplaces, these marketplaces can serve people who want to buy and sell a wide range of products and services. Firms such as **AvantGo** already provide PDAs with downloads of Web site contents, news, restaurant reviews, and maps. Users can create accounts with AvantGo that permit AvantGo to send these downloads to their wireless PDAs, telephones, or other mobile devices automatically. AvantGo has more than 10 million subscribers worldwide. The company earns revenue by selling ads that appear with the downloaded content. Avant-Go's home page appears in Figure 6-12.

FIGURE 6-12 AvantGo home page

Intelligent Software Agents

Some companies provide Web sites that help users find products and services for sale on the Web. These sites use **intelligent software agents** (also called **software robots**, or **bots**), which are programs that search the Web and find items for sale that meet a buyer's specifications.

Some software agents are focused on a particular category of product, such as **Best Book Buys**, which searches more than 20 online bookstores for the best prices on books. In addition to obtaining price information, researchers are developing other software agents that track ratings of buyer and seller reputations. In much the same way that eBay makes reputation reports available to its bidders and sellers about each other, more general software agents can create and search databases of all kinds of buy-sell transactions on the Web. The **MIT Media Lab Software Agents Group** and the **Carnegie Mellon Intelligent Software Agents Lab** have been leaders in the development of intelligent software agents. The **BotSpot** Web site is a good source of information about software agents and includes links to downloadable bot programs. Simon is one of the best shopping agents currently available. In addition to finding product item matches, software agents such as Simon can find the lowest price for an item. You can find Simon at the **mySimon** Web site.

Virtual Communities

A **virtual community**, also called a **Web community** or an **online community**, is a gathering place for people and businesses that does not have a physical existence. Howard Rheingold described the characteristics of these communities in his 1993 book, *The Virtual Community*, which has become recognized as the definitive book on the subject. Virtual communities exist on the Internet today in various forms, including Usenet newsgroups, chat rooms, and Web sites. These communities offer people a way to connect with each other and discuss common issues and interests. The social interaction in these communities can be considerable and many sociologists believe that the communication and relationship-forming activities that occur online are similar to those that occur in physical communities.

One form of virtual community with which you might be familiar is the **virtual learning community**. Many colleges and universities now offer courses that use distance learning platforms such as **Blackboard** or **WebCT** for student-instructor interaction. These distance learning platforms include tools such as bulletin boards, chat rooms, and drawing boards that allow students to interact with their instructors and each other in ways that are similar to the interactions that might occur in a physical classroom setting. Some open-source software projects are devoted to the development of virtual learning communities, including **Moodle** and JA-SIG **uPortal**.

In addition to fulfilling the social interaction needs of individuals, virtual communities can help companies, their customers, and their suppliers plan, collaborate, transact business, and interact in ways that benefit all of them.

Another approach to electronic commerce using virtual communities is the **Google Answers** site. Google Answers gives people a place to ask questions that are then answered by an expert (called a Google Answers Researcher) for a fee. Google administers a test to determine which members of the community qualify to become Google Answers Researchers. The questioner sets the fee (there is a minimum fee of $2.50) and determines whether an answer is sufficient before authorizing the payment of the fee. Most questions posted to date have been answered for fees between $10 and $200. Members of the community who are not Google Answers Researchers are also permitted to answer questions, but they do not collect a fee. Many of the community members who are active answer providers have gone on to take the test and become Google Answers Researchers. When a question is answered, the question and answer appear on the Google Answers site. The Google Answers home page appears in Figure 6-13.

Early Web Communities

One of the first Web communities was the **WELL**. The WELL, which is an acronym for "whole earth 'lectronic link," predates the Web. It began as a series of dialogs among the authors and readers of the *Whole Earth Review* in 1985. Most WELL members were originally from the San Francisco Bay area, and the influence of that area's counterculture heritage is a significant part of the WELL's ambiance. Members of the WELL pay a monthly fee to participate in its forums and conferences. The WELL has been home to many important researchers and participants in the growth of the Internet and the Web. Its membership also includes noted writers and artists. In 1999, the e-zine publisher *Salon.com* bought the WELL and has maintained the sense of community that had existed there for 14

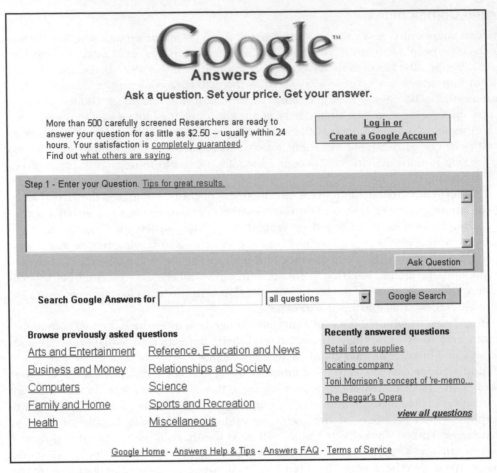

FIGURE 6-13 Google Answers page

years. Access to the WELL community now requires a $10 per month subscription. A premium subscription that includes an "@well.com" e-mail address and the ability to start your own conferences within the WELL costs $15 per month.

As the Web emerged in the mid-1990s, its potential for creating new virtual communities was quickly exploited. In 1995, Beverly Hills Internet opened a virtual community site that featured two Webcams aimed down Hollywood streets and links to entertainment information Web sites. The theme of this community was the formation of digital cities around the focus of the Webcams. The founders of Beverly Hills Internet wanted to create a sense of community and thought that the Webcams would help accomplish that goal. Their hope was that people would be attracted by the Webcam images and want to add their own contributions, thus becoming members of a virtual neighborhood. Members were given free space on the Web site to create pages within these virtual cities on which to add their contributions. As it turned out, the Webcams never did attract much traffic, but the offer of free Web space did. The first of these digital cities were created around Webcams in the Los Angeles area and therefore were named for Los Angeles-area communities.

As the site grew to include more geographic areas, it changed its name to GeoCities. GeoCities earned revenue by selling advertising on members' Web pages and pop-up pages that appeared whenever a visitor accessed a member's site. GeoCities grew rapidly and was purchased in 1999 by Yahoo! for $5 billion.

Other similar sites became virtual communities. Tripod was founded in 1995 in Massachusetts and offered its participants free Web page space, chat rooms, news and weather updates, and health information pages. Like GeoCities, Tripod sold advertising on its main pages and on participants' Web pages. The search engine site **Lycos** purchased Tripod in 1998 for $58 million.

Theglobe.com, also started in 1995, was the outgrowth of a class project at Cornell University. The students who created the site included bulletin boards, chat rooms, discussion areas, and personal ads. They then sold advertising to support the site's operation. Later additions included news feeds, an online art gallery, and shopping pages. Although Theglobe.com offered free Web page space, it did not emphasize that feature to the same extent as competing virtual communities. Theglobe.com turned down several offers to purchase its community during its lifetime. The company experienced declines in advertising revenues during the economic slowdown of 2000 and finally closed in 2001.

Web Community Consolidation

Virtual communities for consumers can succeed as money-making propositions if they offer something sufficiently valuable to justify a charge for membership. For example, people joining the WELL community obtain access to a very interesting set of existing members who frequent the WELL's discussion areas. These areas are open only to members. Thus, WELL owner *Salon.com* can charge a subscription fee for access to the WELL community. As you learned in the previous section, most virtual communities have been unable to support themselves and have either closed or been sold to companies such as Yahoo! or sites that have other revenue-generating activities that they can provide to the purchased community.

Web Communities in the Second Wave of Electronic Commerce

In the early days of the Internet, virtual communities were an essential part of online experience for the small number of people who regularly used the medium. As the Internet and Web grew, some of these communities grew, but others found that their purpose as a place for sharing the new experiences of online communication began to fade. In the second wave of electronic commerce, a new phenomenon in online communication began. People who were now using the Internet no longer found a common bond in the fact that they were using the Internet. Multiple common bonds joined people with all types of common interests. The Internet was no longer the focus of the community, but was simply a tool that enabled the communication among members of the community.

In addition to Web sites and chat communities built on common interests, the second wave saw the introduction of new virtual communities that existed for the sole purpose of community. One of the first of these was **craigslist**, an information resource for San Francisco area residents that was created in 1995 by WELL member Craig Newmark. That community has grown to include information for most major cities in the United

States and in several other countries. The site is operated by a not-for-profit foundation, and all postings other than help-wanted ads are free.

The craigslist Web site was an early pioneer, but significant growth in profit-focused virtual community sites emerged during the second wave of electronic commerce. For example, the virtual community site **Friendster** was founded by Jonathan Abrams in 2002. Other sites followed, including **LinkedIn** and **Tribe.net**. These virtual communities have become useful tools for persons who want to make new local friends, establish acquaintances before moving to a new location, obtain advice of various kinds, or who are looking for a job.

The idea behind these sites is that people are invited to join by existing members who think they would be valuable additions to the community. The site provides a directory that lists members' locations, interests, and qualities. The directory does not disclose the name or contact information of members, however. A member can offer to communicate with any other member, but the communication does not occur until the intended recipient approves the contact (usually after reviewing the sender's directory information).

In addition to searching the directory of the community, new members can work through friends they have established in the community (perhaps starting with the person who invited them to join). By gradually building up a set of connections, members can develop contacts within the community that can prove valuable later. Although all of these virtual community sites are still fairly new, they show great promise for recreating (on a much larger scale) the essence of the original Internet communities. Strategies that build on a combination of virtual communities and other activities are called Web portal revenue models, discussed in the next section.

Web Portal Revenue Models

By the late 1990s, virtual communities were selling advertising to generate revenue. Search engine sites and Web directories were also selling advertising to generate revenue. Beginning in 1998, a wave of purchases and mergers occurred among these sites. The new sites that emerged still used an advertising-only revenue generation model and included all the features offered by virtual community sites, search engine sites, Web directories, and other information-providing and entertainment sites. These portals, which you first learned about in Chapter 3, are so named because their goal is to be every Web surfer's doorway to the Web.

Advertising-Supported Web Portals

Some Web observers believe that Web portal sites could be the great revenue-generating businesses of the future. They argue that adding portal features to existing sites or converting sites to portals can be a wise business strategy. They believe that combining Web communities' sense of belonging with search engine and Web directory tools will yield Web sites with high degrees of stickiness that will be extremely attractive to advertisers.

One rough measure of stickiness is how long each user spends at the site. Figure 6-14 lists the most popular sites on the Web based on the number of users who accessed the sites during the month of August 2004. The information in Figure 6-14 is adapted from

Nielsen//NetRatings reports and shows sites grouped by owner. For example, the numbers for Microsoft include activity on all sites operated by Microsoft, including people with Hotmail accounts checking their e-mail and MSN subscribers using that Web portal's services. The numbers shown include people who accessed the sites from their computers at home and at work. (*Note*: People who have broadband access at work and not at home often use their at-work computers for personal business during nonwork hours.)

Owner	Minutes per unique visitor	Millions of unique visitors
Time Warner	270	113
Yahoo!	147	117
Microsoft	101	136
eBay	98	59
Google	25	80
U.S. Government	22	54
InterActiveCorp	18	40
Amazon	17	40

Adapted from reports for August, 2004 published by
Nielsen//NetRatings at http://www.nielsennetratings.com/

FIGURE 6-14   Stickiness of popular Web sites

Nielsen//NetRatings determines site popularity by measuring the number of unique visitors to a site. The leading sites often have more than 100 million unique visitors per month. In the figure, the site owners are not ranked by popularity, but by the average number of minutes that users spent on the sites.

Because Web portals ask their members to provide demographic information about themselves, the potential for targeted marketing is very high. Industry observers are now predicting that Web portals could be one of the success stories of the second wave of electronic commerce. The top three Web site owners listed in Figure 6-14 draw a significant number of their visitors at their Web portal sites. High visitor counts can yield high advertising rates for these sites. In the boom years of the first wave, Web portals were able to obtain up-front cash payments from advertisers, which is very unusual for any kind of advertising sale. For example, the Excite search engine site paid Netscape (now a part of Time Warner) a $70 million advance fee for two-year rights to a prominent advertising location on its Netcenter Web portal site. Other portal sites have negotiated advertising deals that included a percentage of sales generated from sales leads on the portal site.

The companies that run Web portals have added sticky features such as chat rooms, e-mail, and calendar functions—often by purchasing the companies that create those features. In addition to buying the virtual community site Tripod, Lycos purchased the online directory WhoWhere? for $133 million. In 1999 alone, Yahoo! spent over $10 billion in cash and stock to expand the range of services available on its Web portal site.

This spending spree ended with the decline in online investment that occurred in 2000–2002 at the end of the first wave of electronic commerce. This downturn created serious doubts about whether even the largest and most well-known Web portals, such as Yahoo! and Excite, could survive. Many smaller portals closed. The future of the advertising-supported Web portal is uncertain at this time. Second wave portal strategies are based less on up-front site sponsorship payments and more on the generation of revenues from continuing relationships with people who use their portal sites. The larger portals that have survived are turning to mixed models that offer more stability in their revenue streams.

Mixed-Revenue Web Portals

One of the most successful Web portals is Time Warner's AOL unit, which has always charged a fee to its users and has always run advertising on its site. Many Web portals that have struggled with their advertising-supported revenue models have been moving toward AOL's strategy.

Yahoo! now charges for the Internet phone service that it originally offered at no cost. It still offers free e-mail accounts, but now sells other features, such as more space to store messages and attached files, to members who pay for the "premium" e-mail service. After years of rapid growth, Yahoo! became unprofitable when Internet advertising dropped suddenly. Terry Semel, a media executive with years of experience working for media giants Warner Brothers, CBS, and Disney, was brought in to run Yahoo! in 2001. Semel cut costs and reorganized the company, then set a goal: by 2004, Yahoo! would derive more than half of its revenue from nonadvertising sources. In 2002, Yahoo! announced a partnership with SBC to sell DSL access, which moved the company even closer to an AOL-style revenue model. Semel's strategy appears to be working. In recent years, Yahoo! has sharply increased its revenues from nonadvertising sources and is once again reporting significant profits.

Other advertising-supported Web portals are following the lead of Yahoo! in a strategy called monetizing eyeballs or monetizing visitors. **Monetizing** refers to the conversion of existing regular site visitors seeking free information or services into fee-paying subscribers or purchasers of services. Many of the portals that are conducting these monetizing campaigns are worried about visitor backlash. They are unsure how many existing visitors will stay and pay for services they had become accustomed to receiving at no cost.

Other examples of Web portals that use a mixed-revenue model are financial information sites **The Motley Fool** and **TheStreet.com**. These sites offer investment advice, stock quotes, and financial planning help. Some of the information is provided at no cost, additional information is available to subscribers who pay no fee but who are required to provide personal information, and even more information is available for a fee.

Recently, more and more industry analysts are predicting the end of the "free Web." Although the largest portal sites should be able to survive using a mixed revenue model, it is unclear how smaller portals will fare.

Internal Web Portals

A growing number of large organizations have built Web portals to provide information to their employees. Internal Web portals run on the intranets you learned about in Chapter 2. These portals can save significant amounts of money by replacing the printing and distribution of paper memos, newsletters, and other correspondence with a Web site. Organizations use internal Web portals to publish employee handbooks, newsletters, and employee benefits information.

These organizations are also finding that the internal portal Web site can become a good way of creating a virtual community among employees who are dispersed over a wide geographic area. For example, a global company could create a question and answer page (similar to the Google Answers page you learned about earlier in this chapter) for all of its networking technicians. Such a page would provide mentoring and informal help functions for the networking technician community within the company.

Many companies are adding wireless connectivity to their internal portals and using this technology to extend the reach of the portal to employees who are traveling, meeting with customers or suppliers, or telecommuting from home. These extended portals are yet another example of a second-wave combination of technology (wireless communications) with a business strategy from the first wave (internal Web portals).

Summary

In this chapter, you learned how companies are now using the Web to do things that they have never done before, such as operating auction sites, creating virtual communities, and serving as Web portals. You learned about the key characteristics of the seven major auction types, and how firms are using online auctions to sell goods to their customers and buy from their suppliers.

Although some specialty sites do conduct significant auction activities, the consumer online auction business is dominated by eBay, at least in the United States. B2B auctions give companies a new and efficient way to dispose of excess inventory, and B2B reverse auctions provide an effective procurement tool under some conditions. A number of businesses offer ancillary services to Web users who participate in online auctions. These businesses include escrow services, auction directories and information sites, auction management software for both sellers and bidders, and auction consignment sites.

The Web's ability to bring together geographically dispersed people and organizations that share narrow interests has encouraged the development of focused virtual communities. Businesses are creating virtual communities with their customers and suppliers and using these communities to sell goods and services. In the second wave of electronic commerce, individuals are increasingly using virtual communities for personal, social, and business-related interactions. Organizations are using mobile commerce to sell goods and services to users of handheld devices such as wireless PDAs and mobile phones. The major Web search engine sites evolved into Web portals by adding virtual communities and related features to their sites' offerings, but a decline in online advertising moved many portal sites to a mixed revenue model in which they have added fee-based services to monetize visitor traffic.

Companies are using internal Web portals to communicate with employees and coordinate work across various organizational units. The integration of wireless communications technologies with internal Web portals is an example of a second-wave combination of existing Web strategy with new technology.

Key Terms

Ascending-price auction

Auction consignment services

Auctioneer

Bid

Bidder

Consumer-to-business

Descending-price auction

Double auction

Dutch auction

English auction

Escrow service

First-price sealed-bid auction

Group purchasing site

Intelligent software agent (software robot or bot)

Liquidation broker

Lock-in effect

Minimum bid

Minimum bid increment

Monetizing

Online community

Open auction (open-outcry auction)

Open-outcry double auction

Private valuation

Proxy bid

Reserve price (reserve)

Reverse auction (seller-bid auction)

Reverse bid	Vickrey auction
Sealed-bid auction	Virtual community
Second-price sealed-bid auction	Virtual learning community
Shill bidder	Web community
Snipe	Winner's curse
Sniping software	Wireless Application Protocol (WAP)
Third-party assurance provider	Yankee auction

Review Questions

RQ1. In approximately 100 words, define the term "reserve price" and explain how the use of a reserve price can affect the progress and outcome of an auction.

RQ2. Identify an industry (or a product within an industry) in which buyers would find reverse auctions to be a useful procurement tool. In about 300 words, explain why your chosen industry or product would be a good candidate for a reverse auction procurement process.

RQ3. In about 300 words, describe the services offered by an online auction escrow service. Name one advantage and one disadvantage of using this type of an escrow service.

RQ4. Some eBay users believe that the use of sniping software is unfair and that eBay should prohibit its use. In an essay of about 200 words, present facts and logical arguments that would convince eBay to prohibit the use of sniping software.

RQ5. Assume you work in the procurement department of a small aerospace parts manufacturer. Your company builds switches and relays used in airplanes to control heating and ventilation systems. The parts your company buys must meet precise specifications and the parts are not generally interchangeable; that is, your company's engineers must work with your suppliers to design specific parts for particular systems. The director of procurement has read about online reverse auctions and is interested in exploring the idea. In approximately 300 words, outline the arguments for and against using online reverse auctions at this company, then conclude with a specific recommendation.

Exercises

E1. Use the Online Companion to examine the projects list at either the **MIT Software Agents Group Projects** site or the **Carnegie Mellon Intelligent Software Agents Lab** site. Choose a software agent technology used in one of these projects and, in approximately 200 words, describe how you could use it in an electronic commerce application.

E2. Midland University, like most metropolitan universities, faces a chronic shortage of parking spaces on campus. Each stakeholder group in the typical university community (these groups include students, faculty, administrators, staff, and visitors) believes its members should have the top priority for parking spaces. You have been assigned to a university task force to study the problem. You decide that an annual online auction of parking spaces conducted on the university's intranet could provide a solution. In about 300 words, describe the elements of an annual online auction for parking spaces at Midland University. Be sure to include provisions for disabled persons and for those university employees who do not have regular access to computers in their typical work environment (such as janitors, physical plant maintenance workers, or gardeners).

E3. Follow the links in the Online Companion for **Auction Consignment Sites** to at least two of the sites and become familiar with the services they offer. Prepare a chart that compares the services offered by two of the sites you visit. Include any important factors that a customer would evaluate when deciding which site to use, but be sure to include a comparison of prices, specific services offered, exclusions and limitations on the services, and guarantees, if any. Summarize your findings in a paragraph or two in which you indicate which site you would recommend to a friend.

E4. Use **mySimon**, **Best Web Buys**, or another Web pricing robot of your choice to find sources for a book or DVD that you want to buy. Evaluate the results provided by the robot in terms of how useful the robot was in helping you plan for your purchase. Summarize your findings in a report of approximately 200 words.

E5. You have been hired as an electronic commerce consultant to Oyster Bay, Inc., a dealer in ocean-going yachts. Oyster Bay maintains offices and marinas in major U.S. East Coast ports. The typical purchaser of an Oyster Bay yacht is a high-income business executive, a retiree, or a person of significant inherited wealth. Oyster Bay salespeople have noted that their customers are increasingly aware of the Web. Prepare a proposal for an Oyster Bay Web portal site. You do not need to design the Web pages, but your proposal should include a detailed list of features that should be included in the site design. Describe each feature in detail, and explain why you believe it should be included. For each feature, note whether it will be supplied by Oyster Bay personnel or purchased from an outside supplier. To learn more about existing yacht sales sites, you can use your favorite search engine or consult the **Carver Yachts**, **Moran Yacht & Ship**, and **YachtWorld.com** links in the Online Companion.

E6. In the chapter, there is a discussion of the stickiness of Web sites that have many visitors. **InterActiveCorp** is a company that often appears on lists of sites that have a large number of visitors or sites that have a high degree of stickiness. Visit the company's site and explore it to learn which Web sites it owns. List the names of the two InterActiveCorp sites that you believe have the highest degree of stickiness and, in about 150 words, explain why.

Cases

C1. Alibaba.com

In 1995, Jack Ma taught English in Hangzhou, China, a city near the economic center of Shanghai. Ma wanted to get into the business world, so he raised $2000 from relatives and friends to start Chinapage.com, one of the first Chinese dot-coms. He followed that experience with a job at the Ministry of Foreign Trade and Economic Cooperation. He grew frustrated with the slow pace of the government bureaucracy and left after a year to start his own company again. He placed an ad on the Internet advertising a language translation service for companies that wanted to do business in China. Within two hours, he had received six e-mailed inquiries. 60 percent of the Chinese economy is manufacturing, and 90 percent of manufacturing companies are small or medium-sized businesses. Ma began collecting information from Chinese manufacturing companies that wanted to do business internationally. He translated and organized the information, then posted it on a B2B Web portal site he named **Alibaba.com.**

Alibaba.com has always concentrated on small and medium-sized businesses (SMBs). Ma believed that global companies spend most of their efforts on doing business with large companies. He sees China (and the rest of Asia) as having a different economic structure than the United States or Europe, where the economies are dominated by large companies. Ma believes that Alibaba.com's true opportunities lie in connecting SMBs around the world with SMBs in China. He argues that SMBs seldom have any sales channels outside of their own country. To compensate, SMBs must travel extensively to meet suppliers and customers at exhibitions or trade fairs. Ma believes that Alibaba.com offers SMBs a reasonably-priced alternative.

Foreign companies interested in buying from Chinese suppliers must register on Alibaba.com (buyer registration is free) before they can access the site's supplier database. Alibaba.com charges Chinese companies a membership fee of several thousand dollars for translating and listing their information. The site also lists foreign suppliers. These suppliers can list a small number of items at no charge, however, most choose to pay a small fee that pays for a credit check and allows them to be listed as TrustPass members on the site. The TrustPass designation provides assurance to Chinese companies that want to buy from these suppliers. By 2001, more than 1 million companies had registered with Alibaba.com. In 2003, the company reported its first profitable year, with net income of $12 million. Many of Alibaba.com's registered members are happy with the results they obtain, as indicated by the annual membership renewal rate, which exceeds 70 percent.

Alibaba.com, like all portal sites, suffered a setback during the 2001-2002 time period, but its fee-based revenue model allowed it to recover more quickly than portals that were dependent on advertising revenue. The company sees future growth in the continued expansion of trade between Chinese manufacturers and the rest of the world. Ma is also optimistic about the portal's potential for helping Chinese businesses connect with other Chinese businesses.

Required:

1. Alibaba.com was an early entrant into the B2B portal market in China. In about 100 words, explain how this might have created a lock-in effect, especially given the types of businesses the site attracts.

2. Alibaba.com currently charges foreign sellers an annual fee of about $400 for a TrustPass membership, but Chinese companies pay $8000 or more for their annual listings as China Gold Suppliers. In about 200 words, explain why the site has different listing charges for the two types of members and critically evaluate this practice.

3. You learned in Chapter 5 that large companies, such as General Electric and Sears, often require suppliers to follow specific rules if they want to do business (such as using EDI or even a specific EDI VAN). Alibaba.com currently focuses on connecting SMBs with each other. In about 200 words, discuss opportunities that might exist for Alibaba.com to become an intermediary in relationships between Chinese SMBs and large global companies such as General Electric and Sears.

4. In 2003, Alibaba.com launched **Taobao.com** to compete in the general consumer online auction market against eBay in China. In about 200 words, describe the advantages

Alibaba.com might have over eBay in this new market, then describe the advantages eBay might have over Alibaba.com. Be sure to include a discussion of lock-in effects where appropriate.

Note: Your instructor might assign you to a group to complete this case and might ask you to prepare a formal presentation of your results to your class.

C2. Old Metamora

Betty Shriver is the owner of Betty's Crystal, a small shop that sells collectible glass figurines. Betty's shop carries many items that she purchased from estate sales and regional auctions, but the shop also sells new crystal figurines from manufacturers such as Baccarat, Lalique, Orrefors, and Swarovski. The shop is located in Metamora, Indiana, which is a popular tourist destination for weekend travelers in the Midwest. The town of **Old Metamora** is a small historic area in a rural setting that is less than a day's drive from seven major metropolitan areas: Chicago, Cincinnati, Columbus, Detroit, Indianapolis, Louisville, and St. Louis.

The shop is very busy on weekends and during the spring and summer months when tourists flock to Metamora. In the early fall, the tourist traffic slows considerably, and in the winter months, the town becomes almost deserted. Two years ago, Betty began to pick up extra business during the off season by auctioning items on **eBay**, **Amazon.com**, and **Yahoo! Auctions**. Not only did the auctions help keep inventory moving during the slow months, but Betty found that she was able to carry a wider selection of items in the store. In the past, she would see unusual items at estate sales and auctions that she feared would not sell quickly in the shop. Now Betty knows that any item that does not sell in the shop can be auctioned online quite easily. Another unexpected benefit of participating in online auctions is that Betty developed relationships with regular buyers of crystal figurines and with people who run collectibles stores in other parts of the country. Every auction involves at least two e-mails (one to confirm the final bid and another to confirm the payment). Many successful bidders also send e-mail messages to Betty when they receive the item with questions about the item, or just to thank Betty for sending the item so quickly. Some of these e-mail exchanges continue with discussions related to crystal figurines and other collectible items.

Betty's online auction experiences prompted her to consider expanding the online portion of her business. She has heard (from other shop owners) that eBay allows people to create online stores within the eBay site and that Amazon.com offers a similar service called zShops. She is also interested in creating a Web site that contains photos and descriptions of popular crystal figurines with additional information about how they are made. Betty also wants to include a list of figurines that are no longer manufactured (which makes them more valuable) and a guide to buying collectible crystal figurines that could help her customers and bidders on her auctions make more informed decisions as they add to their collections. She believes that such a site could attract a large number of people interested in crystal figurines. She wants to find ways to direct these site visitors to her auctions and her proposed Web store. Betty has hired you as a consultant to build on her ideas and to help her develop an expansion strategy for her online business activities.

Required:

1. Search for information about Amazon.com's zShops and eBay Stores on the Web and in your library that will help you make a recommendation to Betty regarding which alternative

would provide the best avenue for her online business expansion. Support your recommendation with relevant facts, including specific costs of operating each type of store and specific benefits that Betty could gain by using one or the other. Summarize your recommendation and supporting facts in a report to Betty of about four double-spaced pages.

2. Outline the elements that Betty should include in a virtual community site that meets her stated goals. For each element, explain why it would help create the community, and describe any difficulties that Betty might encounter in building and maintaining that element. Summarize the virtual community outline in a report to Betty of about four double-spaced pages.

Note: Your instructor might assign you to a group to complete this case and might ask you to prepare a formal presentation of your results to your class.

For Further Study and Research

Atanasov, M. 2001. "Going Out of Business Since 1903," *Smart Business*, 14(7), July, 72–76.

Bagozzi, R. and U. Dholakia. 2002. "Intentional Social Action in Virtual Communities," *Journal of Interactive Marketing*, 16(2), Spring, 2–21.

Belson, K., R. Hof, B. Elgin. 2001. "How Yahoo! Japan Beat eBay at Its Own Game," *Business Week*, June 4, 58.

Bieber, M., D. Engelbart, R. Furuta, R. Hiltz, R. Starr, J. Noll, J. Preece, E. Stohr, M. Turoff, and B. Van de Walle. 2002. "Toward Virtual Community Knowledge Evolution," *Journal of Management Information Systems*, 18(4), Spring, 11–35.

Borzo, J. 2004. "Using Online Networking, Job Seekers Turn Friendship into Employment," *The Wall Street Journal*, September 13, R14.

Cassady, R. 1967. *Auctions and Auctioneering*, Berkeley, CA: University of California Press.

Catterall, M. and P. Maclaran. 2002. "Researching Consumers in Virtual Worlds: A Cyberspace Odyssey," *Journal of Consumer Behavior*, 1(3), February, 228–237.

Chang, A. 2003. "Hospitals Auction Nursing Shifts Online," *The Boston Globe*, December 28, A28.

Chen, K. and K. Qiu Haixu. 2004. "Chinese E-Commerce Sites Allow Small Firms to Reach Wider Base," *The Wall Street Journal*, February 25, A12.

Cohen, A. 2001. "The Sniper King," *On Magazine*, May. (http://www.onmagazine.com/on-mag/magazine/article/1,9985,105315-1,00.html)

Davydov, M. 2003. "The Portal Reborn," *Intelligent Enterprise*, 6(17), October 30 20–26.

Deprez, F., J. Rosengren, and V. Soman. 2002. "Portals for all Platforms," *McKinsey Quarterly*, January, 92–99.

Dobrzynski, J. 2000. "F.B.I. Opens Investigation of eBay Bids," *The New York Times*, June 7, 1.

The Economist. 1997. "Going, Going..." May 31, 61.

The Economist. 2001. "We Have Lift-Off" February 3, 69–71.

Elgin, B. 2002. "Can Yahoo Make 'em Pay?" *Business Week*, September 9, 92–94.

Foster, K. 2003. "Captains of Portal Destiny," *Line56: The E-Business Executive Daily*, June 10. (http://www.line56.com/articles/default.asp?articleid=4719)

Freeman, L. 2000. "Blue Mountain Arts: Mark Rinella," *Advertising Age*, June 26, S20.

Ghawi, D. and G. Schneider. 2004. "New Approaches to Online Procurement," *Proceedings of the Academy of Information and Management Sciences*, 8(2), October, 25–28.

Gilbert, J. and A. Kerwin. 1999. "Newspapers Carve Slice of Auction Pie," *Advertising Age*, 70(26), June 21, 32–34.

Greengard, S. 2003. "Portals Shape the Promise of the Internet," *Internet World*, April 1, 26–31.

Grimes, W. 2004. "Just Browsing: That Invisible Hand Guides the Game Of Ticket Hunting," *The New York Times*, June 18, E1.

Gross, N. 1999. "Building Global Communities: How Business Is Partnering with Sites that Draw Together Like-Minded Consumers," *Business Week*, March 22, EB42.

Hafner, K. 2004. "With Internet Fraud Up Sharply, eBay Attracts Vigilantes," *The New York Times*, March 20, A1.

Hannon, D. 2001. "Navy Supply Uncovers Benefits of Auctions," *Purchasing*, 130(12), June 21, 17.

InternetWeek, 2003. "U.S. Air Force Builds Portal," January 26. (http://www.internetweek.com/story/showArticle.jhtml?articleID=6406488)

Kawakami, S. 2003. "China's Visionary B2B," *J@pan Inc.*, May, 14–16.

Kenczyk, M. and V. Reitz. 2001. "Reverse Auctions Are Risky Models for Buying Custom Parts," 73(6), March 22, 148.

Kennedy, J. 1998. "Radio Daze," *Technology Review*, 101(6), November–December, 68–71.

Lechner, U. and J. Hummel. 2002. "Business Models and System Architectures of Virtual Communities: From a Sociological Phenomenon to Peer-to-Peer Architectures," *International Journal of Electronic Commerce*, 6(3), Spring, 41–52.

Lee, J. 2003. "U.S. and States Join to Fight Internet Auction Fraud," *The New York Times*, May 1. (http://www.nytimes.com/2003/05/01/technology/01ONLI.html)

Lloyd, J. 2001. "eBay Founder Pierre Omidyar: His Devotion to Community Created a Global Auction House," *Investor's Business Daily*, August 20, A4.

Ma, M. 1999. "Agents in E-Commerce," *Communications of the ACM*, 42(3), March, 78–80.

Managing Human Resources Information Systems. 2002. "How GM Designed Its Award-Winning Employee Portal," August, 1–14.

Mangalindan, M. 2002. "Yahoo, MSN Plan Broadband Attack on AOL," *The Wall Street Journal*, July 25, B1.

Margulius, D. 2003. "Portal to Higher Learning: JA-SIG Gives Schools an Invaluable Educational Portal," *InfoWorld*, 25(44), November 10, 48.

McGuire, D. 2000. "Auction Sites Stay Popular Despite Fraud Warnings—Study," *Newsbytes*, September 25. (http://www.nbnn.com/)

McLean, B. 1999. "Sothebys.com," *Fortune*, 139(3), February 15, 200.

Mears, J. 2002. "Portals Power Business User Profits," *Network World*, 19(28), July 15, 19-20.

Metz, C. 2004. "Social Networking: Make Contact," *PC Magazine*, 23(1), January 20, 131–136.

The New York Times. 2004. "eBay to Double Some Fraud Coverage," October 1, C3.

Norris, F. 2004, "Google's Offering Proves Stock Auctions Can Really Work," *The New York Times*, August 23, C6.

Norris, G. and D. Duray. 2002. "The Outside-In Portal," *Intelligent Enterprise*, 5(13), August 12, 32–35.

Petersen, A. 1999. "Some Places to Go When You Want to Feel Right at Home: Communities Focus on People Who Need People," *The Wall Street Journal*, January 6, B6.

Petrecca, L. and B. Snyder. 1998. "Auction Universe Puts in $10 Mil Bid for Customers," *Advertising Age*, 43(8), October 26, 8.

Purchasing. 2001. "What Top Supply Execs Say About Auctions," 130(12), June 21, S2-S3.

Quan, J. 1999. "Risky Business," *Rolling Stone*, March 4, 91–92.

Rheingold, H. 1993. *The Virtual Community: Homesteading on the Electronic Frontier.* New York: HarperCollins.

Robins, W. 2000. "Auctions.com Now a Dot-Goner," *Editor & Publisher*, August 28, 6.

Roth, D. 2000. "Meet eBay's Worst Nightmare," *Fortune*, June 26, 199–202.

Rozanski, H. and G. Bollman. 2002. "The Great Portal Payoff: All Consuming Behavior," *strategy+business*, September, 1–5.

Sanborn, S. 2001. "Reverse Auctions Make a Bid for Business," *InfoWorld*, 23(12), March 19, 32.

Schiffman, B. 2001. "A Community That Stays Together, Pays Together," *Forbes*, August 28. (http://www.forbes.com/technology/ecommerce/2001/08/28/0828yahoo.html)

Schonfeld, E. 2002. "eBay's Secret Ingredient," *Business 2.0*, 3(3), March, 52-58.

Schuyler, N. 2000. "Going... Going... Gotcha!" *PC World*, October 1, 181.

Shalo, S. 2002. "Virtual Community Generates Real ROI," *Pharmaceutical Executive*, 22(3), March, 118.

Solomon, S. 1999. "Go Sell It on the Mountain," *Inc*, December, 48–62.

Steiner, I. 2003. "Auction Drop-Off Stores Offer Consignment Services to Non-eBayers," *Auctionbytes Update*, November 2. (http://www.auctionbytes.com/cab/abu/y203/m11/abu0106/s02)

Sternstein, A. 2002. "Movie Reviews to Go," *Business Week Online*, June 21. (http://www.businessweek. com/technology/content/jun2002/tc20020621_6242.htm)

Sullivan, M. 2002. "Online, Itty-bitty Auctions Are Besting the Big Boys," *Forbes*, 170(12), December 9, 228.

Tedeschi, B. 2000. "Creating Marketplaces for Business-to-Business Transactions," *The New York Times*, January 24, C10.

Thaler, R. 1994. *The Winner's Curse: Paradoxes and Anomalies of Economic Life.* Princeton, NJ: Princeton University Press.

Vickrey, W. 1961. "Counterspeculation, Auctions, and Competitive Sealed Tenders," *Journal of Finance*, 16(1), March, 8–37.

The Wall Street Journal. 2000. "Art of Antiquing Turned on Its Head by Online Auctions," August 8, B12B.

Wang, S. 1999. "Analyzing Agents for Electronic Commerce," *Information Systems Management*, 16(1), Winter, 40–48.

Ward, E. 1999. "How to Build Community on Your Site and Participate in Others," *Business Marketing*, June 1, 24.

Wingfield, N. 2001. "Andale Hitches Its Wagon to eBay's Fortunes," *The Wall Street Journal*, September 29, B11.

Wingfield, N. 2004. "Taking on eBay," *The Wall Street Journal*, September 13, R10.

Yoo, W-S., K-S. Suh, and M-B. Lee. 2002. "Exploring the Factors Enhancing Member Participation in Virtual Communities," *Journal of Global Information Management*, 10(3), July–September, 55–71.

CHAPTER 7

THE ENVIRONMENT OF ELECTRONIC COMMERCE: LEGAL, ETHICAL, AND TAX ISSUES

LEARNING OBJECTIVES

In this chapter, you will learn about:

- Laws that govern electronic commerce activities
- Laws that govern the use of intellectual property by online businesses
- Online crime, terrorism, and warfare
- Ethics issues that arise for companies conducting electronic commerce
- Conflicts between companies' desire to collect and use data about their customers and the privacy rights of those customers
- Taxes that are levied on electronic commerce activities

INTRODUCTION

In 1999, **Dell Computer** and Micron Electronics (now doing business as **MPC Computers**), two

companies that sell personal computers through their Web sites, agreed to settle U.S. Federal Trade

Commission (FTC) charges that they had disseminated misleading advertising to their existing and

potential customers. The advertising in question was for computer leasing plans that both companies had

offered on their Web sites. The ads stated the price of the computer along with a monthly payment.

Unfortunately for Dell and Micron, stating the monthly payment without disclosing full details of the lease plan is a violation of the Consumer Leasing Act of 1976. This law is implemented through a federal regulation that was written and that is updated periodically by the Federal Reserve Board. This regulation, called Regulation M, was designed to require banks and other lenders to fully disclose the terms of leases so that consumers would have enough information to make informed financing choices when leasing cars, boats, furniture, and other goods.

Both Dell and Micron had included the required information on their Web pages, but FTC investigators noted that important details of the leasing plans, such as the number of payments and the fees due at the signing of the lease, were placed in a small typeface at the bottom of a long Web page. A consumer who wanted to determine the full cost of leasing a computer would need to scroll through a number of densely filled screens to obtain enough information to make the necessary calculations.

In the settlement, both companies agreed to provide consumers with clear, readable, and understandable information in their lease advertising. The companies also agreed to record-keeping and federal monitoring activities designed to ensure their compliance with the terms of the settlement.

Dell and Micron are computer manufacturers. It apparently did not occur to them that they needed to become experts in Regulation M, generally considered to be a banking regulation. Companies that do business on the Web expose themselves, often unwittingly, to liabilities that arise from today's business environment. That environment includes laws and ethical considerations that may be different from those with which the business is familiar. In the case of Dell and Micron, they were unfamiliar with the laws and ethics of the banking industry. The banking industry has a different culture than that of the computer industry—it is unlikely that a bank advertising manager would have made such a mistake.

As you will learn in this chapter, Dell and Micron are by no means the only Web businesses that have run afoul of laws and regulations. As new and existing companies open online operations, they become subject to unfamiliar laws and different ethical frameworks much more rapidly than in the physical world.

THE LEGAL ENVIRONMENT OF ELECTRONIC COMMERCE

Businesses that operate on the Web must comply with the same laws and regulations that govern the operations of all businesses. If they do not, they face the same set of penalties—fines, reparation payments, court-imposed dissolution, and even jail time for officers and owners—that any business faces.

Businesses operating on the Web face two additional complicating factors as they try to comply with the law. First, the Web extends a company's reach beyond traditional boundaries. As you learned in Chapter 1, a business that uses the Web immediately becomes an international business. Thus, a company can become subject to many more laws more quickly than a traditional brick-and-mortar business based in one specific physical location. Second, the Web increases the speed and efficiency of business communications. As you learned in Chapters 3 and 4, customers often have much more interactive and complex relationships with online merchants than they do with traditional merchants. Further, the Web creates a network of customers who often have significant levels of interaction with each other. Web businesses that violate the law or breach ethical standards can face rapid and intense reactions from many customers and other stakeholders who become aware of the businesses' activities.

In this section, you will learn about the issues of borders, jurisdiction, and Web site content and how these factors affect a company's ability to conduct electronic commerce. You will also learn about legal issues that arise when the Web is used in the commission of crimes, terrorist acts, and even the conduct of war.

Borders and Jurisdiction

Territorial borders in the physical world serve a useful purpose in traditional commerce: They mark the range of culture and reach of applicable laws very clearly. When people travel across international borders, they are made aware of the transition in many ways. For example, exiting one country and entering another usually requires a formal examination of documents, such as passports and visas. In addition, both the language and the currency usually change upon entry into a new country. Each of these experiences, and countless others, are manifestations of the differences in legal rules and cultural customs in the two countries. In the physical world, geographic boundaries almost always coincide with legal and cultural boundaries. The limits of acceptable ethical behavior and the laws that are adopted in a geographic area are the result of the influences of the area's dominant culture. The relationships among a society's culture, laws, and ethical standards appear in Figure 7-1.

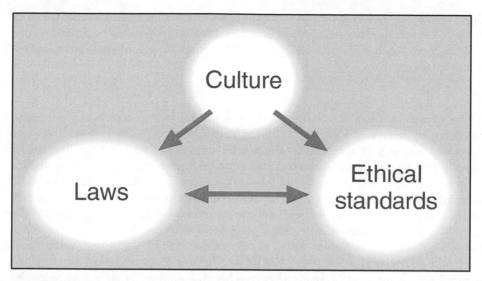

FIGURE 7-1 Culture helps determine laws and ethical standards

The geographic boundaries on culture are logical; for most of our history, humans have been unable to travel great distances to learn about other cultures. In recent years, however, some countries decided that times have indeed changed, and people can travel easily from one country to another within a geographic region. One example is the European Union (EU), which allows free movement within the EU for citizens of member countries. Most of the EU countries (Great Britain being a notable exception) have even formed the European Money Union and use a common currency (the euro) instead of their former individual currencies (for example, French francs, German marks, and Italian lire). Legal scholars define the relationship between geographic boundaries and legal boundaries in terms of four elements: power, effects, legitimacy, and notice.

Power

Power is a form of control over physical space and the people and objects that reside in that space, and is a defining characteristic of statehood. For laws to be effective, a government must be able to enforce them. Effective enforcement requires the power both to exercise physical control over residents, if necessary, and to impose sanctions on those who violate the law. The ability of a government to exert control over a person or corporation is called **jurisdiction**.

Laws in the physical world do not apply to people who are not located in or do not own assets in the geographic area that created those particular laws. For example, the United States cannot enforce its copyright laws on a citizen of Japan who is doing business in Japan and owns no assets in the United States. Any assertion of power by the United States over such a Japanese citizen would conflict with the Japanese government's recognized monopoly on using force with its citizens. Japanese citizens who bring goods into the United States to sell, however, are subject to applicable U.S. copyright laws. A Japanese Web site that offers delivery of goods into the United States is, similarly, subject to applicable U.S. laws.

The Environment of Electronic Commerce: Legal, Ethical, and Tax Issues

The level of power asserted by a government is limited to that which is accepted by the culture that exists within its geographic boundaries. Ideally, geographic boundaries, cultural groupings, and legal structures all coincide. When they do not, internal strife and civil wars can erupt.

Effects

Laws in the physical world are grounded in the relationship between physical proximity and the **effects**, or impact, of a person's behavior. Personal or corporate actions have stronger effects on people and things that are nearby than on those that are far away. Government-provided trademark protection is a good example of this. For instance, the Italian government can provide and enforce trademark protection for a business named Casa di Baffi located in Rome. The effects of another restaurant using the same name are strongest in Rome, somewhat less in geographic areas close to Rome, and even less in other parts of Italy. That is, the effects diminish as geographic distance increases. If someone were to open a restaurant in Kansas City and call it Casa di Baffi, the restaurant in Rome would experience few, if any, negative effects from the use of its trademarked name in Kansas City because it would be so far away and because so few people would be potential customers of both restaurants. Thus, the effects of the trademark violation are controlled by Italian law because of the limited range within which such a violation has an effect.

The characteristics of laws are determined by the local culture's acceptance or rejection of various kinds of effects. For example, certain communities in the United States require that houses be built on lots that are at least five acres. Other communities prohibit outdoor advertising of various kinds. The local cultures in these communities make the effects of such restrictions acceptable.

When businesses begin operations online, the traditional measures of effects—and the laws that have been developed using those measures over many years—do not work very well. For example, France has a law that prohibits the sale of Nazi memorabilia. The people of France have considered this to be a reasonable law for many years. United States laws do not include a similar prohibition. When U.S.-based online auction sites began hosting auctions of Nazi memorabilia, those sites were in compliance with U.S. laws. However, because of the international nature of the Web, these auctions were available to people around the world, including residents of France. The French government ordered Yahoo! Auctions to stop these auctions. Yahoo! argued that it was in compliance with U.S. law, but the French government insisted that the effects of those Yahoo! auctions extended to France and thus violated French law. To avoid protracted legal actions over the jurisdiction issue, Yahoo! decided that it would no longer carry such auctions (*Note:* If you search in Yahoo! auctions using terms such as "Nazi," you might find some items available. These items, which include coins and stamps, are not considered Nazi memorabilia under French law.)

Legitimacy

Most people agree that the legitimate right to create and enforce laws derives from the mandate of those who are subject to those laws. In 1970, the **United Nations** passed a resolution that affirmed this idea of governmental legitimacy. The resolution made clear that the people residing within a set of recognized geographic boundaries are the ultimate source of legitimate legal authority for people and actions within those boundaries. Thus, **legitimacy** is the idea that those subject to laws should have some role in formulating them.

Some cultures allow their governments to operate with a high degree of autonomy and unquestioned authority. China and Singapore are countries in which national culture permits the government to exert high levels of unchecked authority. Other cultures, such as those of the Scandinavian countries, place strict limits on governmental authority.

The levels of authority and autonomy with which governments of various countries operate varies significantly from one country to another. Online businesses must be ready to deal with a wide variety of regulations and levels of enforcement of those regulations as they expand their businesses to other countries. This can be difficult for smaller businesses that operate on the Web.

Notice

Physical boundaries are a convenient and effective way to announce the ending of one legal or cultural system and the beginning of another. The physical boundary, when crossed, provides **notice** that one set of rules has been replaced by a different set of rules. Notice is the expression of such a change in rules. People can obey and perceive a law or cultural norm as fair only if they are notified of its existence. Borders provide this notice in the physical world. The legal systems of most countries include a concept called constructive notice. People receive **constructive notice** that they have become subject to new laws and cultural norms when they cross an international border, even if they are not specifically warned of the changed laws and norms by a sign or a border guard's statement. Thus, ignorance of the law is not a sustainable defense, even in a new and unfamiliar jurisdiction.

This presents particular problems for online businesses, because they may not know that customers from another country are accessing their Web sites. Thus, the concept of notice—even constructive notice—does not translate very well to online business.

Jurisdiction on the Internet

Defining, establishing, and asserting jurisdiction are much more difficult on the Internet than they are in the physical world, mainly because traditional geographic boundaries do not exist. For example, a Swedish company that engages in electronic commerce may have a Web site that is entirely in English and a URL that ends in ".com," thus not indicating to customers that it is a Swedish firm. The server that hosts this company's Web page could be in Canada, and the people who maintain the Web site might work from their homes in Australia.

If a Mexican citizen buys a product from the Swedish firm and is unhappy with the goods received, that person might want to file a lawsuit against the seller firm. However, the world's physical border-based systems of law and jurisdiction do not help this Mexican citizen determine where to file the lawsuit. The Internet does not provide anything like the obvious international boundary lines in the physical world. Thus, the four considerations that work so well in the physical world—power, effects, legitimacy, and notice—do not translate very well to the virtual world of electronic commerce.

Governments that want to enforce laws regarding business conduct on the Internet must establish jurisdiction over that conduct. A **contract** is a promise or set of promises between two or more legal entities—people or corporations—that provides for an exchange of value (goods, services, or money) between or among them. A **tort** is an intentional or negligent action taken by a legal entity that causes harm to another legal entity. People or corporations that wish to enforce their rights based on either contract or tort law must file their claims in courts with jurisdiction to hear their cases. A court has sufficient jurisdiction in a matter if it has both subject-matter jurisdiction and personal jurisdiction.

Subject-Matter Jurisdiction

Subject-matter jurisdiction is a court's authority to decide a particular type of dispute. For example, in the United States, federal courts have subject-matter jurisdiction over issues governed by federal law (such as bankruptcy, copyright, patent, and federal tax matters), and state courts have subject-matter jurisdiction over issues governed by state laws (such as professional licensing and state tax matters). If the parties to a contract are both located in the same state, a state court has subject-matter jurisdiction over disputes that arise from the terms of that contract. The rules for determining whether a court has subject-matter jurisdiction are clear and easy to apply. Few disputes arise over subject-matter jurisdiction.

Personal Jurisdiction

Personal jurisdiction is, in general, determined by the residence of the parties. A court has personal jurisdiction over a case if the defendant is a resident of the state in which the court is located. In such cases, the determination of personal jurisdiction is straightforward. However, an out-of-state person or corporation can also voluntarily submit to the jurisdiction of a particular state court by agreeing to do so in writing or by taking certain actions in the state.

One of the most common ways that people voluntarily submit to a jurisdiction is by signing a contract that includes a statement, known as a **forum selection clause**, that the contract will be enforced according to the laws of a particular state. That state then has personal jurisdiction over the parties who signed the contract regarding any enforcement issue that arises from the terms of that contract. Figure 7-2 shows a portion of the contract that governs site visitors' activities on the **Qpass** site. Qpass sells software to wireless system and network operating companies. The first paragraph shown includes the site's forum selection clause. The second paragraph clarifies that site visitors are subject to their own jurisdictions' laws in addition to the jurisdiction specified in the forum selection clause.

In the United States, individual states have laws that can create personal jurisdiction for their courts. The details of these laws, called **long-arm statutes**, vary from state to state, but generally create personal jurisdiction over nonresidents who transact business or commit tortious acts in the state. For example, suppose that an Arizona resident drives recklessly while in California and, as a result, causes a collision with another vehicle that is driven by a California resident. Due to the driver's tortious behavior in the state of California, the Arizona resident can expect to be called into a California court. In other words, California courts have personal jurisdiction over the matter.

Businesses should be aware of jurisdictional considerations when conducting electronic commerce over state and international lines. In most states, the extent to which these

> These terms of use shall be governed by and
> construed in accordance with the laws of the State of
> Washington, without regard to its conflict of laws
> rules. Any legal action arising out of this Agreement
> shall be litigated and enforced under the laws of the
> State of Washington. In addition, you agree to
> submit to the jurisdiction of the courts of the State of
> Washington, and that any legal action pursued by
> you shall be within the exclusive jurisdiction of the
> courts of King County in the State of Washington.
>
> You are responsible for complying with the laws of
> the jurisdiction from which you are accessing this site
> and you agree that you will not access or use the
> information on this site in violation of such laws.
> Unless expressly stated otherwise herein, any
> information submitted by you through this site shall
> be deemed non-confidential and non-proprietary.

FIGURE 7-2   Forum selection clause on the Qpass Web site

laws apply to companies doing business over the Internet is unclear. Because these procedural laws were written before electronic commerce existed, their application to Internet transactions continues to evolve as more and more disputes arise from online commercial transactions. The trend in this evolving law is that the more business activities a company conducts in a state, the more likely it is that a court will assert personal jurisdiction over that company through the application of a long-arm statute.

One exception to the general rule for determining personal jurisdiction occurs in the case of tortious acts. A business can commit a tortious act by selling a product that causes harm to a buyer. The tortious act can be negligent, in which the seller unintentionally provides a harmful product, or it can be an intentional tort, in which the seller knowingly or recklessly causes injury to the buyer. The most common examples of business-related intentional torts are defamation, misrepresentation, fraud, and theft of trade secrets. Although case law is rapidly developing in this area also, courts tend to invoke their respective states' long-arm statutes much more readily in the case of tortious acts than in other business cases. If the matter involves an intentional tort or a criminal act, courts will assert jurisdiction more liberally.

Jurisdiction in International Commerce

Jurisdiction issues that arise in international business are even more complex than the rules governing personal jurisdiction across state lines within the United States. The exercise of jurisdiction across international borders is governed by treaties between the countries engaged in the dispute. In general, U.S. courts determine personal jurisdiction for foreign companies and people in much the same way that these courts interpret the long-arm statutes in domestic matters. Non-U.S. corporations and individuals can be sued in U.S. courts if they conduct business or commit tortious acts in the United States. Similarly, foreign courts can enforce decisions against U.S. corporations or individuals through the U.S. court system if those courts can establish jurisdiction over the matter.

Courts asked to enforce the laws of other nations sometimes follow a principle called **judicial comity**, which means that they voluntarily enforce other countries' laws or judgments out of a sense of comity, or friendly civility. However, most courts are reluctant to serve as forums for international disputes. Also, courts are designed to deal with weighing evidence and making findings of right and wrong. International disputes often require diplomacy and the weighing of costs and benefits. Courts are not designed to do cost-benefit evaluations and cannot engage in negotiation and diplomacy. Thus, courts (especially U.S. courts) prefer to have the executive branch of the government (primarily the State Department) negotiate international agreements and resolve international disputes.

Jurisdictional issues are complex and change rapidly. Any business that intends to conduct electronic commerce should consult an attorney who is well versed in these procedural issues. However, there are a number of resources online that can be useful to nonlawyers who want to do preliminary investigation of a legal topic such as jurisdiction. The Harvard Law School's **Berkman Center for Internet & Society** Web site includes links to many current Internet-related legal issues. The **UCLA Online Institute for Cyberspace Law and Policy** contains an archive of legal reference materials published between 1995 and 2002.

Contracting and Contract Enforcement in Electronic Commerce

Any contract includes three essential elements: an offer, an acceptance, and consideration. The contract is formed when one party accepts the offer of another party. An **offer** is a commitment with certain terms made to another party, such as a declaration of willingness to buy or sell a product or service. An offer can be revoked as long as no payment, delivery of service, or other consideration has been accepted. An **acceptance** is the expression of willingness to take an offer, including all of its stated terms. **Consideration** is the agreed upon exchange of something valuable, such as money, property, or future services. When a party accepts an offer based on the exchange of valuable goods or services, a contract has been created. An **implied contract** can also be formed by two or more parties that act as if a contract exists, even if no contract has been written and signed.

People enter into contracts on a daily, and often hourly, basis. Every kind of agreement or exchange between parties, no matter how simple, is a type of contract. For example, every time a consumer buys an item at the supermarket, the elements of a valid contract are met:

- The store offers an item at a stated price.
- The consumer accepts this offer by indicating a willingness to buy the product for the stated price.
- The store exchanges its product for another valuable item: the consumer's payment.

Contracts are a key element of traditional business practice, and they are equally important on the Internet. Offers and acceptances can occur when parties exchange e-mail messages, engage in electronic data interchange (EDI), or fill out forms on Web pages. These Internet communications can be combined with traditional methods of forming contracts, such as the exchange of paper documents, faxes, and verbal agreements made over the telephone or in person. An excellent resource for many of the laws concerning contracts, especially as they pertain to U.S. businesses, is the Cornell Law School Web site, which includes the full text of the **Uniform Commercial Code (UCC)**.

When a seller advertises goods for sale on a Web site, that seller is not making an offer, but is inviting offers from potential buyers. If a Web ad were a legal offer to form a contract, the seller could easily become liable for the delivery of more goods than it has available to ship. When a buyer submits an order, which is an offer, the seller can accept that offer and create a contract. If the seller does not have the ordered items in stock, the seller has the option of refusing the buyer's order outright or counteroffering with a decreased amount. The buyer then has the option to accept the seller's counteroffer.

Making a legal acceptance of an offer is quite easy to do in most cases. When enforcing contracts, courts tend to view offers and acceptances as actions that occur within a particular context. If the actions are reasonable under the circumstances, courts tend to interpret those actions as offers and acceptances. For example, courts have held that various actions—including mailing a check, shipping goods, shaking hands, nodding one's head, taking an item off a shelf, or opening a wrapped package—are all, in some circumstances, legally binding acceptances of offers. Although the case law is limited regarding acceptances made over the Internet, it is reasonable to assume that courts would view clicking a button on a Web page, entering information in a Web form, or downloading a file to be legally binding acceptances.

Written Contracts on the Web

In general, contracts are valid even if they are not in writing or signed. However, certain categories of contracts are not enforceable unless the terms are put into writing and signed by both parties. In 1677, the British Parliament enacted a law that specified the types of contracts that had to be in writing and signed. Following this British precedent, every state in the United States today has a similar law, called a **Statute of Frauds**. Although these state laws vary slightly, each Statute of Frauds specifies that contracts for the sale of goods worth over $500 and contracts that require actions that cannot be completed within one year must be created by a signed writing. Fortunately for businesses and people who want to form contracts using electronic commerce, a writing does not require either pen or paper.

Most courts will hold that a **writing** exists when the terms of a contract have been reduced to some tangible form. An early court decision in the 1800s held that a telegraph transmission was a writing. Later courts have held that tape recordings of spoken words, computer files on disks, and faxes are writings. Thus, the parties to an electronic commerce contract should find it relatively easy to satisfy the writing requirement. Courts have been similarly generous in determining what constitutes a signature. A **signature** is any symbol executed or adopted for the purpose of authenticating a writing. Courts have held names on telegrams, telexes, faxes, and Western Union Mailgrams to be signatures. Even typed names or names printed as part of a letterhead have served as signatures. It is reasonable to assume that a symbol or code included in an electronic file would constitute a signature. As you will learn in Chapter 10, the United States now has a law that explicitly makes digital signatures legally valid for contract purposes.

Firms conducting international electronic commerce do not need to worry about the signed writing requirement in most cases. The main treaty that governs international sales of goods, Article 11 of the United Nations Convention on Contracts for the International Sale of Goods (CISG), requires neither a writing nor a signature to create a legally binding acceptance. You can learn more about the CISG and related topics in international commercial law at the **Pace University School of Law CISG Information** Web site.

Warranties on the Web

Most firms conducting electronic commerce have little trouble fulfilling the requirements needed to create enforceable, legally binding contracts on the Web. One area that deserves attention, however, is the issue of warranties. Any contract for the sale of goods includes implied warranties. A seller implicitly warrants that the goods it offers for sale are fit for the purposes for which they are normally used. If the seller knows specific information about the buyer's requirements, acceptance of an offer from that buyer may result in an additional implied warranty of fitness, which suggests that the goods are suitable for the specific uses of that buyer. Sellers can also create explicit warranties by providing a specific description of the additional warranty terms. It is also possible for a seller to create explicit warranties, often unintentionally, by making general statements in brochures or other advertising materials about product performance or suitability for particular tasks.

Sellers can avoid some implied warranty liability by making a warranty disclaimer. A **warranty disclaimer** is a statement declaring that the seller will not honor some or all implied warranties. Any warranty disclaimer must be conspicuously made in writing, which means it must be easily noticed in the body of the written agreement. On a Web page, sellers can meet this requirement by putting the warranty disclaimer in larger type, a bold font, or a contrasting color. To be legally effective, the warranty disclaimer must be stated obviously and must be easy for a buyer to find on the Web site. Figure 7-3 shows a portion of an **Apple Computer** warranty page that includes its warranty disclaimer. The warranty disclaimer is printed in uppercase letters to distinguish it from other text on the page.

WARRANTY COVERAGE

Apple's warranty obligations are limited to the terms set forth below: Apple Computer, Inc. ("Apple") warrants this hardware product against defects in materials and workmanship for a period of ONE (1) YEAR from the date of original retail purchase. If a defect exists, at its option Apple will (1) repair the product at no charge, using new or refurbished replacement parts, (2) exchange the product with a product that is new or which has been manufactured from new or serviceable used parts and is at least functionally equivalent to the original product, or (3) refund the purchase price of the product. A replacement product/part assumes the remaining warranty of the original product or ninety (90) days from the date of replacement or repair, whichever provides longer coverage for you. When a product or part is exchanged, any replacement item becomes your property and the replaced item becomes Apple's property. When a refund is given, your product becomes Apple's property.

EXCLUSIONS AND LIMITATIONS

This Limited Warranty applies only to hardware products manufactured by or for Apple that can be identified by the "Apple" trademark, trade name, or logo affixed to them. The Limited Warranty does not apply to any non-Apple hardware products or any software, even if packaged or sold with Apple hardware. Non-Apple manufacturers, suppliers, or publishers may provide their own warranties. Software distributed by Apple under the Apple brand name (including, but not limited to system software) is not covered under this Limited Warranty. Refer to the Apple Software License Agreement for more information. Apple and its Authorized Service Providers are not liable for any damage to or loss of any programs, data, or other information stored on any media, or any non-Apple product or part not covered by this warranty. Recovery and reinstallation of system and application software and user data are not covered under this Limited Warranty. This warranty does not apply: (a) to damage caused by accident, abuse, misuse, misapplication, or non-Apple products; (b) to damage caused by service (including upgrades and expansions) performed by anyone who is not an Apple Authorized Service Provider; (c) to a product or a part that has been modified without the written permission of Apple; or (d) if any Apple serial number has been removed or defaced

THIS WARRANTY AND REMEDIES SET FORTH ABOVE ARE EXCLUSIVE AND IN LIEU OF ALL OTHER WARRANTIES, REMEDIES AND CONDITIONS, WHETHER ORAL OR WRITTEN, EXPRESS OR IMPLIED. APPLE SPECIFICALLY DISCLAIMS ANY AND ALL IMPLIED WARRANTIES, INCLUDING, WITHOUT LIMITATION, WARRANTIES OF MERCHANTABILITY AND FITNESS FOR A PARTICULAR PURPOSE. IF APPLE CANNOT LAWFULLY DISCLAIM IMPLIED WARRANTIES UNDER THIS LIMITED WARRANTY, ALL SUCH WARRANTIES, INCLUDING WARRANTIES OF MERCHANTABILITY AND FITNESS FOR A PARTICULAR PURPOSE ARE LIMITED IN DURATION TO THE DURATION OF THIS WARRANTY. No Apple reseller, agent, or employee is authorized to make any modification, extension, or addition to this warranty.
APPLE IS NOT RESPONSIBLE FOR DIRECT, SPECIAL, INCIDENTAL OR CONSEQUENTIAL DAMAGES RESULTING FROM ANY BREACH OF WARRANTY OR CONDITION, OR UNDER ANY OTHER LEGAL THEORY, INCLUDING BUT NOT LIMITED TO LOST PROFITS, DOWNTIME, GOODWILL, DAMAGE TO OR REPLACEMENT OF EQUIPMENT AND PROPERTY, ANY COSTS OF RECOVERING, REPROGRAMMING, OR REPRODUCING ANY PROGRAM OR DATA STORED IN OR USED WITH APPLE PRODUCTS, AND ANY FAILURE TO MAINTAIN THE CONFIDENTIALITY OF DATA STORED ON THE PRODUCT. APPLE SPECIFICALLY DOES NOT REPRESENT THAT IT WILL BE ABLE TO REPAIR ANY PRODUCT UNDER THIS WARRANTY OR MAKE A PRODUCT EXCHANGE WITHOUT RISK TO OR LOSS OF PROGRAMS OR DATA. Some states and provinces do not allow the exclusion or limitation of incidental or consequential damages or exclusions or limitations on the duration of implied warranties or conditions, so the above limitations or exclusions may not apply to you. This warranty gives you specific legal rights, and you may also have other rights that vary by state or province.

warranty disclaimer text is capitalized for emphasis

FIGURE 7-3   Apple computer warranty disclaimer

Authority to Form Contracts

As explained previously in this section, a contract is formed when an offer is accepted for consideration. Problems can arise when the acceptance is issued by an imposter or someone who does not have the authority to bind the company to a contract. In electronic commerce, the online nature of acceptances can make it relatively easy for identity forgers to pose as others.

Fortunately, the Internet technology that makes forged identities so easy to create also provides the means to avoid being deceived by a forged identity. In Chapter 10, you will learn how companies and individuals can use digital signatures to establish identity in online transactions. If the contract is for any significant amount, the parties should require each other to use digital signatures to avoid identity problems. In general, courts will not hold a person or corporation whose identity has been forged to the terms of the contract; however, if negligence on the part of the person or corporation contributed to the forgery, a court may hold the negligent party to the terms of the contract. For example, if a company was careless about protecting passwords and allowed an imposter to enter the company's system and accept an offer, a court might hold that company responsible for fulfilling the terms of that contract.

Determining whether an individual has the authority to commit a company to an online contract is a greater problem than forged identities in electronic commerce. This issue, called **authority to bind**, can arise when an employee of a company accepts a contract and the company later asserts that the employee did not have such authority. For large transactions in the physical world, businesses check public information on file with the state of incorporation, or ask for copies of corporate certificates or resolutions, to establish the authority of persons to make contracts for their employers. These methods are available to parties engaged in online transactions; however, they can be time consuming and awkward. You will learn about some good electronic solutions, such as digital signatures and certificates from a certification authority, in Chapter 10.

Terms of Service Agreements

Many Web sites have stated rules that site visitors must follow, although most visitors are not aware of these rules. If you examine the home page of a Web site, you will often find a link to a page titled "Terms of Service," "Conditions of Use," "User Agreement," or something similar. If you follow that link, you find a page full of detailed rules and regulations, most of which are intended to limit the Web site owner's liability for what you might do with information you obtain from the site. These contracts are often called **terms of service (ToS)** agreements even when they appear under a different title. In most cases, a site visitor is held to the terms of service even if that visitor has not read the text or clicked a button to indicate agreement with the terms. The visitor is bound to the agreement by simply using the site. The first few sections of the Amazon.com terms of service agreement appear in Figure 7-4, which shows the top of Amazon's Conditions of Use page.

316

amazon.com.   🛒 VIEW CART | WISH LIST | (YOUR ACCOUNT) | HELP

| WELCOME | YOUR STORE | BOOKS | ELECTRONICS | TOYS & GAMES | SOFTWARE | CELL PHONES & SERVICE | COMPUTER & VIDEO GAMES | ▶ SEE MORE STORES |

Help > Privacy & Security > Privacy Notice > Conditions of Use

Conditions of Use

explains that these terms of service apply to all site visitors

Welcome to Amazon.com. Amazon.com and its affiliates provide their services to you subject to the following conditions. **If you visit or shop at Amazon.com, you accept these conditions.** Please read them carefully. In addition, when you use any current or future Amazon.com service (e.g., Friends & Favorites, e-Cards, Auctions, and Honor System) or visit or purchase from any business affiliated with Amazon.com, whether or not included in the Amazon.com Web site, you also will be subject to the guidelines and conditions applicable to such service or business.

PRIVACY

Please review our Privacy Notice, which also governs your visit to Amazon.com, to understand our practices.

ELECTRONIC COMMUNICATIONS

When you visit Amazon.com or send e-mails to us, you are communicating with us electronically. You consent to receive communications from us electronically. We will communicate with you by e-mail or by posting notices on this site. You agree that all agreements, notices, disclosures and other communications that we provide to you electronically satisfy any legal requirement that such communications be in writing.

COPYRIGHT

All content included on this site, such as text, graphics, logos, button icons, images, audio clips, digital downloads, data compilations, and software, is the property of Amazon.com or its content suppliers and protected by United States and international copyright laws. The compilation of all content on this site is the exclusive property of Amazon.com and protected by U.S. and international copyright laws. All software used on this site is the property of Amazon.com or its software suppliers and protected by United States and international copyright laws.

TRADEMARKS

AMAZON.COM; AMAZON.COM BOOKS; EARTH'S BIGGEST BOOKSTORE; IF IT'S IN PRINT, IT'S IN STOCK; MOODMATCHER; 1-CLICK; and other marks indicated on our site are registered trademarks of Amazon.com, Inc. or its subsidiaries, in the United States and other countries. EARTH'S BIGGEST SELECTION, PURCHASE CIRCLES, SHOP THE WEB, ONE-CLICK SHOPPING, AMAZON.COM ASSOCIATES, AMAZON.COM MUSIC, AMAZON.COM VIDEO, AMAZON.COM TOYS, AMAZON.COM ELECTRONICS, AMAZON.COM e-CARDS, AMAZON.COM AUCTIONS, zSHOPS, CUSTOMER BUZZ, AMAZON.CO.UK, AMAZON.DE, BID-CLICK, GIFT-CLICK, AMAZON.COM ANYWHERE, AMAZON.COM OUTLET, BACK TO BASICS, BACK TO BASICS TOYS, NEW FOR YOU, AMAZON HONOR SYSTEM, PAYPAGE, UNPAY, and other Amazon.com graphics, logos, page headers, button icons, scripts, and service names are trademarks or trade dress of Amazon.com, Inc. or its subsidiaries. Amazon.com's trademarks and trade dress may not be used in connection with any product or service that is not Amazon.com's, in any manner that is likely to cause confusion among customers, or in any manner that disparages or discredits Amazon.com. All other trademarks not owned by Amazon.com or its subsidiaries that appear on this site are the property of their respective owners, who may or may not be affiliated with, connected to, or sponsored by Amazon.com or its subsidiaries.

PATENTS

One or more United States and international patents apply to this site, including without limitation: U.S. Patent Nos. 5,715,399; 5,727,163; 5,826,258; 5,960,411; 5,963,949; and 5,999,924.

LICENSE AND SITE ACCESS

Amazon.com grants you a limited license to access and make personal use of this site and not to download (other than page caching) or modify it, or any portion of it, except with express written consent of Amazon.com. This license does not include any resale or commercial use of this site or its contents; any collection and use of any product listings, descriptions, or prices; any derivative use of this site or its contents; any downloading or copying of account information for the benefit of another merchant; or any use of data mining, robots, or similar data gathering and extraction tools. This site or any portion of this site may not be reproduced, duplicated, copied, sold, resold, visited, or otherwise exploited for any commercial purpose without express written consent of Amazon.com. You may not frame or utilize framing techniques to enclose any trademark, logo, or other proprietary information (including images, text, page layout, or form) of Amazon.com and our affiliates without express written consent. You may not use any meta tags or any other "hidden text" utilizing Amazon.com's name or trademarks without the express written consent of Amazon.com. Any unauthorized use terminates the permission or license granted by

FIGURE 7-4   Amazon.com conditions of use page

317

The Environment of Electronic Commerce: Legal, Ethical, and Tax Issues

USE AND PROTECTION OF INTELLECTUAL PROPERTY IN ONLINE BUSINESS

Online businesses must be careful in their use of intellectual property. **Intellectual property** is a general term that includes all products of the human mind. These products can be tangible or intangible. Intellectual property rights include the protections afforded to individuals and companies by governments through governments' granting of copyrights and patents, and through registration of trademarks and service marks. Online businesses must take care to avoid deceptive trade practices, making false advertising claims, engaging in defamation or product disparagement, and violations of intellectual property rights by using unauthorized content on their Web sites or in their domain names.

Web Site Content Issues

A number of legal issues can arise regarding the Web page content of electronic commerce sites. The most common concerns involve the use of intellectual property that is protected by other parties' copyrights, patents, trademarks, and service marks.

Copyright Infringement

A **copyright** is a right granted by a government to the author or creator of a literary or artistic work. The right is for the specific length of time provided in the copyright law and gives the author or creator the sole and exclusive right to print, publish, or sell the work. Creations that can be copyrighted include virtually all forms of artistic or intellectual expression—books, music, artworks, recordings (audio and video), architectural drawings, choreographic works, product packaging, and computer software. In the United States, works created after 1977 are protected for the life of the author plus 70 years. Works copyrighted by corporations or not-for-profit organizations are protected for 95 years from the date of publication or 120 years from the date of creation, whichever is earlier.

The idea contained in an expression cannot be copyrighted. It is the particular form in which an idea is expressed that creates a work that can be copyrighted. If an idea cannot be separated from its expression in a work, that work cannot be copyrighted. For example, mathematical calculations cannot be copyrighted. A collection of facts can be copyrighted, but only if the collection is arranged, coordinated, or selected in a way that causes the resulting work to rise to the level of an original work. For example, the Yahoo! Web Directory is a collection of links to URLs. These facts existed before Yahoo! selected and arranged them into the form of its directory. However, most copyright lawyers would argue that the selection and arrangement of the links into categories probably makes the directory copyrightable.

In the past, many countries (including the United States) required the creator of a work to register that work to obtain copyright protection. U.S. law still allows registration, but registration is no longer required. A work that does not include the words "copyright" or "copyrighted," or the copyright symbol ©, and that was created after 1977, is copyrighted automatically by virtue of the copyright law unless the creator specifically released the work into the public domain.

Most Web pages are protected by copyright because they arrange the elements of words, graphics, and HTML tags in a way that creates an original work. This creates a potential problem because of the way the Web works. As you learned in Chapter 2, when a Web client requests a page, the Web server sends an HTML file to the client. Thus, a copy of the HTML file (along with any graphics or other files needed to render the page) resides on the Web client computer. Most legal experts agree that this copying is a fair use of the copyrighted Web page. The U.S. copyright law includes an exemption from infringement actions for fair use of copyrighted works. The **fair use** of a copyrighted work includes copying it for use in criticism, comment, news reporting, teaching, scholarship, or research. The law's definition of fair use is intentionally broad and can be difficult to interpret. When you make fair use of a copyrighted work, you must be careful to provide a citation to the original work to avoid charges of plagiarism. The **University of Texas Crash Course in Copyright** is a particularly helpful source of information on making fair use determinations.

Copyright law has always included elements, such as the fair use exemption, that make it difficult to apply. The Internet has made this situation worse because it allows the immediate transmission of exact digital copies of many materials. In the case of digital music, the Napster site provided a network that millions of people used to trade music files that they had copied from their CDs and compressed into MPEG version 3 format, commonly referred to as MP3. This constituted copyright violation on a grand scale, and a group of music recording companies sued Napster for facilitating the violations.

Napster argued that it had only provided the "machinery" used in the copyright violations—much as electronics companies manufacture and sell VCRs that might be used to make illegal copies of videotapes—and had not itself infringed on any copyrights. Both the U.S. District Court and the Federal Appellate Court held that Napster was guilty of vicarious copyright infringement, even though it did not directly violate any music recording companies' copyrights. An entity becomes liable for **vicarious copyright infringement** if it is capable of supervising the infringing activity and obtains a financial benefit from the infringing activity. Napster failed to monitor its network even though it could have done so. It also profited (by selling advertising on its Web site) indirectly from the infringement. Thus Napster was held liable even though Napster itself did not transfer any copies. The courts ordered that Napster be shut down. In late 2001, Napster agreed to pay $26 million in damages for copyright infringement to a group of music publishing associations and began working on relaunching the site with agreements in place to pay copyright holders for the music that would be downloaded in the future. After Napster filed for bankruptcy in 2002, software company **Roxio** bought all of Napster's intellectual property, including its name and Web site, for about $5 million. Roxio launched a new **Napster** site in October 2003. The site now offers legal music downloads to subscribers.

With the growth in popularity of portable music devices such as Apple's iPod, the demand for music in the MP3 (and similar) formats has continued to increase. The companies that sell music downloads, such as the new Napster site and Apple's iTunes site, each have different rules and restrictions that come with the downloaded files. Some sites allow one copy to be installed on a portable music device. Others allow a limited number of copies to be installed. Still others allow unlimited copies, but only if the devices on which the copies are installed are owned by the person who downloaded the file.

The legality of the common practice of copying files from music CDs and placing those files on a portable music device (or onto another CD) is unclear in many cases. This type of copying is governed in the United States by the fair use provisions of the copyright laws, which you learned about earlier in this chapter. The fair use provisions as they relate to copying music tracks are, at best, unclear and difficult to interpret. Some lawyers would argue that a person has the right under the fair use provisions to make a backup copy of a music CD track, but other lawyers would disagree. A person who makes one copy for a portable music device, a second copy for a computer, and a third copy on a CD for backup purposes would be less likely to be protected under the fair use provisions, but some lawyers would argue that all three are protected uses.

Patent Infringement

A **patent** is an exclusive right granted by the government to an individual to make, use, and sell an invention. In the United States, patents on inventions protect the inventor's rights for 20 years. A patent on the design for an invention provides protection for 14 years. To be patentable, an invention must be genuine, novel, useful, and not obvious given the current state of technology. In the early 1980s, companies began obtaining patents on software programs that met the terms of the U.S. patent law. However, most firms that develop software to use in Web sites and for related transaction processing have not found the patent law to be very useful. The process of obtaining a patent is expensive and can take several years. Most developers of Web-related software believe that the technology in the software could become obsolete before the patent protection is secured.

One type of patent has been of interest to companies engaging in electronic commerce. A U.S. Court of Appeals ruled in 1998 that patents could be granted on "methods of doing business." The **business process patent**, which protects a specific set of procedures for conducting a particular business activity, is quite controversial. In addition to the Amazon.com patent on its 1-Click purchasing method (which you read about in Chapter 4), other Web businesses have obtained business process patents. The Priceline.com "name your own price" price-tendering system, About.com's approach to aggregating information from many different Web sites, and Cybergold's method of paying people to view its Web site have each received business process patents.

The ability of companies to enforce their rights under these patents is not yet clear. Many legal experts and business researchers believe that the issuance of business process patents grants the recipients unfair monopoly power and is an inappropriate extension of patent law. In 1999, Amazon.com sued Barnes & Noble for using a process on its Web site that was similar to the 1-Click method. The case was settled out of court in 2002, but the terms of the settlement were not disclosed. The U.S. Supreme Court has not yet ruled on any cases involving business process patents. To read an interesting discussion of both sides of the business process patent issue that includes exchanges between Jeff Bezos, founder of Amazon.com, and book publisher Tim O'Reilly, see the article posted at **My Conversation with Jeff Bezos**.

Trademark Infringement

A **trademark** is a distinctive mark, device, motto, or implement that a company affixes to the goods it produces for identification purposes. A **service mark** is similar to a trademark, but it is used to identify services provided. In the United States, trademarks and service marks can be registered with state governments, the federal government, or both. The name (or a part of that name) that a business uses to identify itself is called a **trade name**. Trade names are not protected by trademark laws unless the business name is the same as the product (or service) name. They are protected, however, under common law. **Common law** is the part of British and U.S. law established by the history of court decisions that has accumulated over many years. The other main part of British and U.S. law, called **statutory law**, arises when elected legislative bodies pass laws, which are also statutes.

The owners of registered trademarks have often invested a considerable amount of money in the development and promotion of their trademarks. Web site designers must be very careful not to use any trademarked name, logo, or other identifying mark without the express permission of the trademark owner. For example, a company Web site that includes a photograph of its president who happens to be holding a can of Pepsi could violate Pepsi's trademark rights. Pepsi can argue that the appearance of its trademarked product on the Web site implies an endorsement of the president or the company by Pepsi.

Domain Names, Cybersquatting, and Name Stealing

Considerable controversy has arisen recently about intellectual property rights and Internet domain names. **Cybersquatting** is the practice of registering a domain name that is the trademark of another person or company in the hopes that the owner will pay huge amounts of money to acquire the URL. In addition, successful cybersquatters can attract many site visitors and, consequently, charge high advertising rates. A related problem, called **name changing**, occurs when someone registers purposely misspelled variations of well-known domain names. These variants sometimes lure consumers who make typographical errors when entering a URL. **Name stealing** occurs when someone posing as a site's administrator changes the ownership of the site's assigned domain name to another site and owner. Name stealing is more of a nuisance than a serious problem because the act can be quickly identified and the ownership of the domain name switched back to the rightful owner before significant damage occurs.

Since 1999, the U.S. Anticybersquatting Consumer Protection Act has prevented businesses' trademarked names from being registered as domain names by other parties. The law provides for damages of up to $100,000 per trademark. If the registration of the domain name is found to be "willful," damages can be as much as $300,000. Recent U.S. cases that were settled out of court illustrate the problem. For example, three cybersquatters made headlines when they tried to sell the URL barrydiller.com for $10 million. Barry Diller, the CEO of USA Networks, sued the trio and won.

Registering a generic name such as Wine.com is very different from registering a trademarked name in bad faith—cybersquatting. Registering a generic name is legal speculation that the name might one day become valuable. Disputes that arise when one person has registered a domain name that is an existing trademark or company name are settled by the **World Intellectual Property Association (WIPO)**. The WIPO began settling domain name disputes in 1999 under its Uniform Domain Name Dispute Resolution Policy (UDRP).

One common type of dispute arises when a business has a trademark that is a common term. If a person obtains the domain name containing that common term, the owner of the trademark must seek resolution at the WIPO. In 2000, Gordon Sumner, who had performed music for more than 20 years as Sting, filed a complaint with the WIPO because a Georgia man obtained the domain name www.sting.com and had reportedly offered to sell it to Sting for $25,000. In more than 80 percent of its cases, the WIPO has held for the trademark name owner; however, in this case, the WIPO noted that the word "sting" was in common and general use and had multiple meanings other than as an identifier for the musician. The WIPO refused to award the domain to Sumner. After the WIPO decision, the two parties came to undisclosed terms and the musician's official Web site is now at www.sting.com.

Many critics have argued that the WIPO UDRP policy has been enforced unevenly and that many of the decisions under the policy have been inconsistent. One problem faced by those who have used the WIPO resolution service is that the WIPO decisions are not appealed to one authority. Instead, the party seeking redress must file suit in a court with the appropriate jurisdiction. No central authority maintains records of all WIPO decisions and appeals. You can learn more about WIPO UDRP decisions by reading the Harvard Law School's **Berkman Center UDRP Opinion Guide**. A complete list of all UDRP decisions with links to the text of each decision appears on the **ICANN UDRP Proceedings** Web pages.

After obtaining a domain name, companies still face the possibility that someone will steal unsuspecting customers by registering a domain name that is a slight variation, or even a misspelling, of a company's well-known domain name. A simple typo in a Web address could lead a Web surfer to LLBaen.com instead of LLBean.com. The Anticybersquatting Consumer Protection Act now helps distinguish between cases that are true cybersquatting and those that are permissible competition. Most businesses agree that the practice of name changing is annoying to affected online businesses and confusing to customers. A company's best defense is to register as many variations in product and company spellings as possible. Unfortunately, there is no complete solution to this problem; as new high-level domains such as .biz become available, the name changing problem recurs.

Perhaps the most flagrant example of domain name abuse is name stealing. Name stealing occurs when someone other than a domain name's owner changes the ownership of the domain name. A **domain name ownership change** occurs when owner information maintained by a public domain registrar is changed in the registrar's database to reflect a new owner's name and business address. This usually happens only when safeguards are not in place. Once domain name ownership is changed, the name stealer can manipulate the site, post graffiti on it, or redirect online customers to other sites selling substandard goods. The main purpose of name stealing is to harass the site owner. The temporary loss of its domain name can cut off a business from its Web site for several days.

Protecting Intellectual Property Online

Several industry trade groups have proposed solutions to the current problems in digital copyright protection, including host name blocking, packet filtering, and proxy servers. All three approaches illustrate how an Internet service provider might try to block access to an entire offending site. However, none of these approaches are really effective in preventing theft or providing identification of property obtained without the copyright holder's permission.

Several methods show promise in the battle to protect digital works, but they only provide partial protection. New and improved methods are continually being developed. One promising technique employs steganography to create a **digital watermark**. The watermark is a digital code or stream embedded undetectably in a digital image or audio file. It can be encrypted to protect its contents, or simply hidden among the bits—digital information—comprising the image or recording. **Verance** is a company that provides, among other products, digital audio watermarking systems to protect audio files on the Internet. Its systems identify, authenticate, and protect intellectual property. Verance's ARIS MusiCode system enables recording artists to monitor, identify, and control the use of their digital recordings.

The audio watermarks do not alter the audio fidelity of the recordings in which they are embedded. The Verance SoniCode product provides verification and authentication tools. SoniCode was originally developed by ARIS Technologies, which is now owned by Verance Corporation. SoniCode can ensure that telephonic conversations have not been altered. The same is true for audiovisual transcripts and depositions. **Blue Spike** produces a watermarking system called Giovanni. Like the SoniCode system, the Giovanni watermark authenticates the copyright and provides copy control. **Copy control** is an electronic mechanism for limiting the number of copies that one can make of a digital work.

A group of more than 180 companies and organizations devoted to providing protection for intellectual property—digital music in this case—is the **Secure Digital Music Initiative (SDMI)** organization. Its members include information technology and consumer electronics companies, security technology firms, Internet service providers, and the music recording industry. SDMI's charter is to develop open, public technology specifications that protect the playing, storing, and distribution of digital music.

Digimarc is another company providing watermark protection systems and software. Its products embed a watermark that allows any works protected by its Digimarc system to be tracked across the Web. In addition, the watermark can link viewers to commerce sites and databases. It can also control software and playback devices. Finally, the imperceptible watermark contains copyright information and links to the image's creator, which enables nonrepudiation of a work's authorship and facilitates electronic purchase and licensing of the work.

Defamation

A **defamatory** statement is a statement that is false and that injures the reputation of another person or company. If the statement injures the reputation of a product or service instead of a person, it is called **product disparagement**. In some countries, even a true and honest comparison of products may give rise to product disparagement. Because the difference between justifiable criticism and defamation can be hard to determine, commercial Web sites should avoid making negative, evaluative statements about other persons or products.

Web site designers should be especially careful to avoid potential defamation liability by altering a photo or image of a person in a way that depicts the person unfavorably. In most cases, a person must establish that the defamatory statement caused injury. However, most states recognize a legal cause of action, called **per se defamation**, in which a court deems some types of statements to be so negative that injury is assumed. For example, the court will hold inaccurate statements alleging conduct potentially injurious to a person's business, trade, profession, or office as defamatory per se—the complaining party need not prove injury to recover damages. Thus, online statements about competitors should always be carefully reviewed before posting to determine whether they contain any elements of defamation.

An important exception in U.S. law exists for statements that are defamatory but that are about a public figure (such as a politician or a famous actor). The law allows considerable leeway for statements that are satirical or that are valid expressions of personal opinion. Other countries do not offer the same protections, so operators of Web sites with international audiences do need to be careful.

Also, recall that defaming or disparaging statements must be false. This protects Web sites that include unfavorable reviews of products or services if the statements made are not false. For example, if a person reads a book and believes it to be terrible, that person can safely post a review on Amazon.com that includes assessments of the book's lack of literary value. Such statements of personal opinion are true statements and thus neither defamatory or disparaging.

Deceptive Trade Practices

The ease with which they can edit graphics, audio, and video files allows Web site designers to do many creative and interesting things. Manipulations of existing pictures, sounds, and video clips can be very entertaining. If the objects being manipulated are trademarked, however, these manipulations can violate the trademark holder's rights. Fictional characters can be trademarked or otherwise protected. Many personal Web pages include unauthorized use of cartoon characters and scanned photographs of celebrities; often, these images are altered in some way. A Web site that uses an altered image of Mickey Mouse speaking in a modified voice is likely to hear from the Disney legal team.

Web sites that include links to other sites must be careful not to imply a relationship with the companies sponsoring the other sites unless such a relationship actually exists. For example, a Web design studio's Web page may include links to company Web sites that show good design principles. If those company Web sites were not created by the design studio, the studio must be very careful to state that fact. Otherwise, it would be easy for a visitor to assume that the linked sites were the work of the design studio.

In general, trademark protection prevents another firm from using the same or a similar name, logo, or other identifying characteristic in a way that would cause confusion in the minds of potential buyers of the trademark holder's products or services. For example, the trademarked name "Visa" is used by one company for its credit card services and another company for its type of synthetic fiber. This use is acceptable because the two products are significantly different. However, the use of very well known trademarks can be protected for all products if there is a danger that the trademark might be diluted. Various state laws define **trademark dilution** as the reduction of the distinctive quality of a trademark by alternative uses. Trademarked names such as "Hyatt," "Trivial Pursuit," and "Tiffany," and the shape of the Coca-Cola bottle have all been protected from dilution by court rulings. A Web site that sells gift-packaged seafood and claims to be the "Tiffany of the Sea" risks a lawsuit from the famous jeweler claiming trademark dilution.

Advertising Regulation

In the United States, advertising is regulated primarily by the **Federal Trade Commission**. The FTC publishes regulations and investigates claims of false advertising. Its Web site includes a number of information releases that are useful to businesses and consumers. The FTC business education campaign publications are available on its Advertising Guidance page, shown in Figure 7-5. These publications include information to help businesses comply with the law.

Any advertising claim that can mislead a substantial number of consumers in a material way is illegal under U.S. law. In addition to conducting its own investigations, the FTC accepts referred investigations from organizations such as the Better Business Bureau. The FTC provides policy statements that can be helpful guides for designers creating electronic commerce Web sites. These policies include information on what is permitted in advertisements and cover specific areas such as these:

- Bait advertising
- Consumer lending and leasing
- Endorsements and testimonials
- Energy consumption statements for home appliances
- Guarantees and warranties
- Prices

Other federal agencies have the power to regulate online advertising in the United States. These agencies include the Food and Drug Administration (FDA), the Bureau of Alcohol, Tobacco, and Firearms (BATF), and the Department of Transportation (DOT). The FDA regulates information disclosures for food and drug products. In particular, any Web site that is planning to advertise pharmaceutical products will be subject to the FDA's drug labeling and advertising regulations. The BATF works with the FDA to monitor and enforce federal laws regarding advertising for alcoholic beverages and tobacco products. These laws require that every ad for such products includes statements that use very specific language. Many states also have laws that regulate advertising for alcoholic beverages and tobacco products. The state and federal laws governing advertising and the sale of firearms are even more restrictive. Any Web site that plans to deal in these products should consult with an attorney who is familiar with the relevant laws before posting any online advertising for such products. The DOT works with the FTC to monitor the advertising of companies over which it has jurisdiction, such as bus lines, freight companies, and airlines.

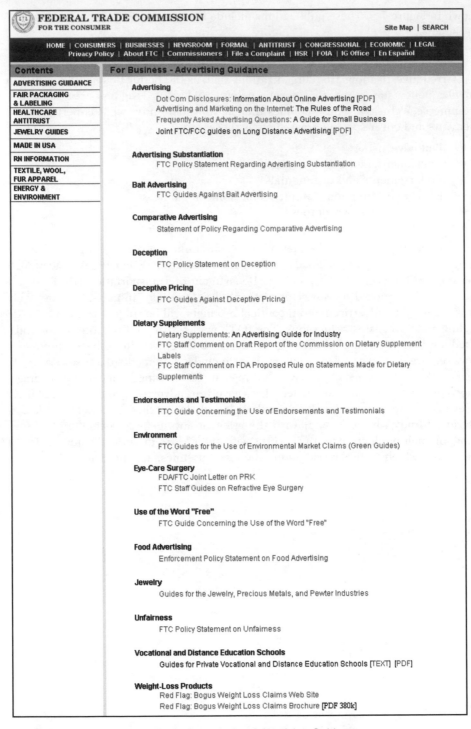

FIGURE 7-5   U.S. Federal Trade Commission Advertising Guidance page

ONLINE CRIME, TERRORISM, AND WARFARE

The Internet has opened up many possibilities for people to communicate and get to know each other better—no matter where in the world they live. The Internet has also opened doors for businesses to reach new markets and create opportunities for economic growth. It is sad that some people in our world have found the Internet to be a useful tool for perpetrating crimes, conducting terrorism, and even waging war.

Online Crime

Crime on the Web includes online versions of crimes that have been undertaken for years in the physical world, including theft, stalking, distribution of pornography, and gambling. Other crimes, such as commandeering one computer to launch attacks on other computers, are new.

Law enforcement agencies have difficulty combating many types of online crime. The first obstacle they face is the issue of jurisdiction. As you learned earlier in this chapter, determining jurisdiction can be tricky on the Internet. Consider the case of a person living in Canada who uses the Internet to commit a crime against a person in Texas. It is unclear which elements of the crime could establish sufficient contact with Texas to allow police there to proceed against a citizen of a foreign country. It is possible that the actions that are considered criminal under Texas and U.S. law might not be considered so in Canada. If the crime is theft of intellectual property (such as computer software or computer files), the questions of jurisdiction become even more complex. You can learn more about online crime issues at the **U.S. Department of Justice Cybercrime** Web site.

Enforcing laws against distribution of pornographic material has also been difficult because of jurisdiction issues. The distinction between legal adult material and illegal pornographic material is, in many cases, subjective and often difficult to make. The U.S. Supreme Court has ruled that state and local courts can draw the line based on local community standards. This creates problems for Internet sales. For example, consider a case in which questionable adult content is sold on a Web site located in Oregon to a customer who downloads the material in Georgia. A difficult question arises regarding which community standards might apply to the sale.

A similar jurisdiction issue arises in the case of online gambling. Many gambling sites are located outside the United States. If people in California use their computers to connect to an offshore gambling site, it is unclear where the gambling activity occurs. Several states have passed laws that specifically outlaw Internet gambling, but the jurisdiction of those states to enforce laws that limit Internet activities is not yet clear.

Another problem facing law enforcement officers is the difficulty of applying laws that were written before the Internet became prevalent to criminal actions carried out on the Internet. For example, most states have stalking laws that provide criminal penalties to people who harass, annoy, or alarm another person in a way that presents a credible threat. Many of these laws are triggered by physical actions, such as physically following the person targeted. The Internet gives a stalker the opportunity to use e-mail or chat room discussions to create the threatening situation. Laws that require physical action on the part of the stalker are not effective against online stalkers. Only a few states have passed laws that specifically address the problem of online stalking.

Online Warfare and Terrorism

Many Internet security experts believe that we are at the dawn of a new age of terrorism and warfare that could be carried out or coordinated through the Internet. A considerable number of Web sites currently exist that openly support or are operated by hate groups and terrorist organizations.

The Internet provides an effective communications network on which many people and businesses have become dependent. Although the Internet was designed from its inception to continue operating while under attack, a sustained effort by a well-financed terrorist group or rogue state could slow down the operation of major transaction-processing centers. As more business communications traffic moves to the Internet, the potential damage that could result from this type of attack increases. You will learn more about security threats and countermeasures for those threats in Chapter 10.

ETHICAL ISSUES

Companies using Web sites to conduct electronic commerce should adhere to the same ethical standards that other businesses follow. If they do not, they will suffer the same consequences that all companies suffer: the damaged reputation and long-term loss of trust that can result in loss of business. In general, advertising or promotion on the Web should include only true statements and should not omit any information that could mislead potential purchasers or wrongly influence their impressions of a product or service. Even true statements have been held to be misleading when the ad omits important related facts. Any comparisons to other products should be supported by verifiable information. The next section explains the role of ethics in formulating Web business policies, such as those affecting visitors' privacy rights and companies' Internet communications with children.

Ethics and Web Business Policies

Web businesses are finding that ethical issues are important to consider when they are making policy decisions. Recall from Chapter 3 that buyers on the Web often communicate with each other. A report of an ethical lapse that is rapidly passed among customers can seriously affect a company's reputation. In 1999, *The New York Times* ran a story that disclosed Amazon.com's arrangements with publishers for book promotions. Amazon.com was accepting payments of up to $10,000 from publishers to give their books editorial reviews and placement on lists of recommended books as part of a cooperative advertising program. When this news broke, Amazon.com issued a statement that it had done nothing wrong and that such advertising programs were a standard part of publisher-bookstore relationships. The outcry on the Internet in newsgroups and mailing lists was overwhelming. Two days later—before most mass media outlets had even reported the story—Amazon.com announced that it would end the practice and offer unconditional refunds to any customers who had purchased a promoted book. Amazon.com had done nothing illegal, but the practice appeared to be unethical to many of its existing and potential customers.

In early 1999, eBay faced a similar ethical dilemma. Several newspapers had begun running stories about sales of illegal items, such as assault weapons and drugs, on the eBay auction site. At this point in time, eBay was listing about 250,000 items each day. Although eBay would investigate claims that illegal items were up for auction on its site, eBay did not actively screen or filter listings before the auctions were placed on the site.

Even though eBay was not legally obligated to screen the items auctioned, and even though screening would be fairly expensive, eBay's executive team decided that screening for illegal and copyright-infringing items would be in the best long-run interest of eBay. The team decided that such a decision would send a signal about the character of the company to its customers and the public in general. The eBay executive team also decided to remove an entire category—firearms—from the site. Not all of eBay's users were happy about this decision—the sale of firearms on eBay, when done properly, was legal. However, the eBay executive team again decided that presenting an overall image of an open and honest marketplace was so important to the future success of eBay that it chose to ban all firearms sales.

An important ethical issue that organizations face when they collect e-mail addresses from site visitors is how the organization limits the use of the e-mail addresses and related information. In the early days of the Web, few organizations made any promises to visitors who provided such information. Today, most organizations state their policy on the protection of visitor information, but many do not. In the United States, organizations are not legally bound to limit their use of information collected through their Web sites. They may use the information for any purpose, including the sale of that information to other organizations. This lack of government regulation that might protect site visitor information is a source of concern for many individuals and privacy rights advocates. These concerns are discussed in the next section.

Privacy Rights and Obligations

The issue of online privacy is continuing to evolve as the Internet and the Web grow in importance as tools of communication and commerce. Many legal and privacy issues remain unsettled and are hotly debated in various forums. The **Electronic Communications Privacy Act of 1986** is the main law governing privacy on the Internet today. Of course, this law was enacted before the general public began its wide use of the Internet. The law was written to update existing law that prevented interception of audio signal transmissions so that any type of electronic transmissions (including, for example, fax or data transmissions) would be given the same protections. In 1986, the Internet was not used to transmit commercially valuable data in any significant amount, so the law was written to deal primarily with interceptions that might occur on leased telephone lines.

In recent years, a number of legislative proposals have been advanced that specifically address online privacy issues, but, thus far, none have withstood constitutional challenges. In July 1999, the FTC issued a report that examined how well Web sites were respecting visitors' privacy rights. Although it found a significant number of sites without posted privacy policies, the report concluded that companies operating Web sites were developing privacy practices with sufficient speed and that no federal laws regarding privacy were required at that time. Privacy advocacy groups responded to the FTC report with outrage and calls for legislation. Thus, the near-term future of privacy regulation in the United States is unclear. The Direct Marketing Association (DMA), a trade association

329

of businesses that advertise their products and services directly to consumers using mail, telephone, Internet, and mass media outlets, has established a set of privacy standards for its members. However, critics note that past efforts by the DMA to regulate its members' activities have been less than successful.

Ethics issues are significant in the area of online privacy because laws have not kept pace with the growth of the Internet and the Web. The nature and degree of personal information that Web sites can record when collecting information about visitors' page-viewing habits, product selections, and demographic information can threaten the privacy rights of those visitors.

The Internet has also changed traditional assumptions about privacy because it allows people anywhere in the world to gather data online in quantities that would have been impossible a few years ago. For example, real estate transactions are a matter of public record in the United States. These transactions have been recorded in county records for many years and have been available to anyone who wanted to go to the county recorder's office and spend hours leafing through large books full of handwritten records. Many counties have made these records available on the Internet, so now a researcher can examine thousands of real estate transaction records in hours without traveling to a single county office. Many privacy experts see this change in the ease of data access to be an important shift that affects the privacy rights of those who participate in real estate transactions. Because the Internet makes such data more readily available to a wider range of people, the privacy previously afforded to the participants in those transactions has been reduced.

Differences in cultures throughout the world have resulted in different expectations about privacy in electronic commerce. In Europe, for example, most people expect that information they provide to a commercial Web site will be used only for the purpose for which it was collected. Many European countries have laws that prohibit companies from exchanging consumer data without the express consent of the consumer. In 1998, the European Union adopted a Directive on the Protection of Personal Data. This directive codifies the constitutional rights to privacy that exist in most European countries and applies them to all Internet activities. In addition, the directive prevents businesses from exporting personal data outside the European Union unless the data will continue to be protected in accordance with provisions of the directive. The European Union and its member countries have consistently exhibited a strong preference for using government regulations to protect privacy. The United States has exhibited an opposite preference. U.S. companies, especially those in the direct mail marketing industry, have consistently and successfully lobbied to avoid government regulation and allow the companies to police themselves.

One of the major privacy controversies in the United States today is the opt-in vs. opt-out issue. Most companies that gather personal information in the course of doing business on the Web would like to be able to use that information for any purpose of their own. Some companies would also like to be able to sell or rent that information to other companies. No U.S. law currently places limits on companies' use of such information. Companies are, in general, also free to sell or rent customer information. An increasing number of U.S. companies do provide a way for customers who would like to restrict use of their personal information to do so. The most common policy used in U.S. companies today is an opt-out approach. In an **opt-out** approach, the company collecting the information assumes that the customer does not object to the company's use of the information unless the customer specifically chooses to deny permission (that is, to opt out of having their information used). In the less common **opt-in** approach, the company collecting the information does not use the information for any other purpose (or sell or rent the information) unless the customer specifically chooses to allow that use (that is, to opt in and grant permission for the use). Figure 7-6 shows an example Web page that presents a series of opt-in choices to site visitors. The Web site will not send any of these three items to a site visitor unless that visitor opts in by checking one or more boxes.

```
Many of our site visitors and customers enjoy receiving our newsletter, periodic notices of
sales and special product offerings, and offers from other companies that we have chosen to
ensure that they will be of interest to our site visitors. Please check the boxes below to add
your e-mail address to our distribution list for any or all of these electronic mailings.

    ☐  Weekly e-mail newsletter

    ☐  Periodic notices of sales and special product offerings

    ☐  Offers from other companies
```

FIGURE 7-6   Sample Web page showing opt-in choices

Figure 7-7 shows the opt-out approach. A Web site that uses the opt-out approach will send all three items to the site visitor unless the site visitor checks the boxes to indicate that the items are not wanted.

```
Many of our site visitors and customers enjoy receiving our newsletter, periodic notices of
sales and special product offerings, and offers from other companies that we have chosen to
ensure that they will be of interest to our site visitors. Please check the boxes below if you do
not wish to be added to our distribution list for any or all of these electronic mailings.

    ☐  Weekly e-mail newsletter

    ☐  Periodic notices of sales and special product offerings

    ☐  Offers from other companies
```

FIGURE 7-7   Sample Web page showing opt-out choices

As you can see, it is easy for site visitors to misread the text and make the wrong choice when deciding whether or not to check the boxes. Sites that use the opt-out approach are often criticized for requiring their visitors to take an affirmative action (checking the empty boxes) to prevent the site from sending items. Another approach to presenting opt-out choices is to use a page that includes checked boxes and instructs the visitor to "uncheck the boxes of the items you do not wish to receive." Most privacy advocates believe that the opt-in approach is preferable because it gives the customer privacy protection unless that customer specifically elects to give up those rights. Most U.S. businesses have traditionally taken the position that they have a right to use the information they collect unless the provider of the information explicitly objects. Some of these companies are changing to the opt-in approach, often at the prodding of privacy advocacy groups.

Until the legal environment of privacy regulation becomes more clear, privacy advocates recommend that electronic commerce Web sites be conservative in their collection and use of customer data. Mark Van Name and Bill Catchings, writing in *PC Week* in 1998, outlined four principles for handling customer data that provide a good outline for Web site administrators even today. These principles are as follows:

- Use the data collected to provide improved customer service.
- Do not share customer data with others outside your company without the customer's permission.
- Tell customers what data you are collecting and what you are doing with it.
- Give customers the right to have you delete any of the data you have collected about them.

A number of organizations are active in promoting privacy rights. You can learn more about current developments in privacy legislation and practices throughout the world by following the links to these organizations' Web sites that appear under the heading **Privacy Rights Advocacy Groups** in the Online Companion.

LEARNING FROM FAILURES

DOUBLECLICK

As you learned in Chapter 4, **DoubleClick** is one of the largest banner advertising networks in the world. DoubleClick arranges the placement of banner ads on Web sites. Like many other Web sites, DoubleClick uses **cookies**, which are small text files placed on Web client computers, to identify returning visitors.

Most visitors find the privacy risk posed by cookies to be acceptable. Visitors to Amazon.com, for example, have Amazon.com cookies placed on their computers so that the Web server at Amazon.com recognizes them when they return. This can be useful, for example, when a visitor who has placed several items in a shopping cart before being interrupted can return to Amazon.com later in the day and find the shopping cart intact—the Web server can read the client's Amazon.com cookie and find the shopping cart from the client's previous session. The Amazon.com server can read only its own cookies; it cannot read the cookies placed on the client computer by any other Web server.

There are two important differences between the Amazon.com scenario and what happens when DoubleClick serves a banner ad. First, the visitor usually does not know that the banner ad is coming from DoubleClick (and thus, does not know that the DoubleClick server could be writing a cookie to the client computer). Second, DoubleClick serves ads through Web sites owned by thousands of companies. As a visitor moves from one Web site to another, that visitor's computer can collect many DoubleClick cookies. The DoubleClick server can read all of its own cookies, gathering information from each one about which ads were served and the sites through which they were served. Thus, DoubleClick can compile a tremendous amount of information about where a visitor has been on the Web.

continued

Even this amount of information collection would not trouble most people. Double-Click can use the cookies to track a particular computer's connections to Web sites, but it does not record any identity information about the owner of that computer. Therefore, DoubleClick accumulates a considerable record of Web activity, but cannot connect that activity with a person.

In 1999, DoubleClick arranged a $1.7 billion merger with Abacus Direct Corporation. Abacus had developed a way to link information about people's Web behavior (collected through cookies such as those placed by DoubleClick's banner ad servers) to the names, addresses, and other information about those people that had been collected in an offline consumer database.

The reaction from online privacy protection groups was immediate and substantial. The FTC launched an investigation, the Internet's privacy issues e-mail lists and chat rooms buzzed with discussions, and, in the end, DoubleClick abandoned its plans to integrate its cookie-generated data with the identity information in the Abacus database. Although DoubleClick is still one of the largest banner advertising networks, it has not met its profitability targets. DoubleClick had been counting on generating additional revenue by using the information in the combined database that it was unable to create.

When the FTC probe concluded two years later, DoubleClick was not charged with any violations of laws or regulations. The lesson here is that a company violates the Internet community's ethical standards at its own peril, even if the transgression does not break any laws.

Communications with Children

An additional set of privacy considerations arises when Web sites attract children and engage in some form of communication with those children. Adults who interact with Web sites can read privacy statements and make informed decisions about whether to communicate personal information to the site. The communication of private information (such as credit card numbers, shipping addresses, and so on) is a key element in the conduct of electronic commerce.

The laws of most countries and most sets of ethics consider children to be less capable than adults in evaluating information sharing and transaction risks. Thus, we have laws in the physical world that prevent or limit children's ability to sign contracts, get married, drive motor vehicles, and enter certain physical spaces (such as bars, casinos, tattoo parlors, and race tracks). Children are considered to be less able (or unable) to make informed decisions about the risks of certain activities. Similarly, many people are concerned about children's ability to read and evaluate privacy statements and then consent to providing personal information to Web sites.

Under the laws of most countries, people under the age of 18 or 21 are not considered adults. However, those countries that have proposed or passed laws that specify differential treatment for the privacy rights of children often define "child" as a person below the age of 12 or 13. This complicates the issue because it creates two classes of nonadults.

In the United States, Congress enacted the Children's Online Protection Act (COPA) in 1998 to protect children from "material harmful to minors." This law was held to be unconstitutional because it unnecessarily restricted access to a substantial amount of material that is lawful, thus violating the First Amendment. Congress was more successful with the **Children's Online Privacy Protection Act of 1998 (COPPA)**, which provides restrictions on data collection that must be followed by electronic commerce sites aimed at children. This law does not regulate content, as COPA attempted to do, so it has not been successfully challenged on First Amendment grounds. In 2001, Congress enacted the Children's Internet Protection Act (CIPA). The CIPA requires schools that receive federal funds to install filtering software on computers in their classrooms and libraries. Filtering software is used to block access to adult content Web sites. In 2003, the Supreme Court held that the CIPA was constitutional.

Companies with Web sites that appeal to nonadults must be careful to comply with the laws governing their interactions with these young visitors. **Disney Online** is a site that appeals primarily to young children. The Disney Online registration page offers three choices to visitors who want to register with the site and receive regular communications and updates. The first registration choice is for adults, a second choice is for "teens," and a third choice is for "kids." The "kids" choice leads to a screen that explains that Disney does not enroll subscribers who are under the age of 13. The Disney.com registration page for "teens" asks for the visitor's name, birthday, and the e-mail address of a parent. Disney uses the birthday to calculate the visitor's age and refuses to enroll the visitor if the age is less than 13. Disney uses the parent's e-mail address to notify parents of their child's registration and to invite them to set up a family account. Family accounts are controlled by parents who can elect to allow family members who are under the age of 13 to use the site. By refusing to enroll any child under age 13 as a site subscriber, Disney Online meets the requirements of the COPPA law. Other sites that appeal to a young audience use similar techniques to limit unsupervised access to their Web pages. For example, Sanrio (the company that produces Hello Kitty and related products) asks for a birthdate before allowing access to its English-language site that is directed at U.S. customers, **Sanriotown**. As shown in Figure 7-8, the site encourages visitors to notify the company that operates the site if they know a child who has gained access to the site in violation of COPPA.

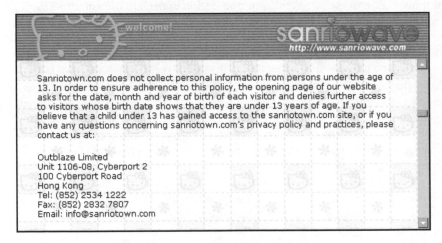

FIGURE 7-8 Sanrio's approach to COPPA compliance

TAXATION AND ELECTRONIC COMMERCE

Companies that do business on the Web are subject to the same taxes as any other company. However, even the smallest Web business can become instantly subject to taxes in many states and countries because of the Internet's worldwide scope. Traditional businesses may operate in one location and be subject to only one set of tax laws for years. By the time those businesses are operating in multiple states or countries, they have developed the internal staff and record-keeping infrastructure needed to comply with multiple tax laws. Firms that engage in electronic commerce must comply with these multiple tax laws from their first day of existence.

An online business can become subject to several types of taxes, including income taxes, transaction taxes, and property taxes. **Income taxes** are levied by national, state, and local governments on the net income generated by business activities. **Transaction taxes**, which include sales taxes, use taxes, excise taxes, and customs duties, are levied on the products or services that the company sells or uses. Customs duties are taxes levied by the United States and other countries on certain commodities when they are imported into the country. **Property taxes** are levied by states and local governments on the personal property and real estate used in the business. In general, the taxes that cause the greatest concern for Web businesses are income taxes and sales taxes.

Nexus

A government acquires the power to tax a business when that business establishes a connection with the area controlled by the government. For example, a business that is located in Kansas has a connection with the state of Kansas and is subject to Kansas taxes. If that company opens a branch office in Arizona, it forms a connection with Arizona and becomes subject to Arizona taxes on the portion of its business that occurs in Arizona. This connection between a taxpaying entity and a government is called **nexus**. The concept of nexus is similar in many ways to the concept of personal jurisdiction discussed earlier in this chapter. The activities that.create nexus in the United States are determined by state law and thus vary from state to state. Nexus issues have been frequently litigated, and the resulting common law is fairly complex. Determining nexus can be difficult when a company conducts only a few activities in or has minimal contact with the state. In such cases, it is advisable for the company to obtain the services of a professional tax advisor.

Companies that do business in more than one country face national nexus issues. If a company undertakes sufficient activities in a particular country, it establishes nexus with that country and becomes liable for filing tax returns in that country. The laws and regulations that determine national nexus are different in each country. Again, companies will find the services of a professional tax lawyer or accountant who has experience in international taxation to be valuable.

U.S. Income Taxes

The **Internal Revenue Service (IRS)** is the U.S. government agency charged with administering the country's tax laws. A basic principle of the U.S. tax system is that any verifiable increase in a company's wealth is subject to federal taxation. Thus, any company whose U.S.-based Web site generates income is subject to U.S. federal income tax. Furthermore, a Web site maintained by a company in the United States must pay federal income tax on income generated outside of the United States. To reduce the incidence of double taxation of foreign earnings, U.S. tax law provides a credit for taxes paid to foreign countries. The IRS Web site's home page appears in Figure 7-9.

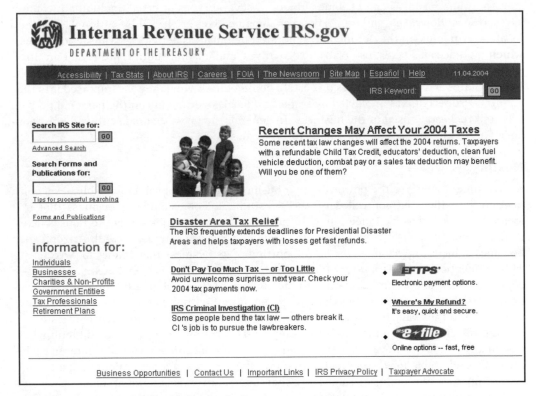

FIGURE 7-9   Internal Revenue Service home page

The IRS site includes links to downloadable tax forms, copies of IRS publications, current tax news, and other useful tax information. The home page offers links to sections of the Web site that are designed to help specific categories of site visitors.

Most states levy an income tax on business earnings. If a company conducts activities in several states, it must file tax returns in all of those states and apportion its earnings in accordance with each state's tax laws. In some states, the individual cities, counties, and other political subdivisions within the state also have the power to levy income taxes on business earnings. Companies that do business in multiple local jurisdictions must apportion their income and file tax returns in each locality that levies an income tax. The number of taxing authorities (which includes states, counties, cities, towns, school districts, water districts, and many other governmental units) in the United States exceeds 30,000.

Companies that sell through their Web sites do not, in general, establish nexus everywhere their goods are delivered to customers. Usually, a company can accept orders and ship from one state to many other states and avoid nexus by using a contract carrier such as FedEx or United Parcel Service to deliver goods to customers.

U.S. State Sales Taxes

Most states levy a transaction tax on goods sold to consumers. This tax is usually called a sales tax. Businesses that establish nexus with a state must file sales tax returns and remit the sales tax they collect from their customers. If a business ships goods to customers in other states, it is not required to collect sales tax from those customers unless the business has established nexus with the customer's state. However, the customer in this situation is liable for payment of a use tax in the amount that the business would have collected as sales tax if it had been a local business.

A **use tax** is a tax levied by a state on property used in that state that was not purchased in that state. Most states' use tax rates are identical to their sales tax rates. In addition to property purchased in another state, use taxes are assessed on property that is not "purchased" at all. For example, lease payments on vehicles are subject to use taxes in most states. The leased vehicle is not purchased (in any state) but when it is used in the lessee's state, it incurs that state's use tax. In the past, few consumers filed use tax returns and few states enforced their use tax laws with regularity. However, an increasing number of states are providing a line on their individual income tax returns that asks people to report and pay their use tax for the year along with their state income taxes. Some states allow taxpayers to estimate their use tax liability; others require an exact statement of the use tax amount.

Larger businesses use complex software to manage their sales tax obligations. Not only are the sales tax rates different in the 7500 U.S. sales tax jurisdictions (which include states, counties, cities, and other sales tax authorities), but the rules about which items are taxable differ. For example, New York's sales tax law provides that large marshmallows are taxable (because they are "snacks"), but small marshmallows are not taxable (because they are "food").

Some purchasers are exempt from sales tax, such as certain charitable organizations and businesses buying items for resale. Thus, to determine whether a particular item is subject to sales tax, a seller must know where the customer is located, what the laws of that jurisdiction say about taxability and tax rate, and the taxable status of the customer.

The sales tax collection process in the United States is largely regarded as a serious problem. Even the Supreme Court, in one of its sales tax decisions more than 10 years ago, stated that the situation is needlessly confusing and encouraged Congress to act. Although a number of bills have been introduced over the years, none has become law. Some large online retailers, such as Amazon.com, have announced that they will begin collecting and remitting sales tax on all sales, even when the sale is delivered into a state with which the company does not have nexus.

Many of the states have joined together through the **National Governor's Association** and the **National Conference of State Legislatures** to create the Streamlined Sales and Use Tax Agreement (SSUTA). The SSUTA simplifies state sales taxes by making the various state tax codes more congruent with each other while allowing each state to set its own rates. Each state must adopt the Agreement, and once a state does adopt it, companies in the state can choose one of several simple procedures for collecting and remitting sales taxes nationwide.

European Union Value Added Taxes

The United States raises most of its revenue through income taxes. Other countries, especially those in the European Union (EU), use transfer taxes to generate most of their revenues. The Value Added Tax (VAT) is the most common transfer tax used in these countries. A VAT is assessed on the amount of value added at each stage of production. For example, if a computer keyboard manufacturer purchased keyboard components for $20 and then sold finished keyboards for $50, the value added would be $30. VAT is collected by the seller at each stage of the transaction. For example, a product that goes through five different companies on its way to the ultimate consumer would have VAT assessed on each of the five sales. In most countries, the VAT is calculated at the time of each intermediate sale and remitted to the country in which that sale occurs.

The EU enacted legislation concerning the application of VAT to sales of digital goods that became effective in mid-2003. Companies based in EU countries must collect VAT on digital goods no matter where in the EU the products are sold. This legislation has attracted the attention of companies based outside of the EU that sell digital goods to consumers based in one or more EU countries. Under the law, non-EU companies that sell into the EU must now register with EU tax authorities and levy, collect, and remit VAT if their sales include digital goods delivered into the EU.

Summary

The legal concept of jurisdiction on the Internet is still unclear and ill defined. The relationship between geographic boundaries and legal boundaries is based on four elements: power, effects, legitimacy, and notice. These four elements have helped governments create the legal concept of jurisdiction in the physical world. Because the four elements exist in somewhat different forms on the Internet, the jurisdiction rules that work so well in the physical world do not always work well in the online world.

As in traditional commerce, contracts are a part of doing business on the Web and are established through various types of offers and acceptances. Any contract for the electronic sale of goods or services includes implied warranties. Many companies include contracts or rules on their Web sites in the form of terms of service agreements. Contracts can be invalidated when one of the parties to the transaction is an imposter; however, forged identities are becoming easier to detect through electronic security tools.

Seemingly innocent inclusion of photographs, whether manipulated or not, and other elements on a Web page can lead to infringement of trademarks, copyrights, or patents; defamation; and violation of intellectual property rights. An international administrative mechanism now exists for resolving domain name disputes that has reduced the need for lengthy and expensive litigation in many cases. Electronic commerce sites must be careful not to imply relationships that do not actually exist. Negative evaluative statements about entities, even when true, are best avoided given the subjective nature of defamation and product disparagement.

Unfortunately, some people use the Internet for perpetrating crimes, advocating terrorism, and even waging war. Law enforcement agencies have found it difficult to combat many types of online crime, and governments are working to create adequate defenses for online war and terrorism.

Web business practices such as collecting information and tracking consumer habits have led to questions of ethics regarding online privacy. Some countries are far more restrictive than others in terms of what type of information collection is acceptable and legal. Companies that collect personal information can use an opt-in policy, in which the customer must take an action to permit information collection, or an opt-out policy, in which the customer must take an action to prevent information collection. Opt-in policies are more protective of customers' privacy rights. Web businesses also must be careful when communicating with children. In general, laws require parental consent be obtained before information is collected from children under the age of 13.

Companies that conduct electronic commerce are subject to the same laws and taxes as other companies, but the nature of doing business on the Web can expose companies to a large number of laws and taxes sooner than traditional companies usually face them. The international nature of all online business further complicates a firm's tax obligations. Although some legal issues are straightforward, others are difficult to interpret and follow because of the newness of electronic commerce and the unsettled nature of applicable law. The large number of government agencies that have jurisdiction and the power to tax makes it essential that companies doing business on the Web understand the potential liabilities of doing business with customers in those jurisdictions.

Key Terms

Acceptance	Nexus
Authority to bind	Notice
Business process patent	Offer
Common law	Opt-in
Consideration	Opt-out
Constructive notice	Patent
Contract	Per se defamation
Cookies	Personal jurisdiction
Copy control	Power
Copyright	Product disparagement
Cybersquatting	Property tax
Defamatory	Service mark
Digital watermark	Signature
Domain name ownership change	Statute of Frauds
Effects	Statutory law
Fair use	Subject-matter jurisdiction
Forum selection clause	Terms of service (ToS)
Implied contract	Tort
Income tax	Trade name
Intellectual property	Trademark
Judicial comity	Trademark dilution
Jurisdiction	Transaction tax
Legitimacy	Use tax
Long-arm statute	Vicarious copyright infringement
Name changing	Warranty disclaimer
Name stealing	Writing

Review Questions

RQ1. In about 100 words, explain why online businesses might have difficulty limiting the effects of their actions to a relatively small geographic area.

RQ2. In about 300 words, describe the differences between subject-matter jurisdiction and personal jurisdiction.

RQ3. The advantages and disadvantages of issuing business process patents have been hotly debated by legal scholars and business people. One compromise proposal advanced by Jeff Bezos, founder of Amazon.com, is to allow the issuance of business patents, but only allow them to be effective for a short time, perhaps two or three years. In about 300 words, present logical and factual arguments that support the issuance of such limited-term business process patents.

RQ4. Define product disparagement. In two or three paragraphs, present an example of product disparagement.

RQ5. In about 300 words, explain the idea of nexus. Why is it an important concept in state and international taxation? In what ways is it similar to jurisdiction?

Exercises

E1. Use **Google** or your favorite Web search engine to obtain a list of Web pages that include the words "privacy statement." Visit the Web pages on the search results list until you find a page that includes the text of a privacy statement. Print the page and turn it in with your answers to the following questions:

- Does the site follow an opt-in or opt-out policy (or is the policy not clearly stated in the privacy statement)?

- Does the privacy statement include a specific provision or provisions regarding the collection of information from children?

- Does the privacy statement describe what happens to the collected personal information if the company goes out of business or is sold to another company? List those provisions.

Write one paragraph in which you evaluate the clarity of the privacy statement.

E2. Use your favorite search engine, the links in the Online Companion for this exercise, and your library to learn more about the Napster lawsuit. Identify the main issues in the case and the principal arguments that could be used by either side.

- In about 300 words, present the case against Napster.

- In about 300 words, present one or more well-reasoned arguments to support Napster's position.

E3. Use **Google** or your favorite search engine to find a Web site (other than Disney or Sanriotown) that is directed to young people. Examine the site to determine how it complies with COPPA. Test the site to ensure that it does not accept information from children under the age of 13. Evaluate the site's compliance with COPPA in a report of about 200 words.

E4. In the United States, a law called the Internet Tax Moratorium (ITM) has been enacted and renewed several times. The purpose of the ITM is to prevent federal, state, or local governments from enacting any new taxes on Internet business activities. Use **Google** or your favorite search engine to learn more about the ITM. In about 300 words, critically evaluate the rationale behind the law and take a position on whether the law should be renewed again.

Cases

C1. Nissan.com

The Nissan Motor Company of Japan had sold its cars in the United States under the brand name Datsun for many years. In the late 1980s, the company changed its branding policy and began selling cars in the U.S. market with the name of Nissan. However, the company did not realize that the Web would become an important marketing tool and did not register the name nissan.com as soon as it became available.

Nissan was not the only auto company to miss an opportunity to register its brand's domain name early. General Motors had registered the domain gm.com in 1992, but it had not registered generalmotors.com. The company had to purchase that name from Gil Vanorder, who had registered it in 1997. Vanorder's site featured a cigar-smoking, uniform-wearing cartoon character named "General John C. Motors." Volkswagen (which had registered vw.com when it first became available) successfully sued Virtual Works (an ISP) to obtain the domain name vw.net. Other auto companies have purchased or sued (with mixed results) to obtain domain names that included their product brand names. DaimlerChrysler was able to purchase dodge.com in 2001 from the London financial software company that had registered it originally. Ford had to sue National A-1 Advertising to obtain the right to use lincoln.com. However, Ford was unsuccessful in its attempts to obtain mercury.com. That name is still used by the New York City information technology services company, Mercury Technologies, that first registered the name.

In 1991, Uzi Nissan formed a company named Nissan Computer Corp. in North Carolina to sell computer hardware and provide related repair and consulting services. Nissan's company also offered networking hardware for sale, along with related services. In 1994, the company registered the name nissan.com. In 1996, the company registered the domain name nissan.net and began offering ISP services to individuals and companies at that Web site.

In 1995, he received a letter from a lawyer representing Nissan Motor Company. The letter requested information about how Nissan was planning to use the domain name nissan.com. Since he was operating a computer company and Nissan was an auto company, Nissan decided there would be no potential confusion in customers' minds about the relationship (or lack thereof) between Nissan Computer and Nissan Motors. Nissan did not respond to the letter. The lawyer did not follow up with any other contact, so Nissan considered the issue closed.

In 2000, Nissan Motors sued Nissan Computer under the U.S. Anticybersquatting Consumer Protection Act for $10 million and the exclusive right to use the names nissan.com and nissan.net. Uzi Nissan argued in court that he was just using his family name (which is a common name in the Middle East) to which he had a basic right, that he had no intent to profit from the name (he was unwilling to sell it to Nissan Motors at any price), and that there was little likelihood that his computer store would be confused in the minds of the consumers with the international auto company of the same name. Nissan Motors argued that its brand name was so well known that any alternative use of the name would be confusing to consumers.

In 2002, opinions issued by the California Superior court and the U.S. Ninth Circuit District Court held that Nissan Computer had not acted in bad faith when it acquired the disputed domain names. However, the court ruled that Nissan Computer could no longer use the domain names for commercial purposes because of the potential confusion it could create in the minds of consumers. Nissan Computer would have to find a different domain name for its business. The court also ordered that Nissan could not place any advertising on his Web sites at nissan.com or nissan.net and prohibited him from placing disparaging remarks or negative commentary about Nissan Motors (or links to such remarks or commentary) on the two sites. The court did not, however, order the transfer of the two domain names to Nissan Motor. The Online Companion includes links to the Web sites operated today by Nissan Computer and Nissan Motors.

Required:

1. U.S. courts sometimes appoint advisors (often called Special Masters) to help them decide cases that involve complex business or technical issues. Assume you are a business advisor to a court that is hearing an appeal of the *Nissan Motor Co. v. Nissan Computer Corp.* case. In about 200 words, explain why Nissan Motors is so concerned about the use of these two domain names and how a monetary damages judgment of $10 million could be justified (if you do not believe that the monetary damages are justified, explain why).

2. In about 200 words, provide an outline of the ethics of the position taken by Uzi Nissan in this dispute.

3. In about 200 words, provide an outline of the ethics of the position taken by Nissan Motors in this dispute.

4. If you believe that the courts' decisions in this case are fair to the parties and the general public, explain why in about 200 words. If you believe that the courts' decisions are not fair, outline a decision (in about 200 words) that you believe would be fair.

Note: Your instructor might assign you to a group to complete this case, and might ask you to prepare a formal presentation of your results to your class.

C2. Ellasaurus Products Enterprises

Ellen Carson is the author and illustrator of a successful series of children's books that chronicle the adventures of Ellasaurus, a four-year-old orange dinosaur. Ellen has done well with the books, but her business advisors have told her that she could earn considerably more money by creating a merchandising business around the Ellasaurus character. Following this advice, she has created Ellasaurus Products Enterprises (EPE), a company that has begun developing and marketing Ellasaurus toys, stuffed animals, coloring books, pajamas, and Halloween costumes.

EPE has had some success in its attempts to get major retailers to stock the Ellasaurus product line, but Ellen is concerned that retailers might not be willing to take on a new and unproven product. She would like to create a Web site through which EPE could sell its merchandise directly to customers. She also sees the Web site as a way to build customer loyalty. Ellen envisions a site with a number of portal features in addition to the product sales. For example, she would like to offer online games, chat rooms, e-mail accounts, and other activities that would promote EPE products and her books.

The Ellasaurus book series appeals to children that are between four and six years old. Ellen expects the EPE product line to appeal to children in about the same age range. Ellen has visited sites such as **Hello Kitty** and **Nick Jr.** , which appeal to similar age groups, to get ideas for the site. She would like the site to be appealing to her main audience, but she would like to obtain registration information from site visitors so EPE can send e-mails with information about new products and Web site features to them.

Ellen plans to limit the Web site's merchandise sales to U.S. residents at first, but she hopes to begin selling internationally within a few years. The site will allow visitors from any country to register and participate in the online portal features.

Required:

1. Ellen will use some copyrighted illustrations from her books on the Web site. She will also include themes from the story lines of her books in some of the games that will be available (free) on the site to registered visitors. Prepare a report of about 300 words in which you discuss at least two intellectual property issues that might arise in the operation of the Web site.

2. In about 200 words, describe the ethical issues that Ellen faces because of the ages of her intended audience members.

3. In about 300 words, outline the laws with which the site must comply when it registers site visitors under the age of 13. Include recommendations regarding how Ellen can best comply with those laws.

4. In about 300 words, describe the sales tax liabilities to which the Web site will be exposed. Assume that Ellen will operate the site from her home office in Michigan and that EPE will manufacture the merchandise in Texas. The merchandise will be warehoused at EPE distribution centers in New Jersey, Ohio, and California.

Note: Your instructor might assign you to a group to complete this case, and might ask you to prepare a formal presentation of your results to your class.

For Further Study and Research

Angwin, J. 2001. "Are Domain Panels the Hanging Judges of Cyberspace?" *The Wall Street Journal*, August 20, B1.

Bagby, J. and F. McCarty. 2003. *The Legal and Regulatory Environment of E-Business.* Cincinnati: Thomson South-Western.

Bodeen, C. 2004. "China Shuts Down Internet Blogs," *Salon.com*, March 19. (http://www.salon.com/news/wire/2004/03/19/blogs2/index.html)

Brilmayer, L. 1989. "Consent, Contract, and Territory," *Minnesota Law Review*, 74(1), 11–12.

Cass, S. 2002. "Nissan v. Nissan," *IEEE Spectrum*, 39(10), October, 53–54.

Clark, P. 2001. "Doubts Cloud DoubleClick's Repositioning," *B to B*, 86(15), August 28, 1–2.

Cope, N. 2000. "A Hit for Jethro Tull in Domain Name Dispute," *The Independent*, July 31, 15.

Crane, E. 2000. "Double Trouble," *Ziff Davis Smart Business*, 13(10), October, 62.

Creed, A. 2001. "E-Trade Swallows $90,000 Fine," *BizReport*, July 10. (http://www.bizreport.com/ article.php?id=1692)

Digital Millennium Copyright Act. 1998. Public Law No. 105-304, 112 Statutes 2860.

Direct Marketing. 2001. "FTC Closes DoubleClick Investigation," 63(12), April, 18.

Federal Trade Commission (FTC). 1999. *Self-Regulation and Privacy Online: A Report to Congress.* Washington: FTC.

Flynn, L. 2000. "Whose Name Is It Anyway? Arbitration Panels Favoring Trademark Holders in Disputes Over Web Names," *The New York Times*, September 4, C3.

Foster, A. 2002. "Computer Crime Incidents at Two California Colleges Tied to Investigation Into Russian Mafia," *Chronicle of Higher Education*, June 24. (http://chronicle.com/free/2002/06/2002062401t.htm)

Greene, S. 2001. "Reconciling Napster with the Sony Decision and Recent Amendments to Copyright Law," *American Business Law Journal*, 39(1), Fall, 57–98.

Greenhouse, L. 2003. "Court Upholds Law to Make Libraries Use Internet Filters," *The New York Times*, June 24, A1.

Hamblen, M. 2003. "Regulatory Requirements Place New Burdens on IT: U.S. Firms Scramble to Comply with EU Tax," *Computerworld*, June 30, 1.

Hardesty, D. 2004. *Electronic Commerce Taxation and Planning, 2004 Update Edition*. Boston: Warren, Gorham & Lamont.

Hardesty, D. 2004. *Sales Tax and Electronic Commerce*. Larkspur, CA: ClickBank.

Harmon, A. 2001. "As Public Records Go Online, Some Say They're Too Public," *The New York Times*, August 24, A1.

Harvard Law Review. 1999. "The Criminalization of Copyright Infringement in the Digital Era," 112(7), May, 1705–1722.

Heckman, J. 2000. "Trademarks Protected Through New Cyber Act," *Marketing News*, 34(1), January 3, 6–7.

Heilemann, J. 2000. "David Boies: The Wired Interview," *Wired*, October. (http://www.wired.com/wired/archive/ 8.10/boies.html)

Hemphill, T. 2000. "DoubleClick and Consumer Online Privacy: An E-Commerce Lesson Learned," *Business & Society Review*, 105(3), Fall, 361–372.

Hirschman, C. 2001. "Prosecuting in the Name of Privacy," *Telephony*, 241(7), August 13, 82.

Hurt, E. 2000. "FTC Wins Internet's Respect," *Business 2.0*, October 13. (http://www.business2.com/content/ channels/technology/2000/10/13/21123)

Hutheesing, N. 2001. "Master of Your Domain," *Forbes*, 167(5), Spring, 60.

Hwang, W. and J. Klosek. 2003. "Taxing the Sale of Digital Goods in Europe," *E-Commerce Law & Strategy*, 20(3), July 11, 1.

Ian, J. 2002. "The Internet Debacle: An Alternative View," *Performing Songwriter Magazine*, May. (http://www.janisian.com/)

Isenberg, D. 2000. "Many Trademarks, But Just One Domain Name," *Internet World*, July 1, 86.

Janal, D. 1999. "Thirty Essential Steps to Take Right Now to Prevent Online Crime," *Communication World*, 16(4), March, 34–36.

Jones, J. 2000. "Protecting Privacy," *InfoWorld*, 22(18), May 1, 40–41.

Journal of Internet Law. 2002. "Computer Firm's Use of Nissan.com Not Bad Faith Under Anticybersquatting Act," 6(1), July, 23.

Kahin, B. and C. Nesson (eds.). 1997. *Borders in Cyberspace*. Cambridge, MA: MIT Press.

Kaplan, C. 2002. "A Libel Suit May Decide E-Jurisdiction," *The New York Times*, May 27. (http://www.nytimes.com/2002/05/27/technology/27ELAW.html)

Keeler, D. 2000. "Taxation Slips Through the Net," *Global Finance*, 14(6), June, 60–61.

Kisiel, R. 2002. "Two Nissans Collide on Information Highway," *Automotive News*, December 16, 1IT–2IT.

Krim, J. 2004. "Justice Department to Announce Cyber-Crime Crackdown: Actions to Include Arrests, Subpoenas," *The Washington Post*, August 25, E5.

Leonard, A. 2002. "Nissan vs. Nissan," *Salon.com*, June 3. (http://www.salon.com/tech/col/leon/2002/06/03/ nissan/index.html)

Lessig, L. 2000. *Code and Other Laws of Cyberspace*. New York: Basic Books.

Liptak, A. 2003. "U.S. Courts' Role in Foreign Feuds Comes Under Fire," *The New York Times*, August 3, 1.

Manjoo, F. 2001. "Fine Print Not Necessarily in Ink," *Wired News*, April 6. (http://www.wired.com/news/business/0,1367,42858,00.html)

345

McClintock, M., N. Maguire, J. Kilby, and D. Barlow. 2000. "Electronic Commerce," *International Tax Review*, July-August, 9–13.

Meller, P. 2000. "Europe Passes Stiff E-Commerce Law," *The Industry Standard*, December 1. (http://www. thestandard.com/article/display/0,1151,20526,00.html)

Miller, R. and G. Jentz. 2002. *Law for E-Commerce*. Cincinnati: West.

Moran, J. and J. Kummer. 2003. "U.S. and International Taxation of the Internet: Part I," *Computer & Internet Lawyer*, 20(4), April, 1–18.

Mueller, M. 2002. *Rough Justice: An Analysis of ICANN's Uniform Dispute Resolution Policy*. Syracuse, NY: Syracuse University Convergence Center. (http://dcc.syr.edu/roughjustice.htm)

Murray, J. 2000. "E-Contracts Present Courts with Special Legal Challenges," *Purchasing*, 129(3), August 24, 119-120.

Network Briefing Daily. 2002. "Amazon Settles 1-Click Patent Dispute," March 8, 3–4.

Nissan Motor Co. v. Nissan Computer Corp., 246 F.3d 675 (9th Cir. 2002).

Oder, N. 2002. "COPA Ruling Offers Mixed Message," *Library Journal*, 127(11), June 15, 15.

Olavsrud, T. 2002. "Supreme Court Partially Lifts Bar on COPA ," *Internet News*, May 13. (http://www.internetnews.com/bus-news/article.php/1121271)

Olin, J. 2001. "Reducing International E-Commerce Taxes," *World Trade*, 14(3), March, 64–66.

Oliva, R. and S. Prabakar. 1999. "Copyright Perils Can Lurk on the Business Web," *Marketing Management*, 8(1), Spring, 54–57.

Pantazis, A. 1999. "Zeran v. America Online, Inc.: Insulating Internet Service Providers from Defamation Liability," *Wake Forest Law Review*, 34(2), Summer, 531–555.

Phillips, D. 2003. "JetBlue Apologizes for Use of Passenger Records," *The Washington Post*, September 20, E1.

Porter, K. and S. Bradley. 1999. *eBay, Inc.* Case #9-700-007. Cambridge, MA: Harvard Business School.

Radcliff, D. 2000. "Domain Name Game," *Computerworld*, 34(24), June 12, 71.

Radcliff, D. 2001. "E-Merchant Beware," *Computerworld*, 35(25), June 18, 42.

Reagle, J. 1999. "The Platform for Privacy Preferences," *Communications of the ACM*, 42(2), February, 48–51.

Rewick, J. 2000. "DoubleClick Finds Its Abacus Unit Nettlesome," *The Wall Street Journal*, October 19, B6.

Richtel, M. 2004. "U.S. Steps Up Push Against Online Casinos by Seizing Cash," *The New York Times*, May 31, C1.

Samborn, H. 2000. "Nibbling Away at Privacy," *ABA Journal*, 86(2), June, 26-27.

Samuelson, P. 1999. "Good News and Bad News on the Intellectual Property Front," *Communications of the ACM*, 42(3), March, 19–24.

Shaller, D. 2000. "E-mail, the Internet, and Other Legal and Ethical Nightmares," *Strategic Finance*, August, 82(2), 48–52.

Smedinghoff, T. (ed.). 1996. *Online Law: The SPA's Legal Guide to Doing Business on the Internet*. Reading, MA: Addison-Wesley Developers Press.

Stellin, S. 2002. "In Fights Over .Com Names, Trademark Owners Usually Win," *The New York Times*, June 24, 4.

Stone, M. 2001. "Court Dismisses Class Action Against eBay," *BizReport*, January 19. (http://www.bizreport.com/ daily/2001/01/20010119-4.htm)

Surowiecki, J. 2003. "Patent Bending," *The New Yorker*, July 14, 36.

Swire, P. and R. Litan. 1998. *None of your Business: World Data Flows, Electronic Commerce, and the European Privacy Directive.* Washington: Brookings Institution Press.

The Economist. 2000. "Business Ethics: Doing Well by Doing Good," 355(8167), April 22, 65–67.

The Economist. 2000. "The Internet's Chastened Child," 357(8196), November 11, 80.

Tynan, D. 2000. "Privacy 2000: In Web We Trust?" *PC World*, 18(6), June, 103–111.

United Nations. 1970. "Declaration on Principles of International Law Concerning Friendly Relations and Cooperation Among States in Accordance with the Charter of the United Nations," *General Assembly Resolution*, #2625, 35th Session.

Van Name, M. and B. Catchings. 1998. "Practical Advice About Privacy and Customer Data," *PC Week*, 15(27), July 6, 38.

Vijayan, J. and K. Ohlson. 2000. "Standards Issue Mars E-Signatures," *Computerworld*, 34(28), July 10, 1, 16.

Warner, M. 2002. "The New Napsters," *Fortune*, 146(3), August 12, 115–116.

Wiley, L. 1999. "Proposed Revisions to European Copyright Laws Cause a Stir," *E Media Professional*, 12(4), April, 16–17.

Wilke, J. 2001. "Twenty States Oppose Airlines' Proposal for Joint Venture in Online Reservations," *The Wall Street Journal*, January 11, A10.

Williams, J., J. Clark, C. Clark, and J. Noe. 1999. "What a Tangled Web: The Legal and Public Relations Dangers of Operating a Web Site," *Information Strategy*, 15(3), Spring, 6–12.

Wingfield, N. 2002. "Napster Boy, Interrupted: Shawn Fanning Discusses Demise of His Brainchild And Future of Online Music," *The Wall Street Journal*, October 1, B1.

Wood, C. 2001. "Collusion in the Air," *PC Magazine*, 20(9), May 8, 199.

Zuckerman, A. 2001. "Somebody Out There Wants to Cheat You!" *World Trade*, 13(3), March, 36–38.

PART 3

TECHNOLOGIES FOR ELECTRONIC COMMERCE

CHAPTER **8**

WEB SERVER HARDWARE AND SOFTWARE

LEARNING OBJECTIVES

In this chapter, you will learn about:

* Web server basics
* Software for Web servers
* E-mail management and spam control issues
* Internet and Web site utility programs
* Web server hardware

INTRODUCTION

As you learned in earlier chapters, **Lands' End** was one of the most successful clothing retailers on the

Web before it was acquired by Sears in 2003. Now, as a division of Sears, Lands' End continues to be a

leader in adding features that attract customers to its Web site and keep those customers coming back.

Behind the scenes at Lands' End, a team of experienced technology professionals implements new Web

page features and performs the many regular maintenance tasks that are necessary to keep the Lands'

End Web site running smoothly.

Lands' End closely monitors the performance of its Web site to make sure that customers have a

consistent experience each time they visit the site. The Web site's technical team works hard to make sure

that site visitors do not notice the Web site's operating characteristics. This goal has not always been easy to attain because the site's traffic volume has, on average, doubled each year since the site opened. Also, regular major improvements to the Lands' End site keep the Web team busy.

Lands' End's specific goals for performance change as Web technologies improve. For example, the site management team has a target for the time it takes one of the site's Web pages to load on a visitor's computer. In the early days of the site, that target was 15 seconds. Today, the target is under five seconds. The Web site's technical team has always taken a conservative approach to operating the site so that the site can meet its performance goals more easily. For example, the technical team specifies the maximum and average sizes of Web pages and graphics files that the content team can use. In addition, the technical team must complete all major changes to the site (including thorough testing) before November 1 each year, prior to the holiday selling season. Lands' End makes over 40 percent of its total annual sales in November and December and does not want to take any chances with Web site changes during that time period.

The server hardware at Lands' End is a mix of **Sun** and **IBM** computers that are managed by another computer that allocates incoming Web traffic. Some of the Web site's advanced features, such as the graphics-intensive My Virtual Model, are created on a separate set of computers. These computers are all located at the Lands' End division headquarters in a small town near Madison, Wisconsin. The computers run a UNIX-based operating system from Sun called Solaris and a version of the Apache Web server software, about which you will learn more in this chapter. Although the Lands' End technical team writes some of the software that it uses to monitor the Web site's performance, the company also uses the services of Keynote Systems. Keynote can measure how fast particular pages load or how rapidly transactions are completed at various times of the day. Keynote can make these measurements at a number of locations around the world.

By paying close attention to the details, the technical team at Lands' End keeps the Web site operating at or above expected levels. When customers become so absorbed in the shopping experience that they do not notice the operation of the site, the technical team has done its job.

WEB SERVER BASICS

This chapter provides background information on the basic technologies used to build Web sites that can support online business operations. It includes a discussion of server software and hardware. It also includes an introduction to software that these sites use to perform utility functions such as site maintenance, diagnostics, and e-mail management. In later chapters, you will learn about software that accomplishes specific electronic commerce functions, such as order entry and processing, content management and delivery, user verification and security, and payment processing.

The main job of a Web server computer is to respond to requests from Web client computers. The three main elements of a Web server are the hardware (computers and related components), operating system software, and Web server software. All three of these elements must work together to provide sufficient capacity in a given situation.

After most companies have decided on the goals they want to accomplish with their Web sites, they begin developing their sites by estimating the number of visitors they expect to have, how many pages those visitors will view during an average visit, how large those pages will be (including graphics and other page elements), and the likely maximum number of simultaneous visitors. The next step is to determine the hardware and software combination that will work best to meet the needs of site visitors.

Types of Web Sites

An important first step in planning a Web server is to determine what the company wants to accomplish with the server. The company must estimate how many visitors will be connecting to the Web site and what types of files (graphics, multimedia, or text) will be delivered through the site. The company must also assess its existing information technology staff. Some companies have a large staff with a depth of experience, while others have a small or relatively inexperienced staff.

Companies create Web sites for a wide variety of reasons and in a wide variety of forms. Each has a different purpose, requires different computer hardware and software, and requires different monetary and personnel resources. Decisions about server hardware and software should be driven by the volume and type of Web activities expected. Types of sites include:

- *Development sites*: Simple sites that companies use to evaluate different Web designs with little initial investment. A development site can reside on an existing PC running Web server software. Multiple testers access the site through their client computers on an existing LAN.

- *Intranets*: Corporate networks that house internal memos, corporate policy handbooks, expense account worksheets, budgets, newsletters, and a variety of other corporate documents.
- *Extranets*: Intranets that allow certain authorized parties outside the company (such as suppliers or strategic partners) to access certain parts of the information stored in the system.
- *Transaction-processing sites*: Commerce sites such as business-to-business and business-to-consumer electronic commerce sites that must be available 24 hours a day, seven days a week. These sites must have spare server computers for handling high traffic volumes that occur periodically. In addition to requiring fast and reliable hardware, transaction-processing sites must run Web and commerce software that is efficient and easily upgraded when site traffic increases.
- *Content-delivery sites*: Sites that deliver content such as news, histories, summaries, and other digital information. Visitors must be able to locate articles quickly with a fast and precise search engine. The content must be presented rapidly on the visitor's screen. In general, these sites must be available 24 hours a day, seven days a week, just like transaction-processing sites. Hardware requirements for content sites are also similar to those of transaction-processing commerce sites.

Web Clients and Web Servers

When people use their Internet connections to become part of the Web, their computers become Web client computers on a worldwide client/server network. Client/server architectures are used in LANs, WANs, and the Web. In a client/server architecture, the client computers typically request services, such as printing, information retrieval, and database access, from the server, which processes the clients' requests. The computers that perform the server function usually have more memory and larger, faster disk drives than the client computers they serve. Recall from Chapter 2 that Web browser software (for example, Microsoft Internet Explorer or Netscape Navigator) is the software that makes computers work as Web clients. This software is also called Web client software.

The Internet connects many different types of computers running different types of operating system software. Because Web software is platform neutral, it lets these computers communicate with each other easily and effectively. This platform neutrality has been (and continues to be) a critical ingredient in the rapid spread and widespread acceptance of the Web. Figure 8-1 shows how the Web's platform neutrality provides multiple interconnections among a wide variety of client and server computers.

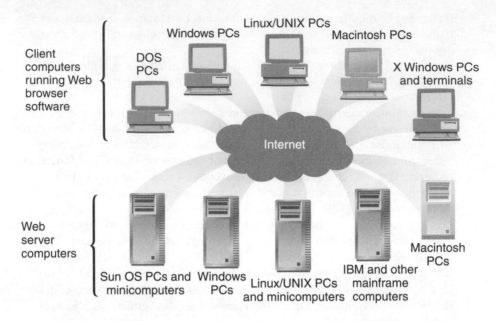

FIGURE 8-1 Platform neutrality of the Web

Dynamic Content

The mix and type of Web pages a system is likely to deliver in response to client requests can affect Web server performance. A **dynamic page** is a Web page whose content is shaped by a program in response to user requests, whereas a **static page** is an unchanging page retrieved from disk. A server delivering mostly static Web pages performs better than the same server delivering dynamic Web pages because static page delivery requires less computing power than dynamic page delivery. The largest performance differences between competing Web server products appear when servers deliver dynamic pages.

Dynamic content is nonstatic information constructed in response to a Web client's request. For example, if a Web client inquires about the status of an existing order by entering a unique customer number or order number into a form, the Web server searches the customer information (or sends a query to the back-end database in a higher tier) and generates a dynamic Web page based on the customer information it found (or that the database management software provided), thus fulfilling the client's request. Assembled from back-end databases and internal data on the Web site, a dynamic page is a specific response to the requester's query.

On a Web site that is a collection of HTML pages, the content on the site can be changed only by editing the HTML in the pages. This is cumbersome and does not allow customized pages to be produced in response to specific queries from site visitors. To create customized pages on the fly, Web sites use one of two basic approaches, server-side scripting or a dynamic page-generation technology.

Server-Side Scripting

The first Web sites to provide dynamic Web pages used an approach called server-side scripting. In **server-side scripting** (also called **server-side includes** or, more generally, **server-side technologies**), programs running on the Web server create the Web pages before sending them back to the requesting Web clients as parts of response messages.

Most server-side technologies are slow, so large Web sites used by online businesses today tend to use dynamic page-generation technologies.

Dynamic Page-Generation Technologies

Microsoft developed a now widely used dynamic page-generation technology called **Active Server Pages (ASP)**. Sun Microsystems developed a similar technology called **JavaServer Pages (JSP)**, and the open-source Apache Software Foundation sponsored a third alternative called **PHP: Hypertext Preprocessor (PHP)**. Yet another alternative is available from Macromedia in its **Cold Fusion** product.

In these page-generation technologies, server-side scripts are mixed with HTML-tagged text to create the dynamic Web page. For example, ASP allows Web programmers to use their choice of programming languages, such as VBScript, Jscript, or Perl. Java, a programming language created by Sun, can be used to produce dynamic pages. Such server-side programs are called **Java servlets**.

The Future of Dynamic Web Page Generation

Many critics of dynamic page-creation technologies note that these approaches do not really solve the problem of dynamic Web page generation. They argue that these dynamic page-creation approaches merely shift the task of creating dynamic pages from people who write HTML code to ASP (or JSP or PHP) programmers.

Several initiatives are under way that are directed at a more comprehensive solution to the dynamic Web page-creation problem. The **Apache Cocoon Project** is one of these initiatives. Apache Cocoon is a Web-development framework that allows programmers to query data that is in XML format and generate output in multiple formats, including HTML. The HTML output option makes Cocoon a useful tool for generating dynamic Web pages. In this approach, the content is stored with XML tags that describe the semantics (meaning) of each content item. The information request is handled by a Java servlet that can read the XML file and select the requested content items using the XML tags in the content file. Instead of creating a Web page, Cocoon can produce a response tailored to the request by applying a style sheet to the data. If a site visitor requests, for example, an Adobe Portable Document Format (PDF) file or a Wireless Markup Language (WML) file for display on a wireless handheld device, a Web site using Cocoon technology can generate the results in those file formats from its XML content files. Many industry experts believe that the Apache Cocoon Project and similar development efforts by Microsoft (the **Microsoft .NET** Framework) and Oracle will provide better ways to generate dynamic Web pages in the future.

Various Meanings of "Server"

All computers that are connected to the Internet and contain documents that their owners have made publicly available through their Internet connections are called Web servers. Unfortunately, the term "server" is used in many different ways by information systems professionals. These multiple uses of the term can be confusing to people who do not have a strong background in computer technology. You are likely to encounter a number of different uses of the word "server."

A **server** is any computer used to provide (or "serve") files or make programs available to other computers connected to it through a network (such as a LAN or a WAN). The software that the server computer uses to make these files and programs available to the other computers is sometimes called **server software**. Sometimes this server software is included as part of the operating system that is running on the server computer. Thus, some information systems professionals informally refer to the operating system software on a server computer as server software, a practice that adds considerable confusion to the use of the term "server."

Some servers are connected through a router to the Internet. As you learned in Chapter 2, these servers can run software, called Web server software, that makes files on those servers available to other computers on the Internet. When a server computer is connected to the Internet and is running Web server software (usually in addition to the server software it runs to serve files to client computers on its own network), it is called a Web server.

Similar terminology issues arise for server computers that perform e-mail processing and database management functions. Recall that the server computer that handles incoming and outgoing e-mail is usually called an e-mail server, and the software that manages e-mail activity on that server is frequently called e-mail server software. The server computer on which database management software runs is often called a **database server**.

Thus, the word "server" is used to describe several types of computer hardware and software, all of which might be found in a typical electronic commerce operation. The only way to determine which server people are talking about when they use the term is from the context or by asking a clarifying question. If you hear a computer technician say "The server is down today," the problem might be in the hardware, the software, or a combination of the two!

Web Client/Server Communication

In Chapter 2, you learned how the Web is software that runs on the Internet. In this section, you will learn more about how Web client and Web server software work. When a person uses a Web browser to visit a Web site, the Web browser (also known as a Web client) requests files from the Web server at the company or organization that operates the Web site. Using the Internet as the transportation medium, the request is formatted by the browser using HTTP and sent to the server computer. A moment later, when the server receives the request, it retrieves the file containing the Web page or other information that the client requested, formats it using HTTP, and sends it back to the client over the Internet.

When the requested information—a file containing the text and markup tags of a Web page, in this instance—arrives at the client computer, the Web browser software determines that the information is an HTML page. It displays the page on the client machine according to the directions defined in the page's HTML code. This process repeats as the client requests, the server responds, and the client displays the result. Sometimes, a single client request results in dozens or even hundreds of separate server responses to locate and deliver information.

A Web page containing many graphics and other objects can be slow to appear in the client's Web browser window because each page element (each graphic or multimedia file) requires a separate request and response.

Two-Tier Client/Server Architecture

The basic Web client/server model is a two-tier model because it has only one client and one server. All communication takes place on the Internet between the client and the server. Of course, other computers are involved in forwarding packets of information across the Internet, but the messages are created and read only by the client and the server computers in a **two-tier client/server architecture**. Figure 8-2 shows how a Web client and a Web server communicate with each other in a two-tier client/server architecture.

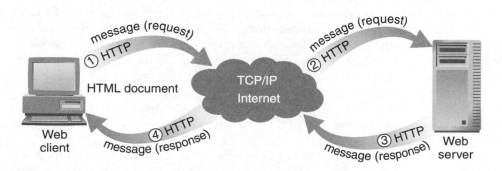

FIGURE 8-2 Message flows in a two-tier client/server network

The message that a Web client sends to request a file or files from a Web server is called a **request message**. A typical request message from a client to a server consists of three major parts:

- Request line
- Optional request headers
- Optional entity body

The **request line** contains a command, the name of the target resource (a filename and a description of the path to that file on the server), and the protocol name and version number. Optional **request headers** can contain information about the types of files that the client will accept in response to this request. Finally, an optional **entity body** is sometimes used to pass bulk information to the server.

When the server receives the request message, it executes the command included in the message (in this case, it sends a particular Web page file back to the client). The server does this by retrieving the Web page file from its disk (or another disk on a network to which it is connected) and then creating a properly formatted **response message** to send back to the client. A server's response consists of three parts that are identical in structure to a request message: a response header line, one or more response header fields, and an optional entity body. In the response, however, each part has a slightly different function than it does in the request. The **response header line** indicates the HTTP version used by the server, the status of the response (whether the server found the file that the client wanted), and an explanation of the status information. Response header fields follow the response header line. A **response header field** returns information describing the server's attributes. The entity body returns the HTML page requested by the client machine.

Three-Tier and N-Tier Client/Server Architectures

Although the two-tier client/server architecture works well for the delivery of Web pages, a Web site that delivers dynamic content and processes transactions must do more than respond to requests for Web pages. A **three-tier architecture** extends the two-tier architecture to allow additional processing (for example, collecting the information from a database needed to generate a dynamic Web page) to occur before the Web server responds to the Web client's request. Higher-order architectures—that is, those that have more than three tiers—are usually called **n-tier architectures**. The third tier often includes databases and related software applications that supply information to the Web server. The Web server can then use the output of these software applications when responding to client requests, instead of just delivering a Web page. Architectures that have four, five, or even more tiers include software applications (just as the three-tier systems), but they also include the databases and database management programs that work with the software applications to generate information that the Web server can turn into Web pages, which it then sends to the requesting client.

A good example of services supported by a database in an n-tier architecture is a catalog-style Web site with search, update, and display functions. Assume that a user requests a display of a company's exotic fruit selections. The client request is formulated into an HTTP message by the Web browser, sent over the Internet to the Web server, and examined by the Web server. The Web server analyzes the request and determines that responding to the request requires the help of the server's database. The server sends a request to the database management software to search for, retrieve, and return all information about exotic fruit in the catalog database. The database information flows back through the database management software system to the server, which formats the response into an HTML document and sends that document inside an HTTP response message back to the client over the Internet.

Three-tier and n-tier systems can track customer purchases stored in shopping carts, look up sales tax rates, keep track of customer preferences, query inventory databases, and keep the company catalog current. Figure 8-3 shows an overview of information flows in a three-tier architecture. Numbers on the flow arrows indicate the order in which the messages flow over the indicated paths.

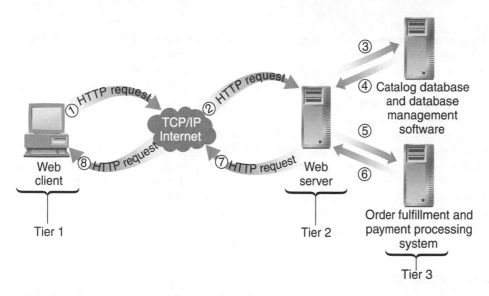

FIGURE 8-3   Message flows in a three-tier client/server network

SOFTWARE FOR WEB SERVERS

Some Web server software can run on only one computer operating system, while some can run on several operating systems. In this section, you will learn about the operating system software used on most Web servers, the Web server software itself, and other programs, such as Internet utilities and e-mail software, that companies often run on Web servers or other computers as part of electronic commerce operations.

Operating Systems for Web Servers

Operating system tasks include running programs and allocating computer resources such as memory and disk space to programs. Operating system software also provides input and output services to devices connected to the computer, including the keyboard, monitor, and printers. A computer must have an operating system to run programs. For large systems, the operating system has even more responsibilities, including keeping track of multiple users logged on to the system and ensuring that they do not interfere with one another.

Most Web servers run on computers that use one of the following operating systems: Microsoft Windows NT Server, Microsoft Windows 2000 Server or Server 2003 products, Linux, or one of several UNIX-based operating systems, such as Solaris or FreeBSD. Many companies believe that **Microsoft server products** are simpler for their information systems staff to learn and use than UNIX-based systems. Other companies worry about the security weaknesses caused by the tight integration between application software and the operating system in Microsoft products. UNIX-based Web servers are more popular, and many users believe that UNIX is a more secure operating system on which to run a Web server.

Linux is an open-source operating system that is easy to install, fast, and efficient. **Open-source software** is developed by a community of programmers who make the software available for download at no cost. Other programmers then use the software, work with it, and improve it. Those programmers can submit their improved versions of the software back to the community. You can learn more about open-source software at the **Open Source Initiative** Web site.

An increasing number of companies that sell computers intended to be used as Web servers include the Linux operating system in default configurations. Although Linux can be downloaded free from the Web, most companies buy it through a commercial distributor. These commercial distributions of Linux include useful additional software, such as installation utilities, and a support contract for the operating system. Commercial Linux distributors that sell versions of the operating system with utilities for Web servers include **Caldera**, **Mandrake**, **Red Hat**, and **SuSE**. **Sun Microsystems** sells Web server hardware along with its UNIX-based operating system, **Solaris**.

Web Server Software

This section describes the most commonly used Web server programs today: Apache HTTP Server, Microsoft Internet Information Server (IIS), and Sun Java System Web Server (JSWS) (often called by its former names, Sun ONE, iPlanet Enterprise Server, and Netscape Enterprise Server). These popularity rankings were accumulated through surveys done by **Netcraft**, a networking consulting company in Bath, England, known throughout the world for its Web server survey. Netcraft continually conducts surveys to tally the number of Web sites in existence and measure the relative popularity of Internet Web server software.

The Netcraft Web server surveys show that the market share of Web server software has stabilized in recent years. Apache generally holds between 65 and 70 percent of the market, and Microsoft's IIS usually holds between 20 and 25 percent of the market. Sun, along with the **National Center for Supercomputing Applications (NCSA)** Web server and a few other products such as Zeus, together account for the remainder of the market. The NCSA Web server was one of the first Web servers developed in the United States. Because it was developed with U.S. government research funds, the NCSA software is available at no cost. You can click the **Netcraft Surveys** link in the Online Companion to check out the latest Web survey results.

According to a *PC Magazine* survey (see the Alwang reference listed in the For Further Study and Research section at the end of the chapter), the market share percentages for intranet Web servers are quite different than for public Web servers. Although the Web server software packages described in this chapter are all top selections among intranet servers, Microsoft IIS and the various Sun servers (JSWS, Sun ONE, iPlanet, Netscape) together account for 75 percent of installed intranet server programs. Other recent surveys show that large company Web sites and high traffic Web sites show a greater use of the Sun servers than the Netcraft Surveys report.

The performance of one Web server differs from that of another based on workload, operating system, and the size and type of Web pages served. *PC Magazine* evaluates computer products regularly. In its tests, some Web server software fared well when delivering static HTML pages, but other Web server software performed better when delivering dynamic Web page content. The differences among servers can be significant; picking the right server for each different business need is critical (see the Machrone reference listed in the For Further Study and Research section). The sections that follow contain descriptions of the main Web server systems used in various electronic commerce applications.

Apache HTTP Server

Apache is an ongoing group software development effort. Rob McCool developed Apache while he was working at the University of Illinois at the **NCSA** in 1994. Several Webmasters from around the world created their own extensions to the server and formed an e-mail group so that they could coordinate their changes (known as "patches") to the system. The system consisted of the original core system with a lot of patches—thus, it became known as "a patchy" server, or simply, "Apache." The Apache Web server is currently available on the Web at no cost as open-source software.

Apache HTTP Server has dominated the Web since 1996 because it is free and performs very efficiently. It is powerful enough that IBM includes it in its WebSphere application server package. Other Web server products, such as Zeus, are based on the Apache open-source code. Currently, Apache is used on 65-70 percent of all Web servers, which means it is more widely used than all other Web server software packages combined. Apache runs on many operating systems (including FreeBSD-UNIX, HP-UX, Linux, Microsoft Windows, SCO-UNIX, and Solaris) and the hardware that supports them.

Microsoft Internet Information Server

Microsoft Internet Information Server (IIS) comes bundled with current versions of Microsoft Windows Server operating systems. IIS is used on many corporate intranets because many companies have adopted Microsoft products as their standard products. Small sites running personal Web pages use IIS, as do some of the largest electronic commerce sites on the Web. Most current surveys estimate that about 20-25 percent of all Web servers run some version of IIS. In recent years, the number of Web sites running IIS has been decreasing. Most industry observers believe this decrease has occurred because IIS has been the victim of several well-publicized security breaches. These security breaches allowed Web servers running IIS to be attacked successfully and defaced. You will learn more about Web server security threats and countermeasures in Chapter 10.

IIS, as a Microsoft product, was originally designed to run only on the Windows NT and Windows 2000 operating systems. It has been released for Microsoft Windows Server 2003 and runs on the Windows XP operating system, but it is not included as a standard part of Windows XP. IIS supports the use of ASP, ActiveX Data Objects, and SQL database queries. IIS also includes the Microsoft FrontPage Web site development tool and other reporting tools. IIS's inclusion of ASP provides an application environment in which HTML pages, ActiveX components, and scripts can be combined to produce dynamic Web pages.

Sun Java System Web Server (Sun ONE, iPlanet, Netscape)

A descendant of the original NCSA Web server program, Sun Java System Web Server (JSWS) was formerly sold under the names Sun ONE, Netscape Enterprise Server, and iPlanet Enterprise Server. When AOL (now Time Warner) purchased Netscape in 1999, the company formed a partnership with Sun Microsystems to support and continue to develop Netscape server products. This partnership was named iPlanet and was operated

361

under a three-year agreement that expired in March 2002. When the partnership ended, iPlanet became a part of Sun because the Web server and electronic commerce software that iPlanet sells are more closely related to Sun's businesses than to Time Warner's businesses.

Sun JSWS is not free, but its licensing fee is reasonable. The fee varies with the processing power of the server on which it is installed, but most Web sites pay between $1400 and $5000 for their licenses. The Sun software runs on many operating systems, including HP-UX, Solaris, and Windows. According to recent estimates, Sun JSWS runs on about 5 percent of all Web servers. However, some of the busiest and best-known sites on the Internet, including BMW, Dilbert, E*TRADE, Excite, Lycos, and Schwab, run (or have run) some version of Sun JSWS. Reports from consulting firms such as Gartner, Inc. show that Sun JSWS is in use at more than 40 percent of all public Web sites and at more than 60 percent of the top 100 enterprise Web sites.

Like most other server programs, Sun JSWS supports dynamic application development for server-side applications. Its ODBC conformance means that Sun JSWS provides connectivity to a number of database products as well.

Finding Web Server Software Information

People who want to know the type of operating system and Web server software that a Web site is running can visit the **Netcraft** Web site. On Netcraft's home page is a link named "What's that site running?" that leads to a page with a search function. Visitors can use that search function to find out what operating system and what Web server software a specific site is now running and what the site ran in the past.

ELECTRONIC MAIL (E-MAIL)

Although the Web, with its interactions between Web servers and clients, is the most important technology used in electronic commerce today, many buyers and sellers also use e-mail to gather information, execute transactions, and perform other tasks related to electronic commerce. E-mail originated in the 1970s on the ARPANET. Although the goals of the ARPANET were to control weapons systems and transfer research files, general communications uses emerged on the network. In 1972, Ray Tomlinson, an ARPANET researcher, wrote a program that could send and receive messages over the network. Today, e-mail is the most popular form of business communication—far surpassing the telephone, conventional mail, and fax in volume.

E-mail Benefits

Not only was e-mail one of the first Internet applications, it was also one reason that many people were originally attracted to the Internet. E-mail conveys messages from one destination to another in a few seconds. Messages can contain simple ASCII text, or they can contain character formatting similar to word-processing programs.

One useful feature of e-mail is that documents, pictures, movies, worksheets, or other information can be sent along with the message itself. These **attachments** are frequently the most important part of the message. A business e-mail message attachment might contain an invoice, a 200-page wholesale catalog, or a set of Web pages that describe the company's products. Many electronic commerce sites use e-mail to confirm the receipt of customer orders and then the shipment of items ordered. Software vendors can also use e-mail to send information about a purchase to the buyer.

E-mail Drawbacks

Despite its many benefits, e-mail does have some drawbacks. One annoyance associated with e-mail is the amount of time that businesspeople spend answering their e-mail today. Researchers have found that most managers can deal with e-mail messages at an average rate of about five minutes per message. Some messages can be deleted within a few seconds, but those are balanced by the e-mails that require the manager to spend much more time finding facts, checking files, making phone calls, and doing other tasks as part of answering e-mail. Researchers have found that most people (not including those people who answer e-mails as a full-time job) begin to resent the time that e-mail consumes when they start getting more than 20 or 30 messages a day. At that point, the average person is spending about two hours a day answering e-mail.

A second major irritation brought by e-mail is the **computer virus**, more simply known as a **virus**, which is a program that attaches itself to another program and can cause damage when the host program is activated. Recall that e-mail messages can carry attachments. Although many of these attached files contain useful information, attached files can contain a virus program or other security threat. Using virus protection software and dealing with e-mailed security threats is a cost that all must bear for the convenience of using e-mail. You will learn more about computer viruses and other threats that can be transmitted through e-mail (and how to control them) in Chapter 10.

Probably the most frustrating and expensive problem associated with e-mail today is the issue of unsolicited commercial e-mail. This nagging problem is discussed in the next section.

Unsolicited Commercial E-mail (UCE, Spam)

Spam, also known as **unsolicited commercial e-mail (UCE)** or **bulk mail**, is electronic junk mail and can include solicitations, advertisements, or e-mail chain letters. The origin of the term "spam" is generally believed to have come from a song performed by British comedy troupe Monty Python about Hormel's canned meat product, SPAM. In the song, an increasing number of people join in repeating the song's chorus: "Spam spam spam spam, spam spam spam spam, lovely spam, wonderful spam . . ." Just as in the song, e-mail spam is a tiresome repetition of meaningless text that eventually drowns out any other attempt at communication.

Besides wasting people's time and their computers' disk space, spam can consume large amounts of Internet capacity. If one person sends a useless e-mail to 100,000 people, that unsolicited mail consumes Internet resources for a few moments that would otherwise be available to other users. Although spam has always been an annoyance, in recent years, companies are increasingly finding it to be a major problem. In addition to consuming bandwidth on company networks and space on e-mail servers, spam distracts employees who are trying to do their jobs and requires them to spend time deleting the unwanted messages. A considerable number of spam messages include content that is offensive to the recipient. Some companies worry that their employees might sue them, arguing that offensive spam they receive while working contributes to sexual harassment by creating a hostile work environment. Figure 8-4 shows the rapid increase in the proportion of all e-mail entering business e-mail servers that is spam. Many researchers who track the growth in spam believe that current trends will continue and that more than 90 percent of all e-mail messages (including messages transmitted to both business and personal users) will be spam before any effective technical solutions can be implemented. Other researchers believe that the growth of spam is showing signs of leveling out.

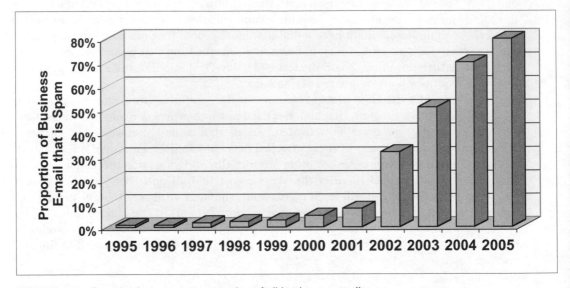

FIGURE 8-4 Growth of spam as a proportion of all business e-mail

Many grassroots and corporate organizations have decided to fight spam aggressively. America Online, for example, has taken an active role in limiting spam through legal channels. In 2005, America Online reported that the amount of spam passing through its servers declined for the first time in five years. The company attributed this success to the effects of lawsuits it has filed against spammers and its improved technical abilities to detect and delete spam e-mails before they are distributed to its users. A number of companies now offer software that organizations can run on their e-mail server computers to limit the amount of spam that gets through to their employees. Although individual users can install client-based spam-filtering programs on their computers or set filters that might be available within their e-mail client software, most companies find it more effective and less costly to eliminate spam before it is downloaded to user computers. These antispam efforts and software products can help limit the annoyance and cost of spam.

Solutions to the Spam Problem

As spam has grown to become a serious problem for all users of e-mail, an increasing number of approaches have been devised or proposed. Some of these approaches require the passing of laws, and some require technical changes in the mail handling systems of the

Internet. Other approaches can be implemented under existing laws and with current technologies, but only if large numbers of organizations and businesses cooperate. A few tactics that reduce spam can be undertaken by individual e-mail users. In the sections below you will learn more about each of these approaches to controlling the spam problem.

Individual User Anti-spam Tactics

One way individuals can limit spam is to reduce the likelihood that a spammer can automatically generate their e-mail addresses. Many organizations create e-mail addresses for their employees by combining elements of each employee's first and last names. For example, small companies often combine the first letter of an employee's first name with the entire last name to generate e-mail addresses for all employees at small companies (larger companies often must use more complex algorithms as they are likely to have both a Jane Smith and a Judy Smith working for them). Any spam sender able to obtain an employee list can generate long lists of potential e-mail addresses using the names on the list. If no employee list is available, the spam sender can simply generate logical combinations of first initials and common names. The cost of sending e-mail is so low that a spammer can afford to send thousands of e-mails to randomly generated addresses in the hope that a few of them are valid. By using an e-mail address that is more complex, such as "xq7yy23@mycompany.com," individuals can reduce the chances that a spammer can randomly generate his or her address. Of course, such an address is hard to remember, which somewhat defeats the purpose of e-mail as a convenient way to communicate.

A second way to reduce spam is to control the exposure of an e-mail address. Spammers use software robots to search the Internet for character strings that include the "@" character (which appears in every e-mail address). These robots search Web pages, discussion boards, chat rooms, and any other online source that might contain e-mail addresses. Again, the spammer can afford to send thousands of messages to e-mail addresses gathered in this way. Even if only one or two people respond, the spammer can earn a profit because the cost of sending e-mail messages is so low.

Some individuals use multiple e-mail addresses to thwart spam. They use one address for display on a Web site, another to register for access to Web sites, another for shopping accounts, and so on. If a spammer starts using one of these addresses, the individual can stop using it and switch to another. Many Web hosting services include a large number (often 100–200) of e-mail addresses as part of their service, so this is a good tactic for people or small businesses with their own Web sites.

These three strategies focus on limiting spammer's access to or use of an e-mail address. Other approaches use one or more techniques that filter e-mail messages based on their contents.

Basic Content Filtering

All content-filtering solutions require software that identifies content elements in an incoming e-mail message that indicate the message is (or is not) spam. The content-filtering techniques differ in which content elements they examine, whether they look for indications that the message is spam or that it is not spam, and how strictly they apply the rules for classifying messages. Most basic content filters examine the e-mail headers (From, To, Subject) and look for indications that the message might be spam. The software that performs the filtering task can be placed on individual users' computers (called **client-level filtering**) or on mail server computers (called **server-level filtering**). Server-level filtering can be implemented on an ISP's mail server, an individual company's mail server, or both. Also, many individuals that have ISP and/or company mail servers that filter their e-mail also install client-level filters on their computers. Spam that gets through one filter can be trapped by another filter.

The most common basic content-filtering techniques are black lists and white lists. A **black list spam filter** looks for From addresses in incoming messages that are known to be spammers. The software can delete the message or put it into a separate mailbox for review. A black list spam filter can be implemented at the individual, organization, or ISP level. Several organizations, such as the **Mail Abuse Protection System** and the **Open Relay Database**, collect black lists and make them available to ISPs and company e-mail administrators. Other groups, such as the **Spamhaus Project**, track known spammers and publish lists of the mail servers they use. Some of these are free services, others charge a fee. The biggest drawback to the black list approach is that spammers frequently change their e-mail servers, which means that a black list must be continually updated to be effective. This updating requires that many organizations cooperate and communicate information about known spammers. In addition to its black list, the Spamhaus Project maintains a list of known spammers on its site. These are individuals and companies who have had their services terminated by an ISP for spam-related violations of an acceptable use policy more than three times. The Spamhaus Project provides detailed information about those on this list to law enforcement agencies.

A **white list spam filter** examines From addresses and compares them to a list of known good sender addresses (for example, the addresses in an individual's address book). A white list filter is usually applied at the individual user level, although it is possible to do the filtering at the organization level if the e-mail administrator has access to all individuals' address books (some companies mandate such access for security purposes). The main drawback to this approach is that it filters out any messages sent by unknown parties, not just spam. Because the number of **false positives** (messages that are rejected but should not have been) can be very high for white list filters, the rejected e-mails are always placed into a review mailbox instead of being deleted.

White list and black list approaches can be used in client-level or server-level filters, but both have serious drawbacks. To overcome these drawbacks, the two approaches are often used together or with other content-filtering approaches to achieve an acceptable level of filtering without an excessive false positive rate.

Challenge-response Content Filtering

One content-filtering technique uses a white list as the basis for a confirmation procedure. This technique, called **challenge-response**, compares all incoming messages to a white list. If the message is from a sender who is not on the white list, an automated e-mail response is sent to the sender. This message (the challenge) asks the sender to reply to the e-mail (the response). The reply must contain a response to a challenge presented in the e-mail.

These challenges are designed so that a human can respond easily, but a computer would have difficulty formulating the response. For example, a good challenge might include a picture of a fruit bowl and would ask the sender to respond with the number of apples in the bowl. This prevents a spammer from setting up a computer that receives challenges and answers them (the program would have difficulty identifying and counting the number of apples). It would be inefficient for a spammer to hire a human to respond to thousands of challenges. To learn more about this technique, you can visit the **CAPTCHA Project** site at Carnegie Mellon University.

One major drawback to challenge-response systems is that they can be abused. For example, a perpetrator could send out thousands of e-mails to recipients that use challenge-response systems. If the perpetrator includes the victim's e-mail as the From address in those e-mails, the victim will be bombarded by the automated challenges sent out by the challenge-response systems of the recipients. What is worse, the potential damage of this tactic becomes greater as more e-mail servers install challenge-response systems.

Another issue with challenge-response systems will arise if they become widespread. Most mail that any individual receives from unknown senders is spam. A challenge-response system sends a challenge message to every unknown sender. That is, for every spam message received, a second e-mail is sent. A challenge-response system thus doubles the amount of useless e-mail messages that must be handled by the Internet's infrastructure. If everyone were to use a challenge-response system, the Internet capacity wasted by spam would approximately double. Because of the drawbacks associated with challenge-response systems, most industry experts agree that they are, at best, a limited short-term solution.

Advanced Content Filtering

Advanced content filters that examine the entire e-mail message can be more effective than basic content filters that only examine the message headers or the IP address of the e-mail's sender. Creating effective content filters can be challenging. For example, a company might want to delete any e-mail message that includes the word "sex." If the company deletes all e-mails containing that character string, they will unintentionally delete all e-mailed orders from customers in the town of Essex.

Many advanced content filters operate by looking for spam indicators throughout the e-mail message. When the filter identifies an indicator in a message, it increases that message's spam "score." Some indicators increase the score more than others. Indicators can be words, word pairs, certain HTML codes (such as the code for the color white, which makes part of the message invisible in most e-mail clients), and information about where a word occurs in the message. Unfortunately, as soon as spam filter vendors identify a good set of indicators, spammers stop including those indicators in their messages.

One type of advanced content filter that is based on a branch of applied mathematics called Bayesian statistics shows some promise of staying one step ahead of the spammers. **Bayesian revision** is a statistical technique in which additional knowledge is used to revise earlier estimates of probabilities. In software that contains a **naïve Bayesian filter** (the most common type in use today), the software begins by not classifying any messages. The user reviews messages and indicates to the software which messages are spam and which are not. The software gradually learns (by revising its estimates of the probability that a message element appears in a spam message) to identify spam messages.

After seeing a few dozen messages classified, the naïve Bayesian filter can successfully classify spam messages about 80 percent of the time. As the filter continues to work, the user reviews its classifications and tells the software when it makes a mistake. After classifying a few hundred messages (and being corrected by the user when it errs), a naïve Bayesian filter typically reaches correct spam classification rates above 95 percent. Although these filters are highly effective and have low false positive rates, they must be trained, which takes time. The training is best done by each individual user because one person's spam can be another person's important message. Having users train their own filters provides the most rapid training and the best results. Most organizations do not currently use naïve Bayesian filters because they require attention by individual e-mail users. However, naïve Bayesian filters can be installed on some client computers (such as those used by people who receive large amounts of e-mail) in organizations that also use other techniques (such as white list or black list filters) at the server level. Most industry observers expect to see naïve Bayesian filtering used more widely in the future as the spam problem worsens and as more e-mail clients include such filters.

A number of researchers presented papers on the development of naïve Bayesian filters in mid-2002. Later that year, an open-source software development project led by John Graham-Cumming released one of the first functional Bayesian filter products for individual users, **POPFile**. POPFile is a program that installs on individual client computers and works with many different e-mail clients (including Microsoft Outlook, Pegasus, and Qualcomm Eudora) to provide content filtering. Because it is open-source software, POPFile is free (although the project team welcomes donations). POPFile does require that e-mail be retrieved using a POP (Post Office Protocol) connection, so it cannot be used with most Web-based e-mail accounts such as Yahoo! or Hotmail. The latest releases of some e-mail client software, such as Qualcomm Eudora and Mozilla Thunderbird, now include naïve Bayesian filtering tools that work the same way POPFile does.

Figure 8-5 shows the training screen from an installation of POPFile. Each message is listed and classified into categories that the software calls buckets. These buckets are configurable by the user; in this case, the user has created two buckets, "spam" and "OK." The user reviews each message and changes the classification if necessary. Each time the user changes a classification, POPFile revises its internal database using a naïve Bayesian algorithm and uses the revised rules to classify new e-mail messages.

Figure 8-6 shows the summary statistics page from a POPFile filter. This page reflects the filter's activity during a recent six-month period on one of the author's e-mail accounts.

Although the filter caught only 30 percent of spam messages when it was installed, within two weeks, it was catching over 90 percent and eventually was more than 99 percent accurate. Note that the number of false positives in the spam category is also quite small. POPFile also includes a feature, called magnets, that allows the user to implement white list and black list filtering. The user can create a magnet that classifies messages based on specific content in the message and does not send the classified message through the

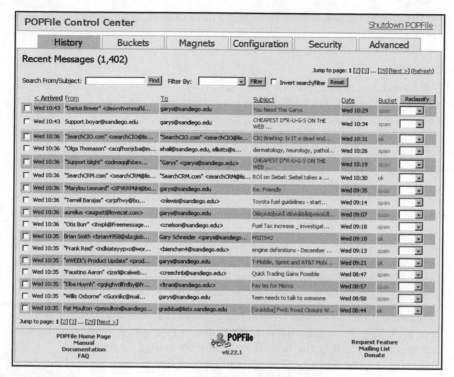

FIGURE 8-5 Training screen in the POPFile naïve Bayesian filter

POPFile Control Center					Shutdown POPFile	
History	**Buckets**	Magnets	Configuration	Security	Advanced	

Classification Accuracy		Messages Classified				Word Counts	
		Bucket	**Classification Count**	**False Positives**	**False Negatives**	**Bucket**	**Word Count**
Messages classified:	35,031	ok	12,764 (36.43%)	79	64	ok	175,732 (47.43%)
Classification errors:	183	spam	22,182 (63.32%)	53	119	spam	194,750 (52.56%)
Accuracy:	99.47%	unclassified	85 (0.24%)	44			
Reset Statistics							

FIGURE 8-6 POPFile summary statistics page

naïve Bayesian filter. In this example, magnets that classify messages as "OK" operate as white list filters and magnets that classify messages as "spam" operate as black list filters.

Naïve Bayesian filters are very effective client-level filters, but they do not work well as server-level filters. The content that is common in one person's spam might be common in another person's valid e-mail; therefore, one user's reclassifications tend to cancel out those of other users. This prevents the filter from building its accuracy to high levels. One good solution for organizations is to use black list filters at the server level combined with white list and naïve Bayesian filters at the client level. The major drawback of any client-level filtering approach is that it requires individual users to update their own filters regularly. Although it takes less time to update a filter than to delete hundreds of spam messages, it still does take time.

Legal Solutions

A number of U.S. jurisdictions have passed laws that provide penalties for the sending of spam. In January, 2004, the U.S. CAN-SPAM law (the law's name is an acronym for "Controlling the Assault of Non-Solicited Pornography and Marketing") went into effect. Researchers who track the amount of spam noted a drop in the percentage of all e-mail that was spam in February and March. A **MessageLabs** study tracked the drop from 62 percent in January to 59 percent in February and 53 percent in March. However, by April, the rate was back up to a new high, 68 percent. It appears that spammers slowed down their activities immediately after the effective date of CAN-SPAM to see if a broad federal prosecution effort would occur. When the threat did not materialize, the spammers went right back to work.

The CAN-SPAM law is the first U.S. federal government effort to legislate controls on spam. It regulates all e-mail messages sent for the primary purpose of advertising or promoting a commercial product or service, including messages that promote the content displayed on a Web site. The law's main provisions include:

- *Misleading address header information*: E-mail headers and routing information, including the originating domain name and e-mail address, must be accurate and must identify the person who sent the e-mail.
- *Deceptive subject headers*: The e-mail's subject line cannot mislead the recipient about the contents or subject matter of the message.
- *Clear and conspicuous notice of message nature*: The e-mail must contain a clear and conspicuous notice that the message is an advertisement or solicitation and that the recipient can opt out of receiving further commercial e-mail from the sender.
- *Physical postal address*: The e-mail must include the sender's valid physical postal address.
- *Mandatory provision of an opt-out mechanism*: The e-mail must include a return e-mail address or another Internet-based response mechanism that allows a recipient to ask not to be sent future e-mail messages. These requests must be honored. The message may include a menu of choices that allows a recipient to opt out of certain types of messages, but one option on the menu must be an option to stop sending all commercial messages of any type.
- *Effectiveness of opt-out mechanism*: Opt-out requests must be honored within 10 business days. Any opt-out mechanism offered must be able to process opt-out requests for at least 30 days after the e-mail is sent. Once an opt-out request has been received, the sender is prohibited from helping any other entity send e-mail to the opt-out address or from having another entity send e-mail on the sender's behalf to that address.
- *Transfer of e-mail addresses*: Once a recipient has submitted an opt-out request, the sender is prohibited from selling or transferring that e-mail address to any other entity.

The law also prohibits misleading address header information in e-mail messages that facilitate an agreed-upon transaction or that update a customer in an existing business relationship. Each violation of a provision of the law is subject to a fine of up to $11,000. Additional fines are assessed for those who violate one of the above provisions and do one or more of the following:

- Harvest e-mail addresses from Web sites or Web services that have published a notice prohibiting the transfer of e-mail addresses for the purpose of sending e-mail.
- Send e-mail messages to addresses that have been generated by combining names, letters, or numbers into multiple combinations and permutations.
- Use scripts or other automated tools to register for multiple e-mail or user accounts that are then used to send commercial e-mail.
- Relay e-mails through a computer or network without the permission of the computer's or network's owner.

As you can see, a successful prosecution could cost the convicted spammer a great deal of money. The law further provides for criminal penalties, including imprisonment, for commercial senders of e-mail who do or conspire to do any of the following:

- Use another person's or entity's computer to send commercial e-mail from or through it without the computer owner's permission.
- Use a computer to relay or retransmit multiple commercial e-mail messages with the intent to deceive or mislead recipients or an Internet access service about the origin of the messages.
- Send multiple e-mail messages that contain false header information.
- Present false identification when registering for multiple e-mail accounts or domain names.
- Falsely represent themselves as owners of multiple IP addresses that are used to send commercial e-mail messages.

You can learn more about the law on the **U.S. Federal Trade Commission CAN-SPAM Law** information pages. The FTC issues new rules from time to time under the law. To obtain current updates on those rules, visit the **U.S. Federal Trade Commission Spam** information pages.

In the months following CAN-SPAM's effective date, several large lawsuits were filed under the Act. Microsoft, AOL, and Earthlink collectively filed six lawsuits and Microsoft filed an additional eight lawsuits on its own shortly thereafter. Beyond these actions, only a few spammers have been prosecuted under the law. The most significant verdict to date was the December 2003 verdict issued by a U.S. District Court in Iowa. That court used a combination of Iowa law and Federal racketeering statutes to order three spammers to pay more than $1 billion to an ISP they had inundated with spam. Despite these few headline-grabbing cases, the large wave of prosecutions that many observers had hoped to see has not yet occurred. The FTC refused to create a do-not-spam list that would have been modeled after its do-not-call list, which has been reasonably successful in limiting marketers' phone calls.

Few industry experts expect CAN-SPAM or similar laws to be effective in stemming the tide of spam. After all, spammers have been violating existing deceptive advertising laws for years. Many spammers use mail servers located in countries that do not have (and that are unlikely to adopt) antispam laws. As you learned in Chapter 7, the issues of jurisdiction can be unclear for businesses that operate online. Even if a plaintiff is successful in court, enforcement of court ordered fines or collection of damages can be difficult. Spammers can also evade cease-and-desist orders because they can move their operations from one server to another in minutes. Many spammers forward mail through servers that they have hijacked (you will learn more about server hijacking in Chapter 10).

Although laws are not likely to stop the most determined spammers, some industry observers hope that laws such as CAN-SPAM will at least enforce constraints on the legitimate marketers that send commercial e-mails. In January 2005, *PC World* conducted a study to determine whether e-mail marketers were complying with the law (see the Spring reference in the For Further Study and Research section at the end of the chapter). The researchers signed up for 100 marketing information mailing lists, then tried to opt out of the mailings. Of the 100 lists, seven sent messages that did not include a valid postal address and two more did not include an opt-out mechanism. When the researchers opted out, 85 of the companies complied fully with the CAN-SPAM law provisions, but eight sent e-mails beyond the 10-day limit and another four had marketing partners that continued to send e-mails. One list sent messages that included an opt-out link that did not work.

Some critics argue that any legal solution to the spam problem is likely to fail until the prosecution of spammers becomes cost-effective for governments. To become cost-effective, prosecutors must be able to identify spammers easily (to reduce the cost of bringing an action against them) and must have a greater likelihood of winning the cases they file (or must see a greater social benefit to winning). The best way to make spammers easier to find is to make technical changes in the e-mail transport mechanism in the Internet's infrastructure.

Technical Solutions

The Internet was not designed to do many of the things it does today. It was not designed to be secure, to process transactions, or to handle billions of e-mail messages. As you learned in Chapter 2, Internet e-mail was an incidental afterthought in a system designed to transfer large files from one researcher to another. As it was originally designed, and as it operates today, the Internet did not include any mechanisms for ensuring that the identity of an e-mail sender would always be known to the e-mail's recipient.

Most industry observers agree that the ultimate solution to the spam problem will come when new e-mail protocols are adopted that provide absolute verification of the source of each e-mail message. This will require all mail servers on the Internet to be upgraded. The new protocols have not yet been written, so this solution is several years away.

Proposals for identification standards have been made by Time Warner's AOL division, Microsoft, Yahoo!, and other companies and organizations. The Internet Engineering Task Force (IETF) working group that has responsibility for e-mail standards has rejected some of these proposals, but has stated its commitment to working out a set of standards that will accomplish sender authentication. You can learn more about current developments on this issue by following the links in the Online Companion in the Additional Information section under the heading **Spam Information Sites**.

WEB SITE AND INTERNET UTILITY PROGRAMS

In addition to Web server software, people who develop Web sites work with a number of utility programs, or tools. TCP/IP supports a wide variety of these utility programs. Some of these programs run on the Web server itself, while others run on the client computers that Web developers use when they are creating Web sites. E-mail was one of the earliest Internet utility programs and it has become one of the most important. In earlier chapters, you learned how companies are using e-mail as a key element in their electronic commerce strategies. In this section, you will learn some of the more significant technical details of how e-mail works. You will learn about several of these programs and see examples of how they work.

Finger and Ping Utilities

Finger is a program that runs on UNIX operating systems and allows a user to obtain some information about other network users. A Finger command yields a list of users who are logged on to a network, or reports the last time a user logged on to the network. Many organizations have disabled the Finger command on their systems for privacy and security reasons. For example, if you send a Finger command to a server at www.microsoft.com, you receive no response. Some e-mail programs have the Finger program built into them, so you can send the command while reading your e-mail.

A program called **Ping**, short for **Packet Internet Groper**, tests the connectivity between two computers connected to the Internet. Ping provides performance data about the connection between Internet computers, such as the number of computers (hops) between them. It sends two packets to the specified address and waits for a reply. Network technicians often use Ping to troubleshoot Internet connections. Many freeware and shareware Ping programs are available on the Internet. You can send out a Ping from a Windows PC by opening an MS-DOS window and typing "ping" followed by the IP address.

Tracert and Other Route-Tracing Programs

Tracert (TRACE RouTe) sends data packets to every computer on the path (Internet) between one computer and another computer and clocks the packets' round-trip times. This provides an indication of the time it takes a message to travel from one computer to another and back, ensures that the remote computer is online, and pinpoints any data traffic congestion. Route-tracing programs also calculate and display the number of hops between computers and the time it takes to traverse the entire one-way path between machines.

Route-tracing programs such as Tracert work by sending a series of packets to a particular destination. Each router along the Internet path between the originating computer and the destination computer reports its IP address and the time it took to reach it. After the program completes its packet transmissions, it displays the number of hops and how much time it took to reach each node and travel the entire path.

Graphical user interface route-tracing programs provide a plot of the packets' route on a map. Network engineers can use route-tracing programs to determine the location of the greatest delays on the Internet. Companies that provide Internet connections to customers often run route-tracing programs to monitor and improve services. Visualware offers its **VisualRoute** route-tracing program for download, trial, and purchase. The site also offers a demonstration of VisualRoute that runs on its Web site so potential customers can test the program without downloading any software. Figure 8-7 shows a route traced from a VisualRoute server in Englewood, Colorado (USA) to a server at SingTel in Milton, Australia using the VisualRoute program.

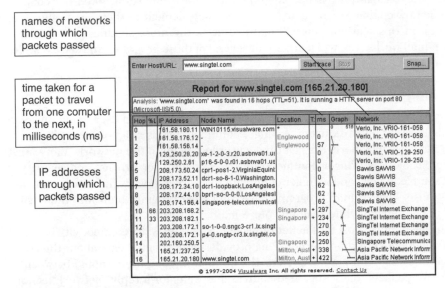

names of networks through which packets passed

time taken for a packet to travel from one computer to the next, in milliseconds (ms)

IP addresses through which packets passed

FIGURE 8-7 Tracing a path between two computers on the Internet

You can see that the packets traveled through 16 computers (that is, the path included 16 hops) and that the path went from Colorado to Virginia to Washington, D.C. to Los Angeles, to Singapore, to Australia. The test message took 422 milliseconds (ms), which is just under one-half of a second, to travel more than half-way around the world.

Telnet and FTP Utilities

Telnet is a program that allows users to log on to a computer that is connected to the Internet. This remote login capability can be useful for running older software that does not have a Web interface. Several Telnet client programs are available as free downloads on the Internet, and Microsoft Windows systems include a Telnet client called Telnet.exe. Telnet lets a client computer give commands to programs running on a remote host. Telnet programs use a set of rules called the **Telnet protocol**. Some Web browsers function as a Telnet client. A user can enter "telnet://" followed by the domain name of the remote host. As more companies place information on Web pages, which are accessible through any Web browser, the use of Telnet will continue to decrease.

The **File Transfer Protocol (FTP)** is the part of the TCP/IP rules that defines the formats used to transfer files between TCP/IP-connected computers. FTP can transfer files one at a time, or it can transfer many files at once. FTP also provides other useful services, such as displaying remote and local computers' directories, changing the current client's or server's active directory, and creating and removing local and remote directories. FTP uses TCP and its built-in error controls to copy files accurately from one computer to another.

Accessing a remote computer with FTP requires that the user log on to the remote computer. A number of FTP client programs exist; however, many people just use their Web browser software. Typing the protocol name, ftp://, before the domain name of the remote computer establishes an FTP connection. Users who have accounts on remote computers can log on to their accounts using the FTP client. FTP establishes contact with the remote computer and logs onto the account on that computer.

An FTP connection to a computer on which the user has an account is called **full-privilege FTP**. Another way to access a remote computer is called anonymous FTP. **Anonymous FTP** allows the user to log on as a guest. By entering the username "anonymous" and an e-mail address as a password, users can read and copy files that are on the remote computer.

Indexing and Searching Utility Programs

Search engines and indexing programs are important elements of many Web servers. Search engines or search tools search either a specific site or the entire Web for requested documents. An indexing program can provide full-text indexing that generates an index for all documents stored on the server. When a browser requests a Web site search, the search engine compares the index terms to the requester's search term to see which documents contain matches for the requested term or terms. More advanced search engine software (such as that used by the popular search engine site Google) uses complex relevance ranking rules that consider things such as how many other Web sites link to the target site. Many Web server software products also contain indexing software. Indexing software can often index documents stored in many different file formats.

Data Analysis Software

Web servers can capture visitor information, including data about who is visiting a Web site (the visitor's URL), how long the visitor's Web browser viewed the site, the date and time of each visit, and which pages the visitor viewed. This data is placed into a Web **log file**. As you can imagine, the file grows very quickly—especially for popular sites with thousands of visitors each day. Careful analysis of the log file can be fruitful and reveal many interesting facts about site visitors and their preferences. To make sense of a log file, you must run third-party Web log file analysis programs. These programs summarize log file information by querying the log file and either returning gross summary information, or accumulating details that reveal how many visitors came to the site per day, hour, or minute, or which hours of the day were peak loading times. Popular Web log file analysis programs include products by **Analog**, **Keylime Software**, **Urchin Web Analytics**, **Web Side Story**, and **WebTrends**. A report generated by the WebTrends Web server log file analyzer software appears in Figure 8-8.

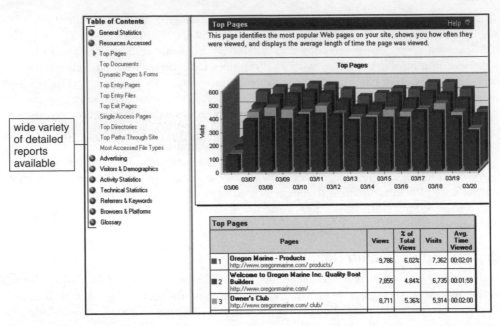

wide variety of detailed reports available

Top Pages   Help

This page identifies the most popular Web pages on your site, shows you how often they were viewed, and displays the average length of time the page was viewed.

Top Pages

Top Pages					
	Pages	Views	% of Total Views	Visits	Avg. Time Viewed
■1	Oregon Marine - Products http://www.oregonmarine.com/ products/	9,786	6.02%	7,362	00:02:01
■2	Welcome to Oregon Marine Inc. Quality Boat Builders http://www.oregonmarine.com/	7,855	4.84%	6,735	00:01:59
■3	Owner's Club http://www.oregonmarine.com/ club/	8,711	5.36%	5,914	00:02:00

FIGURE 8-8   WebTrends log file analysis

Link-Checking Utilities

Dedicated site management tools include a standard set of features, starting with link checking. A **link checker** examines each page on the site and reports on any URLs that are broken, seem broken, or are in some way incorrect. It can also identify orphan files. An **orphan file** is a file on the Web site that is not linked to any page. Other important site management features include script checking and HTML validation. Some management tools can locate error-prone pages and code, list broken links, and e-mail maintenance results to site managers.

On the company Web site, it is important to regularly check links that point to pages both within and outside the corporate Web site. Some Web server software does contain link-checking features. A **dead link**, when clicked, displays an error message rather than a Web page. Maintaining a site that is free of dead links is vital because visitors who encounter too many dead links on a site might jump to another site. Web-browsing customers are just a click away from going to a competitor's site if they become annoyed with an errant Web link.

Some Web site development and maintenance tools, such as Macromedia's Dreamweaver, include link-checking features. Most link-checking programs, however, run as separate programs. One of these link-checking programs, **Elsop LinkScan**, is available in a demo version as a free download. The results of the link checker either appear in a Web browser or are e-mailed to a recipient. Besides checking links, Web site validation programs sometimes check spelling and other structural components of Web pages.

Checkers that run on a company's own Web site are also available. Commercial site checkers, such as **Big Brother** software from **Watchfire**, produce more comprehensive results and more detailed site analyses than do the free products. The Watchfire Linkbot products include an enterprise version that is a complete Web site analysis tool; it scans a site for broken links and many other potential problems and then generates reports detailing any errors it finds. Figure 8-9 shows a report produced by the Watchfire Linkbot enterprise product.

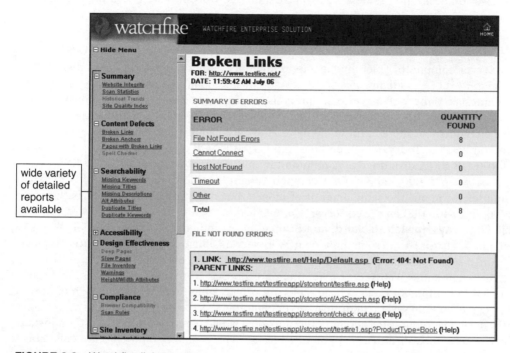

wide variety of detailed reports available

FIGURE 8-9 Watchfire link report

LinxCop is one of several reverse link checkers available. A **reverse link checker** checks on sites with which a company has entered a link exchange program (which you learned about in Chapter 4) and ensures that link exchange partners are fulfilling their obligation to include a link back to the company's Web site.

Remote Server Administration

With **remote server administration**, a Web site administrator can control a Web site from any Internet-connected computer. Although all Web sites provide administrative controls—most through a workstation computer on the same network as the server computer or through a Web browser—it is convenient for an administrator to be able to fix the server from wherever he or she happens to be. For example, an administrator can install **Web Site Garage** on any Internet-connected Windows computer and monitor and change anything on the Web site from that computer. **NetMechanic** offers a variety of link-checking, HTML troubleshooting, site-monitoring, and other programs that can be useful in managing the operation of a Web site.

WEB SERVER HARDWARE

Now that you have learned about Web server and Internet utility software, your next step is to learn about Web server hardware. Companies use a wide variety of computer brands, types, and sizes to host electronic commerce operations. Some small companies can run Web sites on desktop PCs. Most electronic commerce Web sites are operated on computers designed for site hosting, however.

Server Computers

Web server computers generally have more memory, larger (and faster) hard disk drives, and faster processors than the typical desktop or notebook PCs with which you are probably familiar. Many Web server computers use multiple processors; very few desktop PCs have more than one processor. Because Web server computers use more capable hardware elements and more of these elements, they are usually much more expensive than workstation PCs. Today, a high-end desktop PC with 2 GB of RAM, a 3.6 GHz processor, a fast 400 GB SATA disk drive, a good monitor, and a complement of DVD/CD-RW drives costs between $2000 and $4000. A company might be able to buy a low-end Web server computer for about the same amount of money, but most companies spend between $6000 and $400,000 on a Web server. Companies that sell Web server hardware, such as **Dell**, **Gateway**, **Hewlett Packard**, and **Sun**, all have configuration tools on their Web sites that allow visitors to design their own Web servers. Figure 8-10 shows three typical midrange server computer configurations available for sale on Sun's Web site.

Although some Web server computers are housed in freestanding cases, most are installed in equipment racks. These racks are usually about 6 feet tall and 19 inches wide. They can each hold several midrange server computers. A recent innovation in server computer design is to put small server computers on a single computer board and then install many of those boards into a rack-mounted frame. These servers-on-a-card are called **blade servers**, and some manufacturers now make them so small that more than 300 of them can be installed in a single 6-foot rack.

Recall that the fundamental job of a Web server is to process and respond to Web client requests that are sent using HTTP. For a client request for a Web page, the server program finds and retrieves the page, creates an HTTP header, and appends the HTML document to it. For dynamic pages, the server uses an architecture with three or more tiers that uses other programs, receives the results from the back-end process, formats the response, and sends the pages and other objects to the requesting client program. IP-sharing, or a virtual server, is a feature that allows different groups to share a single Web server's IP address. A **virtual server** or **virtual host** is a feature that maintains more than one server on one machine. This means that different groups can have separate domain names, but all domain names refer to the same physical Web server.

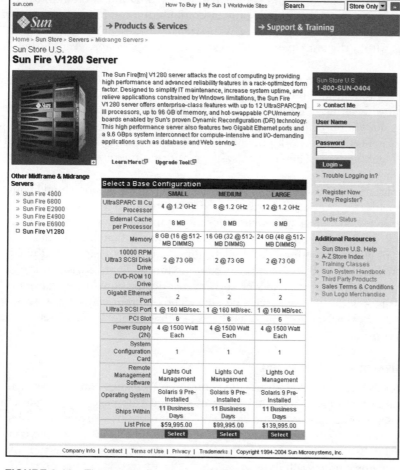

FIGURE 8-10   Three typical mid-range Sun server computer configurations

Web Server Performance Evaluation

Benchmarking Web server hardware and software combinations can help in making informed decisions for a system. **Benchmarking**, in this context, is testing that is used to compare the performance of hardware and software.

Elements affecting overall server performance include hardware, operating system software, server software, connection speed, user capacity, and type of Web pages being delivered. When evaluating Web server performance, a company should know exactly what factors are being measured and ensure that these are important factors relative to the expected use of the Web server. Another factor that can affect a Web server's performance is the speed of its connection. A server on a T3 connection can deliver Web pages to clients much faster than on a T1 connection.

The number of users the server can handle is also important. This can be difficult to measure because results are affected by the server's line speed, the clients' line speeds, and the sizes of the Web pages delivered. Two factors to evaluate when measuring a server's Web page delivery capability are throughput and response time. **Throughput** is the number of HTTP requests that a particular hardware and software combination can process in a unit of time. **Response time** is the amount of time a server requires to process one request. These values should be well within the anticipated loads a server can experience, even during peak load times.

One way to choose Web server hardware configurations is to run tests on various combinations, remembering to consider the system's scalability. Of course, you need to have the hardware and software set up to do this, so it is difficult to evaluate potential configurations that you have not yet purchased. Independent testing labs such as **Mindcraft** test software, hardware systems, and network products for users. Its site contains reports and statistics comparing combinations of application server platforms, operating systems, and Web server software products.

Anyone contemplating purchasing a server that will handle heavy traffic should compare standard benchmarks for a variety of hardware and software configurations. Customized benchmarks can give Web managers guidelines for modifying file sizes, cache sizes, and other parameters. Web managers should run benchmarks regularly. Benchmarks are not as meaningful for small Web sites with much smaller numbers of daily visitors. In the latter case, a focus on Web design and site navigation can maximize clients' satisfaction.

Companies that operate more than one Web server must decide how to configure servers to provide site visitors with the best service possible. The different ways that servers can be connected to each other and to related hardware, such as routers and switches, are called **server architectures**.

Web Server Hardware Architectures

Earlier in this chapter, you learned that electronic commerce Web sites can use two-tier, three-tier, or n-tier architectures to divide the work of serving Web pages, administering databases, and processing transactions. Some electronic commerce sites are so large that more than one computer is required within each tier. For example, large electronic commerce Web sites must deliver millions of individual Web pages and process thousands of customer and vendor transactions each day.

Administrators of these large Web sites must plan carefully to configure their Web server computers, which can number in the hundreds or even thousands, to handle the daily Web traffic efficiently. These large collections of servers are called **server farms** because the servers are often lined up in large rooms, row after row, like crops in a field. One approach, sometimes called a **centralized architecture**, is to use a few very large and fast computers. A second approach is to use a large number of less powerful computers and divide the workload among them. This is sometimes called a **distributed architecture** or, more commonly, a **decentralized architecture**.

Each approach has benefits and drawbacks. The centralized approach requires expensive computers and is more sensitive to the effects of technical problems. If one of the few servers becomes inoperable, a large portion of the site's capability is lost. Thus, Web sites with centralized architectures must have adequate backup plans. Any server problem, no matter how small, can threaten the operation of the site. The decentralized architecture spreads that risk over a large number of servers. If one server becomes inoperable, the site can continue to operate without much degradation in capability. The smaller servers used in the decentralized architecture are less expensive than the large servers used in the centralized approach. That is, the total cost of 100 small servers is usually less than the cost of one large server with the same capacity as the 100 small servers. However, the decentralized architecture does require additional hubs or switches to connect the servers to each other and to the Internet. Most large decentralized sites use load-balancing systems, which do cost additional money, to assign the workload efficiently.

LEARNING FROM FAILURES

WEB SERVERS AT EBAY

Online auction site **eBay** is very popular, as you have learned in earlier chapters. Indeed, it is so popular that its Web servers deliver more than 150 million pages per day. These pages are a combination of static HTML pages and dynamically generated Web pages. The dynamic pages are created from queries run against eBay's Oracle database, in which it keeps all of the information about all auctions that are under way or have closed within the most recent 30 days. With millions of auctions under way at any moment, this database is extremely large. The combination of a large database and high transaction volume makes eBay's Web server operation an important part of the company's success and a potential contributor to its failure. The servers at eBay failed more than 15 times during the first five years (1995–2000) of the company's life. The worst series of failures occurred during May and June of 2000, when the site went down four times. One of these failures kept the site offline for more than a day—a failure that cost eBay an estimated $5 million. The company's stock fell 20 percent in the days following that failure.

At that point, eBay decided it needed to make major changes in its approach to Web server configuration. Many of eBay's original technology staff had backgrounds at Oracle, a company that has a tradition of selling large databases that run on equally large servers. Further, the nature of eBay's business—any visitor might want to view information about any auction at any time—led eBay management initially to implement a centralized architecture with one large database residing on a few large database server computers. It made sense also to use similar hardware to serve the Web pages generated from that database.

In mid-2000, following the worst site failure in its history, eBay decided to move to a decentralized architecture. This was a tremendous challenge because it meant that the single large auction information database had to be replicated across groups, or clusters, of Web and database servers. However, eBay realized that using just a few large servers had made it too vulnerable to the failure of those machines. Once eBay completed the move to decentralization, it found that adding more capacity was easier. Instead of installing and configuring a large server that might have represented 15 percent or more of the site's total capacity, clusters of six or seven smaller machines could be added that represented less than one percent of the site's capacity. Routine periodic maintenance on the servers also became easier to schedule.

The lesson from eBay's Web server troubles is that the architecture should be carefully chosen to meet the needs of the site. Web server architecture choices can have a significant effect on the stability, reliability, and, ultimately, the profitability of an electronic commerce Web site.

Web Server Hardware and Software

Load-Balancing Systems

A **load-balancing switch** is a piece of network hardware that monitors the workloads of servers attached to it and assigns incoming Web traffic to the server that has the most available capacity at that instant in time. In a simple load-balancing system, the traffic that enters the site from the Internet through the site's router encounters the load-balancing switch, which then directs the traffic to the Web server best able to handle the traffic. Figure 8-11 shows a basic load-balancing system.

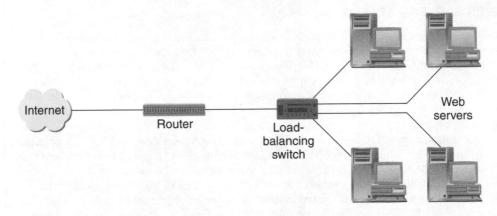

FIGURE 8-11   A load-balancing system in a decentralized architecture

In more complex load-balancing systems, the incoming Web traffic, which may enter from two or more routers on a larger Web site, is directed to groups of Web servers dedicated to specific tasks. In the sample complex load-balancing system that appears in Figure 8-12, the Web servers have been gathered into groups of servers that handle delivery of static HTML pages, servers that coordinate queries of an information database, servers that generate dynamic Web pages, and servers that handle transactions. Load-balancing switches and the software that helps them do their work cost roughly between $10,000 and $50,000, and include products such as **E-Load**, **Loadrunner**, **ServerIron**, and **Silkperformer**.

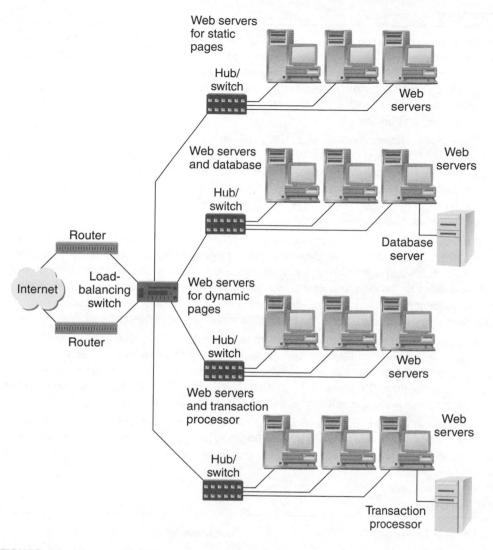

FIGURE 8-12 Complex load balancing

Summary

The Web uses a client/server architecture in which the client computer requests a Web page and a server computer that is hosting the requested page locates and sends a page back to the client. For simple HTTP requests, a two-tier architecture works well. The first tier is the client computer and the second tier is the server. More complicated Web interactions, such as electronic commerce, require the integration of databases and payment-processing software in a three-tier or higher (n-tier) architecture.

Operating systems commonly used on Web server computers include Microsoft server operating systems, UNIX-based operating systems, such as SunOS, FreeBSD, and Linux. The most popular Web server programs are Apache HTTP Server, Microsoft Internet Information Server, and Sun Java System Web Server. Web server computers also run a variety of utility programs such as Finger, Ping, Tracert, e-mail server software, Telnet, and FTP. Most Web servers also have software that helps with link checking and remote server administration tasks.

The problem of unsolicited commercial e-mail (spam) has grown dramatically in recent years. Content filters, particularly naïve Bayesian filters, are becoming available to deal with the problem. An increasing number of organizations are using a combination of server-level filters and client-level filters to reduce spam to tolerable levels. New laws designed to punish spammers have not stemmed the tide of spam, which most industry experts believe will persist until new e-mail protocols are adopted that provide absolute authentication of e-mail senders' identities.

Web server hardware is also an important consideration in the design of an online business site. The server computer must have enough memory to serve Web pages to all site visitors and enough disk space to store the Web pages and the databases that store the elements of dynamically generated Web pages. Large Web sites that have many Web server computers use load-balancing hardware and software to manage their high-activity volumes.

The operating system, connection speed, user capacity, and the type of pages that the site serves affect overall Web server performance. Benchmarking software and consulting firms that use it can help companies evaluate specific combinations of Web server hardware, software, and operating systems.

Key Terms

Active Server Pages (ASP)	Database server
Anonymous FTP	Dead link
Attachment	Decentralized architecture
Bayesian revision	Distributed architecture
Benchmarking	Dynamic content
Black list spam filter	Dynamic page
Blade server	Entity body
Centralized architecture	False positive
Challenge-response	File Transfer Protocol (FTP)
Client-level filtering	Finger
Computer virus	Full-privilege FTP
	JavaServer Pages (JSP)
	Java servlet

Link checker

Load-balancing switch

Log file

Naïve Bayesian filter

National Center for Supercomputing
Applications (NCSA)

N-tier architecture

Open-source software

Orphan file

PHP: Hypertext Preprocessor (PHP)

Ping (Packet Internet Groper)

Remote server administration

Request header

Request line

Request message

Response header field

Response header line

Response message

Response time

Reverse link checker

Server

Server architecture

Server farm

Server-level filtering

Server-side scripting (server-side includes or
server-side technologies)

Server software

Spam (unsolicited commercial e-mail or bulk
mail)

Static page

Telnet

Telnet protocol

Three-tier architecture

Throughput

Tracert

Two-tier client/server architecture

Virtual host

Virtual server

Virus

White list spam filter

Review Questions

RQ1. Compare the two- and three-tier Web client/server architectures and indicate the role of each tier in each architecture. Which architecture is the most likely candidate for an electronic commerce site? Explain why in about 100 words.

RQ2. Describe and briefly discuss two important measures of a Web site's performance.

RQ3. Beginning with the links provided in the Online Companion, locate more information about two of the three Web servers discussed in the chapter: Apache, Microsoft IIS, and Sun JSWS. Write approximately 250 words about each of the two servers you choose. Include descriptions of six features for each Web server and indicate the computer platforms and operating systems on which each runs.

RQ4. In one paragraph, outline the main differences between a desktop PC and a computer that would be suitable as a Web server for a small Web site.

RQ5. Using the Web or your school library, find articles that discuss the types of server hardware used by at least one electronic commerce site. Outline that site's architecture and approach to server hardware and software in an essay of about 300 words.

Exercises

E1. You are the information systems director for Abbon Laboratories, a biotechnology research firm with about 100 employees. Alice Stampler, Abbon's president, is aware that Abbon's incoming e-mail includes a great deal of spam and has always complained about the time it takes her to delete it. More importantly, she is concerned about the time wasted by the company's employees. Just thinking about those expensive Ph.D. researchers spending time deleting half of their e-mail drives her to distraction. Alice just read a story in a business magazine about naïve Bayesian spam filters. Being a Ph.D. herself, she is fascinated by the prospect that sophisticated mathematics might solve the company's spam problem. Alice asks you to find out all you can about naïve Bayesian filters and present a short report (about 200 words) in which you evaluate the technique and whether it can work for Abbon. Alice envisions one filter installed on the e-mail server that would screen all e-mail as it enters the company's network. You can use your library, your favorite search engines, or the links in the Online Companion under the heading **Naïve Bayesian Filters** to do your research.

E2. You created a Web site for International Paper Products and Pulp, complete with links to other pages on your site and to pages on the Internet. Bob Pardee, your supervisor, wants you to check periodically that the links on the corporate site are still valid. Instead of purchasing and installing a link-checking program, you decide to investigate online link checkers (Web sites that allow you to enter a Web site's root or home address and then check all the links that emanate from that site). Use **Link Check** or **Elsop LinkScan Quick Check** to check the links on any site of your choice. Print a few pages of the report and be prepared to turn them in to your instructor. Be patient. The program takes a little time to complete its work—especially on a Web page that has a large number of links.

E3. In researching Web server computers, you find that many companies that sell these computers offer a configuration option for controlling computers' disk drives called "RAID." Using the Web and your library, investigate the purpose of RAID controllers. Learn what these controllers do and how they do it. Summarize your findings in a 600-word briefing report suitable for presentation to a nontechnical manager.

Cases

C1. Microsoft and the People's Republic of China

Software piracy has been a major challenge for software makers such as Microsoft that want to sell software in the global marketplace. Laws that protect intellectual property vary from country to country, and the laws in many countries provide little or no protection. Governments in developing countries are reluctant to increase the protections afforded by their intellectual property laws because they see no point in passing laws that protect the profits of foreign corporations by imposing higher costs on their struggling local businesses and citizens. In the late 1990s, after years of holding firm on its global pricing, Microsoft began to offer significant discounts on its software to governments, small businesses, and individuals in developing countries. It also provided discounts on Windows operating systems software that was installed in new personal computers manufactured in developing countries. Microsoft donated software licenses to schools in developing countries. Just as these efforts were beginning to show some results, however, Microsoft faced a new threat to its global market position—open source software.

Open source operating system software, such as Linux, gives governments and businesses in developing countries a way to avoid paying any server software licensing fees to Microsoft. In 2000, the Brazilian state of Penambuco became the first governmental entity to pass a law that requires the use of open source software on all computers used for state business. Shortly thereafter, the Brazilian state of Rio Grande do Sul passed a similar law that requires the use of open source software in all of the state's offices and in all privately operated utilities. In 2003, IBM realized the potential for open source consulting business in the country and opened several centers for the development of Linux-based application software in Brazil. Microsoft, concerned about a Latin American open source domino effect, embarked on a public relations campaign in the region that included increased advertising spending and donations to public schools. In 2002, Peru was considering passing a law that would require public schools to use open source software. Microsoft founder Bill Gates flew to Peru and, with great public fanfare, donated $550,000 to the schools that would have been affected by the legislation. In 2004, Microsoft announced that it would donate $1 billion in cash and software over five years through the United Nations Development Program to not-for-profit organizations in 45 countries.

Most industry observers believe that Microsoft's largest non-U.S. market today is the People's Republic of China (PRC). Although the PRC generates about $300 million in licensing revenue for Microsoft, more than 90 percent of all Microsoft products used in China today are pirated. Bootleg copies of the company's latest products can be purchased on the street for a few dollars. Thus, Microsoft believes that converting users to paid licenses could increase its PRC licensing revenues tremendously. As the PRC moves from being a less-developed country toward becoming a major economic power in the world, Microsoft sees an opportunity to increase its licensing revenue in the country. In the past, Microsoft has used a global anti-piracy strategy that relied on identifying users of pirated software and threatening those users with legal action, but the company is changing its approach in developing markets such as Latin America and Asia. In the PRC, Microsoft's near-term goal is to develop a market for full-price software licenses that includes large business and government customers. Its new approach focuses more on recruiting major PRC business organizations as customers and less on sending threatening letters to users of pirated Microsoft software.

In developing its business in the PRC, Microsoft faces a number of challenges. Juliet Wu, former general manager of Microsoft China, published a book in 2000 that was highly critical of the company. The book was widely read in the PRC and received many good reviews. PRC officials have often criticized Microsoft for many things ranging from high prices to the company's use of Taiwanese programmers (the PRC does not officially recognize Taiwan as an independent nation separate from the PRC). Government officials in the PRC are also concerned about security. Microsoft has always maintained that the code to its software products is a trade secret and has refused to allow its publication or distribution. Companies that develop software that runs on Microsoft Windows, for example, must sign a non-disclosure agreement with Microsoft to obtain information they need about how Windows operates so they can make their software compatible with it. Many PRC officials believe that Microsoft, as a U.S.-based company, might include secret back doors into its software that would allow the U.S. government to enter PRC government computers undetected in a time of international conflict or war. At a very basic level, the ideology of the PRC's socialist government is a polar opposite to the highly competitive capitalist principles that have driven Microsoft to success. But the greatest challenge that Microsoft must overcome in the PRC is the attraction of open source software.

Open source programs' code is public; thus it cannot have secret back doors that can remain undetected in the code. The PRC is training 600,000 engineers each year and is expected to soon become the second-largest semiconductor manufacturer in the world. These semiconductors are increasingly being used to manufacture Chinese computers, many of which ship with a Linux operating system. In 1999, PRC officials announced a Chinese version of Linux called Red Flag. The PRC's national lottery and post office system all use Linux operating systems. The Province of Guangdong's accounting system runs on Linux-based computers. IBM is the primary consultant on a multi-year project to put the PRC's social security system onto Linux-based systems. As programming jobs in the United States dry up, some PRC Linux software developers have even been able to hire back Chinese programmers who had gone to U.S. companies. In 2003, the Procurement Center of the State Council issued an edict requiring that any computer purchased by the government after 2004 must be delivered with PRC-produced software only.

In the face of these challenges, Microsoft has worked hard to deliver its message that open source software can result in higher total costs because even though it is free, it requires more effort to install, maintain, and update than Microsoft products. In large organizations, this effort results in extra hours worked and thus, extra costs. Microsoft also argues that open source software's publicly available program code makes it a greater security risk. According to Microsoft, attackers can easily learn how any open source program works and develop strategies for attacking the software when it is running on publicly accessible computers, such as Web servers.

Required:

1. Assume that you are on the staff of a PRC legislator. Outline the arguments that you would use to support a law that required all government agencies to use only open source software on their Web servers.

2. Assume that you are working for the marketing department of Microsoft China. Develop a detailed list of briefing points that would help your salespeople convince top executives of large PRC companies to use Windows operating system software on their Web servers.

3. Assume that you are working for the business systems analysis department in IBM's PRC division, which offers both Microsoft Windows and Linux consulting services to PRC businesses and government offices. Develop a checklist that IBM analysts could use in consulting projects that could help advise clients as they make a choice between Windows or Linux operating system software for their Web servers.

4. Companies such as RedHat, Novell (with its SuSE distribution), and SCO (with its Caldera distribution) offer Linux operating system software for sale. Although Linux is available at no cost from various sources, these companies charge a fee for installation and configuration help. They also offer service contracts to help users maintain and upgrade the software on a continuing basis. Briefly outline the strategies that these companies might use to expand their market share in the PRC.

Note: Your instructor might assign you to a group to complete this case, and might ask you to prepare a formal presentation of your results to your class.

C2. Random Walk Shoes

Amy Lawrence, the owner of Random Walk Shoes, has asked you to help her as she launches her company's first Web site. In college, Amy was a business major with an artistic bent. She helped to pay her way through college by decorating sneakers with her hand-painted designs. Her business grew through word of mouth and through her participation in crafts fairs. By the time she earned her degree, Amy was running a successful business from her dorm room.

Amy expanded her sales efforts to include crafts fairs in nearby towns. She hired two college students to work for her, and she convinced several area gift shops to stock samples of her merchandise. The gift shops were not an ideal retail outlet for her products, however. Most people who want to buy decorated sneakers want to choose specific designs or have special designs created just for them. Customers also want to choose the specific shoes on which the design is placed. One of Amy's student workers suggested that she consider selling her products on the Web.

Realizing that the Web would give Random Walk Shoes a chance to reach a much wider audience and would allow customers to choose design-shoe combinations, Amy began gathering information and developing estimates about her planned Web activity. She bought a digital camera and took several hundred pictures of shoes, designs, and shoe-design combinations. She then hired a local Web designer to create sample pages for the Web site, including catalog pages that contained the digital images.

When the Web designer had completed a prototype of the site, Amy worked with the designer to calculate page sizes (including the images). The average page size was 85 KB. Amy and her employees then navigated the prototype site several hundred times to develop an estimate of how many pages an average visitor would download. They concluded that an average site visitor would visit 72 pages during each visit. Amy worked with the Web designer to develop estimates of activity they expect to occur on the Web site during its first two years of operation. These estimates include:

- The database of Web page information (including the images) will require about 80 MB of disk space.
- The database management software itself will require about 300 MB of disk space.
- The shopping cart software will require about 50 MB of disk space.
- About 6000 customers will visit the site during the first month, and site traffic will grow about 20 percent each month during the first two years.
- The site should accommodate a peak traffic load of 1000 visitors at one time.

Amy wants to include features on the site that are similar to those found on competing sites (a list of links to businesses that sell customized shoes on the Web is included in the Online Companion for your reference). Amy wants the site to provide a good experience for visitors. If the site is successful, it will generate sufficient revenue to allow an upgrade after two years. However, she does not want to spend more money than is necessary to get the site up and keep it running for the next two years.

Required:

1. Determine the features and capacities (RAM, disk storage, processor speed) that Amy should include in the Web server computer she will need for her site. Summarize your purchase recommendation in a one-page memorandum to Amy. You may include information from vendors' sites (such as **Dell**, **Hewlett Packard**, or **Sun**) as an appendix to your memorandum.

2. Consider the advantages and disadvantages of each major operating system that Amy might use on the new Web server computer. In a one-page memorandum to Amy, make a specific recommendation and support it with facts and a logical argument. If you do not believe that one operating system is clearly superior for this application, explain why.

3. Consider the advantages and disadvantages of each major Web server software package for accomplishing the goals that Amy has for this site. In a one-page memorandum to Amy, make a specific recommendation regarding which Web server software package she should use. Provide an explanation that supports your recommendation.

Note: Your instructor might assign you to a group to complete this case, and might ask you to prepare a formal presentation of your results to your class.

For Further Study and Research

Abualsamid, A. 2001. "Dishing Up Dynamic Content," *Network Computing*, 12(8), April 16, 90–92.

Alwang, G. 1998. "Internet Web Servers," *PC Magazine*, 17(9), May 5, 184–208.

Andrews, P. 2003. "Courting China," *U.S. News & World Report*, November 24, 44–45.

Ante, S. 2001. "Big Blue's Big Bet on Free Software," *Business Week*, December 10, 78–79.

Baran, N. 2001. "Load Testing Web Sites," *Dr. Dobb's Journal: Software Tools for the Professional Programmer*, 26(3), March, 112–116.

Business Week Online. 2004. "China and Linux: Microsoft, Beware!" November 15. (http://www.businessweekonline.com)

Cheng, H. and I. Bose. 2004. "Performance Models of a Proxy Cache Server: The Impact of Multimedia Traffic," *European Journal of Operational Research*, 154(1), April, 218–229.

Drucker, D. 2000. "Going Once, Going Twice, It's Sun," *InternetWeek*, May 22, 18.

Dustin, E., J. Rashka, and D. McDiarmid. 2002. *Quality Web Systems*. Boston: Addison-Wesley.

Dyck, T. and J. Rapoza. 2001. "IIS: Stay or Switch?" *eWeek*, October 29, 61–64.

The Economist. 2001. "Stealing Each Other's Clothes: Sun's Battle with IBM Raises Questions About Its Long Term Strategy," October 13, 61–63.

Epstein, J. 2004. "Standing Up to Redmond," *Latin Trade*, 12(6), June, 19.

Galli, P., 2004. "New IBM Unit to Target Emerging Markets," *eWeek*, 21(30), July 26, 9–10.

Gaudin, S. 2004. "Record Broken: 82% of U.S. E-mail Is Spam," *Internet News*, May 5. (http://www.internetnews.com/stats/article.php/3349921)

Glassman, M. 2003. "Fortifying the In Box as Spammers Lay Siege," *The New York Times*, July 31. (http://www.nytimes.com/2003/07/31/technology/circuits/31basi.html)

Graham, P. 2002. "A Plan for Spam," Paul Graham, July. (http://www.paulgraham.com/spam.html)

Graham, P. 2003. "Better Bayesian Filtering," Paul Graham, January. (http://www.paulgraham.com/better.html)

Gralla, P. 2002. "Making a List," *PC Magazine*, 21(10), May 21, 62–63.

Gross, G. 2004. "Judge Awards ISP $1 Billion in Spam Damages," *Computerworld*, December 20. (http://www.computerworld.com/governmenttopics/government/legalissues/story/0,10801,98421,00.html)

Harbaugh, L. 2000. "Balancing Act," *InternetWeek*, January 24, 26–30.

Information Week. 2004. "AOL Reports Big Drop in Spam," December 27. (http://www.informationweek.com/story/showArticle.jhtml?articleID=56200528)

Keizer, G. 2005. "CAN-SPAM Can't Slam Spam," *Information Week*, January 4. (http://www.informationweek.com/story/showArticle.jhtml?articleID=56900503)

Kopytoff, V. 2004. "Spam Mushrooms Despite a New Federal Law," *The San Francisco Chronicle*, September 2, C1.

Krim, J. 2003. "Spam's Cost To Business Escalates: Bulk E-Mail Threatens Communication Arteries," *The Washington Post*, March 13, A1.

Krim, J. 2004. "E-Mail Authentication Will Not End Spam, Panelists Say," *The Washington Post*, November 11, E01.

Kuo, J. 2001. "Work-Ready Linux," *InternetWeek*, September 10, 23–25.

Lee, M. 2004. "Don't Give Up on E-mail," *Network Computing*, 15(12), June 24, 20.

Lee, Y. 2000. "Low-Cost Dedicated Servers," *Web Techniques*, 5(7), July, 88–89.

Lipschutz, R. 1999. "Internet in a Box," *PC Magazine*, 18(12), June 22, 189–202.

Machrone, W. 2000. "Picking the Right Server Is Key," *PC Magazine*, 19(10), May 23, 52.

MacVittie, L. 2001. "IPlanet Goes Where No Server's Gone Before," *Network Computing*, 12(13), June 25, 96–99.

Mears, J. and D. Dubie. 2003. "Users Banking on Blades," *Network World*, 20(40), October 6, 1–2.

Montalbano, E. 2001. "Sun Sets on IPlanet Alliance," *Computer Reseller News*, August 27, 12.

Morgan, C. 2000. "Web Content Management," *Computerworld*, 34(17), April 24, 72.

Nelson, D. 2003. "Defending Your Site Against Spam," *O'Reilly Network*, June 26. (http://www.oreillynet.com/ pub/a/linux/2003/06/26/blocklist.html)

PC Magazine. 2001. "Servers," 20(14), August 12, 118.

PC World, 2005. "Spam Law Test,". 23(1), January, 20–22.

Petreley, N. 2001. "The Cost of Free IIS," *Computerworld*, 35(43), October 22, 49.

Popovich, K. 2002. "Blade Servers Boast Dual CPUs, Internal SCSI Drives," *eWeek*, 19(34), August 26, 7–8.

Roberts, P. 2004. "IETF Panel Deals Setback to Microsoft's Spam Proposal," *Computerworld*, 38(38), September 20, 14.

Roberts-Witt, R. 2001. "The Internet Business," *PC Magazine*, 20(5), March 6, 8–17.

Roberts-Witt, R. 2001. "Web Server Brawn," *PC Magazine*, 20(10), May 22, 144–151.

Sarrel, M. 2002. "The Top 100 (Undiscovered) Web Sites: What We Found," *PC Magazine*, February 26. (http://www.pcmag.com/article/0,2997,s=25087&a=21937,00.asp)

Schafer, S., 2004. "Microsoft's Cultural Revolution," *Newsweek*, June 28, E10-12.

Schroeder, E. 1999. "The Internet Rises and Sets with Sun," *PC Week*, March 8.

Schuchart, S. 2001. "IBM's Small Change," *Network Computing*, 12(9), April 30, 51–58.

Schwartz, J. 2001. "Update: How the NYSE Crashed," *InternetWeek*, June 8. (http://www.internetwk.com/story/INW20010608S0006)

Shankland, S. 2002. "Sun to Announce Leap into Linux," *CNET News.com*, August 5. (http://news.com.com/2100-1001-940277.html)

Shankland, S., M. Kane, and R. Lemos. 2001. "How Linux Saved Amazon Millions," *CNET News.com*, October 30. (http://news.cnet.com/news/0-1003-200-7720536.html)

391

Shen, X. 2005. "Intellectual Property and Open Source: A Case Study of Microsoft and Linux in China," *International Journal of IT Standards & Standardization Research*, 3(1), January-June, 21–43.

Shonfeld, E. 2002. "To Preserve Social Capital Get Big Boxes and Some Really Mean Software," *Business* 2.0, 3(3), March, 57.

Stauffer, T. 2000. "New Features and Fast Load Times Keep LandsEnd.com Humming," *Publish*, November. (http://www.publish.com/features/0011/feature3.html)

Ulfelder, S. 2004. "Spam-busters,"*Network World*, 21(12), March 22, 69–71.

Vijayan, J. 2001. "Sun Attempts to Woo Users Away from IIS," *Computerworld*, 35(42), October 15, 28.

Wagner, M. and T. Kemp. 2001. "What's Wrong with eBay?" *InternetWeek*, January 15, 1–2.

Xinhua, 2004. "Microsoft Teams Up with China's Leading Server and Solutions Supplier, November 9. (http://www.xinhua.org)

Yager, T. 2002. "Server Blades," *InfoWorld*, 24(29), July 22, 36–37.

CHAPTER 9

ELECTRONIC COMMERCE SOFTWARE

LEARNING OBJECTIVES

In this chapter, you will learn about:

- Finding and evaluating Web hosting services
- Basic functions of electronic commerce software
- Advanced functions of electronic commerce software
- Electronic commerce software for small and midsize businesses
- Electronic commerce software for midsize to large businesses
- Electronic commerce software for large businesses that have an existing information technology infrastructure

INTRODUCTION

In 1996, Australian Phillip Merrick started **webMethods** in Fairfax, Virginia. Merrick wanted to create a company that could exploit a new technology called XML (about which you learned in Chapter 2) in helping get B2B electronic commerce off the ground. Since then, the company has installed software in more than 1200 of the world's largest organizations that helps those organizations conduct electronic commerce with their suppliers. **Covisint**, the auto industry procurement portal you learned about in Chapter 6, uses webMethods software to integrate its Oracle database system with its Commerce One procurement and auction software, its Supply Solution supply chain execution software, and a variety of other vendors' software products.

Businesses on the Internet can face challenges when trying to exchange information with each other—information such as invoices and inventory tracking information—using XML. With XML and webMethods software, a manufacturing company's order can be translated into a Web page that both the manufacturer's software and the Web server can understand. One of webMethods' largest customers, Dun & Bradstreet, compiles financial and credit information. It uses webMethods software to translate data from proprietary systems into a common format that any Dun & Bradstreet customer's computer can understand. Dun & Bradstreet's customers save money when they use webMethods software instead of writing their own customized programs to interpret Dun & Bradstreet data. Dun & Bradstreet benefits because it no longer has to worry about supporting many different financial data and credit information formats in its regional data centers; webMethods software takes care of translating the different formats into a single form. As you will learn in this chapter, companies that engage in online business activities often combine software and tools from different vendors to accomplish their goals. Although small companies can sometimes use a single vendor to supply all their electronic commerce software, most larger companies need to integrate a number of software products, each of which performs a particular task or process particularly well.

WEB HOSTING ALTERNATIVES

When companies need to incorporate electronic commerce components, they may opt to run servers in-house; this is called **self-hosting**. This is the option used most often by large companies. Other companies, especially midsize and smaller companies, often decide that a third-party Web hosting service provider is a better choice than self-hosting. Many small Web stores use a third-party host provider for both Web services and electronic commerce functions, particularly when the Web site is small or the company sells a limited number of products.

As you learned in Chapter 2, a number of companies, called Internet service providers (ISPs), are in the business of providing Internet access to companies and individuals. Many of these companies offer Web hosting services as well. To distinguish themselves from companies that provide only Internet access services, these hosting service firms sometimes call themselves something other than ISPs. Because the hosting services they offer are designed to help companies conduct electronic commerce, these hosting service firms sometimes call themselves **commerce service providers (CSPs)**. These firms often offer Web server management and rent application software (such as databases, shopping carts, and content management programs) to businesses; thus, these companies also sometimes call themselves **managed service providers (MSPs)** or **application service providers (ASPs)**. Despite the increasing variety of acronyms, many companies that provide some or all of these additional services still call themselves ISPs.

Service providers offer clients hosting arrangements that include shared hosting, dedicated hosting, and co-location. **Shared hosting** means that the client's Web site is on a server that hosts other Web sites simultaneously and is operated by the service provider at its location. With **dedicated hosting**, the service provider makes a Web server available to the client, but the client does not share the server with other clients of the service provider. In both shared hosting and dedicated hosting, the service provider owns the server hardware and leases it to the client. The service provider is responsible for maintaining the Web server hardware and software, and provides the connection to the Internet through its routers and other network hardware. In a **co-location** (also spelled **collocation** and **colocation**) service, the service provider rents a physical space to the client to install its own server hardware. The client installs its own software and maintains the server. The service provider is responsible only for providing a reliable power supply and a connection to the Internet through its routers and other networking hardware. You can find service providers by looking in your local telephone directory or by using a Web directory such as **The List**, which appears in Figure 9-1.

FIGURE 9-1   The List Web host directory

The **HostIndex** site provides a convenient collection of Web pages that compare Web hosts. **TopHosts.com** and **HostSearch** also provide comprehensive link collections to companies researching Web hosting alternatives and services. Major Web directories can be helpful sources; the **Google Directory of Web Host Directories** is especially comprehensive.

When making Web server hosting decisions, a company should ask whether the hardware platform and software combination can be upgraded when the traffic on its Web site increases. A company's Web server requirements are directly related to its electronic commerce transaction volume and Web site traffic. The best hosting services provide Web server hardware and software combinations that are **scalable**, which means they can be adapted to meet changing requirements when their clients grow.

BASIC FUNCTIONS OF ELECTRONIC COMMERCE SOFTWARE

The size and objectives of electronic commerce sites vary greatly; thus, a variety of software and hardware products are used to build those sites. At the inexpensive end of the spectrum of electronic commerce solutions are choices such as externally hosted stores that provide software tools to build an online store on a host's site. At the other end of the range are sophisticated electronic commerce software suites that can handle high transaction volumes and include a broad assortment of features and tools.

The type of electronic commerce software an organization needs depends on several factors. One of the most important factors is the expected size of the enterprise and its projected traffic and sales. A high-traffic electronic commerce site with thousands of catalog inquiries each minute requires different software than a small online shop selling a dozen items. Another determining factor is budget. Creating an online store can be much less expensive than building a chain of retail stores. The start-up cost of an electronic commerce operation can be much lower than the cost of creating a brick-and-mortar sales and distribution channel that includes warehouses and multiple retail outlets. A traditional store requires a physical location with leases, employees, utility payments, and maintenance. The cost of creating the infrastructure for an online business can be much lower.

Another early decision is whether the company should use an external host or host the electronic commerce site in-house. Companies that have an existing information technology (IT) staff of programmers, Web designers, and network engineers are more likely to choose an in-house hosting approach. If a company does not have or cannot easily hire people with the skills required to set up and maintain an electronic commerce site, it can outsource all or part of the job to a service provider. Companies that are located outside of major metropolitan areas and want to host sites themselves must also consider whether their Internet connections are sufficient. In many cases, these companies find that they are not close enough to a major Internet access point or that their connections do not have sufficient bandwidth to handle large volumes of traffic efficiently. Even if these companies have employees with sufficient skills, they might decide to use a service provider to host their electronic commerce sites.

The specific duties that electronic commerce software performs range from a few fundamental operations to a complete solution—from catalog display to fulfillment notification. All electronic commerce solutions must at least provide:

- A catalog display
- Shopping cart capabilities
- Transaction processing

Larger and more complex electronic commerce sites also use software that adds other features and capabilities to the basic set of commerce tools. These additional software components can include:

- Middleware that integrates the electronic commerce system with existing company information systems that handle inventory control, order processing, and accounting
- Enterprise application integration
- Web services
- Integration with enterprise resource planning (ERP) software
- Supply chain management (SCM) software
- Customer relationship management (CRM) software
- Content management software
- Knowledge management software

Tools required by all electronic commerce sites are described in the following sections. The more advanced functions used by larger sites are covered later in this chapter.

Catalog Display

A catalog organizes the goods and services being sold. To further organize its offerings, a retailer may break them down into departments. As in a physical store, merchandise in an online store can be grouped within logical departments to make locating an item, such as a camping stove, simpler. Web stores often use the same department names as their physical counterparts. In most physical stores, each product is kept in only one place. A Web store has the advantage of being able to include a single product in multiple categories. For example, running shoes can be listed as both footwear and athletic gear.

A small commerce site can have a very simple static catalog. A **catalog** is a listing of goods and services. A **static catalog** is a simple list written in HTML that appears on a Web page or a series of Web pages. To add an item, delete an item, or change an item's listing, the company must edit the HTML of one or more pages. Larger commerce sites are more likely to use a dynamic catalog. A **dynamic catalog** stores the information about items in a database, usually on a separate computer that is accessible to the server that is running the Web site itself. A dynamic catalog can feature multiple photos of each item, detailed descriptions, and a search tool that allows customers to search for an item and determine its availability. The software that implements a dynamic catalog is often included in larger electronic commerce software packages; however, some companies write their own software to link their existing databases of product information to their Web sites.

Most of the Web stores you read about in earlier chapters are large, well-known sites. These sites include many features and have a professional look. Figure 9-2 shows the Web page of a small electronic commerce site that sells guitars and other musical instruments.

This site uses simple, inexpensive electronic commerce software and has a clean look with few features beyond those necessary to make sales.

FIGURE 9-2 Small electronic commerce site

Small Web stores that sell fewer than 30 or 40 items, such as the store shown in Figure 9-2, need only a simple list of products or categories. Organization of the items is not particularly important. Companies that offer only a small number of items can provide a photo of each item on the Web page that is a link to more information about the product. A static catalog is sufficient for their needs. Larger electronic commerce sites require the more sophisticated navigation aids and better product organization tools that are a part of dynamic catalogs.

Good sites give buyers alternative ways to find products. Besides offering a well-organized catalog, large sites with many products can provide a search engine that allows customers to enter descriptive search terms, such as "men's shirts," so they can quickly find the Web page containing what they want to purchase. Remember, the most important rule of all commerce is: Never stand in the way of a customer who wants to buy something.

Shopping Cart

In the early days of electronic commerce, shoppers selected items they wanted to purchase by filling out online forms. Using text box and list box form controls to indicate their choices, users entered the quantity of an item in the quantity text box, the SKU (stock-keeping unit) or product number in another text box, and the unit price in yet another text box. This system was awkward for ordering more than one or two items at a time.

One problem with forms-based shopping was that shoppers had to write down product codes, unit prices, and other information about the product before going to the order form, which was inevitably on another page. Another problem was that customers sometimes forgot whether they had clicked the submit button to send in their orders. As a result, they either sent the same order twice (pressing the submit button when they had already done so) or thought they had submitted the order when they really had not (consequently failing to submit the order). The forms-based method of shopping was confusing and error prone.

Figure 9-3 illustrates the problems that shoppers faced with forms-based ordering systems. First, many customers found it difficult to remember the exact spelling of the coffee names. Second, customers had to enter the coffee prices, which were located on a different Web page, in the text boxes. Thus, the customers needed to either write down or memorize the prices.

Shipping Information

- ◉ Ship the order to me.
- ○ Ship to the address below.

Name:	Otis Toadvine
Address:	1212 Pond Place
City, State, Zip:	Lincoln, Ne 47906

Order Information *(Please complete entire row)*

Coffee Name	Type	Grind	Qty.	Price
Zimbabwe	Caffeinated	Whole Bean	2 lbs	
Yemen Mocha	Caffeinated	Whole Bean	4 lbs	
Kopi Luwak	Caffeinated	Perc	1/2 lb	

text boxes that must be filled out

S&H: 4.95

Total:

Comments

NOTE: *Please do not enter your credit card number.*
We will call back to verify the order and receive the credit card number.

Credit Card Information

Name on credit card	Card Type	Expires

FIGURE 9-3 Using a form to enter an order

The forms-based method of ordering has given way to electronic shopping carts. Today, shopping carts are a standard of electronic commerce. As you learned in Chapter 4, a shopping cart, also sometimes called a shopping bag or shopping basket, keeps track of the items the customer has selected and allows customers to view the contents of their carts, add new items, or remove items. To order an item, the customer simply clicks that item.

All of the details about the item, including its price, product number, and other identifying information, are stored automatically in the cart. If a customer later changes his or her mind about an item, he or she can view the cart's contents and remove the unwanted items. When the customer is ready to conclude the shopping session, the click of a button executes the purchase transaction. Figure 9-4 shows a typical shopping basket page at a site that sells computer equipment.

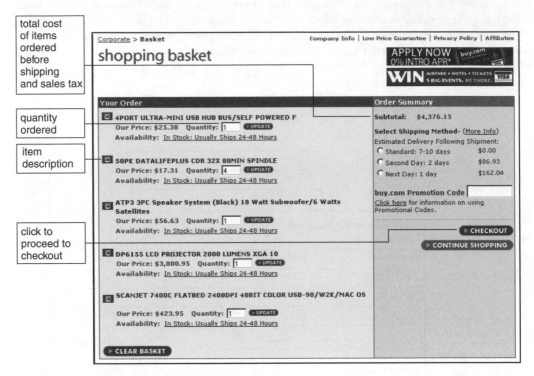

FIGURE 9-4   Typical shopping basket page

Clicking the Checkout button usually displays a screen that asks for billing and shipping information and that confirms the order. As you can see from the figure, the shopping cart software keeps a running total of each type of item. The shopping cart calculates a total as well as sales tax and shipping costs.

Shopping cart software at some Web commerce sites allows the customer to fill a shopping cart with purchases, put the cart in virtual storage, and come back days later to confirm and pay for the purchases. A number of companies, including **BIZNET Internet Services**, **CartIt!**, **SalesCart**, and **WebGenie Software**, sell shopping cart software that sellers can add to their Web sites. These software packages range in price from a few hundred dollars to several thousand dollars, plus an ongoing monthly fee. The shopping cart software sold by SalesCart works with several different Web server software platforms, as shown in Figure 9-5.

Because the Web is a stateless system—unable to remember anything from one transmission or session to another—shopping cart information must be stored explicitly for the

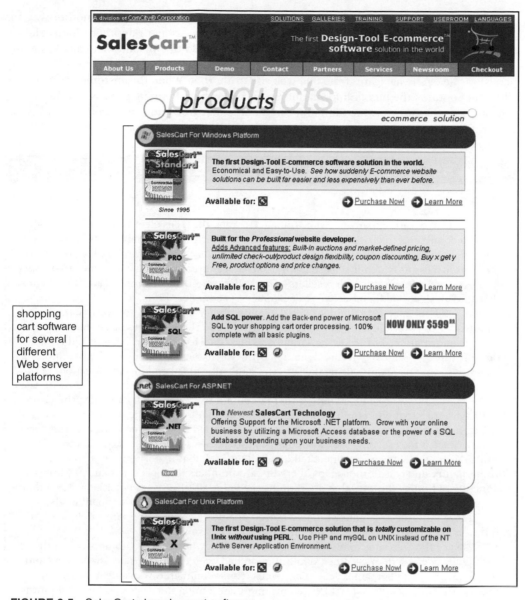

shopping cart software for several different Web server platforms

FIGURE 9-5   SalesCart shopping cart software

shopper to retrieve later. Furthermore, it must distinguish one shopper from another so that the purchases are not mixed up. One way to uniquely identify users and store information about their choices is to create and store cookies, which, as you learned in earlier chapters, are bits of information stored on a client computer. When a customer returns to a site that issued a particular cookie, the shopping software reads either the cookie from the customer's computer or the database record from the merchant's server.

If a shopper's browser does not allow storage of cookies, sites can use another way to preserve shopping cart information from one browser session to another. Some electronic commerce software packages, such as **ShopSite**, do this by automatically assigning a shopper a temporary number. The number is added to the end of the shopper's URL and persists as he or she navigates from one Web site to another. When the customer returns, the URL still contains the bits of information about his or her shopping cart. When the customer closes the browser, the temporary number is discarded and thus cannot be reused, even if the customer later reopens the browser and returns to the same Web site.

LEARNING FROM FAILURES

PDG Software

PDG Software is a company based in Tucker, Georgia, that sells electronic commerce software to companies that operate small and midsize electronic commerce Web sites. PDG sells shopping cart software, auction software, shopping mall software, and a number of other packages. Although it sells some of its software directly to the companies that use it, most of its sales are through resellers—firms that use PDG software as part of Web sites that they design, build, and deliver to customers as complete units.

In April 2001, an attacker discovered a vulnerability in the PDG software that allowed an intruder to enter the shopping cart and open the file that contained customer names, contact information, and credit card numbers. PDG developed a patch that would repair the software the same day it found out about the intrusions. PDG posted the patch on its Web site so that companies using the software could download and install the patch. Both PDG and the FBI issued press releases immediately to warn users of the problem with the shopping cart software and encourage them to obtain the patch. Unfortunately, the users of the software that had purchased it as part of a complete electronic commerce Web site were, in many cases, unaware that their sites included the PDG shopping cart software.

Because it took so long—several months, in some cases—to find and contact the companies using the software, online offenders had an excellent opportunity to exploit this vulnerability and collect thousands of credit card numbers. In most cases such as this, the difficulty of finding the sites that are running the vulnerable software helps slow down the attackers. Unfortunately, in this case, the intruder who discovered the opening also found that entering a specific word in a search engine's search expression would instantly return a list of the thousands of sites running the PDG software.

Most of the Web sites found out about the problem when their customers called them, suspicious because their credit card information had been compromised. The lesson from this failure is that companies that operate electronic commerce Web sites must know the source of the software used in creating and maintaining their sites and must monitor news about the security of that software.

Transaction Processing

Transaction processing occurs when the shopper proceeds to the virtual checkout counter by clicking a checkout button. Then the electronic commerce software performs any necessary calculations, such as volume discounts, sales tax, and shipping costs. At checkout, the customer's Web browser software and the seller's Web server software both switch into a secure state of communication. You will learn more about how Web clients and servers establish these secure communication states in the next two chapters.

Transaction processing can be the most complex part of the online sale. Computing taxes and shipping costs are important parts of this process, and site administrators must continually check tax rates and shipping tables to make sure they are current. Some software enables the Web server to obtain updated shipping rates by connecting directly to shipping companies to retrieve information.

Other calculation complications include provisions for coupons, special promotions, and time-sensitive offers; for example, "purchase a round-trip ticket before the end of the month and receive a 50 percent discount." Some shopping cart software designed for small and midsize companies provides connections to accounting software so that Web sales can be entered simultaneously in the company's accounting system. In larger companies, the integration of the Web site's transaction processing into the accounting and operation-control systems of the company can be very complex. The next section discusses some of the advanced functions that larger companies look for in electronic commerce software.

ADVANCED FUNCTIONS OF ELECTRONIC COMMERCE SOFTWARE

In this section, you will learn about the features that larger companies need in their electronic commerce software. Although there are exceptions, such as Amazon.com and Buy.com, most large companies that have electronic commerce operations also have substantial business activity that is not related to electronic commerce. Thus, integrating electronic commerce activities into the company's other operations is very important.

Middleware

Larger companies usually establish the connections between their electronic commerce software and their existing accounting system by using a type of software called middleware. Some large companies that have sufficient IT staff write their own **middleware**; however, most companies purchase middleware that is customized for their businesses by the middleware vendor or a consulting firm. Thus, most of the cost of middleware is not the software itself, but the consulting fees needed to make the software work in a given company. Making a company's information systems work together is called **interoperability** and is an important goal of companies when they install middleware.

The total cost of a middleware implementation can range from $50,000 to several million dollars, depending on the complexity of the company's underlying operations and its existing information systems. Major middleware vendors include **BEA Systems**, **Broadvision**, **Digital River**, and **IBM Tivoli Systems**. As the market for this type of software has matured, the companies that provide this software have worked to build products that can integrate software throughout the enterprise with company Web sites. The BEA Integration Projects Web page appears in Figure 9-6.

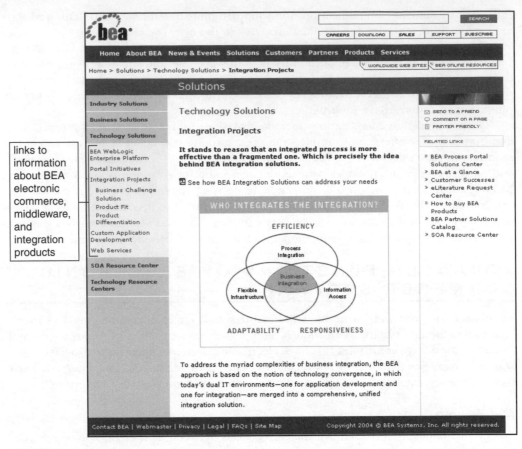

links to information about BEA electronic commerce, middleware, and integration products

FIGURE 9-6   BEA Technology Solutions page

Enterprise Application Integration and Databases

A program that performs a specific function, such as creating invoices, calculating payroll, or processing payments received from customers, is called an **application program**, **application software** or, more simply, an **application**. An **application server** is a computer that takes the request messages received by the Web server and runs application programs that perform some kind of action based on the contents of the request messages. The actions that the application server software performs are determined by the rules used in the business. These rules are called **business logic**. An example of a business rule is: When a customer logs in, check the password entered against the password file in the database.

In many organizations, the business logic is distributed among many different applications that are used in different parts of the organization. In recent years, many IT departments have devoted significant resources to the creation of links among these scattered applications so that the organization's business logic can be interconnected. This activity is called **application integration** or **enterprise application integration**. The integration is accomplished by programs that transfer information from one application to another. For example, a program might transfer information from order entry systems in several different divisions to a single accounts receivable and sales system that integrates all enterprise-wide sales activity. In many cases, the data formats in the various programs are different and the transfer programs must edit and reformat the data before transferring it. Increasingly, programmers are using XML data feeds to move data from one application to another in enterprise integration implementations.

Application servers are usually grouped into two types: page-based and component-based systems. **Page-based application systems** return pages generated by scripts that include the rules for presenting data on the Web page with the business logic. Common page-based server systems include Macromedia ColdFusion, JavaServer Pages (JSP), Microsoft Active Server Pages (ASP), and PHP: Hypertext Preprocessor (PHP). These page-based systems work quite well for small and midsize Web sites. Because they combine the page presentation logic with the business logic, however, they can be difficult to revise and update. Larger businesses often prefer to use a **component-based application system** that separates the presentation logic from the business logic. Each component of logic is created in its own module. This makes updating and changing elements of the system much easier—especially on large electronic commerce sites that are built and maintained by teams of programmers. The most common component-based systems in use today are **Enterprise JavaBeans (EJBs)**, **Microsoft Component Object Model (COM)**, and the Object Management Group **Common Object Request Broker Architecture (CORBA)**.

Application servers usually obtain the business logic information they use to build Web pages from databases. A **database manager** is software that stores information in a highly structured way. The structure of the database makes it easy for the database manager software to retrieve the information stored in the database. Smaller electronic commerce sites can use low-cost databases such as Microsoft Access. Larger sites need the power of more expensive database management software such as IBM DB2, Microsoft SQL Server, or Oracle. These database management software packages can be quite expensive. Typical installations cost between $5000 and $200,000. Companies with very large databases that have operations in many locations must make their data available to users in those locations. Large information systems that store the same data in many different physical locations are called **distributed information systems**, and the databases within those systems are called **distributed database systems**. The complexity of these systems leads to their high cost.

Most companies that can afford it do use commercial database products; however, an increasing number of companies and other organizations are beginning to use MySQL, which was developed and is maintained by a community of programmers on the Web. Similar to the Linux operating system you learned about in earlier chapters, **MySQL** is open-source software that can be downloaded and used at no cost. The term **open source** is used to describe such software because the source code of the software is freely available, or "open." The MySQL home page appears in Figure 9-7.

Except for small sites offering only a few products, companies should consider database support as they evaluate electronic commerce software. Most Web stores selling many products use a database that stores product information, including size, color, type, and price details. Usually, the database that serves an online store is the same one that is used by the existing corporate clients. It is better to have one database serving two separate entities because it eliminates parallel but distinct databases—something companies should

FIGURE 9-7　MySQL page

avoid if possible. If a company has existing inventory and product databases, then it should evaluate only electronic commerce software that supports these systems.

Web Services

Companies are beginning to extend the idea of application server systems so that these programs can communicate across organizational boundaries. Although a generally accepted definition has not yet evolved, many IT professionals define **Web services** as a combination of software tools that let application software in one organization communicate with other applications over a network by using a specific set of standard protocols known by their acronyms: SOAP, UDDI, and WSDL (these protocols are described below). Another definition of Web services that IT professionals use is: a self-contained, modular unit of application logic that provides some business functionality to other applications through an Internet connection.

What Web Services Can Do

Companies are using Web services to offer improved customer service and reduce costs. In some companies, Web services are used to provide the XML data feeds that flow from one application to another in enterprise application integration efforts. In other applications,

Web services provide data feeds between two different companies. J.P. Morgan Chase & Co., a major investment bank, uses Web services in its investment information portal. The Web services pull information, such as general economic forecasts, financial analyses of specific companies, industry forecasts, and financial markets results into continually updated online reports that customers can obtain on the J.P. Morgan Chase portal site. The bank's customers could obtain all of this information themselves, but the aggregation is a service that the bank provides. The information flow in this case is from the bank to its customers.

Nationwide Building Society, a mortgage company in Swindon, England, uses a Web services tool to automate its communications with mortgage application service companies. These service companies obtain information from consumers who want mortgages and then forward the information in a prescribed XML format to Nationwide. The Nationwide Web services software reformats the submission and submits it to Nationwide's enterprise computer system. When a lending decision has been reached, the Web services tool conveys the decision back to the mortgage application service company. This Web services approach has reduced costs and decreased turnaround time for loan decisions at Nationwide.

CUNA Mutual Group sells services to credit unions throughout the United States from its headquarters in Madison, Wisconsin. These services include everything from check clearing to construction management. CUNA provides many of its services by running programs on old computer systems that have been in operation for years. Instead of reprogramming everything so it could be accessible on the Web, CUNA created a Web services layer that takes information from the old computer systems and generates Web pages that its customers can use to obtain those services.

How Web Services Work

A key element of the Web services approach is that programmers can write software that accesses these units of business application logic without knowing the details of how each unit is implemented. Web services can be mixed and matched with other Web services to execute a complex business transaction. Thus, Web services allow programs written in different languages on different platforms to communicate with each other and accomplish transaction processing and other business tasks.

The common format of this machine-to-machine communication was originally HTML; however, most newer Web services implementations use XML. As you learned in Chapter 2, organizations can use XML to mark up content with agreed upon sets of descriptive tags. As Web services become more fully implemented, businesses will be able to connect their operations quickly and cheaply. Thus businesses will be able to reduce transaction costs and improve customer service at the same time. Customers and employees will find it easier to access companies' Web resources from a variety of devices such as PDAs and mobile phones.

The first Web services were information sources. The Web services model allowed programmers to incorporate these information sources into software applications. For example, a company that wanted to collect all of its financial management information into one spreadsheet could use Web services to obtain bank account and loan balances, stock portfolio holdings, and current interest rates on financial instruments. If this information is available through Web services, the spreadsheet program can use those services to update itself automatically. Some of the information might be available as a Web service at no cost; other information access might require a subscription. But Web services can make automated access of the information much easier.

A more advanced example would be a company that uses purchasing software to help manage that activity. That software can use Web services to obtain price information from a variety of vendors. After the purchasing agent reviews the price and delivery information and authorizes the purchase, the software can submit the order and track it until the shipment is received. On the other side of this transaction, the vendor's software can use Web services (in addition to providing price and delivery information) to check the buyer's credit and contract with a freight company to handle the shipment.

SOAP, UDDI, and WSDL

Three rule sets (usually called protocols or specifications) let programs work with the formatted (using XML or HTML) data flows to accomplish the communication that makes Web services work. The **Simple Object Access Protocol (SOAP)** is a message-passing protocol that defines how to send marked up data from one software application to another across a network. You can see the full SOAP specification and learn more about SOAP at the **W3C SOAP Page**.

The characteristics of the logic units that make up specific Web services are described using the **Web Services Description Language (WSDL)**. Today, programmers can use the information in a WSDL description to modify an application program so it can connect to a Web service. When Web services become more complex, WSDL descriptions allow programs to configure themselves to connect to multiple Web services. You can learn more about WSDL and related topics at the **W3C Web Services Activity** pages.

Programmers (and, eventually, the programs themselves) need to find the location of Web services before they can interpret their characteristics (described in WSDL) or communicate with them (using SOAP). The set of protocols that identify locations of Web services and their associated WSDL descriptions is called the **Universal Description, Discovery, and Integration (UDDI) specification**. The **UDDI.org** Web site is a good source of information about this specification and includes the current UDDI Business Registry, which includes a catalog of currently available Web services.

A number of major software vendors have embraced the idea of Web services in new technology initiatives such as **Microsoft .NET** and the **Sun Java 2 Platform, Enterprise Edition**. Many companies that have used Web services to accomplish application integration have found it to be less expensive to implement than older approaches that required programmers to write or adapt multiple middleware software programs. Merrill Lynch was able to use Web services to implement an integration project for $30,000 that would have cost $800,000 using its older application integration approach.

The Future of Web Services

The idea behind Web services is a major change in the way business does computing. The IT industry has historically resisted standards and has frequently used programming languages that cannot communicate with each other. For years, large businesses and other organizations have hired armies of programmers to write middleware software to integrate their hodgepodge of programs for financial management, inventory control, marketing, and other functions. The idea of connecting software within an organization is still revolutionary—connecting software across organizational boundaries is even more revolutionary. Some industry analysts report companies are using Web services in 25 percent of all current data integration projects.

Despite the promise of Web services, there are some potential pitfalls. Much of the data in Web services applications is stored and transmitted in XML format. Because there are so many variations of XML in use today, it is critical that data-providing and data-using partners agree on which XML implementation to use. As Web services become more commonplace, individual companies' software applications will become more dependent on them. This means that Web services must include quality of service and service level specifications on which applications developers at each company can rely. At present, there are no Web services management standards or history of best practices. This lack of standards means that each Web services subscriber needs a detailed agreement (specifying service levels, quality of service standards, and so on) with each Web services provider. Security can be a problem with Web services. By its very nature, a Web services data feed connects directly into a company's internal applications, bypassing any security features installed at the company's perimeter (you will learn more about perimeter security defenses in Chapter 10). These are not insurmountable issues, but they do prevent Web services from being a simple matter.

Despite the hurdles that must be overcome, some companies have begun to implement Web services successfully. For example, the **MSN Money** site buys stock quotes from the Interactive Data Corporation through its ComStock Web service. An MSN Money stock quote page with the **ComStock** Web services acknowledgment appears in Figure 9-8.

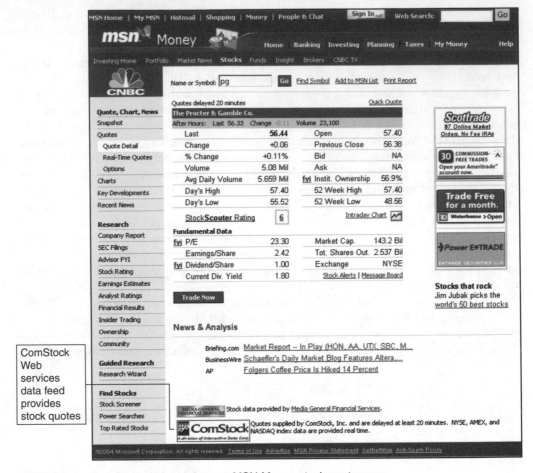

ComStock Web services data feed provides stock quotes

FIGURE 9-8   ComStock Web services on MSN Money stock quote page

Integration with ERP Systems

Larger firms have extranets and intranets requiring tools and capabilities different from those needed to implement simpler electronic commerce Web sites. In the case of large B2B interactions, both the buyer and seller have complex systems. In general, B2B sites require security tools not standard in B2C systems, such as encryption and authentication, as well as signed receipt notices.

Many B2B Web sites must be able to connect to existing information systems such as enterprise resource planning software. **Enterprise resource planning (ERP)** software packages are business systems that integrate all facets of a business, including accounting, logistics, manufacturing, marketing, planning, project management, and treasury functions. The major ERP vendors include **Baan**, **Oracle**, **PeopleSoft** (now a part of Oracle), and **SAP**. A typical installation of ERP software costs between $2 million and $25 million; thus, companies that are already running these systems have made a significant investment in them and expect their electronic commerce sites to integrate with them. Figure 9-9 shows a typical architecture for a B2B Web site that connects to several existing information systems, including the ERP system within the company and its trading partners' systems through EDI connections.

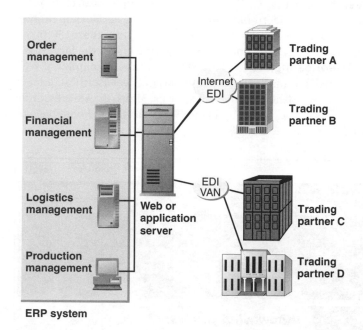

ERP system

FIGURE 9-9   ERP system integration with EDI

ELECTRONIC COMMERCE SOFTWARE FOR SMALL AND MIDSIZE COMPANIES

In this section you will learn about software that small and medium-sized businesses can use to implement online business Web sites. In most cases, these companies can create a Web site that stands alone in its business activities and does not need to be coordinated completely with the business' other activities.

Basic Commerce Service Providers

Using a service provider's shared or dedicated hosting services instead of building an in-house server or using a co-location service means that the staffing burden shifts from the company to the Web host. CSPs have the same advantages as ISP hosting services, including spreading the cost of a large Web site over several "renters" hosted by the service. The biggest single advantage—low cost—occurs because the host provider has already purchased the server and configured it. The host provider has to worry about keeping it working through lightning storms and power outages.

CSPs offer free or low-cost electronic commerce software for building electronic commerce sites that are then kept on the CSP's server. Services in this category usually cost less than $20 per month, and the software is built into the CSP's site, allowing companies to immediately begin building and storing a storefront using the Web interface of the software. These services are designed for small online businesses selling only a few items (usually no more than 50) and having relatively low transaction volumes (fewer than 20 transactions per day). **ValueWeb**, operating since 1996, is an example of a CSP. ValueWeb offers businesses comprehensive electronic commerce hosting services including shared hosting, dedicated hosting, and co-location services. Hosting more than 150,000 Web sites for customers in over 130 countries, ValueWeb has become one of the largest Web hosting companies in the world. **ProHosting.com** and **Interland** are other examples of Web hosting companies serving the small and midsize company market. Because these companies offer a variety of services, they might be called ISPs, CSPs, MSPs, or ASPs by different users, depending on the service they are seeking. Figure 9-10 shows the home page of Interland, which outlines its CSP offerings.

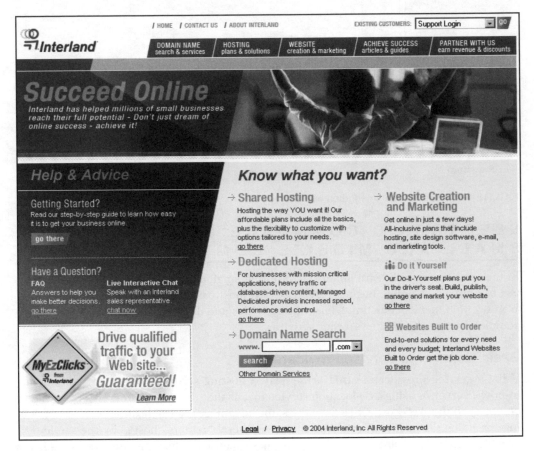

FIGURE 9-10   Interland Web hosting services home page

TopHosts.com (see Figure 9-11) features a comprehensive presentation about CSPs and hosting issues. This site contains hundreds of links and much good information.

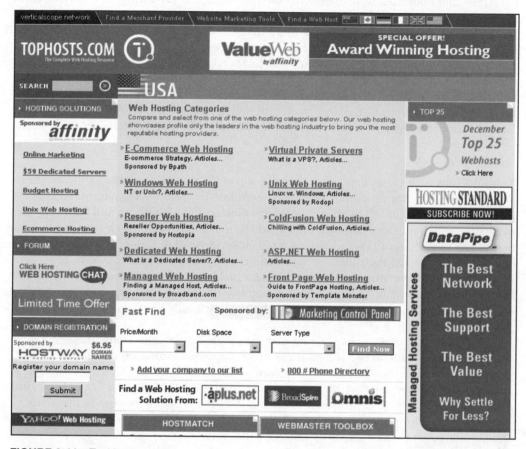

FIGURE 9-11   TopHosts.com home page

Mall-Style Commerce Service Providers

Mall-style CSPs provide small businesses with an Internet connection, Web site creation tools, and little or no banner advertising clutter. Web hosts in this group charge a monthly fee that is often higher than that of lower-end providers, and may also charge one-time setup fees. Some of these providers also charge a percentage of or fixed amount for each customer transaction. These Web hosts also provide high-quality tools, storefront templates, an easy-to-use interface, and quick Web page-generation capabilities and page maintenance.

Mall-style CSPs provide shopping cart software or the ability to use another vendor's shopping cart software. They also furnish customer payment processing so that customers can choose to purchase their goods and services with a credit card or other form of payment. The CSP processes the acceptance and authorization of credit cards on behalf of the merchant. Another benefit is that because they are paying a monthly fee to the CSP, sites do not have to display any Web banners, which can be unattractive and distracting. The

Electronic Commerce Software

fourth benefit of the mall-style CSPs is that they provide higher-quality Web store building and maintenance tools than do the basic CSPs.

CSPs that offer mall-style commerce services include **eBay Stores** and **Yahoo! Store**. Another CSP that began as a mall-style service is **Bigstep**, but it no longer uses the mall structure. All three of these CSPs offer Web site construction tools that can be used by small and midsize businesses to take their businesses online.

You can learn how capable these Web hosting services are by trying them. Many of these services have a 30-day trial period. The eBay Stores service does not offer a free trial, but the charges are minimal and are based on the number of products listed and graphic images used. Creating an eBay test store with only a few items costs less than $20. The Yahoo! Store and Bigstep services are described in the following sections.

Yahoo! Store

Yahoo! Store is the most expensive of this group of Web hosts. It serves as the business Web host not only for thousands of small storefronts, but also for big names, such as the Kennedy Space Center Space Shop, The Sharper Image, PalmPilotGear, and *Rolling Stone* magazine, which all have Yahoo! Stores in addition to their regular Web sites. These companies use Yahoo! Store as an additional retail outlet targeted at Yahoo! site visitors. Other companies do not sell their main products directly to consumers on the Web. These companies can use Yahoo! Store as a retail outlet for incidental merchandise. The **WD-40 Gift Shop** is an example of this use of Yahoo! Store. WD-40 sells its main products through large retailers and hardware stores, but it sells WD-40 baseball caps and polo shirts in its Yahoo! Store gift shop.

Merchants can create, change, and maintain their Yahoo! storefronts through a Web browser. Yahoo! holds all the stores' pages in a proprietary format on its own site. Yahoo! processes payment transactions on a secure server. When a merchant signs up for a Yahoo!-hosted storefront, its URL is a subdomain of the Yahoo! URL. For example, if a business's username is BoldImpressions, then the URL that customers visit is http://store.yahoo.com/BoldImpressions, which is a Yahoo! subdomain. For an extra fee, a Yahoo! Store can be redirected to the company's existing URL.

To update and maintain the store, the merchant logs on with a store owner username and password and then uses the online tools to edit pages, add or delete catalog items, and perform other functions. Merchants can use Manager, a Yahoo!-supplied software product, to examine statistical information, set up acceptable payment methods, select preferred customer shipping methods, and change global site settings. Yahoo! site visitors enter the online mall through the Yahoo! Shopping link on the Yahoo! main page. The stores are organized by category in the mall, as you can see on the Yahoo! Shopping entry page shown in Figure 9-12.

links to Yahoo! Store sites of major retailers

links to Yahoo! Store sites organized by category

FIGURE 9-12   Yahoo! Shopping entry page

Bigstep

Bigstep has received many industry awards for its CSP offering. Bigstep provides two different storefront packages that can meet the electronic commerce software needs of a range of small and midsize businesses. The Bigstep home page, which shows the two levels of service offered, appears in Figure 9-13.

To create a store, merchants must register with Bigstep. The registration process identifies the user with an e-mail address and password. As with Yahoo!, the store's URL is a subdomain of Bigstep unless the merchant pays an additional fee. After logging on with a store owner e-mail address and password, the merchant can create and manage the electronic store.

FIGURE 9-13 Bigstep home page

Bigstep's reports provide data-mining capabilities that search through site data collected in log files. **Data mining**—looking for hidden patterns in data—can help businesses find customers with common interests and discover previously unknown relationships among the data. Reports can indicate problematic pages in a store's design where, for example, a large number of customers get stuck and then leave the Web site. Other facts that Bigstep reports can reveal include the number of pages an average customer must load and display before locating the merchandise he or she wants. If customers have to load too many pages, they might become impatient and leave without making a purchase.

If a merchant has a brick-and-mortar store, Bigstep's built-in map locator can display the location of the store. Additional Bigstep store features include automatic calculation of taxes and shipping, collection of customer data, merchant e-mail notification of sales, and customer e-mail confirmation when products ship.

Estimated Operating Expenses for a Small Web Business

The following table shows an estimate of the first-year expenses that a small business owner might incur to put a store on the Web. The estimate assumes that the Web site will offer fewer than 50 different items for sale. The total omits payment processing charges, which might average 50 cents per transaction and 2 percent of each sale's total. The costs shown are averages. Depending on which hosting service and electronic commerce software options are chosen, the actual costs could be somewhat lower or considerably higher.

Operating Costs	Cost Estimate
Initial site setup fee	$ 200
Annual maintenance fee (12 x $100)	1200
Domain name registration	70
Scanner for photo conversion or digital camera	300
Photo editing software	100
Occasional HTML and site design help	400
Merchant credit card setup fee	200
Total first-year cost	**$2470**

The preceding costs are typical, but they can vary because different Web hosting sites charge varying fees for various services. Additional payment processing fees can run into hundreds and thousands of dollars, but those fees occur only when a site makes sales. A reasonable guideline for payment processing fees that would be charged to a new merchant opening a business on the Web is about 3 percent of gross sales. Thus, if a site's annual gross sales are $50,000, then the payment processing fees should be approximately $1500. That estimate would include both the per-transaction fixed costs and the percentage of total sales costs charged by most merchant credit card processing agencies.

Contrast the preceding costs with comparable estimated costs for self-hosting a Web site. Setup and Web site maintenance costs include equipment, communications, physical location, and staff. Equipment—a server and networking gear—has a one-time cost ranging from $3000 to $20,000. A T1 connection or fraction thereof (see Chapter 2) costs from $1200 to $12,000 per year. A server must be housed in a room that is both secure and convenient to communications access. The cost to secure a room, properly air-condition it, and install a chemical fire extinguishing system can be nearly $5000 a year. A self-hosted system requires a staff of experts well versed in a variety of Web programming and scripting languages, electronic commerce packages, and database management systems. Technicians will likely be required to monitor and maintain equipment. Minimum staff costs range from $50,000 to $100,000 annually. In total, annual operating costs for self-hosting approach $60,000 to $100,000 or more the first year. Costs for subsequent years will be about the same. Companies should carefully compare self-host cost estimates with the fees charged by various hosting services.

The costs previously discussed are for a small electronic commerce site. Costs for larger sites are much more difficult to estimate. The cost of integrating the Web site with the existing systems of the company is often the largest element of the total cost. Midsize businesses typically incur start-up costs ranging from $100,000 to $500,000 and recurring annual costs of about half that amount. Large businesses typically spend between $1 million and $50 million to launch an electronic commerce site and then spend another 50 percent of the launch cost every year to operate, maintain, and improve the site. You will learn more about managing the costs of Web site implementation and management in Chapter 12.

Next, you will learn about midrange electronic commerce packages. Midrange packages are suitable for running larger businesses. These software packages have more features, are capable of handling more inventory items and types of transactions, and thus are more expensive than the template-driven CSP offerings described above.

ELECTRONIC COMMERCE SOFTWARE FOR MIDSIZE TO LARGE BUSINESSES

This section includes a discussion of software that midsize and large companies can use to implement electronic commerce features on their Web sites. It also includes an outline of Web site development tools that can be used for that purpose and an overview of three specific midrange electronic commerce software products that are representative of the types of products available.

These midrange packages allow the merchant to have explicit control over merchandising choices, site layout, internal architecture, and remote and local management options. In addition, the midrange and basic electronic commerce packages differ on price, capability, database connectivity, software portability, software customization tools, and computer expertise required of the merchant.

Web Site Development Tools

Although they are more often used for creating small business sites, it is possible to construct the elements of a midrange electronic commerce Web site using the Web page creation and site management tools you learned about in Chapter 2. For example, recent versions of **Macromedia Dreamweaver** all include integrated development environments. Experienced Web designers using this tool can create the elements of dynamic Web pages as easily as static Web pages.

Other Web page design tools, such as **Microsoft FrontPage**, can also be used to build the framework of a functional midrange electronic commerce site. The remaining elements of the dynamic pages needed to create catalog, customer service, and transaction-processing pages can be added with development tools such as Microsoft's **Visual Studio .NET** product.

After creating the Web site with these development tools, the designer can add purchased software elements, such as shopping carts and content management software, to the site. The final step is to create the middleware that connects the site to the company's existing product and transaction-processing databases.

Buying and using midrange electronic commerce software is significantly more expensive than using one of the CSPs described in the previous section, with annual costs ranging from $2000 to $50,000. Midrange software traditionally offers connectivity to database systems that store catalog information. Having the catalog stored in a database simplifies updates and changes. Several of the midrange systems provide connections into existing inventory and ERP systems. This can yield savings because there is no need to run duplicate inventory systems, and the cost of the existing systems is spread across several software systems.

Three midrange electronic commerce systems are described in this section. They are representative of the whole group, yet are different from one another in important ways. The systems are Intershop Enfinity MultiSite, WebSphere Commerce Suite by IBM, and Commerce Server 2002 by Microsoft.

Intershop Enfinity

Intershop Enfinity MultiSite provides search and catalog capabilities, electronic shopping carts, online credit card transaction processing, and the ability to connect to existing back-end business systems and databases. Intershop Enfinity MultiSite has setup wizards and good catalog and data management tools. It provides many built-in storefront templates. Management and editing of a storefront are done through a Web browser—either locally at the server or remotely through any Internet connection. The products inventory management module tracks inventory levels and allows merchants to view the quantity of items available, create a list of inventory transactions, and enter new products into the inventory. Discount rules are also easy to enter. Merchants define the business rules for a discount and dates during which special discounts apply. Bundled with the software is a database management system. Alternatively, Enfinity can work with DB2 (IBM's relational database) or Oracle databases. The software includes an automated e-mail facility that can send order confirmations to customers. Enfinity includes support for secure transactions. A wide variety of site and customer reports are available to track Web page visits and customer activities.

IBM WebSphere Commerce Professional Edition

IBM produces the **WebSphere Commerce Professional Edition**, which is a family of electronic commerce packages. IBM WebSphere is a set of software components that provides software suitable for midsize to large businesses to sell goods and services on the Internet. It includes catalog templates, setup wizards, and advanced catalog tools to help companies create attractive and efficient electronic commerce sites. WebSphere Commerce Professional Edition can be used both for business-to-business and business-to-consumer applications and provides a smooth connection to existing corporate systems, such as inventory databases and procurement systems.

WebSphere Commerce products run on many different operating systems. Merchants can begin with a small store and then move up to a bigger, more capable store as necessary. A wizard leads the merchant through the process of creating a starter store. Once that is up and working, more functionality can be added by executing commands and writing code. With the basic pages built, the merchant can populate the catalog with products, prices, and product pictures. The WebSphere Commerce Professional Edition also accommodates electronic download products, such as audio tracks or software.

WebSphere offers a large collection of functions, utility programs, and commands that allow a merchant to create a customized online store experience. However, JavaScript, Java, or C++ expertise is required. Typical of commerce programs in this class, WebSphere can connect to existing databases and other legacy systems through DB2 or Oracle databases. A single store or several different stores can be administered from the same browser-based interface. A large number of midrange electronic commerce sites use WebSphere software. Enough IT professionals are involved in installing, maintaining, and customizing WebSphere that a magazine, *WebSphere Advisor*, is devoted to it.

The system has all the standard electronic commerce features, including tools for a shopping cart, e-mail notifications upon sale completion, secure transaction support, promotions and discounting, shipment tracking, links to legacy accounting systems, and browser-based local and remote administration. WebSphere Commerce Professional Edition costs $80,000 per processor. The less powerful Professional Entry Edition of the software costs $20,000 per processor.

Microsoft Commerce Server 2002

Microsoft **Commerce Server 2002** allows businesses to sell products or services on the Web using tools such as user profiling and management, transaction processing, product and service management, and target audience marketing. Commerce Server 2002 is not an out-of-the-box solution. Wizards help users build a site in several steps, but program code must be written to make the software meet specific user needs. The Microsoft Visual Studio .NET tools, bundled with Commerce Server 2002, allow companies to customize the sites they build.

Like other midrange electronic commerce software, Commerce Server 2002 has tools that help companies engage the customer (through marketing and advertising), complete an order, and analyze the sales information after the sale. Commerce Server 2002 also includes tools for advertising, promotions, cross-selling, and customer targeting and personalization.

Commerce Server 2002 provides many predefined reports for analyzing site activities and product sales data. Commerce Server 2002 can grow with increasing business demands. The system provides several storefront templates, wizards for setting up and initializing a store, and database connections. In addition, Commerce Server 2002 provides a shopping cart, confirms completed sales transactions by e-mail, and supports secure transactions. It can connect to existing accounting systems, and the administrator can oversee the site through a Web browser.

ELECTRONIC COMMERCE SOFTWARE FOR LARGE BUSINESSES

Larger businesses require many of the same advanced capabilities as midsize firms, but the larger firms need to handle higher transaction loads. In addition, they need dedicated software applications to handle specific elements of their online business. In this section, you will learn about electronic commerce software that has higher transaction-load capability, and you will learn about software that accomplishes specific tasks in large businesses, such as customer relationship management, supply chain management, content management, and knowledge management.

The distinction between midrange and large-scale electronic commerce software is much clearer than the one between basic systems and midrange systems. The telltale sign is price. Other elements, such as extensive support for business-to-business commerce, also indicate that the software is in this category. Commerce software in this class is sometimes called **enterprise-class software**. The term "enterprise" is used in information systems to describe a system that serves multiple locations or divisions of one company and encompasses all areas of the business or enterprise. Enterprise-class electronic commerce software provides tools for both B2B and B2C commerce. In addition, this software interacts with a wide variety of existing systems, including database, accounting, and ERP systems. As electronic commerce has become more sophisticated, large companies have demanded that their Web sites and supporting information infrastructure do more things.

The cost of these enterprise systems for large companies ranges from $200,000 for basic systems to $10 million and more for comprehensive solutions.

Enterprise-Class Electronic Commerce Software

Enterprise-class electronic commerce software running large online organizations usually requires several dedicated computers—in addition to the Web server system and any necessary firewalls. Examples of enterprise-class products that can be used to run a large online business with high transaction rates include **IBM WebSphere Commerce Business Edition, Oracle E-Business Suite**, and **Broadvision One-To-One Commerce**.

Enterprise-class software typically provides good tools for linking to and supporting supply and purchasing activities. A large part of B2B commerce is ordering supplies from trading or business partners and issuing the appropriate documents, such as purchase orders. For a selling business, e-business software provides standard electronic commerce activities, such as secure transaction processing and fulfillment, but it can also do more. For instance, it can interact with the firm's inventory system and make the proper adjustments to stock, issue purchase orders for needed supplies when they reach a critically low point, and generate other accounting entries in ERP, legacy accounting, or file systems. In contrast, both basic and midrange electronic commerce packages usually require an administrator to check inventory manually and place orders explicitly for items that need to be replenished.

In B2C situations, customers use their Web browsers to locate and browse a company's catalog. For electronic goods (software, research papers, music tracks, and so on), customers can download the items directly from the site, or they can complete order forms and have the hard-copy versions of the products shipped to them. The Web server is linked to back-end systems, including a database management system, a merchant server, and an application server. The database usually contains millions of rows of information about products, prices, inventory, user profiles, and user purchasing history. The history provides a way to recommend to a user on a return visit related items that he or she might wish to purchase. A merchant server houses the e-business system and key back-end software. It processes payments, computes shipping and taxes, and sends a message to the fulfillment department when it must ship goods to a purchaser. Figure 9-14 shows a typical enterprise-class electronic commerce architecture.

As you learned in Chapter 4, companies are storing data about site visitors in large databases and analyzing it to improve their relationships with those customers. These clickstreams track the path a visitor takes through a Web site, including which pages were viewed, the amount of time spent on each page, and the sequence in which pages were viewed. Thus, large electronic commerce sites must include customer relationship management software. In Chapter 5, you learned how companies are using the Web to integrate their supply chains. As a result, enterprise-class commerce Web sites must include or work with supply chain management software.

In Chapter 6, you learned about companies that were building business portal sites to engage their customers and suppliers. A significant part of that strategy is providing useful, fresh content to attract site visitors to the portal. This need has given rise to software that automatically manages and rotates content on Web sites. Some companies have even developed software that helps them manage the knowledge that exists in their businesses. An enterprise-class Web site often includes several of these types of software packages in

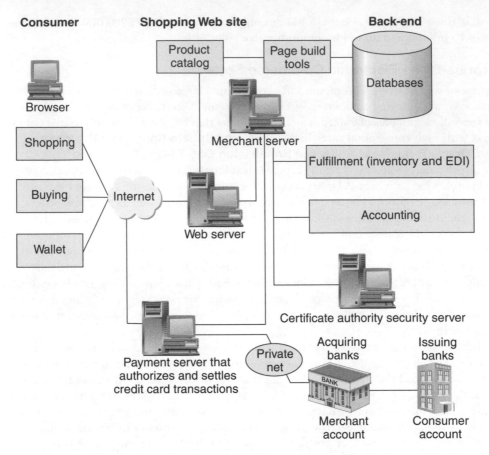

FIGURE 9-14 Typical enterprise-class electronic commerce architecture

its design. The next four sections discuss software that works with electronic commerce software in large companies to help those companies achieve all of their electronic commerce objectives.

Customer Relationship Management Software

You learned about the philosophy and techniques of customer relationship management (CRM) in Chapter 4. The goal of CRM is to understand each customer's specific needs and then customize a product or service to meet those needs. The idea is that a customer whose needs are being met exactly is willing to pay more for the goods or services that are meeting those needs. Although companies of all sizes can practice CRM techniques, large companies can afford to buy and implement expensive software products that automate many of CRM's principles.

Customer relationship management (CRM) software must obtain data from operations software that conducts activities such as sales automation, customer service center operations, and marketing campaigns. The software must also gather data about customer activities on the company's Web site and any other points of contact the company has with its existing and potential customers. CRM software uses this data to help managers conduct analytical activities, such as gathering business intelligence, planning marketing strategies, customer behavior modeling, and customizing the products and services to meet the needs of specific customers or categories of customers. In its most basic form, CRM uses information about customers to sell them more (or more profitable) goods or services. More advanced CRM is about delivering extremely attractive and positive experiences regularly to customers. CRM can be very important in maintaining customer loyalty in businesses where the purchase process is long and complex. Companies that design and install custom machinery, software products, or office workflow systems often find themselves involved in these types of long and complex processes. CRM software can help maintain positive and consistent contacts with multiple employees at the purchasing company.

Some companies create their own CRM software using outside consultants and their own IT staffs. In recent years, software vendors have increased the quality and variety of their offerings and today, most large companies are likely to buy a CRM software package. **Siebel Systems** offers a series of leading comprehensive CRM software packages. Siebel was the first company to specialize in CRM software and it has a large share of the market. Other major software firms have created products that compete with Siebel's, including **Oracle CRM**, **PeopleSoft CRM Analytics**, and **MySAP CRM**. Prices for these systems start around $50,000 (on average, about $2000 per user); large implementations can cost millions of dollars.

In the early days of CRM software implementation (approximately 1996 through 2000), companies spent many millions of dollars to buy CRM systems that promised to monitor and improve relationships with existing customers. Most of these systems were focused on giving companies the information they needed to identify changing customer preferences and respond very quickly to those changes. By responding quickly, companies hoped that they would be able to gain sales that might otherwise be lost to competitors that could respond better to the new customer preferences. In addition to gaining sales, the use of CRM software would help retain customers and reduce the need to spend money on marketing to find new customers. The goal was to instantly make available perfect information about all customer behaviors from all customer-interaction points throughout the company.

Most companies did not realize these benefits and CRM software sales dropped from 2000 through 2003. Many industry analysts were announcing that CRM was just another business fad that was dying as quickly as it had become fashionable. Starting in 2003, however, CRM software sales began growing again. Companies had learned from the bad experiences in which they invested large amounts of money to revamp their customer interaction strategies completely. Those companies became less likely to view CRM software as a tool for changing their overall customer strategy, and instead began using CRM software to solve smaller and more specific problems. For example, a cable company might use CRM to track service outages and repair team responses in real time, but would not expect the CRM system to calculate the profitability of on-demand video services on a continual basis.

One of the most popular targets for these new focused CRM applications has been call center operations. By examining problems that arise in their call centers, many companies have identified specific applications where CRM software can improve response times, accuracy, and effectiveness.

Supply Chain Management Software

Supply chain management (SCM) software helps companies to coordinate planning and operations with their partners in the industry supply chains of which they are members. SCM software performs two general types of functions: planning and execution. Most companies that sell SCM software offer products that include both components, but the functions are quite different. SCM planning software helps companies develop coordinated demand forecasts using information from each participant in the supply chain. SCM execution software helps with tasks such as warehouse and transportation management. The two major firms offering SCM software are **i2 Technologies** and **Manugistics**.

The i2 Technologies product, RHYTHM, includes components that manage demand planning, supply planning, and demand fulfillment. The demand planning module includes proprietary algorithms customized for specific industry markets that examine customers' buying patterns and generate continually updated forecasts. The supply planning module coordinates distribution logistics, inventory-level forecasting, collaborative procurement, and supply allocations. The demand fulfillment module handles the execution elements, including order management, customer verification, backlog control, and order fulfillment.

The Manugistics SCM product includes a constraint-based master planning module that controls the other elements of the system. These other elements include modules for transportation management, replenishment management, manufacturing planning, scheduling, purchase planning, and materials control.

The cost of SCM software implementations varies tremendously depending on how many locations (retail stores, wholesale warehouses, distribution centers, and manufacturing plants) are in the supply chain. For example, a retailer with 500 stores might pay between $4 million and $10 million for an SCM package that includes both planning and execution functions, but a wholesaler with only three or four distribution centers might be able to install a good SCM product for $1 million.

Content Management Software

Most electronic commerce software comes with wizards and other automated helpers that create template-driven pages, such as home pages, about pages, and contact pages. But most businesses want to customize Web pages with company and product pictures and text. Content management software should be tested before committing to it. The testing should ensure that company employees find the software's procedures for performing regular maintenance (for example, adding new categories of products and new items to existing product pages) to be straightforward. The software should also facilitate typical content creation tasks, such as adding sale-item specials.

Large companies are finding new ways to use the Web to share information among their employees, customers, suppliers, and partners. **Content management software** helps companies control the large amounts of text, graphics, and media files that have become a key part of doing business. With the rise of wireless devices, such as mobile phones, hand-held computers, and personal digital assistants (PDAs), content management has become even more important.

Companies that need many different ways to access corporate information—for example, product specifications, drawings, photographs, or lab test results—often choose to manage the information and access to that information using content management software. The three leading companies that provide these tools are **Documentum**, **Vignette**, and **webMethods**. Content management software costs between $200,000 and $500,000, but it can cost three or four times that much to customize, configure, and implement. Figure 9-15 shows the Documentum site.

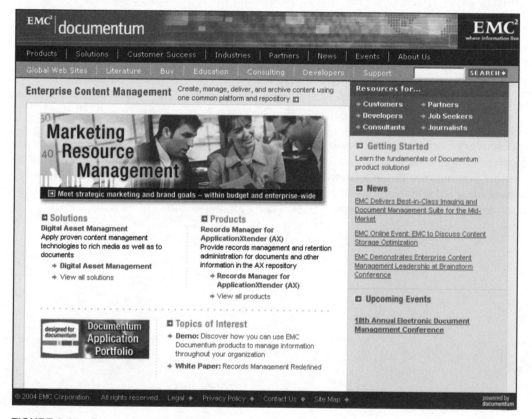

FIGURE 9-15 Documentum content management Web site

Knowledge Management Software

An increasing number of large companies have achieved cost savings by using content management software. Most content management software is designed to help companies manage information that, until recently, was stored in paper reports, schedules, analyses, and memos. Although the cost reductions that can be obtained by moving mountains of paper into an electronic format are significant, some companies have begun to understand that the true value of those documents is in the information contained in them. Thus, they began the search for systems that would help them manage the knowledge itself, rather than the documentary representations of that knowledge. The software that has been developed to meet that goal is called **knowledge management (KM) software**.

KM software helps companies do four main things: collect and organize information, share the information among users, enhance the ability of users to collaborate, and preserve the knowledge gained through the use of information so that future users can benefit from the learning of current users. KM software includes tools that read electronic documents (in formats such as Microsoft Word or Adobe PDF), scanned paper documents, e-mail messages, and Web pages. KM software often includes powerful search tools that use proprietary semantic and statistical algorithms to help users find the content, human experts, and other resources that can aid them in their research and decision-making tasks.

Most early KM software required companies to build a centralized knowledge repository before the software could provide users any real benefits. The building of these repositories required major investments of time and money, and often disrupted the regular flow of work. More recently developed KM systems are less obtrusive and allow the collection of knowledge elements to flow as a natural by-product of the normal interactions users have with information.

The major software vendors have KM software offerings, including **IBM Lotus Discovery Server** and **Microsoft SharePoint Technologies**. Smaller companies have also entered the market with innovative KM software and technologies. Two of the more interesting products are Entopia **Quantum** and Mirror Worlds Technologies **Scopeware**. Total costs for a KM software implementation, including hardware, software licenses, and consultant fees, typically range from $50,000 to $1 million or more.

Summary

In this chapter, you learned about electronic commerce software for small, midsize, and large businesses and the functions provided by each software type. The electronic commerce software a company chooses depends on its size, objectives, and budget, and requires making some major decisions. A company must first choose between paying a service provider to host the site and self-hosting. External hosting options include shared hosting, dedicated hosting, and co-location. Many hosting companies offer comprehensive services to merchants such as databases, shopping carts, and content management in addition to basic Web hosting services.

Key elements of all electronic commerce software include catalogs, shopping carts, and transaction-processing capabilities. An important new way for companies to get their information systems to work across organizational boundaries is the implementation of Web services.

Small enterprises that are just starting an electronic commerce initiative might use a commerce service provider (CSP). Basic CSP and mall-style hosting services for small businesses provide a range of standard features, including tools for quickly creating storefronts, catalogs, and transaction processing. These packages are usually wizard- and template-driven.

If a company already has computing equipment and staff in place, purchasing a midrange electronic commerce software package provides more control over the site and allows for expansion. Midrange software can interact with database software to create dynamic catalogs and shopping carts and handle order processing.

Large enterprises that have high transaction rates, B2B partnerships, or a large investment in ERP and other existing information systems need to invest in larger, more customizable systems that can provide needed features and flexibility. These packages can include customer relationship management, supply chain management, content management, and knowledge management capabilities, or they can work with dedicated software that performs these functions.

Key Terms

Application integration

Application program (application)

Application server

Application service providers (ASPs)

Application software (application)

Business logic

Catalog

Co-location (collocation, colocation)

Commerce service providers (CSPs)

Component-based application system

Content management software

Customer relationship management (CRM) software

Data mining

Database manager

Dedicated hosting

Distributed database systems

Distributed information systems

Dynamic catalog

Enterprise application integration

Enterprise-class software

Enterprise resource planning (ERP)

Interoperability

Knowledge management (KM) software

Managed service providers (MSPs)

Middleware

Open source

Page-based application system

Scalable

Self-hosting

Shared hosting

Simple Object Access Protocol (SOAP)

Static catalog

Supply chain management (SCM) software

Transaction processing

Universal Description, Discovery and Integration (UDDI) specification

Web services

Web Services Description Language (WSDL)

Review Questions

RQ1. Provide a brief definition of the term "middleware." In one or two paragraphs, explain why middleware can be difficult to write and test.

RQ2. Using your library or the Web, find an article that describes a successful application of Web services. In about 200 words, discuss how the company that implemented the Web services application overcame the lack of standards for such applications.

RQ3. List two disadvantages of hosting an electronic commerce site on a host that is free or available at a very low cost. What is missing from such a host's services that would make an online entrepreneur's job more difficult?

RQ4. In about 200 words, describe the differences between basic electronic commerce software and midrange electronic commerce software. Discuss at least four differences and give examples of each type of software.

RQ5. What are the characteristics of large firms conducting both B2B and B2C transactions that require more robust and capable electronic commerce systems? Consider the volume and types of transactions and store maintenance activities that differ between a small storefront operation and, for example, an Amazon.com-caliber store.

RQ6. Visit the product Web sites to learn more about two of the knowledge management software products discussed in the chapter. In a report of about 300 words addressed to the president of a local university, explain how that university could benefit from an implementation of knowledge management software.

Exercises

E1. Your friend Faye Borthick wants to set up a small Web site devoted to gardening. She believes her many years of experience in gardening give her an understanding of the kinds of gardening tools, fertilizers, soil amendment products, herbicides, pesticides, and plants that appeal to the serious gardener. Right now Faye doesn't want to sell anything, although she might change her mind in the future. She merely wants to display pages of plant photography, write and store short how-to papers for novice gardeners, and provide links to other gardening tips on the Web. She wants your advice on whether to self-host the Web site or use an ISP (or CSP) to start her endeavor. Use **The List** or the **TopHosts** sites to locate information on the cost of using a service provider to host a Web site. Then, estimate what a small Web site might cost in terms of the minimal configuration of hardware and software. Estimate the design and development costs and the annual maintenance costs. Then, select one of the Web server programs. Estimate the cost of a Web connection. Write a 200-word summary of everything you think Faye needs to know to use either of the two options (she builds it or she uses a service provider) for creating her site.

E2. Annette Jackson owns a small crafts store in central Missouri. She wants to expand her store's reach outside the region to increase her profits and simultaneously reduce her inventory. Annette has been watching her teenage daughter, Kelly, use the Internet to order music CDs and books. After learning from Kelly how simple it is to order from online stores, Annette decided that she needs to create an online store. She asked you to do a little research on how much it might cost in the first year to create a simple store with a catalog of about 100 items. Annette wants you to investigate two CSP offerings and report back to her what you find. Because her store is small, limit your research to basic commerce and mall-style services. You might want to begin your research with sites such as **Freemerchant.com** or **Bigstep**. Annette would like to consider the following information for the two CSP offerings you examine:

- Costs: initial setup fee, monthly fee, and transaction fees
- Amount of disk space the CSP would provide for Annette's 100-item store
- Existence of a search engine within each store
- Promotion and marketing opportunities
- Customer communications capabilities, such as automated e-mail confirmation of orders
- Shopping cart or other order entry mechanism
- Storefront-building wizards for creating a new store
- Security provisions for transactions
- Nature of the domain names available (subdomain of the site or not)
- Upload capabilities for product names, descriptions, images, and costs (can they be uploaded from files or databases, or must the merchant enter each item individually?)
- Existence of an online user manual for the merchant

Produce a report of about 500 words summarizing your findings.

E3. Write a 400-word report summarizing the costs and features of any enterprise-class commerce package for large businesses. You can review a product mentioned in the chapter or one of your own choosing. Pick seven characteristics of the software package and describe them in detail in your report. The Online Companion includes links to several vendors of these products under the Exercise 3 heading.

E4. Review the material in Chapter 4 on customer relationship management (CRM). Then visit the Web sites of two or more of the providers of CRM software discussed in this chapter. In about 300 words, critically evaluate one of the CRM software packages by comparing what it accomplishes to the goals of CRM.

Cases

C1. Ingersoll-Rand Club Car Division

Ingersoll-Rand is a $9 billion diversified manufacturing company that sells its products worldwide. Its well-known brands include Ingersoll-Rand tools and portable power generators, Bobcat construction equipment, Thermo King refrigerated transport systems, Dexter and Schlage locks, and ARO industrial fluids equipment. The company's Club Car division manufactures and sells a variety of small electric cart vehicles to golf courses and industrial users. The division also sells a rough-terrain version designed for farmers, ranchers, construction workers, and recreational users.

In 2001, the Club Car division was experiencing a sales decline. The downturn in the general economy was affecting golf courses, which, in turn, were reducing the size and frequency of their golf cart orders. Club Car had a general sense that this major market segment was causing their revenues to decline, but their information systems were not providing enough data about exactly which sales were being most affected by the economic downturn.

Club Car sales managers relied on their sales representatives for information about likely future sales. Sales forecasting was a matter of judgment, guesswork, and a few spreadsheet software models scattered throughout the regional sales offices. The sales representatives had little influence on how the carts were customized for particular customer segments or for individual customers.

The company decided it needed better information about all of its sales and marketing activities, so it spent more than $2 million to install a comprehensive CRM system. This system was designed to automate the entire customer sales cycle: prospect evaluation, proposal writing, product configuration, and order entry. However, the users at Club Car division found the new system difficult to use and therefore were reluctant to spend much time learning how to use it. Thus, the promised benefits of improved productivity and more detailed reports were not forthcoming. Sales managers did not see the ultimate benefits that the system might provide. Salespeople found that the new system was requiring them to spend time entering data into the system rather than seeing customers. The order entry staff found the system to be cumbersome and unfamiliar.

When Club Car's president realized that the CRM system was not delivering on its promise, he had the management team go back and re-examine the key elements in the division's customer relationships and asked them to choose one or two issues that needed attention. The management team identified two major issues. First, the order entry process required the time of salespeople and order entry staff, but it did not include any interaction with customers. Second, the division was not producing accurate and timely sales forecasts.

In 2002, Club Car division re-launched its CRM efforts and focused on these two problem areas. The new effort included the sales representatives in redesigning the order entry process. The division was able to reduce the data entry time and effort required, especially the time of salespeople. Salespeople do have remote access to the system, so they can work on-site with customers to configure the carts to the customers' exact specifications. Salespeople can obtain pricing information and explore various alternatives with customers while they are at the customer's site. They can also examine manufacturing schedules and provide more accurate delivery date estimates. All of this remote, real-time information access helps salespeople close deals and increase sales volume and profitability.

Sales forecasts are more accurate now because the information about sales orders is automatically collected when the sales representatives close sales at the customers' sites. The CRM system combines this real-time sales order information with general industry information on cart demand, cart replacement cycles, and economic trends in their customers' industries. The increased accuracy of sales forecasts allows the company to create more stable production schedules, which means that more customers receive their carts on the delivery date they were promised.

Required:

1. List the types of information that Club Car division's new CRM system makes available to sales representatives in the field. For each type of information, briefly explain how salespeople's remote access to that type of information can help them close sales on their customers' sites.

2. In the CRM re-launch, Club Car division focused on two CRM elements. In about 200 words, explain why this approach would work better, in general, than implementing a comprehensive CRM system that could track all of the division's sales activities and related information in real time.

3. In about 200 words, explain how Club Car division might use Web services in its CRM system.

Note: Your instructor might assign you to a group to complete this case, and might ask you to prepare a formal presentation of your results to your class.

C2. Web Services for State Government

You are a member of the Web site management team of a state government. You have worked on all of the state's Web sites from time to time and have managed the launch of four major sites and the redesign and relaunch of two others. Some of the Web sites on which you have worked include electronic commerce features such as order acceptance, payment processing, and purchasing.

You report to Anne Nelson, the state's CIO. Anne asked you to lead a project to explore the potential uses of Web services in carrying out state government activities. She scheduled a formal briefing at which you will present an overview of Web services technology. You will also outline specific applications of Web services technologies to specific tasks that the state either currently performs or that it might perform in the future.

Anne knows that the state has many current and potential applications that could use Web services technologies, so she asked you to focus on four specific areas of state government in your briefing. At the briefing, you will address the directors of four state departments: the Attorney General's Department of Corporation Records, the Tax Administration and Collection Department, the Department of Motor Vehicles, and the Department of Fish and Wildlife Management.

The Attorney General's Department of Corporation Records maintains the official records of corporations chartered by the state or holding licenses to do business in the state. In addition to the original charter or license, companies must file annual reports that include the names and addresses of corporate directors and officers, the amount of company stock issued or redeemed during the year, and the current address of the company.

The Tax Administration and Collection Department is responsible for accepting income tax, personal property tax, and sales tax return filings of companies and individuals. The department also processes payments of these taxes and authorizes the State Treasurer to issue refunds that are due to taxpayers who have overpaid their taxes. This department currently provides tax forms and instructions in Adobe PDF format on its Web site. It also maintains an extensive frequently asked questions (FAQ) list on the site.

The Department of Motor Vehicles issues driver's license renewals and vehicle registration renewals (for cars, trucks, and boats) and accepts auto dealerships' monthly reports of vehicles purchased or sold on its Web site. The site also includes extensive collections of information about motor vehicle laws and administrative rulings that visitors can review to ensure they are in compliance.

The Department of Fish and Wildlife Management provides downloadable applications for hunting and fishing licenses on its site. Current hunting and fishing license holders can renew their licenses and pay their annual fees on the Web site. Companies that have state-issued permits to undertake logging or mining operations can file their monthly activity reports on the department's Web site, too.

Anne suggests that you review current IT trade publications (both in print and on the Web) to learn more about Web services applications that have been implemented in government agencies. She also recommends that you examine a number of other state Web sites to see how they are performing these tasks.

Required:

1. Prepare a briefing report of about four double-spaced pages in which you describe Web services technology in a way that will be understandable to the four department directors. These directors are experienced administrators, but they are not technology experts.

2. Prepare a briefing report that outlines opportunities for the use of Web services in each department. Include about three double-spaced pages for each department.

3. Prepare an analysis of costs and benefits for each major application of Web services that you identify. In this setting, a benefit can arise from an increase in revenue, a reduction in expense, an improvement in the quality of service provided, or an increase in the speed with which a service is provided. This report should be directed to Anne and should include an implementation recommendation (whether the state should implement or should not implement) for each Web service application you identified.

Note: Your instructor might assign you to a group to complete this case, and might ask you to prepare a formal presentation of your results to your class.

For Further Study and Research

Abate, C. 2002. "Going Once, Going Twice... Sold!" *Smart Business*, 15(4), May, 72–76.

Atanasov, M. 2001, "The ASP Trap," *Ziff Davis Smart Business*, 14(7), July, 58–63.

Bailor, C. 2004. "Ten Technologies That Are Reinventing the CRM Industry," *CRM Magazine*, 8(12), December, 44–48.

Caulfield, B., S. Finch, M. Maier, D. Orenstein, R. Tate, and O. Thomas. 2002. "The Who's Who of E-Business," *Business 2.0*, January. (http://www.business2.com/articles/mag/0,1640,35691|2,FF.html)

Clyman, J. 2002. "Server's Advantage," *PC Magazine*, 21(1), January, 106–108.

Desai, G., E. Sanchez, and J. Fenner. 2001. "Web Application Servers Come of Age," *Network Computing*, 12(15), July 23, 63–71.

DiSabatino, J. 2002. "Knowledge Management Users Pick Smaller Vendors," *Computerworld*, 36(28), July 8, 17.

Dyck, T. 2001. "Web Server Brains," *PC Magazine*, 20(10), May 22, 124–132.

Dyck, T. 2002. "Web Services Impact," *eWeek*, September 16, 39–51.

Epstein, J. 2003. "Getting on Track with Web Services," *Darwin Magazine*, May 1. (http://www.darwinmag.com/read/050103/webserve.html)

Ferguson, G. 2002. "Have Your Objects Call My Objects," *Harvard Business Review*, 80(6), June, 138–143.

Fingar, P. 2002. "Web Services Among Peers," *Internet World*, January, 21.

Gambhir, S. and M. Muchmore, 2001. "Secrets," *PC Magazine*, 20(19), 130–131.

Gladwin, L. 2001. "Covisint Focuses on Tech Integration," *Computerworld*, 35(27), July 2, 10.

Guernsey, L. 2003. "On the Web, Without Wasting Time," *The New York Times*, May 6, G10.

Hall, M. 2003. "Web Services' Sharp Edge," *Computerworld*, 37(20), May 19, 34.

Hane, P. 2002. "Entopia Ships New Release of Quantum," *Information Today*, 19(8), September, 7.

Homan, D. 2002. "Look Past Products to Web Services' True Promise," *InformationWeek*, August 26, 44–45.

Information Security. 2001. "Crackers Steal Credit Card Data from Shopping Carts," May, 34.

Ismail, A., S. Patil, and S. Saigal. 2002. "When Computers Learn to Talk: A Web Services Primer," *McKinsey Quarterly*, Special Edition (Issue 2), June, 70–78.

Keizer, G. 2002. "Software Reviews: Yahoo Store," *Fortune*, 144(10), 190.

Kennedy, D. 2004. "What Lawyers Need to Know About the Open Source Licenses," *Journal of Internet Law*, 7(8), February, 3–10.

Kumar, K. 2001. "Technology for Supporting Supply Chain Management," *Communications of the ACM*, 44(6), June, 6–9.

Lohr, S. 2003. "Competitors Shape Strategy to Gain Edge in Web Services," *The New York Times*, February 3, C1.

Maamar, Z., E. Dorion, and C. Daigle. 2001. "Toward Virtual Marketplaces for E-Commerce Support," *Communications of the ACM*, 44(12), December, 35–38.

MacSweeney, G. 2002. "Web Services: Here To Stay?" *Insurance & Technology*, 27(10), September, 53–55.

Macvittie, L. 2004. "Choosing the Right Web Server," *Network Computing*, 15(17), September 2, 76–77.

Miller, M. 2002. "Web Services: Can't We All Just Get Along?" *PC Magazine*, 21(6), March 26, 7.

Morse, G. 2003. "Plumbing Web Connections," *Harvard Business Review*, 81(9), September, 18–19.

Nielsen, J. 2001. "The End of Homemade Websites," *Alertbox*, October. (http://www.useit.com/alertbox/ 20011014.html)

Pallato, J. 2002. "Power Play: Oracle9i Turns Up the Heat on BEA WebLogic, IBM WebSphere," *Internet World*, January, 58.

Pallato, J. 2002. "Web Services Deliver," *Internet World*, October, 32–37.

Rigby, D. and D. Ledingham. 2004. "CRM Done Right," *Harvard Business Review*, 82(11), November, 118–127.

Roberts-Witt, S. 2001. "The Internet Business," *PC Magazine*, 20(5), March 6, 8–18.

Rosenberg, A. 2004. "Which CRM Is Right for You?" *Call Center Magazine*, 17(12), December, 28–35.

Rubenking, N. 2001. "Hidden Messages," *PC Magazine*, 20(10), May 22, 86–88.

Schuff, D. and R. Saint Louis. 2001. "Centralization vs. Decentralization of Application Software," *Communications of the ACM,* 44(6), June, 88–94.

Schultz, K. 2001. "SCM Turned Inside Out: Vendors Incorporate Collaboration into the Supply Chain," *InternetWeek*, June 25, 25–30.

Siebel Systems. 2004. *Ingersoll-Rand Maximizes Customer Focus*, San Mateo, CA: Siebel Systems. (http://www.siebel.com/downloads/case_studies/)

Smetannikov, M. 2001. "MSPs Face Hard Questions About Software," *Interactive Week*, 8(34), September 3, 35.

Tristram, C. 2001. "The Next Computer Interface," *Technology Review*, 104(10), December, 52–59.

Ulfelder, S. 2001. "The Web's Last Gap," *Computerworld*, 35(25), June 18, 48–49.

Vizard, M. 2002. "Web Services Are Delivering Savings," *InfoWorld*, 24(33), August 19, 8.

Walker, L. 2001. "Web Services: High Stakes Amid the Hype," *The Washington Post*, October 18, E1.

Whiting, R. 2002. "Oracle Clarifies License Options," *InformationWeek*, September 2, 32.

CHAPTER 10

ELECTRONIC COMMERCE SECURITY

LEARNING OBJECTIVES

In this chapter, you will learn about:

- Online security issues
- Security for client computers
- Security for the communication channels between computers
- Security for server computers
- Organizations that promote computer, network, and Internet security

INTRODUCTION

In 2002, the U.S. Congress held hearings to review the federal government's computer security status. The results were not encouraging. The **General Accounting Office (GAO)** summarized its previous two years' work in reviewing security at 24 government agencies. According to the GAO, 16 of those agencies had failed completely in their computer security efforts, and all 24 had at least one major security weakness.

Most of the security problems identified by the GAO did not involve sophisticated technological issues, nor did they require large amounts of money to resolve. The most prevalent security weaknesses stemmed from inadequate employee training and awareness and failure to keep software updated with the latest security patches available. The most common problem was failure to enforce basic standards for access control, such as rotating passwords periodically and having employees maintain the confidentiality of their passwords.

In many of the agencies, readily available security patches for well-known vulnerabilities had not been applied to system software. The GAO noted that more than 90 percent of all successful attacks on U.S. government agency systems had exploited known vulnerabilities for which a patch was available but had not been installed. The GAO concluded that by simply adhering to their own existing policies, these agencies could improve their level of computer security significantly. In many cases, the agencies had not made any person responsible for monitoring vulnerabilities and for ensuring that available solutions were applied. The GAO report emphasized that this state of affairs was unacceptable, especially in the wake of the terrorist attacks of September 11, 2001.

When businesses began using computers 50 years ago, security was accomplished by using physical controls over access to the computers. Alarmed doors and windows, guards, security badges to admit people to sensitive areas, and surveillance cameras were the tools used to secure computers. Back then, interactions between people and computers were limited to terminals (which had no internal processing capabilities) connected directly to large mainframe computers. There were no other connections to computers, and there were very few networks of computers (and those few networks did not extend outside the organization in which they existed). Computer security meant dealing with the few people who had access to terminals or physical access to the computer room. In many computer installations of the day, people ran programs by submitting decks of punched cards that were fed into card readers. The card readers translated the punched holes in the cards into electrical impulses that were processed by the computer. The computer printed out the results when it was finished running the program. When program submitters returned to the computer operations center (often the next day; computers were not very fast then), they could pick up the printouts and reclaim their punched card decks from the input/output clerk. Security was a pretty simple matter.

Both the population of computer users and the methods to access computing resources have increased tremendously since those early years of computing. Millions of people now have access to computing power over both private and public networks that connect millions of computers. It is no longer a simple matter to determine who is using a computing resource. A user in South Africa could be using a computer in California. New security tools and methods have evolved and are employed today to protect computers and the electronic assets they store. The transmission of valuable information, such as electronic receipts, purchase orders, payment data, and order confirmations, has drastically increased the need for security and new automatic methods to deal with security threats.

Data security measures date back to the time of the Roman Empire, when Julius Caesar coded information to prevent enemies from reading secret war and defense plans carried by his Roman legions. Many modern electronic security techniques were developed for wartime use. The U.S. Department of Defense was the main driving force behind early security requirements and more recent advances. In the late 1970s, the Defense Department formed a committee to develop computer security guidelines for handling classified information on computers. The result of that committee's work was *Trusted Computer System Evaluation Criteria*, known in defense circles as the "Orange Book" because its cover was orange. It spelled out rules for mandatory access control—the separation of confidential, secret, and top secret information—and established criteria for certification levels for computers ranging from D (not trusted to handle multiple levels of classified documents at once) to A1 (the most trustworthy level).

This early security work has been helpful because it provided a basis for electronic commerce security research. This research today provides commercial security products and practical security techniques. This early work also helped current security efforts by developing formal approaches to security analysis and evaluation, including the explicit evaluation and management of risk.

ONLINE SECURITY ISSUES OVERVIEW

In the early days of the Internet, one of its most popular uses was electronic mail. Despite e-mail's popularity, people have often worried that a business rival might intercept e-mail messages for competitive gain. Another fear was that employees' nonbusiness correspondence might be read by their supervisors, with negative repercussions. These were significant and realistic concerns.

Today, the stakes are much higher. The consequences of a competitor having unauthorized access to messages and digital intelligence are now far more serious than in the past. Electronic commerce, in particular, makes security a concern for all users. A typical worry of Web shoppers is that their credit card numbers might be exposed to millions of people as the information travels across the Internet. Recent surveys show that more than 80 percent of all Internet users have at least "some concern" about the security of their credit card numbers in electronic commerce transactions. This echoes the fear shoppers have expressed for many years about credit card purchases over the phone.

Consumers are now more comfortable giving their credit card numbers and other information over the phone, but many of those same people fear providing that same information on a Web site. As you learned in Chapter 7, people are concerned about personal information they provide to companies over the Internet. Increasingly, people doubt that these companies have the willingness and the ability to keep customers' personal information confidential. This chapter examines security in the context of electronic commerce, presenting an introduction to important security problems and some solutions to those problems.

Computer security is the protection of assets from unauthorized access, use, alteration, or destruction. There are two general types of security: physical and logical. **Physical security** includes tangible protection devices, such as alarms, guards, fireproof doors, security fences, safes or vaults, and bombproof buildings. Protection of assets using nonphysical means is called **logical security**. Any act or object that poses a danger to computer assets is known as a **threat**.

Managing Risk

Countermeasure is the general name for a procedure, either physical or logical, that recognizes, reduces, or eliminates a threat. The extent and expense of countermeasures can vary, depending on the importance of the asset at risk. Threats that are deemed low risk and unlikely to occur can be ignored when the cost to protect against the threat exceeds the value of the protected asset. For example, it would make sense to protect from tornadoes a computer network in Oklahoma City, where there is significant and regular tornado activity, but not to protect a similar network in Los Angeles, where tornadoes are rare. The risk management model shown in Figure 10-1 illustrates four general actions that an organization could take, depending on the impact (cost) and the probability of the physical threat. In this model, a tornado in Oklahoma would be in quadrant II, whereas a tornado in Southern California would be in quadrant IV.

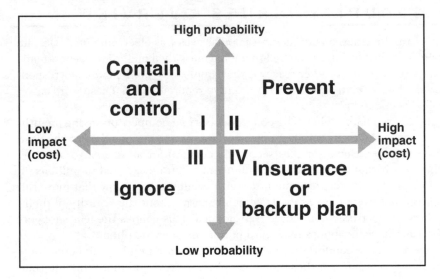

FIGURE 10-1 Risk management model

The same sort of risk management model applies to protecting Internet and electronic commerce assets from both physical and electronic threats. Examples of the latter include impostors, eavesdroppers, and thieves. An **eavesdropper**, in this context, is a person or device that can listen in on and copy Internet transmissions. People who write programs or manipulate technologies to obtain unauthorized access to computers and networks are called **crackers** or **hackers**.

A cracker is a technologically skilled person who uses their skills to obtain unauthorized entry into computers or network systems—usually with the intent of stealing information or damaging the information, the system's software, or even the system's hardware. Originally, the term hacker was used to describe a dedicated programmer who enjoyed writing complex code that tested the limits of technology. Although the term hacker is still used in a positive way—even as a compliment—by computer professionals (who make a strong distinction between the terms hacker and cracker), the media and the general public usually use the term to describe those who use their skills for ill purposes. Some IT people also use the terms **white hat hacker** and **black hat hacker** to make the distinction between good hackers and bad hackers.

To implement a good security scheme, organizations must identify risks, determine how to protect threatened assets, and calculate how much to spend to protect those assets. In this chapter, the primary focus in risk management protection is on the central issues of identifying the threats and determining the ways to protect assets from those threats, rather than on the protection costs or value of assets.

Computer Security Classifications

Computer security is generally classified into three categories: secrecy, integrity, and necessity (also known as denial of service). **Secrecy** refers to protecting against unauthorized data disclosure and ensuring the authenticity of the data source. **Integrity** refers to preventing unauthorized data modification. **Necessity** refers to preventing data delays or denials (removal). Secrecy is the best known of the computer security categories. Every month, newspapers report on break-ins to government computers or theft and use of stolen credit card numbers that are used to order goods and services. Integrity threats are reported less frequently and, thus, may be less familiar to the public. For example, an integrity violation occurs when an Internet e-mail message is intercepted and its contents are changed before it is forwarded to its original destination. In this type of integrity violation, which is called a **man-in-the-middle exploit**, the contents of the e-mail are often changed in a way that negates the message's original meaning. Necessity violations take several forms, and they occur relatively frequently. Delaying a message or completely destroying it can have grave consequences. Suppose that a message sent at 10:00 a.m. to an online stockbroker includes an order to purchase 1000 shares of IBM at market price. If the stockbroker does not receive the message (because an attacker delays it) until 2:30 p.m. and IBM's stock price has increased by $3, the buyer loses $3000.

Security Policy and Integrated Security

Any organization concerned about protecting its electronic commerce assets should have a **security policy** in place. A security policy is a written statement describing which assets to protect and why they are being protected, who is responsible for that protection, and which behaviors are acceptable and which are not. The policy primarily addresses physical security, network security, access authorizations, virus protection, and disaster recovery. The policy develops over time and is a living document that the company and security officer must review and update at regular intervals.

Both defense and commercial security guidelines state that organizations must protect assets from unauthorized disclosure, modification, or destruction. However, military security policy differs from commercial policy because military applications stress separation of multiple levels of security. Corporate information is usually classified as either "public" or "company confidential." The typical security policy concerning confidential company information is straightforward: Do not reveal company confidential information to anyone outside the company.

The first step an organization must take in creating a security policy is to determine which assets to protect from which threats. For example, a company that stores its customers' credit card numbers might decide that those numbers are an asset that must be protected from eavesdroppers. Then, the organization must determine who should have access to various parts of the system. Next, the organization determines what resources are available to protect the assets identified. Using the information it has acquired, the organization develops a written security policy. Finally, the organization commits resources to building or buying software, hardware, and physical barriers that implement the security policy. For example, if a security policy disallows any unauthorized access to customer information, including credit card numbers and credit history, then the organization must either create or purchase software that guarantees end-to-end secrecy for electronic commerce customers.

A comprehensive plan for security should protect a system's privacy, integrity, and availability (necessity), and authenticate users. When these goals are used to create a security policy for an electronic commerce operation, they should be selected to satisfy the list of requirements shown in Figure 10-2. These requirements provide a minimum level of acceptable security for most electronic commerce operations.

Requirement	Meaning
Secrecy	Prevent unauthorized persons from reading messages and business plans, obtaining credit card numbers, or deriving other confidential information.
Integrity	Enclose information in a digital envelope so that the computer can automatically detect messages that have been altered in transit.
Availability	Provide delivery assurance for each message segment so that messages or message segments cannot be lost undetectably.
Key management	Provide secure distribution and management of keys needed to provide secure communications.
Nonrepudiation	Provide undeniable, end-to-end proof of each message's origin and recipient.
Authentication	Securely identify clients and servers with digital signatures and certificates.

FIGURE 10-2 Requirements for secure electronic commerce

The Network Security Library, which is sponsored by GFI Software (a company that sells security and messaging software), is a good source for information about security policies. The Network Security Library includes a number of white papers that provide guidance on how to craft a workable security policy. **Information Security Policy World** is another Web site that provides information about security policy matters.

Although absolute security is difficult to achieve, organizations can create enough barriers to deter most intentional violators. With good planning, organizations can also reduce the impact of natural disasters or terrorist acts. Integrated security means having all security measures working together to prevent unauthorized disclosure, destruction, or modification of assets. A security policy covers many security concerns that must be addressed by a comprehensive and integrated security plan. Specific elements of a security policy address the following points:

- *Authentication*: Who is trying to access the electronic commerce site?
- *Access control*: Who is allowed to log on to and access the electronic commerce site?
- *Secrecy*: Who is permitted to view selected information?
- *Data integrity*: Who is allowed to change data?
- *Audit*: Who or what causes specific events to occur, and when?

In this chapter, you will explore these security policy issues with a focus on how they apply to electronic commerce in particular. The electronic commerce security topics in this chapter are organized to follow the transaction processing flow, beginning with the consumer and ending with the Web server (or servers) at the electronic commerce site. Each logical link in the process includes assets that must be protected to ensure security: client computers, the communication channel on which the messages travel, and the Web servers, including any other computers connected to the Web servers.

SECURITY FOR CLIENT COMPUTERS

Client computers, usually PCs, must be protected from threats that originate in software and data that are downloaded to the client computer from the Internet. In this section, you will learn that active content delivered over the Internet in dynamic Web pages can be harmful. Another threat to client computers can arise when a malevolent server site masquerades as a legitimate Web site. Users and their client computers can be duped into revealing information to those Web sites. This section explains these threats, describes how they work, and outlines some protection mechanisms that can prevent or reduce the threats they pose to client computers.

Cookies

The Internet provides a type of connection between Web clients and servers called a stateless connection. In a **stateless connection**, each transmission of information is independent; that is, no continuous connection (also called an **open session**) is maintained between any client and server on the Internet. Earlier in this book, you learned that cookies are small text files that Web servers place on Web client computers to identify returning visitors. Cookies also allow Web servers to maintain continuing open sessions with Web clients. An open session is necessary to do a number of things that are important in online business activity. For example, shopping cart and payment processing software both need an open session to work properly. Early in the history of the Web, cookies were devised as a way to maintain an open session despite the stateless nature of Internet connections. Thus, cookies were invented to solve the stateless connection problem by saving information about a Web user from one set of server-client message exchanges to another.

There are two ways of categorizing cookies: by time duration and by source. The two kinds of time duration cookie categories include **session cookies**, which exist until the Web client ends the connection (or "session"), and **persistent cookies**, which remain on the client computer indefinitely. Electronic commerce sites use both kinds of cookies. For example, a session cookie might contain information about a particular shopping visit and a persistent cookie might contain login information that can help the Web site recognize visitors when they return to the site on subsequent visits. Each time a browser moves to a different part of a merchant's Web site, the merchant's Web server asks the visitor's computer to send back any cookies that the Web server stored previously on the visitor's computer.

Another way of categorizing cookies is by their source. Cookies can be placed on the client computer by the Web server site, in which case they are called **first-party cookies**, or they can be placed by a different Web site, in which case they are called **third-party cookies**. A third-party cookie originates on a Web site other than the site being visited. These third-party Web sites usually provide advertising or other content that appears on the Web site being viewed. The third-party Web site providing the advertising is often interested in tracking responses to their ads by visitors who have already seen the ads on other sites. If the advertising Web site places its ads on a large number of Web sites, it can use persistent third-party cookies to track visitors from one site to another. Earlier in this book, you learned about DoubleClick and similar online ad placement services that perform this function.

The most complete way for Web site visitors to protect themselves from revealing private information or being tracked by cookies is to disable cookies entirely. The problem with this approach is that useful cookies are blocked along with the others, requiring visitors to enter information each time they revisit a Web site. The full resources of some sites are not available to visitors unless their browsers are set to allow cookies. For example, most distance learning software used by schools to deliver online courses does not work properly in student Web browsers unless cookies are enabled.

Web users can accumulate large numbers of cookies as they browse the Internet. Most Web browsers have settings that allow the user to refuse only third-party cookies or to review each cookie before it is accepted. Some browsers, such as **Netscape Navigator**, **Mozilla**, **Mozilla Firefox**, and **Opera**, provide comprehensive cookie management functions. Figure 10-3 shows the dialog box that can be used to manage stored cookies in the Mozilla Firefox Web browser.

Another approach is to use one of the many third-party programs, called **cookie blockers**, that prevent cookie storage selectively. Some of these programs, such as **WebWasher**, plug into a browser and allow users to block cookies from the Web servers that load advertising banners into Web pages. Other cookie blocking programs, such as **Cookie Pal**, allow cookies to be filtered by Internet (IP) address, allowing in the "good"

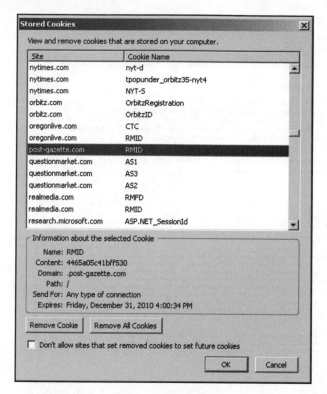

FIGURE 10-3 Mozilla Firefox dialog box for managing stored cookies

cookies and denying storage to all others. **Cookie Crusher** is another program that controls cookies before they are stored on a user's hard drive.

WebSideStory provides software that Web site managers can use to analyze Internet traffic at their sites. The company also sells a reporting service to Web sites that provides information about who visits their sites and what sites the visitors came from. WebSideStory's HitBox software collects and warehouses data from Web site visitors remotely, securely, and anonymously. The company does allow Web site visitors to examine the cookies that it reads from their computers. Figure 10-4 shows an example of the information that WebSideStory obtained from a cookie its HitBox software had placed on one of the author's computers.

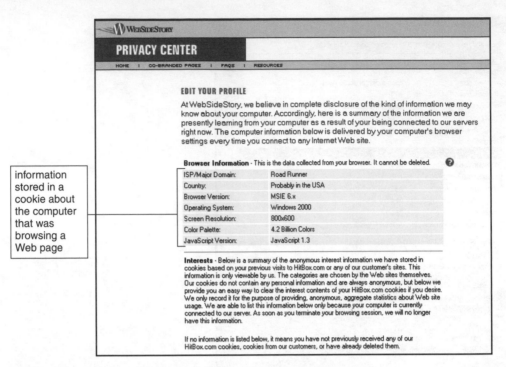

information stored in a cookie about the computer that was browsing a Web page

FIGURE 10-4 Information stored in a cookie on a client computer

Web Bugs

Some advertisers send images (from their third-party servers) that are included on Web pages, but are too small to be visible. A **Web bug** is a tiny graphic that a third-party Web site places on another site's Web page. When a site visitor loads the Web page, the Web bug is delivered by the third-party site, which can then place a cookie on the visitor's computer. A Web bug's only purpose is to provide a way for a third-party Web site (the identity of which is unknown to the visitor) to place cookies from that third-party site on the visitor's computer. The Internet advertising community sometimes calls Web bugs "clear GIFs" or "1-by-1 GIFs" because the graphics can be created in the GIF format with a color value of "transparent" and can be as small as 1 pixel by 1 pixel.

Active Content

Until the debut of executable Web content, Web pages could do little more than display content and provide links to related pages with additional information. The widespread use of active content has changed the situation. **Active content** refers to programs that are embedded transparently in Web pages and that cause action to occur. For example, active content can display moving graphics, download and play audio, or implement Web-based spreadsheet programs. Active content is used in electronic commerce to place items into a shopping cart and compute a total invoice amount, including sales tax, handling, and shipping costs. Developers use active content because it extends the functionality of HTML and moves some data processing chores from the busy server machine to the user's client computer. Unfortunately, because active content elements are programs that run on the client computer, active content can damage the client computer. Thus, active content can pose a threat to the security of client computers.

Active content is provided in several forms. The best-known active content forms are cookies, Java applets, JavaScript, VBScript, and ActiveX controls. Other ways to provide Web active content include graphics, Web browser plug-ins, and e-mail attachments.

JavaScript and VBScript are **scripting languages**; they provide scripts, or commands, that are executed. An **applet** is a small application program. Applets typically run within the Web browser. Active content is launched in a Web browser automatically when that browser loads a Web page containing active content. The applet downloads automatically with the page and begins running. Depending on how the browser's security settings are configured, the browser might open a warning dialog box, such as the one shown in Figure 10-5, announcing the active content and asking the user for permission to open that content.

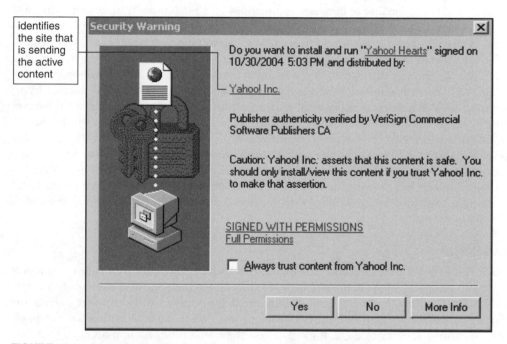

identifies the site that is sending the active content

FIGURE 10-5   Dialog box asking for permission to open a Java applet

Because active content modules are embedded in Web pages, they can be completely transparent to anyone browsing a page containing them. Crackers intent on doing mischief to client computers can embed malicious active content in these seemingly innocuous Web pages. This delivery technique is called a Trojan horse. A **Trojan horse** is a program hidden inside another program or Web page that masks its true purpose. The Trojan horse could snoop around a client computer and send back private information to a cooperating Web server—a secrecy violation. The program could alter or erase information on a client computer—an integrity violation. Zombies are equally threatening. A

zombie is a program that secretly takes over another computer for the purpose of launching attacks on other computers. Zombie attacks can be very difficult to trace to their creators.

Java Applets

Java is a programming language developed by Sun Microsystems that is used widely in Web pages to provide active content. The Web server sends the Java applets along with Web pages requested by the Web client. In most cases, the Java applet's operation will be visible to the site visitor; however, it is possible for a Java applet to perform functions that would not be noticed by the site visitor. The client computer then runs the programs within its Web browser. Java can also run outside the confines of a Web browser. Java is platform independent; that is, it can run on many different computers. This "develop once, deploy everywhere" feature reduces development costs because only one program needs to be developed for all operating systems.

Java adds functionality to business applications and can handle transactions and a wide variety of actions on the client computer. That relieves an otherwise busy server-side program from handling thousands of transactions simultaneously. Once downloaded, embedded Java code can run on a client's computer, which means that security violations can occur. To counter this possibility, a security model called the **Java sandbox** has been developed. The Java sandbox confines Java applet actions to a set of rules defined by the security model. These rules apply to all untrusted Java applets. **Untrusted Java applets** are those that have not been established as secure. When Java applets are run within the constraints of the sandbox, they do not have full access to the client system. For example, Java applets operating in the sandbox cannot perform file input, output, or delete operations. This prevents secrecy (disclosure) and integrity (deletion or modification) violations. You can follow the Online Companion link to the **Java Security Page** maintained by the Center for Education and Research in Information and Assurance (CERIAS) to learn more about Java applet security.

JavaScript

JavaScript is a scripting language developed by Netscape to enable Web page designers to build active content. Despite the similar-sounding names, JavaScript is based only loosely on Sun's Java programming language. Supported by popular Web browsers, JavaScript shares many of the structures of the full Java language. When a user downloads a Web page with embedded JavaScript code, it executes on the user's (client) computer.

Like other active content vehicles, JavaScript can be used for attacks by executing code that destroys the client's hard disk, discloses the e-mail stored in client mailboxes, or sends sensitive information to the attacker's Web server. JavaScript code can also record the URLs of Web pages a user visits and capture information entered into Web forms. For example, if a user enters credit card numbers while reserving a rental car, a JavaScript program could copy the credit card number. JavaScript programs, unlike Java applets, do not operate under the restrictions of the Java sandbox security model.

Unlike Java applets, a JavaScript program cannot commence execution on its own. To run an ill-intentioned JavaScript program, a user must start the program. For example, a site with a retirement income calculator might require a visitor to click a button to see a retirement income projection. Once the user clicks the button, the JavaScript program starts and does its work.

ActiveX Controls

An **ActiveX** control is an object that contains programs and properties that Web designers place on Web pages to perform particular tasks. ActiveX components can be constructed using many different programming languages, but the most common are C++ and Visual Basic. Unlike Java or JavaScript code, ActiveX controls run only on computers with Windows operating systems.

When a Windows-based Web browser downloads a Web page containing an embedded ActiveX control, the control is executed on the client computer. Other ActiveX controls include Web-enabled calendar controls and Web games. The **ActiveX** page at Download.com contains a comprehensive list of ActiveX controls.

The security danger with ActiveX controls is that once they are downloaded, they execute like any other program on a client computer. They have full access to all system resources, including operating system code. An ill-intentioned ActiveX control could reformat a user's hard disk, rename or delete files, send e-mails to all the people listed in the user's address book, or simply shut down the computer. Because ActiveX controls have full access to client computers, they can cause secrecy, integrity, or necessity violations. The actions of ActiveX controls cannot be halted once they begin execution. Most Web browsers can be configured to provide a notice when the user is about to download an ActiveX control. Figure 10-6 shows an example of the warning issued when Internet Explorer detects an ActiveX control.

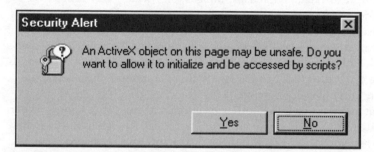

FIGURE 10-6 Internet Explorer ActiveX Control warning message

Graphics and Plug-Ins

Graphics, browser plug-ins, and e-mail attachments can harbor executable content. Some graphics file formats have been designed specifically to contain instructions on how to render a graphic. That means that any Web page containing such a graphic could be a threat because the code embedded in the graphic could cause harm to a client computer. Similarly, browser **plug-ins**, which are programs that enhance the capabilities of browsers, handle Web content that a browser cannot handle. Plug-ins are normally beneficial and perform tasks for a browser, such as playing audio clips, displaying movies, or animating graphics. Apple's QuickTime, for example, is a plug-in that downloads and plays movies stored in a special format.

Plug-ins can also pose security threats to a client computer. Users download these plug-in programs and install them so their browsers can display content that cannot be included in HTML tags. Popular plug-ins include Macromedia's Flash Player and Shockwave Player, Apple's QuickTime Player, and RealNetworks' RealPlayer.

In 1999, *The New York Times* revealed that RealNetworks had been using its RealPlayer plug-in to gather information surreptitiously from users. Downloaded and installed easily from the Internet, RealPlayer was recording user information such as the RealPlayer user's name, e-mail address, country, ZIP code, computer operating system, and other details. RealPlayer used the Internet connection to send the information it had gathered back to RealNetworks. Soon after the discovery, and after considerable public embarrassment, RealNetworks issued a statement that a software patch was available for all current users. The patch prevents the RealNetworks software from collecting and transmitting user information.

Many plug-ins execute commands buried within the media being manipulated. This opens the door to the possibility that someone intent on doing harm could embed commands within a seemingly innocuous video or audio clip. The ill-intentioned commands hidden within the object that the plug-in is interpreting could damage a client computer by erasing some (or all) of its files.

Viruses, Worms, and Antivirus Software

The potential dangers lurking in e-mail attachments get a lot of news coverage and are the most familiar to the general population. E-mail attachments provide a convenient way to send nontext information over a text-only system—electronic mail. Attachments can contain word-processing files, spreadsheets, databases, images, or virtually any other information you can imagine. Most programs, including Web browser e-mail programs, display attachments by automatically executing an associated program; for example, the recipient's Excel program reads an attached Excel workbook file and opens it, or Word opens and displays a Word document. Although this activity itself does not cause damage, Word and Excel macro viruses inside the loaded files can damage a client computer and reveal confidential information when those files are opened.

A virus is software that attaches itself to another program and can cause damage when the host program is activated. A worm is a type of virus that replicates itself on the computers that it infects. Worms can spread quickly through the Internet. A **macro virus** is a type of virus that is coded as a small program, called a macro, and is embedded in a file. You have probably read about or have personally experienced recent examples of e-mail attachment-borne virus attacks.

E-mail attachments containing viruses and other malicious software are reported daily. Some of the most famous in recent years include the ILOVEYOU virus, also known as the "love bug," and its variants. The ILOVEYOU virus was eventually traced to a 23-year-old computer science student who lived in the Philippines. The virus spread through the Internet with amazing speed as an e-mail message. It infected the computer of anyone who opened the e-mail attachment and clogged e-mail systems with thousands of copies of the useless e-mail message. The virus spread quickly because it automatically sent itself to as many as 300 addresses stored in a computer's Microsoft Outlook address book. Besides replicating itself explosively through e-mail, the virus caused other harm, destroying digital music and photo files stored on the target computers. The ILOVEYOU virus also searched for other users' passwords and forwarded that information to the original perpetrator. Within days, the virus spread to 40 million computers in more than 20 countries and caused an estimated $9 billion in damages—most of it in lost worker productivity.

In 2001, the incidences of virus and worm attacks increased. With more than 40,000 reported security violations occurring that year, the parade of attacks included Code Red and Nimda virus-worm combinations, each affecting millions of computers and costing billions of dollars to clean up. Both Code Red and Nimda are examples of a **multivector virus**, so called because they can enter a computer system in several different ways (vectors). Even though Microsoft issued security patches that should have stopped the Code Red virus-worm, it continued to propagate throughout the Internet in 2002. Both the original Code Red virus and a variant called Code Red 2 infected thousands of new computers during the year.

New virus-worm combinations also appeared in 2002 and 2003, including a version of the Code Red virus called Bugbear. Bugbear was spread through Microsoft Outlook e-mail clients. The person receiving the e-mail did not even have to click on an attachment to run the malicious code—Bugbear started itself through a security loophole in the connection between Outlook and the Internet Explorer browser. Of course, Microsoft issued a security patch for the browser, but many users did not install the patch (or, in many cases, even know about it). When launched, Bugbear first checked to see if the computer was running antivirus software. **Antivirus software** detects viruses and worms and either deletes them or isolates them on the client computer so they cannot run. If antivirus software existed on the system, Bugbear attempted to destroy it. Then it installed a Trojan horse program on the computer that let attackers access the computer through the Internet and upload or download files at will. (Bugbear was difficult to eliminate from an infected computer because it gives its own files a randomly generated name; thus, the virus files have different names on every infected computer.) Bugbear would then send out e-mail messages with attachments that would infect the recipients. It did not create its own e-mail messages, but took previously sent e-mail messages that were on the computer and resent them to different addresses. This often fools recipients because the e-mail messages have subject headers that seem normal and do not hint that the e-mail might contain a virus.

Symantec and **McAfee**, among other companies, keep track of viruses and sell antivirus software. You can follow the links in the Online Companion to those companies to find descriptions of thousands of viruses. Antivirus software is only effective if the antivirus data files are kept current. The data files contain virus-identifying information that is used to detect viruses on a client computer. Because people generate new viruses by the hundreds every month, users must be vigilant and update their antivirus data files regularly so that the newest viruses are recognized and eliminated. Some Web e-mail systems, such as Yahoo! Mail, let users scan attachments using antivirus software before downloading e-mail. In these cases, the antivirus software is run by the Web site and the user does not need to take any action to keep the software updated.

MICROSOFT INTERNET INFORMATION SERVER

As you learned in Chapter 8, Internet Information Server (IIS) is Microsoft's Web server software. Microsoft supplies versions of the IIS software with its Windows server operating systems that are suitable for use in operating electronic commerce Web sites.

In August 2001, Microsoft faced an uncomfortable situation that many U.S. manufacturing companies have experienced with recalled, defective products—Microsoft executives stood by at a news conference while a U.S. government official announced to gathered reporters that there was a serious flaw in a Microsoft product. The director of the FBI's National Infrastructure Protection Center was warning reporters that the Code Red worm, which was spreading through the Internet for the third time in as many weeks, was a serious threat to the continued operation of the Internet. A **worm** is a type of virus that replicates itself on the computers that it infects.

The Code Red worm exploits a vulnerability in the Microsoft IIS Web server software. When the worm was first identified, Microsoft rapidly made a patch available on its Web site. Microsoft also announced that Web server installations that had kept current with all of the updates and patches that Microsoft had issued would not be subject to attack by the worm.

Many Microsoft customers were outraged by these statements, noting that Microsoft had issued more than 40 software patches in the first half of 2001 and 100 or more patches in each of several prior years. IIS users complained that keeping the software current was virtually impossible and called for Microsoft to deliver software that was more secure when first installed.

Many IIS users began to consider switching to other Web server software. Gartner, Inc., a major IT consulting firm, recommended to its clients that they seriously consider alternatives to IIS for their critical Web server installations. Many industry observers and software engineers agree that Microsoft was a victim of its own success. It had created a very popular and complex piece of software. It is extremely difficult to ensure that no bugs exist in complex software products, and the popularity of the software made it an attractive target for crackers—one worm could bring down many of the servers operating on the Internet. These two factors, plus the likelihood that many IIS servers would not have all of the available security upgrades installed, combined to make it an irresistible target for a worm creator.

Microsoft has struggled to gain the confidence of large corporate IT departments. The company has worked hard in recent years to establish the reputation of its operating system software as reliable and trustworthy. The Code Red worm attack on its Web server software was a major setback in its reputation-building effort. You can review the **Microsoft Security Pages** through the link in the Online Companion to see how Microsoft is still trying to establish that its software is secure in the face of continuing cracker and virus-writer attacks that are both regular and frequent.

Digital Certificates

One way to control threats from active content is to use digital certificates. A **digital certificate** or digital ID is an attachment to an e-mail message or a program embedded in a

Web page that verifies that the sender or Web site is who or what it claims to be. In addition, the digital certificate contains a means to send an encrypted message—encoded so others cannot read it—to the entity that sent the original Web page or e-mail message. In the case of a downloaded program containing a digital certificate, the encrypted message identifies the software publisher (ensuring that the identity of the software publisher matches the certificate) and indicates whether the certificate has expired or is still valid. The digital certificate is a **signed** message or code. Signed code or messages serve the same function as a photo on a driver's license or passport. They provide proof that the holder is the person identified by the certificate. Just like a passport, a certificate does not imply anything about either the usefulness or quality of the downloaded program. The certificate only supplies a level of assurance that the software is genuine. The idea behind certificates is that if the user trusts the software developer, signed software can be trusted because, as proven by the certificate, it came from that trusted developer.

Digital certificates are used for many different types of online transactions, including electronic commerce, electronic mail, and electronic funds transfers. A digital ID verifies a Web site to a shopper and, optionally, identifies a shopper to a Web site. Web browsers or e-mail programs exchange digital certificates automatically and invisibly when requested to validate the identity of each party involved in a transaction.

Figure 10-7 displays the digital certificate owned by Amazon.com. Whenever a browser indicates that it has established secure communication with a Web site; that is, when a lock appears in the browser's status line, the user can double-click the lock to display the Web site's digital certificate.

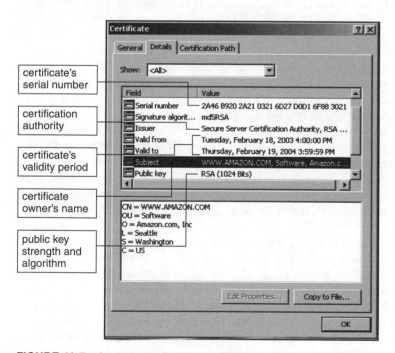

FIGURE 10-7   Amazon.com's digital certificate

A digital certificate for software is an assurance that the software was created by a specific company. The certificate does not attest to the quality of the software, just to the identity of the company that published it. Digital certificates are issued by a **certification authority (CA)**. A CA can issue digital certificates to organizations or individuals. A CA requires entities applying for digital certificates to supply appropriate proof of identity. Once the CA is satisfied, it issues a certificate. Then, the CA signs the certificate, and its stamp of approval is affixed in the form of a public encryption key, which "unlocks" the certificate for anyone who receives the certificate attached to the publisher's code.

Digital certificates cannot be forged easily. A digital certificate includes six main elements, including:

- Certificate owner's identifying information, such as name, organization, address, and so on
- Certificate owner's public key (you will learn more about public and private keys later in this chapter)
- Dates between which the certificate is valid
- Serial number of the certificate
- Name of the certificate issuer
- Digital signature of the certificate issuer

A **key** is simply a number—usually a long binary number—that is used with the encryption algorithm to "lock" the characters of the message being protected so that they are undecipherable without the key. Longer keys usually provide significantly better protection than shorter keys. In effect, the CA is guaranteeing that the individual or organization that presents the certificate is who or what it claims to be.

Identification requirements vary from one CA to another. One CA might require a driver's license for individuals' certificates; others might require a notarized form or fingerprints. CAs usually publish identification requirements so that any Web user or site accepting certificates from each CA understands the stringency of that CA's validation procedures. There are only a small number of CAs because the certificates issued are only as trustworthy as the CA itself and only a few companies have decided to build the reputation needed to be a successful seller of digital certificates. Two of the most commonly used CAs are **Thawte** and **VeriSign**, but other companies such as **Entrust** and **Equifax Secure** also offer CA services. The digital certificate for Amazon.com (information about this certificate appears in the dialog box shown in Figure 10-7) was issued by VeriSign. As you examine the certificates of various Web sites, you will notice that many of them indicate that the issuer is "RSA Data Security," which is the division of VeriSign that issues many of its digital certificates.

Certificates are classified as low, medium, or high assurance, based largely on the identification requirements imposed on certificate seekers. The fees charged by CAs vary with the level of assurance provided; higher levels of assurance are more expensive. For example, VeriSign provides certificate issuing and revocation services and offers several classes of certificates—from Class 1 through Class 4—that are differentiated by assurance level, which is the confidence level one can assume based on the process the CA uses to verify the owner's identity. Class 1 certificates are the lowest level and bind e-mail addresses and associated public keys. Class 4 certificates apply to servers and the server organizations. Requirements for Class 4 certificates are significantly greater than those for Class 1.

VeriSign's Class 4 certificate, for example, offers assurance of the individual's identity and that person's relationship to the specified company or organization.

Digital certificates expire after a period of time (often one year). This built-in limit provides protection for both users and businesses. Limited-duration certificates guarantee that businesses and individuals must submit their credentials for reevaluation periodically. The expiration date appears in the certificate itself and in the dialog boxes that browsers display when a Web page or applet that has a digital certificate is about to be opened. Certificates become invalid on their expiration dates or when they are intentionally revoked by the CA. If the CA determines that a Web site has begun delivering malicious code, it will refuse to issue new certificates to that site and revoke any existing certificates it might already have obtained.

Steganography

The term **steganography** describes the process of hiding information (a command, for example) within another piece of information. This information can be used for malicious purposes. Frequently, computer files contain redundant or insignificant information that can be replaced with other information. This other information resides in the background and is undetectable by anyone without the correct decoding software. Steganography provides a way of hiding an encrypted file within another file so that a casual observer cannot detect that there is anything of importance in the container file. In this two-step process, encrypting the file protects it from being read, and steganography makes it invisible.

Many security analysts believe that the terrorist organization Al Qaeda used steganography to hide attack orders and other messages in images that it had its confederates post on Web sites. Messages hidden using steganography are extremely difficult to detect. This fact, combined with the fact that there are millions of images on the Web, makes the use of steganography by global terrorist organizations a deep concern of governments and security professionals. The Online Companion includes a link to a site with more information about **Steganography and Digital Watermarking**.

Physical Security for Clients

In the past, physical security was a major concern for large computers that ran important business functions such as payroll or billing; however, as networks (including intranets and the Internet) have made it possible to control important business functions from client computers, concerns about physical security for client computers have become greater. Many of the physical security measures used today are the same as those used in the early days of computing; however, some interesting new technologies have been implemented as well.

Devices that read fingerprints are now available for personal computers. These devices, which cost less than $200, provide a much stronger protection than traditional password approaches. In addition to fingerprint readers, companies can use other biometric security devices that are more accurate and, of course, cost more. A **biometric security device** is one that uses an element of a person's biological makeup to perform the identification. These devices include writing pads that detect the form and pressure of a person writing a signature, eye scanners that read the pattern of blood vessels in a person's retina or the color levels in a person's iris, and scanners that read the palm of a person's hand (rather than just one fingerprint) or that read the pattern of veins on the back of a person's hand.

COMMUNICATION CHANNEL SECURITY

The Internet serves as the electronic connection between buyers (in most cases, clients) and sellers (in most cases, servers). The most important thing to remember as you learn about communication channel security is that the Internet was not designed to be secure. Although the Internet has its roots in a military network, that network was not designed to include any significant security features. It was designed to provide redundancy in case one or more communications lines were cut. In other words, the goal of the Internet's packet-switching design was to provide multiple alternative paths on which critical military information could travel. The military always sends sensitive information in an encrypted form so that the content of messages traveling over any network—even if intercepted—remain secret. The security of messages traversing the military predecessors to the Internet was provided by software that operated independently of the network to encrypt messages. As the Internet developed, it did so without any significant security features that became a part of the network itself.

Today, the Internet remains largely unchanged from its original, insecure state. Message packets on the Internet travel an unplanned path from a source node to a destination node. A packet passes through a number of intermediate computers on the network before reaching its final destination. The path can vary each time a packet is sent between the same source and destination points. Because users cannot control the path and do not know where their packets have been, it is possible that an intermediary can read the packets, alter them, or even delete them. That is, any message traveling on the Internet is subject to secrecy, integrity, and necessity threats. This section describes these problems in more detail and outlines several solutions for those problems.

Secrecy Threats

Secrecy is the security threat that is most frequently mentioned in articles and the popular media. Closely linked to secrecy is privacy, which also receives a great deal of attention. Secrecy and privacy, though similar, are different issues. Secrecy is the prevention of unauthorized information disclosure. **Privacy** is the protection of individual rights to nondisclosure. The **Privacy Council**, which helps businesses implement smart privacy and data practices, created an extensive Web site surrounding privacy—covering both business and legal issues. Secrecy is a technical issue requiring sophisticated physical and logical mechanisms, whereas privacy protection is a legal matter. A classic example of the difference between secrecy and privacy is e-mail.

A company might protect its e-mail messages against secrecy violations by using encryption (you will learn more about encryption later in this chapter). In encryption, a message is encoded into an unintelligible form that only the proper recipient can convert back into the original message. Secrecy countermeasures protect outgoing messages. E-mail privacy issues address whether company supervisors should be permitted to read employees' messages randomly. Disputes in this area center around who owns the e-mail messages: the company, or the employees who sent them. The focus in this section is on secrecy, preventing unauthorized persons from reading information they should not be reading.

One significant threat to electronic commerce is theft of sensitive or personal information, including credit card numbers, names, addresses, and personal preferences. This kind of theft can occur any time anyone submits information over the Internet because it is easy for an ill-intentioned person to record information packets (a secrecy violation) from the Internet for later examination. The same problems can occur in e-mail transmissions. Software applications called **sniffer programs** provide the means to record information that passes through a computer or router that is handling Internet traffic. Using a sniffer program is analogous to tapping a telephone line and recording a conversation. Sniffer programs can read e-mail messages and unencrypted Web client-server message traffic such as user logins, passwords, and credit card numbers.

Periodically, security experts find electronic holes, called **backdoors**, in electronic commerce software. These can be left open accidentally by the software developer, or they can be left open intentionally. Either way, content is exposed to secrecy threats. A backdoor allows anyone with knowledge of the existence of the backdoor to cause damage by observing transactions, deleting data, or stealing data. In 2000, the Cart32 shopping cart software made by McMurtrey/Whitaker & Associates was found to have a backdoor through which credit card numbers could be obtained by anyone with a backdoor password. The company quickly supplied a patch to eliminate the backdoor. Although the backdoor resulted from a programming error and not from intentional efforts, the consequences were serious for merchants that used the software—their customers' credit card numbers were available to hackers around the world.

Credit card number theft is an obvious problem, but proprietary corporate product information or prerelease data sheets mailed to corporate branches can be intercepted and passed along easily, too. Confidential information can be considerably more valuable than information about credit cards, which usually have spending limits. Stolen corporate information can be worth millions of dollars.

Here is an example of how an online eavesdropper might obtain confidential information. Suppose a user logs on to a Web site that contains a form with text boxes for name, address, and e-mail address. When the user fills out those text boxes and clicks the Submit button, the information is sent to the Web server for processing. Some Web servers obtain and track that data by collecting the text box responses and placing them at the end of the server's URL (which appears in the address box of the user's Web browser). This long URL (with the text box responses appended) is included in all HTTP request and response messages that travel between the user's browser and the server.

So far, no violations have occurred. Suppose, however, that the user decides not to wait for a response from the server. Instead, the user visits another Web site. The server at this second Web site might be set up to collect Web demographics. If it is, it logs the URL from which the user just came by capturing it from the HTTP request message that the browser sends. Web sites use this URL logging technique for the completely legitimate purpose of identifying sources of customer traffic. However, any employee at the second site who has access to the server log can read the part of the URL that includes the information entered into those text boxes on the first site, thus obtaining that user's confidential information.

Web users continually reveal information about themselves when they use the Web. This information includes IP addresses and the type of browser being used. Such data exposure is a secrecy breach. Several Web sites offer an anonymous browser service that hides personal information from sites visited. One of these sites, **Anonymizer**, provides a measure of secrecy to Web surfers who use the site as a portal (the beginning site from which they visit other sites). Anonymizer places its address on the front end of any URLs that the user visits. This shield reveals only the Anonymizer Web site URL to other Web sites that the user visits. This can make anonymous Web surfing possible, but tedious, because each URL that the user wants to visit must be typed in the text box on the Anonymizer home page. To make the process easier, Anonymizer and other companies provide browser plug-in software that users can download and install for an annual subscription fee. Figure 10-8 shows Anonymizer's home page.

FIGURE 10-8 Anonymizer home page

Integrity Threats

An integrity threat, also known as **active wiretapping**, exists when an unauthorized party can alter a message stream of information. Unprotected banking transactions, such as deposit amounts transmitted over the Internet, are subject to integrity violations. Of course, an integrity violation implies a secrecy violation because an intruder who alters information can read and interpret that information. Unlike secrecy threats, where a viewer simply sees information he or she should not, integrity threats can cause a change in the actions a person or corporation takes because a mission-critical transmission has been altered.

Cybervandalism is an example of an integrity violation. **Cybervandalism** is the electronic defacing of an existing Web site's page. The electronic equivalent of destroying property or placing graffiti on objects, cybervandalism occurs whenever someone replaces a Web site's regular content with his or her own content. Recently, several cases of Web page defacing involved vandals replacing business content with pornographic material and other offensive content.

Masquerading or **spoofing**—pretending to be someone you are not, or representing a Web site as an original when it is a fake—is one means of disrupting Web sites. **Domain name servers (DNSs)** are the computers on the Internet that maintain directories that link domain names to IP addresses. Perpetrators can use a security hole in the software that runs on some of these computers to substitute the addresses of their Web sites in place of the real ones to spoof Web site visitors.

For example, a hacker could create a fictitious Web site masquerading as www.widgets.com by exploiting a DNS security hole that substitutes his or her fake IP address for Widgets.com's real IP address. All subsequent visits to Widgets.com would be redirected to the fictitious site. There, the hacker could alter any orders to change the number of widgets ordered and redirect shipment of those products to another address. The integrity attack consists of altering an order and passing it to the real company's Web server. The Web server is unaware of the integrity attack and simply verifies the consumer's credit card number and passes on the order for fulfillment. Major electronic commerce sites that have been the victims of masquerading attacks in recent years include Amazon.com, AOL, eBay, and PayPal. Some of these schemes combine spam with spoofing. The perpetrator sends millions of spam e-mails that appear to be from a respectable company. The e-mails contain a link to a Web page that is designed to look exactly like the company's site. The victim is encouraged to enter username, password, and sometimes even credit card information. These exploits, which capture confidential customer information, are called **phishing expeditions**. The most common victims of phishing expeditions are users of online banking and payment system (such as PayPal) Web sites. You will learn more about the phishing problem and the measures banks and other companies are taking to combat it in Chapter 11.

Necessity Threats

The purpose of a **necessity threat**, also known by other names such as a delay, denial, or denial-of-service (DoS) threat, is to disrupt normal computer processing, or deny processing entirely. A computer that has experienced a necessity threat slows processing to an intolerably slow speed. For example, if the processing speed of a single ATM transaction slows from one or two seconds to 30 seconds, users will abandon ATMs entirely. Similarly, slowing any Internet service drives customers to competitors' Web or commerce sites—possibly discouraging them from ever returning to the original commerce site. In other words, slower processing can render a service unusable or unattractive. For example, an online newspaper that reports three-day-old news is worth very little.

DoS attacks remove information altogether, or delete information from a transmission or file. One documented denial attack caused selected PCs that have Quicken (an accounting program) installed to divert money to the perpetrator's bank account. The denial attack denied money from its rightful owners. In another famous DoS attack against high-profile electronic commerce sites such as Amazon.com and Yahoo!, the attackers used zombie computers to send a flood of data packets to the sites. This overwhelmed the sites' servers and choked off legitimate customers' access. Prior to the attack, perpetrators located vulnerable computers and loaded them with the software that attacked the commerce sites. The Internet Worm attack of 1998, which disabled thousands of computer systems that were connected to the Internet, was the first recorded example of a DoS attack.

Threats to the Physical Security of Internet Communications Channels

The Internet was designed from its inception to withstand attacks on its physical communication links. Recall from Chapter 2 that the main purpose of the U.S. government research project that led to the development of the Internet was to provide an attack-resistant technology for coordinating military operations. Thus, the Internet's packet-based network design precludes it from being shut down by an attack on a single communications link on that network.

However, an individual user's Internet service can be interrupted by destruction of that user's link to the Internet. Few individual users have multiple connections to an ISP. However, larger companies and organizations (and ISPs themselves) often do have more than one link to the main backbone of the Internet. Typically, each link is purchased from a different network access provider. If one link becomes overloaded or unavailable, the service provider can switch traffic to another network access provider's link to keep the company, organization, or ISP (and its customers) connected to the Internet.

Threats to Wireless Networks

As you learned in Chapter 2, networks can use wireless access points (WAPs) to provide network connections to computers and other mobile devices within a range of several hundred feet. If not protected, a wireless network allows anyone within that range to log in and have access to any resources connected to that network. Such resources might include any data stored on any computer connected to the network, networked printers, messages sent on the network, and, if the network is connected to the Internet, free access to the Internet. The security of the connection depends on the Wireless Encryption Protocol (WEP), which is a set of rules for encrypting transmissions from the wireless devices to the WAPs.

Companies that have large wireless networks are usually careful to turn on WEP in devices, but smaller companies and individuals who have installed wireless networks in their homes often do not turn on the WEP security feature. Many WAPs are shipped to buyers with a default login and password already set. Companies that install these WAPs sometimes fail to change that login and password. This has given rise to a new avenue of entry into networks.

In some cities that have large concentrations of wireless networks, attackers drive around in cars using their wireless-equipped laptop computers to search for accessible networks. These attackers are called **wardrivers**. When wardrivers find an open network (or a WAP that has a common default login and password), they sometimes place a chalk mark on the building so that other attackers will know that an easily entered wireless network is nearby. This practice is called **warchalking**. Some warchalkers have even created Web sites that include maps of wireless access locations in major cities around the world. Companies can avoid becoming targets by simply turning on WEP in their access points and changing the logins and passwords to something other than the manufacturers' default settings.

In 2002, Best Buy was using wireless point-of-sale (POS) terminals in some of its 1900 stores. The wireless POS terminals could be moved easily from one area of the store to another, and they helped Best Buy handle large customer flows better than it could using only fixed POS terminals. Unfortunately, Best Buy failed to enable WEP on these terminals. A customer who had just purchased a wireless card for his laptop decided to launch a sniffer utility program on the laptop in his car in the parking lot. The customer was able to intercept data from the POS terminals, including transaction details and what he said looked like credit card numbers. Best Buy stopped using the wireless POS terminals when the story appeared on several Web sites and wire services.

Encryption Solutions

Encryption is the coding of information by using a mathematically based program and a secret key to produce a string of characters that is unintelligible. The science that studies encryption is called **cryptography**, which comes from a combination of the two Greek words *krypto* and *grapho*, which mean "secret" and "writing," respectively. That is, cryptography is the science of creating messages that only the sender and receiver can read.

Cryptography is different from steganography, which makes text undetectable to the naked eye. Cryptography does not hide text; it converts it to other text that is visible, but does not appear to have any meaning. What an unauthorized reader sees is a string of random text characters, numbers, and punctuation.

Encryption Algorithms

The program that transforms normal text, called **plain text**, into **cipher text** (the unintelligible string of characters) is called an **encryption program**. The logic behind an encryption program that includes the mathematics used to do the transformation from plain text to cipher text is called an **encryption algorithm**. There are a number of different encryption algorithms in use today. Some have been developed by the U.S. government and others have been developed by IBM and other commercial enterprises. You can learn more about the development of encryption algorithms, including an evaluation of currently available algorithms, by consulting a Web security textbook (see, for example, the Mackey reference in the For Further Study and Research section at the end of this chapter).

Messages are encrypted just before they are sent over a network or the Internet. Upon arrival, each message is decoded, or **decrypted**, using a **decryption program**—a type of encryption-reversing procedure. Encryption algorithms are considered so vitally important to preserving security within the United States that the National Security Agency has control over their dissemination. Some encryption algorithms are considered so important that the U.S. government has banned publication of details about them. Currently, it is illegal for U.S. companies to export some of these encryption algorithms. Web pages containing software whose distribution is restricted include warnings about U.S. export laws. The Freedom Forum Online contains a number of articles on lawsuits and legislation surrounding encryption export laws. Critics consider publication restrictions a freedom of speech issue. If you are interested in reading more about the latest arguments in the ongoing debates over freedom of speech and export law, search the **Freedom Forum** using the keyword "encryption" as the search term.

One property of encryption algorithms is that someone can know the details of the algorithm and still not be able to decipher the encrypted message without knowing the key that the algorithm used to encrypt the message. The resistance of an encrypted message to attack attempts depends on the size (in bits) of the key used in the encryption procedure. A 40-bit key is currently considered to provide a minimal level of security. Longer keys, such as 128-bit keys, provide much more secure encryption. A sufficiently long key can help make the security unbreakable.

The type of key and associated encryption program used to lock a message, or otherwise manipulate it, subdivide encryption into three functions:

- Hash coding
- Asymmetric encryption
- Symmetric encryption

Hash Coding

Hash coding is a process that uses a **hash algorithm** to calculate a number, called a **hash value**, from a message of any length. It is a fingerprint for the message because it is almost certain to be unique for each message. Good hash algorithms are designed so that the probability of two different messages resulting in the same hash value, which would create a **collision**, is extremely small. Hash coding is a particularly convenient way to tell whether a message has been altered in transit because its original hash value and the hash value computed by the receiver will not match after a message is altered.

Asymmetric Encryption

Asymmetric encryption, or **public-key encryption**, encodes messages by using two mathematically related numeric keys. In 1977, Ronald Rivest, Adi Shamir, and Leonard Adleman invented the RSA Public Key Cryptosystem while they were professors at MIT. Their invention revolutionized the way sensitive information is exchanged. In their system, one key of the pair, called a **public key**, is freely distributed to the public at large—to anyone interested in communicating securely with the holder of both keys. The public key is used to encrypt messages using one of several different encryption algorithms. The second key—called a **private key**—belongs to the key owner, who keeps the key secret. The owner uses the private key to decrypt all messages received.

Here is an overview of how asymmetric encryption system works: If Herb wants to send a message to Allison, he obtains Allison's public key from any of several well-known public places. Then, he encrypts his message to Allison using her public key. Once the message is encrypted, only Allison can read the message by decrypting it with her private key.

Because the keys are unique, only one secret key can open the message encrypted with a corresponding public key, and vice versa. Reversing the process, Allison can send a private message to Herb using Herb's public key to encrypt the message. When he receives Allison's message, Herb uses his private key to decrypt the message and then read it. If they are sending e-mail to one another, the message is secret only while in transit. Once a message is downloaded from the mail server and decoded, it is stored in plain text on the recipient's machine for all to view.

One of the most popular technologies used to implement public-key encryption today is called **Pretty Good Privacy (PGP)**. PGP was invented in 1991 by Phil Zimmerman, who charged businesses for use of PGP, but allowed individuals to use PGP at no cost. PGP is a set of software tools that can use several different encryption algorithms to perform public-key encryption. The PGP business was purchased by Network Associates in 1997 and sold back to the product's developers, who formed PGP Corporation in 2002. Today, individuals can download free versions of PGP for personal use from the **PGP Corporation** site and from the **PGP International** site. Individuals can use PGP to encrypt their e-mail messages to protect them from being read if they are intercepted on the Internet. The PGP Corporation site sells licenses to businesses that want to use the technology to protect business communication activities.

Symmetric Encryption

Symmetric encryption, also known as **private-key encryption**, encodes a message with one of several available algorithms that use a single numeric key, such as 456839420783, to encode and decode data. Because the same key is used, both the message sender and the message receiver must know the key. Encoding and decoding messages using symmetric encryption is very fast and efficient. However, the key must be guarded. If the key is made public, then all messages sent previously using that key are vulnerable, and both the sender and receiver must use new keys for future communication.

It can be difficult to distribute new keys to authorized parties while maintaining security and control over the keys. The catch is that to transmit *anything* privately, it must be encrypted. This includes the new, secret key. Another significant problem with private keys is that they do not scale well in large environments such as the Internet. Each pair of users on the Internet who wants to share information privately must have their own private key. That results in a huge number of key-pair combinations, similar to a telephone system of private lines without switching stations. Enabling 12 people to have a private key pair between all pairs (or private telephone lines between each pair) would require 66 private keys. In general, n individual Internet clients require $(n(n-1))/2$ private key pairs.

In secure environments such as the defense sector, using private-key encryption is simpler, and it is the prevalent method to encode sensitive data. Distribution of classified information and encryption keys is straightforward in the defense sector. It requires guards (two-person control) and secret transportation plans. The **Data Encryption Standard (DES)** is a set of encryption algorithms adopted by the U.S. government for encrypting sensitive or commercial information. It is the most widely used private-key encryption system. However, the DES private-key size is increased periodically because individuals are using increasingly fast computers to break messages encoded with shorter keys. In 1999, for example, the Electronic Frontier Foundation's Deep Crack key breaker used 100,000 PCs on the Internet to break a DES-encrypted test message in under 23 hours (see the **EFF DES Cracker Project** for more information).

Today, the U.S. government uses a stronger version of the Data Encryption Standard, called **Triple Data Encryption Standard (Triple DES or 3DES)**. Triple DES offers good protection because it cannot be cracked even with today's supercomputers. Experts expect that it will continue to be extremely difficult to crack for the next several years. However, the U.S. government's **National Institute of Standards and Technology** (NIST) has developed a new encryption standard designed to keep government information secure. The new standard is called the **Advanced Encryption Standard (AES)**. In February 2001, the NIST announced that the four-year development process had been successful and that two cryptography researchers from Belgium had created the algorithm chosen for AES. The algorithm's name is Rijndael (pronounced "rain doll"); you can learn more about the development process and the algorithm at the NIST's **AES Algorithm (Rijndael)** Web site.

Comparing Asymmetric and Symmetric Encryption Systems

Public-key (asymmetric) systems provide several advantages over private-key (symmetric) encryption methods. First, the combination of keys required to provide private messages between enormous numbers of people is small. If n people want to share secret information with one another, then only n unique public-key pairs are required—far fewer than an equivalent private-key system. Second, key distribution is not a problem. Each person's public key can be posted anywhere and does not require any special handling to distribute. Third, public-key systems make implementation of digital signatures possible. This means that an electronic document can be signed and sent to any recipient with nonrepudiation. That is, with public-key techniques, it is not possible for anyone other than the signer to produce the signature electronically; in addition, the signer cannot later deny signing the electronic document.

Public-key systems have disadvantages. One disadvantage is that public-key encryption and decryption are significantly slower than private-key systems. This extra time can add up quickly as individuals and organizations conduct commerce on the Internet. Public-key systems do not replace private-key systems, but serve as a complement to them. Public-key systems are used to transmit private keys to Internet participants so that additional, more efficient communication can occur in a secure Internet session. Figure 10-9 shows a graphical representation of the hashing, private-key, and public-key encryption methods: Figure 10-9a shows hash coding; Figure 10-9b depicts private-key encryption; and Figure 10-9c illustrates public-key encryption.

Several encryption algorithms exist that can be used with secure Web servers. The U.S. government approves the use of several of these inside the United States. Electronic commerce Web servers can accommodate most of these algorithms because they must be able to communicate with a wide variety of Web browsers.

The **Secure Sockets Layer (SSL)** system developed by Netscape Communications and the Secure Hypertext Transfer Protocol (S-HTTP) developed by CommerceNet are two protocols that provide secure information transfer through the Internet. SSL and S-HTTP allow both the client and server computers to manage encryption and decryption activities between each other during a secure Web session.

SSL and S-HTTP have different goals. SSL secures connections between two computers, and S-HTTP sends *individual* messages securely. Encryption of outgoing messages

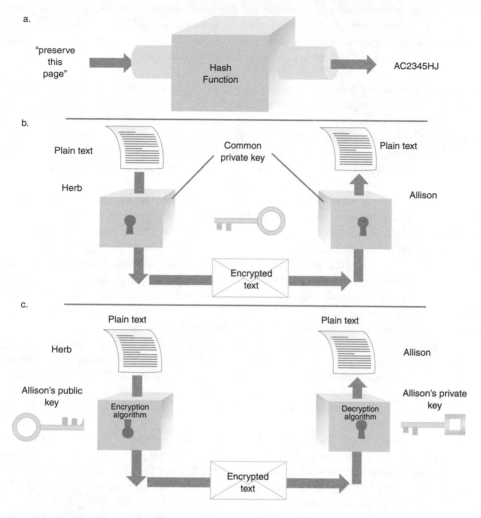

a.

"preserve this page" → Hash Function → AC2345HJ

b.

Plain text

Herb

Common private key

Plain text

Allison

Encrypted text

c.

Plain text

Herb

Allison

Allison's public key

Encryption algorithm

Plain text

Allison's private key

Decryption algorithm

Encrypted text

FIGURE 10-9   (a) hash coding, (b) private-key, and (c) public-key encryption

and decryption of incoming messages happen automatically and transparently with both SSL and S-HTTP.

Secure Sockets Layer (SSL) Protocol

SSL provides a security "handshake" in which the client and server computers exchange a brief burst of messages. In those messages, the level of security to be used for exchange of digital certificates and other tasks is agreed upon. Each computer identifies the other. After identification, SSL encrypts and decrypts information flowing between the two computers. This means that information in both the IITTP request and any HTTP response is encrypted. Encrypted information includes the URL the client is requesting, any forms containing information the user has completed (which might include a credit card number), and HTTP access authorization data, such as usernames and passwords. In short, *all* communication between SSL-enabled clients and servers is encoded. When SSL encodes everything flowing between the client and server, an eavesdropper receives only unintelligible information.

SSL can secure many different types of communication between computers in addition to HTTP. For example, SSL can secure FTP sessions, enabling private downloading and uploading of sensitive documents, spreadsheets, and other electronic data. SSL can secure Telnet sessions in which remote computer users can log on to corporate host machines and send their passwords and usernames. The protocol that implements SSL is HTTPS. By preceding the URL with the protocol name HTTPS, the client is signifying that it would like to establish a secure connection with the remote server.

Secure Sockets Layer allows the length of the private session key generated by every encrypted transaction to be set at a variety of bit lengths (such as 40-bit, 56-bit, 128-bit, and 168-bit). A **session key** is a key used by an encryption algorithm to create cipher text from plain text during a single secure session. The longer the key, the more resistant the encryption is to attack. A Web browser that has entered into an SSL session indicates that it is in an encrypted session (most browsers use an icon in the browser status bar). Once the session is ended, the session key is discarded permanently and not reused for subsequent secure sessions.

Here is how SSL works with an exchange between a client and an electronic commerce site: Remember that SSL has to authenticate the commerce site and encrypt any transmissions between the two computers. When a client browser sends a request message to a server's secure Web site, the server sends a hello request to the browser (client). The browser responds with a client hello. The exchange of these greetings, or the handshake, allows the two computers to determine the compression and encryption standards that they both support.

Next, the browser asks the server for a digital certificate—proof of identity. In response, the server sends to the browser a certificate signed by a recognized certification authority. The browser checks the serial number and certificate fingerprint on the server certificate against the public key of the CA stored within the browser. Once the CA's public key is verified, the endorsement is verified. That action authenticates the Web server.

Both the client and server agree that their exchanges should be kept secure because they involve transmitting credit card numbers, invoice numbers, and verification codes over the Internet. To implement secrecy, SSL uses public-key (asymmetric) encryption and private-key (symmetric) encryption. Although public-key encryption is handy, it is slow compared to private-key encryption. That is why SSL uses private-key encryption for nearly all its secure communications. Because it uses private-key encryption, SSL must have a way to get the key to both the client and server without exposing it to an eavesdropper. SSL accomplishes this by having the browser generate a private key for both to share. Then the browser encrypts the private key it has generated using the server's public key. The server's public key is stored in the digital certificate that the server sent to the browser during the authentication step. Once the key is encrypted, the browser sends it to the server. The server, in turn, decrypts the message with its private key and exposes the shared private key.

From this point on, public-key encryption is no longer used. Instead, only private-key encryption is used. All messages sent between the client and the server are encrypted with the shared private key, also known as the session key. When the session ends, the session key is discarded. A new connection between a client and a secure server starts the entire process all over again, beginning with the handshake between the client browser and the server. The client and server can agree to use 40-bit encryption or a 128-bit encryption. The client and server also agree on which specific encryption algorithm to use.

Figure 10-10 illustrates the SSL handshake that occurs before a client and server exchange private-key encoded business information for the remainder of the secure session.

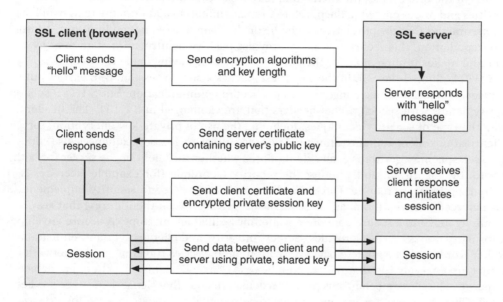

FIGURE 10-10   Establishing an SSL session

You can determine details about an SSL site by visiting Netcraft's Web Server Survey site. Use the Online Companion link **What's that SSL site running?** to visit the site and follow the instructions provided there.

Secure HTTP (S-HTTP)

Secure HTTP (S-HTTP) is an extension to HTTP that provides a number of security features, including client and server authentication, spontaneous encryption, and request/response nonrepudiation. The protocol was developed by **CommerceNet**, a consortium of organizations interested in promoting electronic commerce. S-HTTP provides symmetric encryption for maintaining secret communications and public-key encryption to establish client/server authentication. Either the client or the server can use S-HTTP techniques separately. That is, a client browser may require security through the use of a private (symmetric) key, whereas the server may require client authentication by using public-key techniques.

The details of S-HTTP security are conducted during the initial negotiation session between the client and server. Either the client or the server can specify that a particular security feature be required, optional, or refused. When one party stipulates that a particular security feature be required, the client or server continues the connection only if the other party (client or server) agrees to enforce the specified security. Otherwise, no secure connection is established. Suppose the client browser specifies that encryption is required to render all communications secret. In such a situation, the transactions of a high-fashion clothing designer purchasing silk from a Far East textile house will remain confidential. Eavesdropping competitors cannot learn which fabrics are featured next season. On the other hand, the textile mill may insist that integrity be enforced so that quantities and prices quoted to the purchaser remain intact. In addition, the textile mill may want assurances that the purchaser is who he or she claims to be, not an imposter. A form of nonrepudiation, this security property provides positive confirmation of an offer by a client and makes it impossible for the client to deny ever having made the offer.

S-HTTP differs from SSL in the way it establishes a secure session. SSL carries out a client/server handshake exchange to set up a secure communication, but S-HTTP sets up security details with special packet headers that are exchanged in S-HTTP. The headers define the type of security techniques, including the use of private-key encryption, server authentication, client authentication, and message integrity. Header exchanges also stipulate which specific algorithms each side supports, whether the client or the server (or both) supports the algorithm, and whether the security technique (for example, secrecy) is required, optional, or refused. Once the client and server agree to security implementations enforced between them, all subsequent messages between them during that session are wrapped in a secure container, sometimes called an envelope. A **secure envelope** encapsulates a message and provides secrecy, integrity, and client/server authentication. In other words, it is a complete package. With it, all messages traveling on the network or Internet are encrypted so that they cannot be read. Messages cannot be altered undetectably because integrity mechanisms provide a detection code that signals a message has been altered. Clients and servers are authenticated with digital certificates issued by a recognized certification authority. The secure envelope includes all of these security features. S-HTTP is no longer used by many Web sites. SSL has become a more generally accepted standard for establishing secure communication links between Web clients and Web servers.

You have learned how encryption provides message secrecy and confidentiality, and you have learned how digital certificates serve to authenticate a server to a client, and vice versa. However, you have not learned how to implement message integrity. The methods that allow you to ensure that an interloper does not change a message in transit appear in the next section.

Ensuring Transaction Integrity with Hash Functions

Electronic commerce ultimately involves a client browser sending payment information, order information, and payment instructions to the Web server and that server responding with a confirmation of the order details. If an Internet interloper alters any of the order information in transit, harmful consequences can result. For instance, the perpetrator could alter the shipment address so that he or she receives the merchandise instead of the original customer. This is an example of an **integrity violation**, which occurs whenever a message is altered while in transit between the sender and receiver.

Although it is difficult and expensive to *prevent* a perpetrator from altering a message, there are security techniques that allow the receiver to *detect* when a message has been altered. When the receiver—a Web server, for example—receives a damaged message, the receiver simply asks the sender to retransmit the message. Apart from being annoying, a damaged message harms no one as long as both parties are aware of the alteration. Harm occurs when unauthorized message changes go undetected by the message's sender and receiver.

A combination of techniques creates messages that are both tamperproof and authenticated. Additionally, those techniques provide the property of nonrepudiation —making it impossible for message creators to claim that the message was not theirs or that they did not send it. To eliminate fraud and abuse caused by messages being altered, two separate algorithms are applied to a message. First, a hash algorithm is applied to the message. Hash algorithms are **one-way functions**, meaning that there is no way to transform the hash value back to the original message. This approach is acceptable because a hash value is compared only with another hash value to see if there is a match—the original, prehash values are never compared with one another.

All encryption programs convert text into a **message digest**, which is a small integer number that summarizes the encrypted information. A hash algorithm uses no secret key; the message digest it produces cannot be inverted to produce the original information; the algorithm and information about how it works are publicly available; and finally, hash collisions are nearly impossible. Once the hash function computes a message's hash value, that value is appended to the message. Suppose the message is a purchase order containing the customer's address and payment information. When the merchant receives the purchase order and attached message digest, he or she calculates a message digest value for the message (exclusive of the original attached message digest). If the message digest value that the merchant calculates matches the message digest attached to the message, the merchant then knows the message is unaltered—that is, no interloper altered the amount or the shipping address information. Had someone altered the information, then the merchant's software would compute a message digest value different from the message digest that the client calculated and sent along with the purchase order.

Ensuring Transaction Integrity with Digital Signatures

Hash functions are not a complete solution. Because the hash algorithm is public and (by design) widely known, anyone could intercept a purchase order, alter the shipping address and quantity ordered, re-create the message digest, and send the message and new message digest on to the merchant. Upon receipt, the merchant would calculate the message digest value and confirm that the two message digest values match. The merchant is fooled into concluding that the message is unadulterated and genuine. To prevent this type of fraud, the sender encrypts the message digest using his or her private key.

An encrypted message digest (message hash value) is called a **digital signature**. A purchase order accompanied by a digital signature provides the merchant with positive identification of the sender and assures the merchant that the message was not altered. Because the message digest is encrypted using a public key, only the owner of the public/private key pair could have encrypted the message digest. Thus, when the merchant decrypts the message with the user's public key and subsequently calculates a matching message digest value, the result is proof that the sender is authentic. Furthermore, matching hash values prove that only the sender could have authored the message (non-repudiation) because only his or her private key would yield an encrypted message that could be decrypted successfully by an associated public key. This solves the spoofing problem.

If necessary, both parties can agree to provide transaction secrecy in addition to the integrity, nonrepudiation, and authentication that the digital signature provides. Simply encrypting the entire string—digital signature and message—guarantees message secrecy. Used together, public-key encryption, message digests, and digital signatures provide a high level of security for Internet transactions. Figure 10-11 illustrates how a digital signature and a signed message are created and sent.

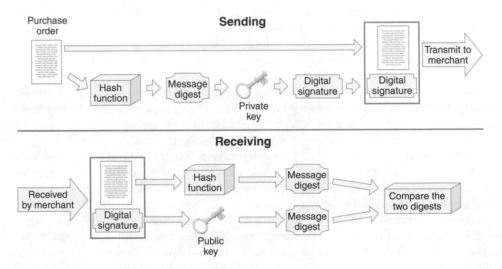

FIGURE 10-11   Sending and receiving a digitally signed message

In 2000, U.S. President Bill Clinton signed a bill that gave digital signatures the same legal status as traditional signatures. Clinton first signed the paper version of the new digital signature legislation with a pen. Then, he signed the electronic version of the bill with a smart card (you will learn about smart cards in Chapter 11) containing his digital signature. After doing so, the name "Bill Clinton" appeared on the screen under the text of the new law entitled *Electronic Signatures in Global and National Commerce Act*. People can now electronically sign all sorts of legal documents, such as online car lease agreements, loan papers, and purchase orders.

The European Union followed closely on the heels of the U.S. legislation and required all of its member countries to enact digital signature laws by mid-2001. Most Canadian provinces had also enacted digital signature legislation by the end of 2001. Other countries have passed or are working toward passing laws that enable the use of digital signatures.

Guaranteeing Transaction Delivery

As you learned earlier in this chapter, denial or delay-of-service attacks remove or absorb resources. Neither encryption nor a digital signature protects information packets from theft or slowdown. However, the Transmission Control Protocol (TCP) half of the TCP/IP pair is responsible for end-to-end control of packets. When it reassembles packets at the destination in the correct order, it handles all the details when packets do not appear. Among TCP's duties are to request that the client computer resend data when packets seem to be missing. That is, no special computer security protocol beyond TCP/IP is required as a countermeasure against denial attacks. TCP/IP builds checks into the data so that it can tell when data packets are altered, inadvertently or otherwise.

SECURITY FOR SERVER COMPUTERS

The server is the third link in the client-Internet-server electronic commerce path between the user and a Web server. Servers have vulnerabilities that can be exploited by anyone determined to cause destruction or acquire information illegally. One entry point is the Web server and its software. Other entry points are any back-end programs containing data, such as a database and the server on which it runs. Although no system is completely safe, the Web server administrator's job is to make sure that security policies are documented and considered in every part of the electronic commerce operation.

Web Server Threats

Web server software, as you learned in Chapter 8, is designed to deliver Web pages by responding to HTTP requests. Although Web server software is not inherently high-risk software, it has been designed with Web service and convenience as the main design goals. The more complex the software, the greater the probability that it contains coding errors or security weaknesses.

A Web server can compromise secrecy if it allows automatic directory listings. The secrecy violation occurs when the contents of a server's folder names are revealed to a Web browser. This happens frequently and is caused when a user enters a URL, such as http://www.somecompany.com/FAQ/, and expects to see the default page in the FAQ directory. The default Web page that the server normally displays is named index.htm or index.html. If that file is not in the directory, a Web server that allows automatic directory listings displays all of the file and folder names in that directory. Then, visitors can click folder names at random and open folders that might otherwise be off limits. Careful site administrators turn off this folder name display feature. If a user attempts to browse a folder where protections prevent browsing, the Web server issues a warning message stating that the directory is not available.

Web servers can compromise security by requiring users to enter a username and password. The username and password can be subsequently revealed when the user visits multiple pages within the same Web server's protected area if the server requires that users reestablish their usernames and passwords for each protected page they visit. This repeated information requirement is necessary because the Web is stateless—it cannot remember what happened during the last transaction. The most convenient way to remember a username and password is to store the user's confidential information in a cookie on his or her computer. That way, the Web server can request confirmation of the data by requesting that the computer send a cookie. Although cookies are not inherently unsafe, a Web server should not ask a Web browser to transmit a cookie in unencrypted form. The **W3C Security FAQ** provides additional information about server security.

One of the most sensitive files on a Web server is the file that holds Web server username and password pairs. If that file is compromised, an intruder can enter privileged areas masquerading as someone else. Such an intruder can obtain usernames and passwords if that information is readily available and not encrypted. Most Web servers store user authentication information in encrypted form.

The passwords that users select can be a threat. Users sometimes select passwords that are guessed easily, such as mother's maiden name, name of a child, a telephone number, or some easily obtained identification number, such as a Social Security number. **Dictionary attack programs** cycle through an electronic dictionary, trying every word in the book as a password. Users' passwords, once broken, may provide an opening for illegal entry into a server that can remain undetected for a long time. To prevent dictionary attacks, many organizations use a dictionary check as a preventive measure in their password assignment software. When a user selects a new password, the password assignment software checks the password against its dictionary and, if it finds a match, refuses to allow the use of that password. An organization's password assignment software dictionary typically includes common words, names (including common pet names), acronyms that are commonly used in the organization, and words or characters (including numbers) that have some meaning for the user requesting the password (for example, employees might be prohibited from using their employee numbers as passwords).

Database Threats

Electronic commerce systems store user data and retrieve product information from databases connected to the Web server. Besides storing product information, databases connected to the Web contain valuable and private information that could damage a company irreparably if disclosed or altered. Most large-scale database systems include security features that rely on usernames and passwords. Once a user is authenticated, select portions of the database become available to that user. However, some databases either store username/password pairs in an unencrypted table, or they fail to enforce security altogether and rely on the Web server to enforce security. If unauthorized users obtain user authentication information, they can masquerade as legitimate database users and reveal or download confidential and potentially valuable information. Trojan horse programs hidden within the database system can also reveal information by changing the access rights of various user groups. A Trojan horse can even remove access controls within a database, giving all users complete access to the data—including intruders.

Other Programming Threats

Web server threats can arise from programs executed by the server. Java or C++ programs that are passed to Web servers by a client, or that reside on a server, frequently make use of a buffer. A **buffer** is an area of memory set aside to hold data read from a file or database. A buffer is necessary whenever any input or output operation takes place because a computer can process file information much faster than the information can be read from input devices or written to output devices. Programs filling buffers can malfunction and overfill the buffer, spilling the excess data outside the designated buffer memory area. This is called a **buffer overrun** or **buffer overflow** error. Usually, this occurs because the program contains an error or bug that causes the overflow. Sometimes, however, the buffer overflow is intentional. The Internet Worm of 1988 was such a program. It caused an overflow condition that eventually consumed all resources until the affected computer could no longer function.

A more insidious version of a buffer overflow attack writes instructions into critical memory locations so that when the intruder program has completed its work of overwriting buffers, the Web server resumes execution by loading internal registers with the address of the main attacking program's code. This type of attack can open the Web server to severe damage because the resumed program—which is now the attacker program—may regain control of the computer, exposing its files to disclosure and destruction by the attacking program. **The Red Hat Linux Buffer Overflow Attacks Web Page** describes the buffer vulnerabilities of Web servers that run on the Linux operating system. Good programming practices can reduce the potential damage from buffer overflows and some computers include hardware that works with the operating system to limit the effects of buffer overflows that are intentionally programmed to create damage.

A similar attack, one in which excessive data is sent to a server, can occur on mail servers. Called a **mail bomb**, the attack occurs when hundreds or even thousands of people each send a message to a particular address. The attack might be launched by a large team of well-organized hackers, but more likely the attack is launched by one or a few hackers who have gained control over others' computers using a Trojan horse virus or some other method of turning those computers into zombies. The accumulated mail received by the target of the mail bomb exceeds the allowed e-mail size limit and can cause e-mail systems to malfunction. Although it is fairly easy to track the people responsible for the attack, it is debilitating nonetheless.

Threats to the Physical Security of Web Servers

Web servers and the computers that are networked closely to them, such as the database servers and application servers used to supply content and transaction-processing capabilities to electronic commerce Web sites, must be protected from physical harm. For many companies, these computers have become repositories of important data (information about customers, products, sales, purchases, and payments). They have also become important parts of the revenue-generating function in many businesses. As key physical resources, these computers and related equipment warrant high levels of protection against threats to their physical security.

As you learned in Chapter 8, many companies use CSPs to host Web sites. Even large companies that own servers and have IT staff to maintain those servers often put the computers in a CSP facility. The security that CSPs maintain over their physical premises (see earlier section on Threats to the Physical Security of Internet Communications Channels) is, in many cases, stronger than the security that a company could provide for computers maintained at its own location.

Companies can take additional steps to protect their Web servers. Many companies maintain backup copies of server contents at a remote location. If the Web server operation is critical to the continuation of the business, a company can maintain a duplicate of the entire Web server physical facility at a remote location. In the case of a natural disaster or a terrorist attack, the Web operations can be switched over in a matter of seconds to the backup location. Examples of mission-critical Web servers that would warrant such a comprehensive (and expensive) level of physical security include airline reservation systems, stock brokerage firm trading systems, and bank account clearing systems.

Some companies rely on their service providers to help with Web server security. Major service providers that offer managed services, such as **Level 3**, **PSINet**, and **Verio Security Services**, often include Web server security as an add-on service. Other companies hire smaller, specialized security service providers to handle security (see Learning From Failures—Pilot Network Services to learn more about one alternative to this approach). Having a service provider handle security usually adds an additional $1000 to $3000 per month to the bandwidth charges. The specialized security firms often charge two to three times more than that for their services.

LEARNING FROM FAILURES

PILOT NETWORK SERVICES

Pilot Network Services began operations in 1993, at the dawn of commercial use of the Internet. Its goal was to build a network that would be secure for electronic commerce activities. It built a network that included its own carefully monitored connections to the Internet and a database of attack signatures. Attack signatures are descriptions of the Internet traffic characteristics that indicate a cracker attack on a Web server. Pilot, as a firm specializing in security services, built an excellent collection of attack signatures and kept it updated much better than other firms that were not security specialists.

Pilot maintained the Web servers for many of its clients, and it used versions of the operating systems and Web server software that it had customized to be especially resistant to attacks. Pilot's engineers meticulously applied patches for all known points of access to the software and worked to identify new, as yet unknown, points of vulnerability—for which they immediately created and applied protective patches. For customers hosting their own servers, Pilot provided the Internet connection through its own secure network. The router between the client's network and Pilot's network and the operating system running the Pilot network were customized to eliminate any known security loopholes.

Pilot had 24/7 monitoring of its network by computer security experts, in addition to the network technicians that any other Web hosting company would provide as part of a managed services offering. Because it offered high-quality services, its fees were considerably higher than the security service charges imposed by other service providers. Typical charges were $6000 per month for the basic connection, plus $4000 per month for each Web server.

Even at these high prices, Pilot had many fans among the Fortune 500. Pilot never had more than 300 customers, but it monitored more than 70,000 individual networks for a customer list that included General Electric, PeopleSoft, Sovereign Bancorp, The Washington Post Company, and many other major accounts. By 1999, Pilot appeared to be doing well. Its revenue had increased more than 80 percent over 1998. News releases were issued regularly announcing new customers.

continued

In late 2000, Pilot's stock price began to fall, along with the stock prices of many companies in Internet-related businesses. Although Pilot's sales were growing, its costs were escalating at an even more rapid rate. The company had never reported a profit, and its annual losses had increased to $21.7 million in 2000. Pilot executives assured its customers that the company was financially sound, but the ability of companies in Internet-related businesses to survive on the promise of future earnings had disappeared. Pilot's ability to raise the cash it needed to continue operating had vanished.

In early 2001, some Pilot customers noticed that the service was failing. Phone calls and e-mails were not being returned quickly. On the afternoon of April 25, 2001, Pilot employees received four e-mails. The first explained that telephones would be disconnected that evening. The second asked all employees to turn in their mobile phones and pagers. The third announced that the chief financial officer had resigned. The final e-mail stated that all employees were out of a job as of 4:30 p.m.

Pilot's clients, many of which found out about the collapse from the Pilot employees who had been servicing their accounts, were in serious trouble. Connections to the Internet vanished with no warning. The companies that had used Pilot to host entire Web operations were in an even worse situation. A group of Pilot customers convinced AT&T (the provider of Pilot's Internet connections) to continue to carry traffic from Pilot, even though Pilot had not paid AT&T. Providian Financial, a major bank holding company and credit card processor, sent its own employees into Pilot operations centers to keep Providian's Web servers operating. Other Pilot customers that were Providian's competitors protested loudly. Most Pilot customers were concerned that their Web servers were suddenly open and vulnerable to attack.

Several of Pilot's competitors tried to raise funding to take over the business, but all of those attempts failed, and on May 9, 2001—two weeks after the collapse—AT&T cut Internet service and Pilot was liquidated. Pilot's former customers were scrambling to hire security staff, find alternative hosting firms, or join forces with other companies to keep their electronic commerce sites operating. The lesson from this failure is that security is a critical part of an electronic commerce operation. It should be handled with the same care that a company would use to protect any physical asset. If any part of the security function is handed over to another company, that company's condition becomes an important concern and must be monitored carefully.

473

Access Control and Authentication

Access control and authentication refers to controlling who and what has access to the Web server. Most people who work with Web servers in electronic commerce environments do not sit at a keyboard connected to the server. Instead, they access the server from a client computer. Recall that authentication is verification of the identity of the entity requesting access to the computer. Just as users can authenticate servers with which they are interacting, servers can authenticate individual users. When a server requires positive identification of a user, it requests that the client send a certificate.

The server can authenticate a user in several ways. First, the certificate represents the user's admittance voucher. If the server cannot decrypt the user's digital signature contained in the certificate using the user's public key, then the certificate did not come from the true owner. Otherwise, the server is certain that the certificate came from the owner.

This procedure prevents fraudulent certificates of "admission" to a secure server. Second, the server checks the timestamp on the certificate to ensure that the certificate has not expired. A server will reject an expired certificate and provide no further service. Third, a server can use a callback system in which the user's client computer name and address are checked against a list of usernames and assigned client computer addresses. Such a system works especially well in an intranet where usernames and client computers are controlled closely and assigned systematically. On the Internet, a callback system is more difficult to manage—particularly if client users are mobile and work from different locations. It is easy to see how certificates issued by trusted CAs play a central role in authenticating client computers and their users. Certificates provide attribution —irrefutable evidence of identity—if a security breach occurs.

Usernames and passwords can also provide some element of protection. To authenticate users using passwords and usernames, the server must acquire and store a database containing rightful users' passwords and usernames. Many Web server systems store usernames and passwords in a file. Large electronic commerce sites usually keep username/ password combinations in a separate database with built-in security features.

The easiest way to store passwords is to maintain usernames in plain text and encrypt passwords using a one-way encryption algorithm. With the plain text username and encrypted password stored, the system can validate users when they log on by checking the usernames they enter against the list of usernames stored in the database. The password that a user enters when he or she logs on to a system is encrypted. Then the resulting encrypted password from the user is checked against the encrypted password stored in the database. If the two encrypted versions of the password match for the given user, the login is accepted. That is why even a system administrator cannot tell you what your forgotten password is on most systems. Instead, the administrator must assign a new temporary password that the user can change to another password.

Passwords are not immune to discovery, and a person truly intent on stealing a password can often figure out a way to do so. Figure 10-12 shows an example of a login box for accessing subscription-only areas of *The Wall Street Journal*'s Web site.

Note that the site visitor can save his or her username and password as a cookie on the client computer, which allows access to subscription areas of the site without entering the username and password on subsequent site visits. The trouble with that system of cookies is that the information might be stored on the client computer in plain text. If the cookie contains login and password information, then that information is visible to anyone who has access to the user's computer.

Web servers often provide access control list security to restrict file access to selected users. An **access control list (ACL)** is a list or database of files and other resources and the usernames of people who can access the files and other resources. Each file has its own access control list. When a client computer requests Web server access to a file or document that has been configured to require an access check, the Web server checks the resource's ACL file to determine if the user is allowed to access that file. This system is especially convenient to restrict access of files on an intranet server so that individuals can only access selected files on a need-to-know basis. The Web server can exercise fine control over resources by further subdividing file access into the activities of read, write, or

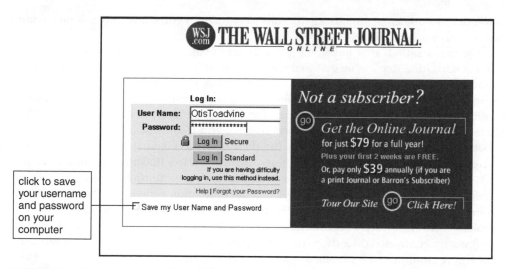

FIGURE 10-12   Logging on with a username and a password

execute. For example, some users may be permitted to read the corporate employee handbook, but not allowed to update or write to the file. Only the human resources (HR) manager would have write access to the employee handbook, and that access privilege is stored along with the HR manager's ID and password in an ACL.

Firewalls

A **firewall** is software or a hardware and software combination that is installed in a network to control the packet traffic moving through it. Most organizations place a firewall at the Internet entry point of their networks. The firewall provides a defense between a network and the Internet or between a network and any other network that could pose a threat. Firewalls have the following characteristics:

- All traffic from inside to outside and from outside to inside the network must pass through it.
- Only authorized traffic, as defined by the local security policy, is allowed to pass through it.
- The firewall itself is immune to penetration.

Those networks inside the firewall are often called **trusted**, whereas networks outside the firewall are called **untrusted**. Acting as a filter, firewalls permit selected messages to flow into and out of the protected network. For example, one security policy a firewall might enforce is to allow all HTTP (Web) traffic to pass back and forth, but disallow FTP or Telnet requests either into or out of the protected network. Ideally, firewall protection should prevent access to networks inside the firewall by unauthorized users, and thus prevent access to sensitive information. Simultaneously, a firewall should not obstruct legitimate users. Authorized employees outside the firewall ought to have access to firewall-protected networks and data files. Firewalls can separate corporate networks from one another and prevent personnel in one division from accessing information from another division of the same company. Using firewalls to segment a corporate network into secure zones serves as a coarse need-to-know filter.

Large organizations that have multiple sites and many locations must install a firewall at each location that has an external connection to the Internet. Such a system ensures an unbroken security perimeter that is effective for the entire corporation. In addition, each firewall in the organization must follow the same security policy. Otherwise, one firewall might permit one type of transaction to flow into the corporate network that another excludes. The result is an unwanted access that is permitted throughout the corporation because one firewall left a small security door open to the entire network.

Firewalls should be stripped of any unnecessary software. Because the firewall computer is used only as a firewall and not as a general-purpose computing machine, only essential operating system software and firewall-specific protection software should remain on the computer. Having fewer software programs on the system should reduce the chances for malevolent software security breaches. Access to a firewall should be restricted to a console physically connected directly to the firewall machine. Otherwise, remote administration of the firewall must be provided, which opens up the possibility of a break in the firewall by an imposter remotely accessing the firewall along the same path that an administrator would use.

Firewalls are classified into the following categories: packet filter, gateway server, and proxy server. **Packet-filter firewalls** examine all data flowing back and forth between the trusted network (within the firewall) and the Internet. Packet filtering examines the source and destination addresses and ports of incoming packets and denies or permits entrance to the packets based on a preprogrammed set of rules.

Gateway servers are firewalls that filter traffic based on the application requested. Gateway servers limit access to specific applications such as Telnet, FTP, and HTTP. Application gateways arbitrate traffic between the inside network and the outside network. In contrast to a packet-filter technique, an application-level firewall filters requests and logs them at the application level, rather than at the lower IP level. A gateway firewall provides a central point where all requests can be classified, logged, and later analyzed. An example is a gateway-level policy that permits incoming FTP requests, but blocks outgoing FTP requests. That policy prevents employees inside a firewall from downloading potentially dangerous programs from the outside.

Proxy server firewalls are firewalls that communicate with the Internet on the private network's behalf. When a browser is configured to use a proxy server firewall, the firewall passes the browser request to the Internet. When the Internet sends back a response, the proxy server relays it back to the browser. Proxy servers are also used to serve as a huge cache for Web pages.

One problem faced by companies that have employees working from home is that the location of computers outside the traditional boundaries of the company's physical site expands the number of computers that must be protected by the firewall. This **perimeter expansion** problem is particularly troublesome for companies that have salespeople using laptop computers to access confidential company information from all types of networks at customer locations, vendor locations, and even public locations, such as airports.

Another problem faced by organizations connected to the Internet is that their servers are under almost constant attack. Crackers spend a great deal of time and energy on attempts to enter the servers of organizations. Some of these crackers use automated programs to continually attempt to gain access to servers. Organizations often install intrusion detection systems as part of their firewalls. **Intrusion detection systems** are designed to monitor attempts to login to servers and analyze those attempts for patterns that might indicate a cracker's attack is underway. Once the intrusion detection system identifies an attack, it can block further attempts that originate from the same IP address until the organization's security staff can examine and analyze the access attempts and determine whether they are an attack.

In addition to firewalls installed on organizations' networks, it is possible to install software-only firewalls on individual client computers. These firewalls are often called **personal firewalls**. The use of personal firewalls, such as **ZoneAlarm**, has become an important tool in the protection of expanded network perimeters for many companies. Many home computer users are installing personal firewalls on their home networks. You can learn more about firewall protection for your home computer at the **Gibson Research Shields Up!** Web site.

ORGANIZATIONS THAT PROMOTE COMPUTER SECURITY

Following the occurrence of the Internet Worm of 1988, a number of organizations were formed to share information about threats to computer systems. These organizations are devoted to the principle that sharing information about attacks and defenses for those attacks can help everyone create better computer security. Some of the organizations began at universities; others were launched by government agencies. In this section, you will learn about some of these organizations and their resources.

CERT

In 1988, a group of researchers met to study the infamous Internet Worm attack soon after it occurred. They wanted to understand how worms worked and how to prevent damage from future attacks of this type. The National Computer Security Center, part of the National Security Agency, initiated a series of meetings to figure out how to respond to future security breaks that might affect thousands of people. Soon after that meeting of security experts in 1988, the U.S. government created the Computer Emergency Response Team and housed it at Carnegie Mellon University in Pittsburgh. The organization is now operated as part of the federally funded Software Engineering Institute at Carnegie Mellon, and it has changed its legal name from the Computer Emergency Response Team (which had been abbreviated to "CERT" by most people who wrote and talked about it) to **CERT**. CERT still maintains an effective and quick communications infrastructure among security experts so that security incidents can be avoided or handled quickly.

Today, CERT responds to thousands of security incidents each year and provides a wealth of information to help Internet users and companies become more knowledgeable about security risks. CERT posts alerts to inform the Internet community about security events, and it is regarded as a primary authoritative source for information about viruses, worms, and other types of attacks.

Other Organizations

CERT is the most prominent of these organizations and has formed relationships, such as the **Internet Security Alliance**, with other industry associations. However, CERT is not the only computer security resource. In 1989, one year after CERT was formed, a cooperative research and educational organization called the Systems Administrator, Audit, Network, and Security Institute was launched. Now known as the **SANS Institute**, this organization includes more than 150,000 members who work in computer security consulting firms and information technology departments of companies as auditors, systems administrators, and network administrators.

Many SANS education and research efforts yield resources such as news releases, research reports, security alerts, and white papers that are available on the Web site at no cost. SANS also sells publications to generate funds that it uses for research and educational programs. The SANS Institute operates the **SANS Internet Storm Center**, a Web site that provides current information on the location and intensity of computer attacks throughout the world. Purdue University's Center for Education and Research in Information Assurance and Security **(CERIAS)** is a center for multidisciplinary research and education in information security. The CERIAS Web site provides resources in computer, network, and communications security and includes a section on information assurance. The **Center for Internet Security** is a not-for-profit cooperative organization devoted to helping companies that operate electronic commerce Web sites reduce the risk of disruptions from technical failures or deliberate attacks on their computer systems. It also provides information to auditors who review such systems and to insurance companies that provide coverage for companies who operate such systems. **Microsoft Security Research Group** is a privately sponsored site that offers free information about computer security issues. For current information about computer security, you can visit **CSO Online**, which carries articles that have appeared in *CSO Magazine* along with other news items related to computer security.

The U.S. government has several Web sites devoted to security enhancement efforts. The **U.S. Department of Justice's Cybercrime** site offers information about computer crimes and intellectual property violations. The U.S. Department of Homeland Security operates the **National Infrastructure Protection Center** (NIPC) Web site, which provides information about threats to U.S. infrastructure, including its computing infrastructure.

Computer Forensics and Ethical Hacking

A small group of firms, endorsed by corporations and security organizations, have the unlikely job of breaking into client computers. Called **computer forensics experts** or **ethical hackers**, these computer sleuths are hired to probe PCs and locate information that can be used in legal proceedings. The field of **computer forensics** is responsible for the collection, preservation, and analysis of computer-related evidence. Ethical hackers are often hired by companies to test computer security safeguards. Links to the Web sites of several companies that offer computer forensics and ethical hacking services are included in the Additional Resources section of the Online Companion for this chapter.

Summary

Electronic commerce is vulnerable to a wide range of security threats. Attacks against electronic commerce systems can disclose or manipulate proprietary information. The three general assets that companies engaging in electronic commerce must protect are client computers, computer communication channels, and Web servers. Key security provisions in each of these parts of the Web client-Internet-Web server linkage are secrecy, integrity, and available service. Threats to commerce can occur anywhere in the commerce chain. News accounts of virus attacks have kept Web users aware of the security risks to client computers. Antivirus software is also an important element in the protection of client computers. More subtle threats are delivered as client-side applets. Java, JavaScript, and ActiveX controls run on client machines and have the potential to breach security. Cookies, if not controlled and used properly, can present threats to client computers.

Communication channels, in general, and the Internet, in particular, are especially vulnerable to attacks. The Internet is a vast network and because no control exists over the nodes through which Internet traffic passes, information sent through the Internet is vulnerable to unauthorized disclosure. This can lead to disclosure of private information, alteration of critical business documents, and theft or loss of important business messages. Encryption provides secrecy, and several forms of encryption are available that use hash functions or other more complex algorithms. They include private-key and public-key techniques. Although public-key encryption eliminates the problem of sharing a secret key, it is much slower than private-key encryption. Private-key encryption is used during most commerce sessions because it is fast and efficient. Integrity protections ensure that messages between clients and servers are not altered. Digital certificates provide both integrity controls and user authentication. A trusted third party such as a certification authority can provide digital certificates to users and organizations. Several Internet protocols, including Secure Sockets Layer and Secure HTTP, use encryption to provide secure Internet transmission capabilities. As wireless networks have grown to become important parts of the data communication infrastructure, security concerns have increased. Although many wireless networks (especially home networks) are installed without security features, wireless encryption methods that make them more secure are available. Most wireless networks installed in businesses today do have wireless encryption.

Web servers are susceptible to security threats. Programs that run on servers have the potential to damage databases, abnormally terminate server software, or make subtle changes in proprietary information. Attacks can come from within the server in the form of programs, or they can come from outside the server. One type of external attack can occur when a message overflows a server's internal storage region and overwrites crucial information. Overwritten information is replaced with either data or instructions that cause other programs on the server to execute. Backup copies of servers provide redundancy in the case of a physical threat to a server. The Web server must be protected from both physical threats and Internet-based attacks on its software. Protections for the server include access control and authentication, provided by username and password login procedures and client certificates. Firewalls can be used to separate trusted inside computer networks and clients from untrusted outside networks, including other divisions of a company's enterprise network system and the Internet.

A number of organizations have been formed to share information about computer security threats and defenses. CERT, the SANS Institute, and similar organizations address security outbreaks by linking knowledgeable security experts. When large security outbreaks occur, the members of these organizations join together and discuss methods to locate and eliminate the threat. Computer forensics firms that undertake attacks against their clients' computers can play an important role in helping identify security weaknesses.

Key Terms

Access control list (ACL)

Active content

Active wiretapping

ActiveX

Advanced Encryption Standard (AES)

Antivirus software

Applet

Asymmetric encryption

Backdoor

Biometric security device

Black hat hacker

Buffer

Buffer overrun (buffer overflow)

Certification authority (CA)

Cipher text

Collision

Computer forensics

Computer forensics expert

Computer security

Cookie blocker

Countermeasure

Cracker

Cryptography

Cybervandalism

Data Encryption Standard (DES)

Decrypted

Decryption program

Dictionary attack program

Digital certificate (digital ID)

Digital signature

Domain name server (DNS)

Eavesdropper

Encryption

Encryption algorithm

Encryption program

Ethical hacker

Firewall

First-party cookies

Gateway server

Hacker

Hash algorithm

Hash coding

Hash value

Integrity

Integrity violation

Intrusion detection system

Java sandbox

JavaScript

Key

Logical security

Macro virus

Mail bomb

Man-in-the-middle exploit

Masquerading (spoofing)

Message digest

Multivector virus

Necessity

Necessity threat (delay, denial, or denial-of-service threat)

One-way function

Open session

Packet-filter firewall

Perimeter expansion

Persistent cookie

Personal firewall

Phishing expeditions

Physical security

Plain text

Plug-ins

Pretty Good Privacy (PGP)

Privacy

Private key

Private-key encryption

Proxy server firewall

Public key

Public-key encryption

Scripting language

Secrecy

Secure envelope

Secure Sockets Layer

Security policy

Session cookie

Session key

Signed (message or code)

Sniffer program

Stateless connection

Steganography

Symmetric encryption

Third-party cookies

Threat

Triple Data Encryption Standard (Triple DES, 3DES)

Trojan horse

Trusted (network)

Untrusted (network)

Untrusted Java applet

Warchalking

Wardrivers

Web bug

White hat hacker

Worm

Zombie

Review Questions

RQ1. In about 200 words, explain why Web sites use cookies. In your answer, discuss the reasons that cookies were first devised and explain where cookies are stored. You can use the links in the Online Companion to help with your research.

RQ2. In about 100 words, describe steganography and explain its connection to the topic of online security. You can use the links in the Online Companion to help with your research.

RQ3. In about 200 words, explain the differences between public-key encryption and private-key encryption. List advantages and disadvantages of each encryption method. Explain which method you would use for e-mail sent from a field sales office to corporate headquarters. Assume that the e-mail regularly includes highly confidential information about upcoming sales opportunities.

RQ4. In about 300 words, describe the security threats that a company will face when it implements a wireless network. Assume that the company occupies the six middle floors in a 12-story office building that is located in a downtown business area between two other buildings of similar height. Briefly explain how the company could reduce the risks it faces.

RQ5. Consider the reasons that programs such as Java applets that run on client machines are considered security threats. In about 200 words, explain how these programs could breach security and compare the security risks posed by JavaScript programs to the risks posed by Java applets

RQ6. Write a 200-word description of computer forensics in general and ethical hacking in particular. In your essay, describe at least one real situation in which computer forensic experts or ethical hackers used their talents to help a company overcome a security weakness.

Exercises

E1. Brought Back Bugs is a used Volkswagen dealer in Lincoln, Nebraska. The dealership hired you to update its Web site. One of the requirements is that the site must display a few banner advertisements showing the week's specials. You decide that active content would be the best technology to automate the placement and rotation of the advertisement. You are also considering using active content to make the content of the banner ad more interesting. You decide to investigate Java, JavaScript, Jscript, and Java applets as alternatives. Use the Online Companion and Web search engines to learn more about these alternatives and write a 300-word summary that describes each and evaluates its use for automating the rotation and placement of banner ads on the Brought Back Bugs Web site.

E2. You are the administrator of a Web server for an electronic commerce site. The site receives about 12,000 visitors per day, maintains a product catalog of about 4000 items, and processes about 2000 sales per day. The average sale amount is $87. The site accepts four major credit cards and it outsources its payment processing for all of the credit cards to another company. In about 200 words, describe the types of threats that could be launched against your Web server, given the types of activity (catalog presentation, order entry, and payment processing) it handles and the volume of those activities. Consult sources on the Internet or in your library to help you complete this exercise.

E3. Write a 300-word paper in which you evaluate the CERT organization. Include information about when it was founded, what groups or people are members, and where it is headquartered. Include in your discussion at least three current security alerts, specifying the name of the virus or attack program, the date the alert was posted, and two sentences about each reported security alert. Use Internet search engines and the CERT Web site to help you locate information.

E4. Your colleague, Andrea Flemington, is responsible for security of your company's electronic commerce Web site. Her manager has sent her, rather unexpectedly, to one of the company's divisions in Sydney. Before leaving, she left a note on your desk asking you to help her with a brief analysis of other electronic commerce sites. In her note, she requested that you conduct a survey of 10 Web sites that employ the Secure Sockets Layer (SSL) protocol to protect transactions. Andrea indicated that you could select six sites yourself, but she wanted you to be sure to include these sites in the survey: www.amazon.com, www.ebay.com, www.fedex.com, and www.gateway.com. Print out the query results from each of the 10 sites for reference as you write a report about your research. Create a table containing 11 rows and five columns. In the leftmost column of the table, list the name (or abbreviation) of each company. Place the following labels in the first row of the table: Company Name, HTTPS Server, SSL Ciphers, Validity Period, and CA's Name. Fill in each row with the appropriate information. Here is an example of the table:

Company Name	HTTPS Server	SSL Ciphers	Validity Period	CA's Name
Sun Microsystems	Netscape Enterprise 3.0	RC4 with MD5 RC2 with MD5 DES with MD5 Triple DES	January 1, 2004 to January 1, 2005	VeriSign

Cases

C1. Bibliofind

Bibliofind was founded in 1996 as one of the first Web sites to specialize in hard-to-find and collectible books. The site featured a powerful search engine for used and rare books. The search engine's database was populated with the results of Bibliofind's daily surveys of a worldwide network of suppliers. Registered site visitors could specify the title for which they were searching, a price range, and whether they were seeking a first edition. The site also allowed visitors to build a wish list that would trigger an e-mail when a specific book on the list became available.

Bibliofind had developed a large customer list, an excellent reputation, and a solid network of rare book dealers, all of which made the company an attractive acquisition for other online bookstores. In 1999, Amazon.com bought Bibliofind, but the company continued to operate its own Web site and conduct its business as it had before the acquisition.

In 2001, Bibliofind's Web site was hacked. The cracker had gained access to the company's Web server and replaced the company's Web pages with defaced versions. Bibliofind shut down its Web site for several days and undertook a complete review of its Web site's security. When the company's IT staff examined the server logs carefully, they found that the Web page hacking was only the tip of the iceberg. Entries in the logs showed that attackers had been accessing Bibliofind's computers for more than four months. Even worse, some of the crackers had been able to go through the Web servers to gain access to the computers that held Bibliofind customer information, including names, addresses, and credit card numbers. That information had been stored in plain text files on Bibliofind's transaction servers.

Bibliofind called in state and federal law enforcement officials to investigate the hacking incidents and sent an e-mail notification to the 98,000 customers whose private information might have been obtained by the crackers. The investigation did not result in any arrests, nor did it determine the identity of the intruders. Many of Bibliofind's customers were very upset when they learned what had happened.

A month after the hacking incident, Amazon.com moved Bibliofind into its zShops online mall. As an Amazon zShop, Bibliofind could process its transactions through Amazon's system and no longer needed to maintain private information about its customers on its computers. Eventually, Bibliofind was closed down. A successful business had been seriously damaged because it failed to maintain adequate security over the customer information it had gathered.

Required:

1. In about 300 words, explain how Bibliofind might have used firewalls to prevent the intruders from gaining access to its transaction servers. Be specific about where the firewalls should have been placed in the network and what kinds of rules they should have used to filter network traffic at each point.

2. In about 200 words, explain how encryption might have helped prevent or lessen the effects of Bibliofind's security breach.

3. In 2003, the State of California enacted a law that requires companies to inform customers whose private information might have been exposed during a security breach like the one that Bibliofind experienced. While the legislation was being debated, businesses argued that the law would encourage nuisance lawsuits. In about 300 words, present arguments for and against this type of legislation.

Note: Your instructor might assign you to a group to complete this case, and might ask you to prepare a formal presentation of your results to your class.

C2. Wilderness Trailhead

Wilderness Trailhead, Inc. (WTI) is a retailer that offers hiking, rock-climbing, and survival gear for sale on its Web site. WTI targets the serious outdoor enthusiast and offers high-quality equipment at competitive prices. The company has been in business on the Web for five years. It has grown rapidly and has been profitable since its first year of operations. WTI offers about 1200 different items for sale and has about 1000 visitors per day at its Web site. Because the company's offerings are specialized and high quality, its average transaction size is much higher than other outdoor equipment stores. WTI makes about 200 sales each day on its site, with an average transaction value of $372.

WTI sells products primarily through its Web site (it does have a small retail outlet store for discontinued items in Bellingham, Washington) to customers in the United States and Canada. WTI ships orders from its two warehouses—one in Vancouver, British Columbia, and a second in Shoreline, Washington.

WTI accepts four major credit cards and processes its own credit card transactions. It stores records of all transactions on a database server that shares a small room with the Web server computer at WTI's main offices in a small industrial park just outside of Bellingham. Harry Bogdosian, the manager of IT for WTI, has become increasingly concerned about the security of the company's Web and database servers as the company has grown.

Required:

1. WTI faces certain risks that arise from its storage of customer credit card numbers on its database server. List at least four specific threats to the database server's security, and identify defenses, deterrents, or countermeasures that might reduce or eliminate the potential damage that could be caused by those threats.

2. Write a security policy for the operation of the WTI database server. Be sure to consider the threats that exist because that server stores customer credit card numbers. You can use the links included in the Online Companion under the heading "Computer Security Policy Resources" to help you as you write this policy.

3. WTI is considering moving its existing Web and database server computers to a CSP in a co-location arrangement. Prepare a two-page outline of the security features that WTI should ask a CSP to provide as part of this co-location service.

Note: Your instructor might assign you to a group to complete this case, and might ask you to prepare a formal presentation of your results to your class.

C3. Materials Equipment

You are an information technology (IT) consultant to Materials Equipment, Inc. (MEI), a major industrial equipment distributor. Its products include materials-handling machinery for assembly lines and product-packaging areas, hydraulic equipment (for moving fluids), hoses, hose fittings, and similar items. MEI has been in business for more than 70 years and sells more than $200 million worth of parts and equipment each year to its 3000 customers. MEI's customers are located all over the world, but most are in the United States, Mexico, Malaysia, China, and Singapore.

Joe Andrejewski, MEI's director of sales, has retained you to help him with a new marketing idea. He has read about other companies that have created Web portal sites for customers, and he is interested in developing a portal site that MEI could operate with three other companies that sell products (such as bearings, seals, hoses, and hose fittings) and services (design, layout, and installation of materials-handling equipment) that are complementary to MEI products. The portal would provide MEI customers with a Web site at which they could buy MEI products, buy the products and services of the three MEI strategic partners, and obtain information about current trends in industrial equipment technologies and the application of those technologies. The portal site would also include a used equipment area in which MEI customers could list equipment for sale. Joe believes that giving customers a convenient way to liquidate old equipment will make it easier for his sales representatives to sell new equipment to those customers.

Joe has put together an internal team to examine the feasibility of the portal site, including key employees from MEI's Sales, Finance, Product Engineering, and IT Services departments. The team has identified several security issues that they want to resolve before they take the portal idea much further. Joe would like you to help the team understand two security technologies—digital certificates and encryption—and how these techniques might be used in MEI's proposed portal site.

Required:

1. Prepare two briefing reports of about 700 words each for the MEI portal team—one about digital certificates and one about encryption. Each report should explain the technology and describe one or two common applications.

2. Assume that the MEI portal project is approved and implemented. Further assume that MEI has decided to require each customer that participates in the portal to obtain a digital certificate. Write a memo of about 500 words addressed to potential participants (MEI customers) in which you explain why they must obtain a digital certificate as a condition of participation.

Note: Your instructor might assign you to a group to complete this case, and might ask you to prepare a formal presentation of your results to your class.

For Further Study and Research

Alexander, S. 2000. "Viruses, Worms, Trojan Horses and Zombies," *Computerworld*, 34(18), May 1, 74.

Anderson, R. and F. Petitcolas. 1998. "On the Limits of Steganography," *IEEE Journal of Selected Areas in Communications*, 16(4), May, 474–481.

Austin, R. and C. Darby. 2003. "The Myth of Secure Computing," *Harvard Business Review*, 81(6), June, 120–126.

Betts, M. 2000. "Digital Signatures Law to Speed Online B-to-B Deals," *Computerworld*, 34(26), June 26, 8.

Butler, R. and A. Goldstein. 2001. "Keeping the Hackers at Bay," *Time*, 158(23), November 26, 87.

Cohen, A. 2001. "When Terror Hides Online," *Time*, November 12, 65.

Colkin, E., A. Gilbert, G. Hulme, M. McGee, and J. Rendleman. 2001. "IT Security and the Law," *Information Week*, November 26, 22–24.

Connell, S. 2004. "Security Lapses, Lost Equipment Expose Students to Possible ID Theft Loss," *The Los Angeles Times*, August 29, B4

Costanzo, C. 2003. "Dealing with Phishing and Spoofing," *American Banker*, 168(184), September 24, 10.

Dacey, R. 2001. *Information Security: IRS Electronic Filing Systems* (GAO-01-306). Washington, D.C.: U.S. General Accounting Office.

Delio, M. 2001. "SirCam: Devious, But Not Sinister," *Wired News*, July 25. (http://www.wired.com/news/ technology/0,1282,45506-2,00.html)

DeMaria, M. 2001. "Symantec Firewall/VPN Devices Secure the Small Office at the Right Price," *Network Computing*, 12(24), November 26, 30–31.

DoD Directive 5215.1 CSC-STD-001-83. 1983. *Department of Defense Trusted Computer System Evaluation Criteria* (the "Orange Book"), Washington, D.C.

Dyck, T. 2001. "How to Shore Up IT Resources to Prevent Attacks in the First Place," *eWeek*, 18(44), November 12, 65–66.

Erlanger, L. 2002. "Defensive Strategies," *PC Magazine*, 21(19), November 5, 70–72.

Evers, J. 2001. "Hackers Get Credit Card Data from Amazon's Bibliofind," *PC World*, March 6. (http://www.pcworld.com/news/article/0,aid,43582,00.asp)

Fratto, M. 2001. "Buyers Guide: Enterprise Firewalls," *Network Computing*, 12(25), December 10, 90–92.

Gallagher, S. 2002. "Best Buy: May Day Mayday for Security," *Baseline*, June 7. (http://www.baselinemag.com/article2/0,3959,687,00.asp)

Gardner, E. 2000. "ZixIt Corp.," *Internet World*, July 1, 28.

Garfinkel, S. and G. Spafford. 2002. *Web Security, Privacy, & Commerce.* Cambridge, MA: O'Reilly.

Glass, B. and D. Fisher. 2004. "Biometrics Security," *PC Magazine*, 23(1), January 20, 66.

Gurley, J. 2001. "From Wired To Wiretapped," *Fortune*, 144(7), October 15, 214–215.

Hancock, B. 2001. "Terrorism and Steganography: Shaken, Not Stirred," *Computers & Security*, 20(2) 110–111.

Harrison, A. 2000. "Advanced Encryption Standard," Computerworld, 34(22), May 29, 57.

Hayes, F. 2002. "Thanks, Warchalkers," *Computerworld*, 36(35), August 26, 56.

Information Technology Association of America (ITAA). 2000. *Intellectual Property Protection in Cyberspace*. Arlington, VA: ITAA.

Katzenheisser, S. and F. Petitcolas (eds.). 1999. *Information Hiding Techniques for Steganography and Digital Watermarking*, Norwood. MA: Artech House.

Krim, J. 2003. "WiFi Is Open, Free and Vulnerable to Hackers: Safeguarding Wireless Networks Too Much Trouble for Many Users," *The Washington Post*, July 27, A1.

Krim, J. 2003. "Microsoft Critic Forced Out, Firm Does Business With Software Giant," *The Washington Post*, September 26, E1.

Kuchinskas, S. 2003. "Lack of Trust Could Impact E-Commerce Sales," *E-Commerce Guide*, December 3. (http://ecommerce.internet.com/news/news/article/0,10375_3115741,00.html)

Kutter, M. 1998. "Watermarking Resistance to Translation, Rotation, and Scaling," *Proceedings of SPIE: Multimedia Systems and Applications*, Vol. 3528, November 1–6, 423–431.

Lohmeyer, D., J. McCrory, and S. Pogreb. 2002. "Managing Information Security," *The McKinsey Quarterly*, June, 12–15.

Mackey, D. 2003. *Web Security for Network and System Administrators*. Boston: Course Technology.

Manes, S. 2001. "Security, Microsoft Style: No Safety Net?" *PC World*, 19(11), November, 210.

Maney, K. 2001. "Osama's Messages Could Be Hiding in Plain Sight," *USA Today*, December 19, 6B.

McCracken, H. 2004. "Microsoft's Security Problem—and Ours," *PC World*, 22(1), January, 25.

McCullagh, D. 2001. "'Secure' U.S. Site Wasn't Very," *Wired News*, July 6. (http://www.wired.com/news/privacy/0,1848,45031,00.html)

Merkow, M., J. Breithaupt, and K. Wheeler. 1998. *Building SET Applications for Secure Transactions*. New York: John Wiley & Sons.

Nerney, C. 2003. "Get It Right, Redmond," *Internet News*, May 12. (http://www.internetnews.com/commentary/ article.php/2205081)

Neuman, B. and T. Tso. 1994. "An Authentication Service for Computer Networks," *IEEE Communications*, 32(9), September, 33–38.

Nielsen, J. 2004. "User Education Is Not the Answer to Security Problems," *Alertbox*, October 25. (http://www.useit.com/alertbox/20041025.html)

Null, C. 2000. "Name Grab," *PC Computing*, 13(4), April, 40–42.

O'Harrow, R. 2001. "Prozac Maker Reveals Patient E-Mail Addresses," *The Washington Post*, July 4, E1.

Oppliger, R. 1997. "Internet Security: Firewalls and Beyond," *Communications of the ACM*, 40(5), May, 92–102.

Palmer, C. 2001. "Ethical Hacking," *IBM Systems Journal*, 40(3), 769–780.

Petreley, N. 2001. "The Cost of Free IIS," *Computerworld*, 35(43), October 22, 49.

Piazza, P. 2003. "Phishing for Trouble," *Security Management*, 47(12), December, 32–33.

Pleas, K. 1999. "Certificates, Keys, and Security," *PC Magazine*, 18(8), April 20, 227–230.

Popovich, K. 2000. "Biometrics Gets Thumbs Up," *eWeek*, 17(24), June 12, 37.

Powell, T. 2004. "Quick Tips for Web Application Security," *Network World*, 21(20), May 17, 50–51.

Regan, K. 2001. "Hack Victim Bibliofind to Move to Amazon," *E-Commerce Times*, April 6. (http://www.ecommercetimes.com/story/8768.html)

Rivest, R. 1992. *The MD5 Message-Digest Algorithm*, IETF RFC 1321.

Rosencrance, L. 2004. "Federal Audit Raises Doubts About IRS Security System," *Computerworld*, 38(36), September 6, 9.

Rothstein, P. 2001. "Disaster Recovery: September 11 Changes Everything," *Information Security*, 4(11), 48–49.

Rubenking, N. 2002. "Securing Web Services," *PC Magazine*, 21(17), October 1, IP01–04.

Rutrell, Y. 2001. "So Many Patches, So Little Time," *InternetWeek*, October 8, 1–2.

Saita, A. 2001. "Deep Digital Cover," *Information Security*, 4(10), October, 22.

Schwartz, J. 2002. "13,000 Credit Reports Stolen by Hacker," *The New York Times*, May 17. (http://www.nytimes.com/2002/05/17/technology/17IDEN.html)

Security Management, 2002. "Government Infosec Gets Failing Grade,"46(2), February, 34–35.

Shipley, G. 2001. "Growing Up with a Little Help from the Worm," *Network Computing*, 12(20), October 1, 39.

Shively, G. 2002. *Network InSecurity*. Newport Beach, CA: PivX Solutions.

Skoudis, E. 2002. "Infosec's Worst Nightmares: The Five Past Attacks That Haunt Us, the Five Fears That Trouble Us," *Information Security*, November. (http://www.infosecuritymag.com/2002/nov/nightmares.shtml)

Skoudis, E. 2005. "Five Malicious Code Myths and How To Protect Yourself in 2005," *SearchSecurity.com*, January 4. (http://searchsecurity.techtarget.com/tip/1,289483,sid14_gci1041736,00.html)

Sterling, B. 2001. "Steganography Goes Digital," *The New York Times*, December 9, 102.

Thompson, H. and J. Whittaker. 2002. "Testing for Software Security," *Dr. Dobb's Journal*, November, 24–32.

Tippett, P. 2001. "The Crypto Myth," *Information Security*, 4(5), May, 38–40.

Trombly, M. 2001. "Wall Street Firms Look Externally for Web Security," *Securities Industry News*, 13(33), August 20, 3–4.

U.S. National Institute of Standards and Technology. 1993. *Data Encryption Standard (DES): Federal Information Processing Standards Publication 46–2*. Gaithersburg, MD: U.S. Computer Systems Laboratory.

Verton, D. 2001. "Microsoft in Hot Seat After Code Red," *Computerworld*, 35(32), August 6, 1–2.

Verton, D. 2002. "Fixes Named Along With Top 20 Holes," *Computerworld*, 36(41), October 7, 1–2.

Verton, D. 2002. "Mapping of Wireless Networks Could Pose Enterprise Risk," *Computerworld*, August 14. (http://computerworld.com/securitytopics/security/story/0,10801,73479,00.html)

Vijayan, J. 2001. "Corporations Left Hanging as Security Outsourcer Shuts Doors," *Computerworld*, 35(18), April 30, 13.

Vijayan, J. 2000. "Possible S & P Security Holes Reveal Risks of E-Commerce," *Computerworld*, 34(22), May 29, 6.

Vijayan, J. 2000. "Analysts: Better to Be Safe Than Sorry With Viruses," *Computerworld*, 34(26), June 26, 20.

Weber, T. 2001. "Why You Need to Install A Firewall—And Why You Shouldn't Have To," *The Wall Street Journal*, July 30, B1.

Yeh, W-H. and J-J. Hwang. 2001. "Hiding Digital Information Using a Novel System Scheme," *Computers & Security*, 20(6), 533–538.

Zetter, K. 2001. "Holey Software!" *PC World*, 19(11), November, 135–140.

CHAPTER **11**

PAYMENT SYSTEMS FOR ELECTRONIC COMMERCE

LEARNING OBJECTIVES

In this chapter, you will learn about:

- The basic functions of online payment systems
- The use of payment cards in electronic commerce
- The history and future of electronic cash
- How electronic wallets work
- The use of stored-value cards in electronic commerce
- Internet technologies and the banking industry

INTRODUCTION

In 1991, a teenager named Max Levchin immigrated from the Ukraine to the United States. Settling in Chicago, Levchin had a burning interest in cryptography. Growing up in a Soviet police state convinced him that the ability to send coded messages that could not be read or intercepted was both important and useful. He majored in computer science at the University of Illinois and spent many hours at the school's **Center for Supercomputing**, pursuing his passion for making and breaking codes. When he graduated in 1998, he wanted to follow the American dream of turning his knowledge into money, so he headed for the heart of the computer industry in Palo Alto, California. Levchin's plan to build the ultimate transmission encryption scheme has not yet panned out, but he has managed to turn his knowledge into a successful business. As cofounder and chief technical officer of **PayPal**, an online payment processing company that you will learn about in this chapter, Levchin has used his expertise in cryptography and computer security to protect the firm from losses that could destroy it.

PayPal, founded in 1999, operates a service that lets people exchange money over the Internet. It has become the most used payment system for clearing auction transactions on eBay. People can also use PayPal to send money to anyone who has an e-mail address and to receive money.

PayPal charges very small fees to business users and no fees at all to individuals, so its profit margins are small. However, it has grown so rapidly that its thin profit margins are realized on a very large number of users. The company has been able to raise $211 million in funding (almost half of that amount after the dot-com bubble burst) and its prospects look good. A single, well-organized, large-scale fraud attack on PayPal, however, could put the company out of business quickly. Levchin's current contribution to the company's success is his development of payment surveillance software that continually monitors PayPal transactions. The software searches millions of transactions as they occur every day and looks for patterns that might indicate fraud. The software notifies PayPal managers immediately when it finds something suspicious.

The software appears to be working very well. Companies that process credit card transactions have experienced much larger fraud occurrence rates on the Web (about 1.13 percent) than in physical stores (about .70 percent). PayPal claims to have kept its fraud rate below .50 percent. As long as PayPal can keep its fraud rate low, it can continue to charge lower transaction fees than its competitors and still make a profit. Some industry observers believe that PayPal's ability to avoid high fraud rates could make it a serious competitor to banks in other areas of financial transaction handling, such as credit card processing.

PayPal's largest customer group has always been the participants (buyers and sellers) on the auction Web site eBay. As you will learn in this chapter, eBay spent three years working to establish its own payments service that could compete effectively with PayPal. In October 2002, eBay finally gave up and bought PayPal for $1.4 billion. PayPal continues to offer payment services under its own name as a division of eBay.

ONLINE PAYMENT BASICS

An important function of electronic commerce sites is the handling of payments over the Internet. Most electronic commerce involves the exchange of some form of money for goods or services. As you learned in Chapter 5, many companies use electronic funds transfers (EFTs) or financial EDI to make online payments. In this chapter, you will learn about a number of online payment alternatives that are available to individual consumers.

Online payment systems for consumer electronic commerce are still evolving. A number of proposals and implementations of payment systems currently compete for dominance. Regardless of format, electronic payments are far cheaper than mailing paper checks. Electronic payments can be convenient for customers and can save companies money. Estimates of the cost of billing one person by mail range between $1 and $1.50. Sending bills and receiving payments over the Internet can drop the transaction cost to an average of 50 cents per bill. The total savings is huge when the unit cost is multiplied by the number of customers who could use electronic payment. For example, a telephone company in a major metropolitan area might have 5 million customers, each of whom receives a bill every month. In one year, a savings of 50 cents on each of those 60 million bills adds up to about $30 million. The environmental impact is also significant. Those 60 million paper bills weigh about 1.7 million pounds. It takes 2200 trees to make that much paper—along with the energy consumed and the wastes generated in the paper-making process.

Today, four basic ways to pay for purchases dominate both traditional and electronic business-to-consumer commerce. Cash, checks, credit cards, and debit cards account for more than 90 percent of all consumer payments in the United States. A small but growing percentage of consumer payments are made by electronic transfer. The most popular consumer electronic transfers are automated payments of auto loans, insurance payments, and mortgage payments made from consumers' checking accounts. Figure 11-1 shows the estimated proportions of the $6.7 trillion in payments projected for 2005 in the United States for all types of consumer commerce, online and offline.

Credit cards are by far the most popular method that consumers use to pay for online purchases. Recent surveys have found that more than 85 percent of worldwide consumer Internet purchases are paid for with credit cards. In the United States, the proportion is about 96 percent.

Another payment medium is limited-use scrip. **Scrip** is digital cash minted by a company instead of by a government. Most scrip cannot be exchanged for cash; it must be

Type	Number of transactions	Dollar value of transactions
Cash	35%	15%
Checks	21%	32%
Credit cards	19%	26%
Debit cards	17%	12%
Electronic transfers	5%	11%
Other	3%	4%

Adapted from Table 1182, *2004-2005 Statistical Abstract of the United States*, Washington, D.C.: U.S. Census Bureau, p. 746.

FIGURE 11-1 Payment methods for all types of U.S. consumer transactions, 2005 projections

exchanged for goods or services by the company that issued the scrip. Scrip is like a gift certificate that is good at more than one store. In the early days of the Web, many experts predicted that scrip would become a popular way of making payments for consumer goods and services online. Unfortunately for many investors and at least two companies (see the Learning from Failures feature), this turned out not to be true. Most current scrip offerings, such as **eScrip**, focus on the not-for-profit fundraising market. This market consists mainly of primary and secondary schools in the United States.

LEARNING FROM FAILURES

Flooz and Beenz

Flooz and Beenz were two pioneers in the business of issuing scrip for use on the Web. The scrip created by these two companies could be bought, traded, and exchanged for merchandise, or discounts on merchandise, at hundreds of Web retailers.

In 1998, Beenz began offering its scrip for sale on its Web site. The scrip was called beenz, and the company's logo included a small kidney bean shape. A number of merchants agreed to accept the beenz scrip and by mid-2000, Beenz had more than a million customers who were accumulating and using beenz to buy merchandise on the Web. Beenz formed a partnership with Columbus Bank and Trust that allowed beenz holders to transfer their beenz value to a debit card that they could use in the physical world.

Flooz began selling its scrip product, flooz, in late 1999. Flooz had overwhelming support from major partners, such as NextCard, and quickly signed an agreement with BarnesandNoble.com in which the bookseller would accept flooz scrip for purchases on its Web site. Flooz undertook major promotional activities, including an $8 million advertising campaign featuring Whoopi Goldberg.

By August 2001, both companies had ceased operations. The idea of using scrip was novel and it did give parents a way to allow their children to make purchases on the Web. However, scrip did not solve any major problems for most online buyers and it required that they learn a new and different technique for making Web payments. Another major barrier to adoption was that neither product meshed very well with existing payment systems.

continued

The lesson from the Flooz and Beenz failures is that any Web product or service must meet a real need of consumers, and it must not require those consumers to learn a new way to do something that they are already comfortable doing. The new product or service must also integrate well with existing systems and practices.

Merchants should offer their customers payment options that are safe, convenient, and widely accepted. The key is to determine which choices work the best for the company and its customers. The information in this chapter will help you make those decisions. Companies such as **Payment Online**, shown in Figure 11-2, sell packages of payment processing services to Web merchants that allow those merchants to accept several different types of payments.

FIGURE 11-2   Payment processing service offerings of Payment Online

You will learn about four different payment technologies in this chapter: payment cards, electronic cash, software wallets, and smart cards (also called stored-value cards). Each technology has unique properties, costs, advantages, and disadvantages. Some are methods that are already popular and widely accepted; others are only beginning to catch on and have an unclear future. All of these electronic payment methods can work well for B2C Web commerce sites.

PAYMENT CARDS

Businesspeople often use the term **payment card** as a general term to describe all types of plastic cards that consumers (and some businesses) use to make purchases. The main categories of payment cards are credit cards, debit cards, and charge cards.

A **credit card**, such as a **Visa** or a **MasterCard**, has a spending limit based on the user's credit history; a user can pay off the entire credit card balance or pay a minimum amount each billing period. Credit card issuers charge interest on any unpaid balance. Many consumers already have credit cards, or are at least familiar with how they work. Credit cards are widely accepted by merchants around the world and provide assurances for both the consumer and the merchant. A consumer is protected by an automatic 30-day period in which he or she can dispute an online credit card purchase. Merchants that already accept credit cards in an offline store can accept them immediately for online payment because they already have established a mechanism for accepting credit card payments. Online credit card purchases are similar to telephone purchases in that the card holder is not present and cannot provide proof of identity as easily as he or she can when standing at the cash register. Online and telephone purchases are often called **card not present transactions** and both require an extra degree of security.

A debit card looks like a credit card, but it works quite differently. Instead of charging purchases against a credit line, a **debit card** removes the amount of the sale from the cardholder's bank account and transfers it to the seller's bank account. Debit cards are issued by the cardholder's bank and usually carry the name of a major credit card issuer, such as Visa or MasterCard, by agreement between the issuing bank and the credit card issuer. By branding their debit cards (with the Visa or MasterCard name), banks ensure that their debit cards will be accepted by merchants who recognize the credit card brand names.

A **charge card**, such as one from **American Express** or **Diner's Club**, carries no spending limit, and the entire amount charged to the card is due at the end of the billing period. Charge cards do not involve lines of credit and do not accumulate interest charges. In the United States, many retailers, such as department stores and oil companies that own gas stations, issue their own charge cards. In the rest of this chapter, the term "payment card" refers to credit cards, debit cards, and charge cards.

Many consumers have concerns about providing their payment card numbers to vendors online, especially when the vendor is unknown to them. To address this concern, several payment card companies now offer cards with disposable numbers. These cards, sometimes called **single-use cards**, give consumers a unique card number that is valid for one transaction only. This prevents an unscrupulous vendor from using the card number to complete unauthorized transactions on the consumer's account or selling the card number to others. In 2000, American Express was the first to offer single-use cards. A few other card issuers followed suit, but the number of companies that offer single-use cards continues to be small. Neither Visa nor MasterCard have required all of their issuing banks to provide single-use cards; the only major issuing banks to do so are MBNA and Citigroup. J.P. Morgan offers a single-use version of its Discover card. In 2004, American Express

494

stopped offering its single-use card, but many industry analysts believe that consumer interest in these types of cards will continue to grow. The problem with single-use cards thus far has been that they require consumers to behave differently and not enough consumers see the benefit of learning how to use this new product. As concerns over stolen credit card numbers increase, this benefit could become compelling.

Advantages and Disadvantages of Payment Cards

Payment cards have several features that make them an attractive and popular choice with both consumers and merchants in online and offline transactions. For merchants, payment cards provide fraud protection. When a merchant accepts payment cards for online payment or for orders placed over the telephone, the merchant can authenticate and authorize purchases using a payment card processing network. For U.S. consumers, payment cards are advantageous because the Consumer Credit Protection Act limits the cardholder's liability to $50 if the card is used fraudulently. Once the cardholder notifies the card's issuer of the card theft, the cardholder's liability ends. Frequently, the payment card's issuer waives the $50 consumer liability when a stolen card is used to purchase goods.

Perhaps the greatest advantage of using payment cards is their worldwide acceptance. Payment cards can be used anywhere in the world, and the currency conversion, if needed, is handled by the card issuer. For online transactions, payment cards are particularly advantageous. When a consumer reaches the electronic checkout, he or she enters the payment card number and his or her shipping and billing information in the appropriate fields to complete the transaction. The consumer does not need any special hardware or software to complete the transaction.

Payment cards have one significant disadvantage for merchants when compared to cash. Payment card service companies charge merchants per-transaction fees and monthly processing fees. These fees can add up, but merchants view them as a cost of doing business. Any merchant that does not accept payment cards for purchases risks losing a significant portion of sales to other merchants that do accept payment cards. The consumer pays no direct transaction-based fees for using payment cards, but the prices of goods and services are slightly higher than they would be in an environment free of payment cards. Most consumers also pay an annual fee for credit cards and charge cards. This annual fee is much less common on debit cards.

Payment cards provide built-in security for merchants because merchants have a higher assurance that they will be paid through the companies that issue payment cards than through the sometimes slow direct invoicing process. To process payment card transactions, a merchant must first set up a merchant account. The series of steps in a payment card transaction is usually transparent to the consumer. Several groups and individuals are involved: the merchant, the merchant's bank, the customer, the customer's bank, and the company that issued the customer's payment card. All of these entities must work together for customer charges to be credited to merchant accounts (and vice versa when a customer receives a payment card credit for returned goods).

Payment Acceptance and Processing

Most people are familiar with the use of payment cards: In a physical store, the customer or a sales clerk runs the card through the online payment card terminal and the card account is charged immediately. The process is slightly different on the Internet, although the purchase and charge processes follow the same rules. Payment card processing has been made easier over the past two decades because Visa and MasterCard, along with MasterCard's international affiliate, **MasterCard International** (formerly known as Europay), have implemented a single standard for the handling of payment card transactions called the **EMV standard** (EMV is derived from the names of the companies: Europay, MasterCard, and Visa).

In a brick-and-mortar store, customers walk out of the store with purchases in their possession, so charging and shipment occur nearly simultaneously. Online stores and mail order stores in the United States must ship merchandise within 30 days of charging a payment card. Because the penalties for violating this law can be significant, most online and mail order merchants do not charge payment card accounts until they ship merchandise. Payment card transactions follow these general steps once the merchant receives a consumer's payment card information, which is usually sent using the SSL encryption technique you learned about in Chapter 10:

1. The merchant authenticates the payment card to ensure it is valid and not stolen.
2. The merchant checks with the payment card issuer to ensure that credit or funds are available and puts a hold on the credit line or the funds needed to cover the charge.
3. Settlement occurs, usually a few days after the purchase, which means that funds travel between banks and are placed into the merchant's account.

Open and Closed Loop Systems

In some payment card systems, the card issuer pays the merchants that accept the card directly and does not use an intermediary, such as a bank or clearinghouse system. These types of arrangements are called **closed loop systems** because no other institution is involved in the transaction. American Express and Discover Card are examples of closed loop systems.

Open loop systems involve three or more parties. Suppose an Internet shopper uses his or her Visa card issued by the First Bank of Woodland to purchase an item from Web Wonders, whose bank account is at the Hackensack Commerce Bank. The banking system includes one or more intermediary banks that coordinate the transfer of funds from the First Bank of Woodland to the Hackensack Commerce Bank. Whenever a third party, such as the intermediary banks in this example, processes a transaction, the system is called an **open loop system**. Systems using Visa or MasterCard are the most visible examples of open loop systems. Many banks issue both cards. Unlike American Express or Discover, neither Visa nor MasterCard issues cards directly to consumers. Visa and MasterCard are **credit card associations** that are operated by the banks who are members in the associations. These member banks, which are also called **customer issuing banks**, issue credit cards to individual consumers and are responsible for establishing customer credit limits.

Merchant Accounts

A **merchant bank** or **acquiring bank** is a bank that does business with sellers (both Internet and non-Internet) that want to accept payment cards. In other words, to process payment cards for Internet transactions, an online merchant must set up a **merchant account**. When the merchant's bank collects credit card receipts on behalf of the merchant from the payment card issuer, it credits their value to the merchant's account.

A merchant must provide business information before the bank will provide an account through which the merchant can process payment card transactions. Typically, a new merchant must supply a business plan, details about existing bank accounts, and a business and personal credit history. The merchant bank wants to be sure that the merchant has a good prospect of staying in business and wants to minimize its risk. An online merchant that appears disorganized is less attractive to a merchant bank than a well-organized online merchant.

The type of business also influences the bank's likelihood of granting the account. In some industries, merchant banks will be reluctant to offer a merchant account because of the type of business; some businesses have a higher likelihood of customers repudiating payment card charges than others. For example, a business that sells a guaranteed weight loss scheme—a business in which many customers might want their money back—will find many merchant banks unwilling to provide an account. The bank assesses the level of risk in the business based on the type of business and the credit information that is provided. Merchant banks must estimate what percentage of sales are likely to be contested by cardholders. When a cardholder successfully contests a charge, the merchant bank must retrieve the money it placed in the merchant account in a process called a **chargeback**. To ensure that sufficient funds are available to cover chargebacks, a merchant bank might require a company to maintain funds on deposit in the merchant account. For example, a new or risky business that plans to make $100,000 in sales each month might be required to keep $50,000 or more on deposit in its merchant account.

One problem facing online businesses is that the level of fraud in online transactions is much higher than either in-person or telephone transactions of the same nature (that is, the same amount and the same type of good or service being purchased). Fewer than 5 percent of all credit card transactions are completed online, but those transactions are responsible for about 50 percent of the total dollar amount of credit card fraud. A Celent Communications study reported in *Credit Card Management* (see the reference in the For Further Study and Research section at the end of this chapter) has projected that online credit card fraud will be over $2 billion by 2007 and will amount to 62 percent of all credit card fraud.

Several third-party Internet and Web-based services are available to handle all the details of processing payment card transactions. The next section discusses payment card processing options for Internet stores.

Processing Payment Cards Online

Software packaged with electronic commerce software can handle payment card processing automatically, or merchants can contract with a third party to handle payment card processing. Several companies, called **payment processing service providers**, offer these services. **InternetSecure**, for example, allows merchants to concentrate on business

497

while it provides secure payment card services. InternetSecure supports payments with Visa and MasterCard for Canadian and United States accounts. The company provides risk management and fraud detection and handles transactions from online merchants using existing, bank-approved payment card processing infrastructure, secure links, and firewalls. InternetSecure notifies the merchant of all approved orders and also supplies authorization codes to buyers of digital content, who can download their purchases upon payment card approval. InternetSecure ensures that the transactions it processes are credited to the correct merchant's account.

First Data provides merchant payment card processing services with the **ICVERIFY**, **PCAuthorize**, and **WebAuthorize** programs. ICVERIFY is intended for small retailers that use Microsoft Windows electronic cash registers and point-of-sale terminal systems. PCAuthorize is intended for smaller online stores (unlike ICVERIFY, it does not require any point-of-sale hardware), and WebAuthorize is for large, enterprise-class merchant sites. PCAuthorize automatically accepts merchant forms containing consumers' payment card information, captures the data, and deposits the amount in the merchant's payment card account. Once the merchant receives payment from a customer through a Web page form, the merchant passes on the Web page with customer payment card information to the processing bank.

Services such as ICVERIFY, PCAuthorize, and WebAuthorize connect directly to a network of banks called the **Automated Clearing House (ACH)** and to credit card authorization companies. You can learn more about ACHs by following the Online Companion links to the **Electronic Payments Network**, **NACHA - The Electronic Payments Association**, **The Clearing House**, and the U.S. Federal Reserve Bank's **FedACH** site. Banks connect to an ACH through highly secure, private leased telephone lines. The merchant sends the card information to a payment card authorization company, which reviews the customer account and, if it approves the transaction, sends the credit authorization to the issuing bank. Then the issuing bank deposits the money in the merchant's bank account through the ACH. The merchant's Web site receives confirmation of the acceptance of the consumer transaction. After receiving notification of acceptance or rejection of the transaction, the merchant Web site confirms the sale to the customer over the Internet. In addition, the merchant site usually sends an e-mail confirmation of the sale to the consumer with details about the purchase price and shipping information. Figure 11-3 is a graphic representation of the process.

Other payment card processing companies include VeriSign's **PayFlow Link** system and InfoSpace's **Authorize.Net**. PayFlow is an online payment system developed by Cyber-Cash that is now operated by VeriSign. Authorize.Net is an online, real-time payment card processing service that allows merchants to link their sites to the Authorize.Net system by simply inserting a small block of HTML code into their transaction page. With Authorize.Net, a customer's order is encrypted and transferred to the Authorize.Net server. The server, in turn, relays the transaction to a bank network through a private leased line. Merchants must have an Authorize.Net account to use the service. Customers are usually not aware that the transaction is being handled by a third-party supplier. Check the Online Companion links for more details about these services.

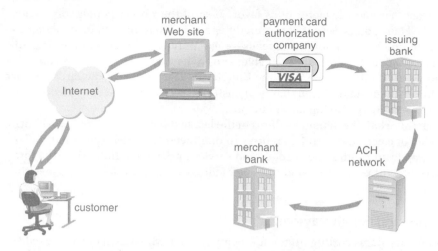

FIGURE 11-3   Processing a payment card transaction

ELECTRONIC CASH

Although credit cards dominate online payments today, electronic cash shows promise for the future. **Electronic cash** (also called e-cash or digital cash) is a general term that describes any value storage and exchange system created by a private (nongovernmental) entity that does not use paper documents or coins and that can serve as a substitute for government-issued physical currency. A significant difference between electronic cash and scrip is that electronic cash can be readily exchanged for physical cash on demand. Because electronic cash is issued by private entities, there is a need for common standards among all electronic cash issuers so that one issuer's electronic cash can be accepted by another issuer. This need has not yet been met. Each issuer has its own standards and electronic cash is not universally accepted, as is government-issued physical currency.

As you learned in the previous section, banks that issue credit cards make money by charging merchants a processing fee on each transaction. This fee ranges from 1 percent to 4 percent of the value of the transaction. Often, banks impose a minimum fee of 20 cents or more per transaction. Many banks charge electronic commerce sites more than similar brick-and-mortar stores—up to $1 more per credit card transaction. The cost of an online transaction can be 50 percent higher than the cost to process the same transaction for a brick-and-mortar retailer.

Many stores that accept credit cards require a minimum purchase amount of $10 or $15. Merchants impose a minimum purchase amount because the bank fees for small purchase amounts would be greater than the profits on those transactions. The same is true for Internet purchases. Small purchases are not profitable for merchants that accept only credit cards for payment. There is a market for small purchases on the Internet—purchases below $10. This is one potentially significant market for electronic cash. With very low fixed costs, electronic cash provides the promise of allowing users to spend, for example, 50 cents for an online newspaper, or 80 cents to send an electronic greeting card.

Electronic cash has another factor in its favor: Most of the world's population do not have credit cards. Many adults cannot obtain credit cards due to minimum income requirements or past debt problems. Children and teens—eager purchasers representing a significant percentage of online buyers—are ineligible, simply because they are too young. People living in most countries other than the United States hold few credit cards because they have traditionally made their purchases in cash. For all of these people, electronic cash provides the solution to paying for online purchases.

Even though there have been many failures in the last few years in electronic cash introductions, the idea of electronic cash just refuses to die. Electronic cash shows particular promise in two applications: the sale of goods and services priced less than $10—the lower threshold for credit card payments—and the sale of all goods and services to those without credit cards.

Micropayments and Small Payments

Internet payments for items costing from a few cents to approximately a dollar are called **micropayments**. Micropayment champions see many applications for such small transactions, such as paying 5 cents for an article reprint or 25 cents for a complicated literature search. However, micropayments have not been implemented very well on the Web yet. Another barrier to micropayments is a matter of human psychology. Researchers have found in a number of studies that many people prefer to buy small value items in fixed-price chunks rather than in individual small increments, even when buying the small increments would cost less money overall. A good example of this behavior is the preference most mobile telephone users have for fixed monthly payment plans over charges based on minutes used. The comfort of knowing the exact amount of the monthly bill is more important to many people than getting the lowest price on the minutes used.

The payments that are between $1 and $10 do not have a generally accepted name (some industry observers use the term micropayment to describe any payment of less than $10); in this book, the term **small payments** will be used to include all payments of less than $10.

Two companies now offer products for handling small payments that use credit cards as an alternative to electronic cash. The logic behind these products is that credit cards are more widely accepted than electronic cash. **Yaga** has targeted its product to large media companies such as Hearst, Time, and Ziff-Davis. These companies want to sell copies of articles from their publications, but the transaction fees charged by credit card processors make such sales unprofitable. Yaga accumulates charges made by an individual and then processes them in one lump sum at the end of a month or longer period. If a site visitor obtained six articles in a month, Yaga allows the site to process a credit card charge once (incurring just one transaction fee) instead of six times. **BitPass** targets smaller content providers—individual authors and musicians—by offering site visitors an account that they can draw against at any BitPass participating site. A customer authorizes BitPass to make a small (usually $3) charge to the customer's credit card to create that customer's BitPass account. The customer can then draw down the BitPass account at participating content vendor sites.

Privacy and Security of Electronic Cash

All electronic payment schemes have issues that must be resolved satisfactorily to allay consumers' fears and give them confidence in the technology. Concerns about electronic payment methods include privacy and security, independence, portability, and convenience. Privacy and security questions are probably the most important issues that have to be addressed with any payment system to be used by consumers. Consumers want to know whether transactions are vulnerable and whether the electronic currency can be copied, reused, or forged.

Electronic cash has unique security problems. Electronic cash should have two important characteristics in common with physical currency. First, it must be possible to spend electronic cash only once, just as with traditional currency. Second, electronic cash ought to be anonymous, just as hard currency is. That is, security procedures should be in place to guarantee that the entire electronic cash transaction occurs only between two parties, and that the recipient knows that the electronic currency being received is not counterfeit or being used in two different transactions. Ideally, consumers should be able to use electronic cash without revealing their identities—this prevents sellers from collecting information about individual or group spending habits. Companies in the electronic cash business include **eCharge** and **Valista**.

Electronic cash has the advantages of being independent and portable. When electronic cash is independent, it is unrelated to any network or storage device. That is, electronic cash is really not free-floating currency if its existence depends on a particular proprietary storage mechanism that is specially designed to hold one type of electronic cash. Electronic cash should ideally be able to pass transparently across international borders and be converted automatically to the recipient country's currency. Electronic cash portability means that it must be freely transferable between any two parties. Credit and debit cards do not possess this property of portability or transferability between every combination of two parties. In a credit card transaction, the payment recipient must already have a merchant account established with a bank. A merchant account is not required for a business to receive electronic cash.

Perhaps the most important characteristic of cash is convenience. If electronic cash requires special hardware or software, it is not convenient for people to use. Chances are good that people will not adopt an electronic cash system that is difficult to use.

Holding Electronic Cash: Online and Offline Cash

Two widely accepted approaches to holding cash exist today: online storage and offline storage. Online cash storage means that the consumer does not personally possess electronic cash. Instead, a trusted third party—an online bank—is involved in all transfers of electronic cash and holds the consumers' cash accounts. Online systems work by requiring merchants to contact the consumer's bank to receive payment for a consumer purchase, which helps prevent fraud by confirming that the consumer's cash is valid. This resembles the process of checking with a consumer's bank to ensure that a credit card is still valid and that the consumer's name matches the name on the credit card.

Offline cash storage is the virtual equivalent of money kept in a wallet. The customer holds it, and no third party is involved in the transaction. Protection against fraud is still a concern, so either hardware or software safeguards must be used to prevent fraudulent or double-spending. **Double-spending** is spending a particular piece of electronic cash twice by submitting the same electronic currency to two different vendors. By the time the same electronic currency clears the bank for a second time, it is too late to prevent the fraudulent act. The encryption techniques used to prevent double-spending are described later in this chapter.

Advantages and Disadvantages of Electronic Cash

Billing for goods and services that customers purchase is part of any business. Traditional billing methods in the brick-and-mortar paradigm are costly and involve generating invoices, stuffing envelopes, buying and affixing postage to the envelopes, and sending the invoices to the customers. Meanwhile, the Accounts Payable Department must keep track of incoming payments, post accounts in the database, and ensure that customer data is current.

Online stores have many of the same payment collection inefficiencies as their brick-and-mortar cousins. Most online customers use credit cards to pay for their purchases. Online auction customers also use conventional payment methods, including checks and money orders. Electronic cash systems, though less popular than other payment methods, provide advantages and disadvantages that are unique to electronic cash.

For the most part, electronic cash transactions are more efficient (and therefore less costly) than other methods, and that efficiency should foster more business, which eventually means lower prices for consumers. Transferring electronic cash on the Internet costs less than processing credit card transactions. Conventional money exchange systems require banks, bank branches, clerks, automated teller machines, and an electronic transaction system to manage, transfer, and dispense cash. Operating this conventional money exchange system is expensive.

Electronic cash transfers occur on an existing infrastructure—the Internet—and through existing computer systems. Thus, the additional costs that users of electronic cash must incur are nearly zero. Because the Internet spans the globe, the distance that an electronic transaction must travel does not affect cost. When considering moving physical cash and checks, distance and cost are proportional—the greater the distance that the currency has to go, the more it costs to move it. However, moving electronic currency from Los Angeles to San Francisco costs the same as moving it from Los Angeles to Hong Kong. Merchants can pay other merchants in a business-to-business relationship, and consumers can pay each other. Electronic cash does not require that one party obtain an authorization, as is required with credit card transactions.

Electronic cash does have disadvantages, and they are significant. Using electronic cash provides no audit trail. That is, electronic cash is just like real cash in that it cannot be easily traced. Because true electronic cash is not traceable, another problem arises: money laundering. **Money laundering** is a technique used by criminals to convert money that they have obtained illegally into cash that they can spend without having it identified as the proceeds of an illegal activity. Money laundering can be accomplished by purchasing goods or services with ill-gotten electronic cash. The goods are then sold for physical cash on the open market.

Just as physical currency can be counterfeited, electronic cash is susceptible to forgery. However, it is much more difficult to forge electronic cash than it is to use a fraudulently obtained credit card number. There are several other potentially damaging digital economic factors that might result from the use of electronic cash. These factors have to do with the expansion of the money supply when banks loan electronic cash on consumer and merchant accounts in traditional bank accounts. You can learn more about these economic factors by following the links to **Understanding the Digital Economy** and **The Economic and Social Impacts of Electronic Commerce** in the Online Companion.

Electronic cash has been successful in some parts of the world, but it has not yet become a global commercial success. Making electronic cash a popular alternative payment system requires wide acceptance and a solution to the problems of multiple electronic cash standards. Customers do not want to have to carry a dozen different brands of electronic cash to be able to purchase goods from a majority of the merchants that accept electronic cash. Establishing electronic cash as a popular payment method requires that a standard be developed for electronic cash disbursement and acceptance—a standard that individual vendors then implement for their individual electronic cash systems. Electronic cash from different vendors must be easily interchangeable so that customers can exchange one cash type for another when needed.

How Electronic Cash Works

To begin using electronic cash, a consumer opens an account with an electronic cash issuer (such as a bank that issues electronic cash or a private vendor of electronic cash, such as PayPal) and presents proof of identity. The consumer can then withdraw electronic cash by accessing the issuer's Web site and presenting proof of identity, such as a digital certificate issued by a certification authority, or a combination of a credit card number and a verifiable bank account number. After the issuer verifies the consumer's identity, it gives the consumer a specific amount of electronic cash and deducts the same amount from the consumer's account. In addition, the issuer might charge a small processing fee. The consumer can store the electronic cash in an electronic wallet (described later in this chapter) on his or her computer, or on a stored-value card (also described later in this chapter). In addition, the consumer can authorize the issuer to make payments to third parties from the electronic cash account.

Providing Security for Electronic Cash

You have already learned about one significant problem with electronic cash: its potential for double-spending. The main deterrent to double-spending is the threat of detection and prosecution. Cryptographic algorithms are the keys to creating tamperproof electronic cash that can be traced back to its origins. A two-part lock provides anonymous security that also signals when someone is attempting to double-spend cash. When a second transaction occurs for the same electronic cash, a complicated process comes into play that reveals the attempted second use and the identity of the original electronic cash holder. Otherwise, electronic cash that is used correctly maintains a user's anonymity. This double-lock procedure protects the anonymity of electronic cash users and simultaneously provides built-in safeguards to prevent double-spending. Figure 11-4 shows a graphic representation of this double-spending detection process using a double-lock system.

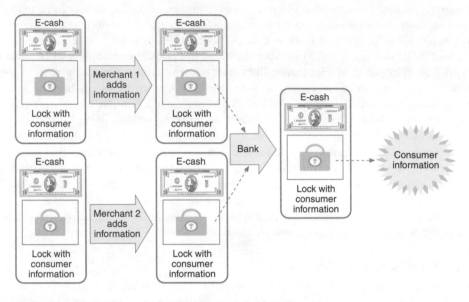

FIGURE 11-4   Detecting double spending of electronic cash

Double-spending can neither be detected nor prevented with truly anonymous electronic cash. **Anonymous electronic cash** is electronic cash that, like bills and coins, cannot be traced back to the person who spent it. One way to be able to trace electronic cash is to attach a serial number to each electronic cash transaction. That way, cash can be positively associated with a particular consumer. That does not solve the double-spending problem, however. Although a single issuing bank could detect whether two deposits of the same electronic cash are about to occur, it is impossible to ascertain who is at fault in such a situation—the consumer or the merchant. Of course, electronic cash that contains serial numbers is no longer anonymous, and anonymity is one reason to acquire electronic cash in the first place. Electronic cash containing serial numbers also raises a number of privacy issues, because merchants could use the serial numbers to track spending habits of consumers.

Creating truly anonymous electronic cash requires a bank to issue electronic cash with embedded serial numbers such that the bank can digitally sign the electronic cash while removing any association of the cash with a particular customer. The process begins when a consumer creates a random serial number that he or she sends to the bank issuing the electronic cash. The bank uses the consumer's random serial number along with the bank's digital signature and sends the random number, electronic cash, and digital signature as one package back to the user. When the user receives the electronic cash bundle, the user extracts the original random serial number and keeps the bank's digital signature. The consumer can now spend the electronic cash, which is digitally signed by the bank. When the consumer spends the electronic cash and the merchant passes it along to the issuing bank, the bank validates the electronic cash because it contains the bank's digital signature. However, the bank cannot determine the identity of the spender. It only knows that the electronic cash is genuine.

Electronic Cash Systems

Electronic cash has not been nearly as successful in the United States as it has been in Europe and Japan. In the United States, most consumers have credit cards, debit cards, charge cards, and checking accounts. These payment alternatives work well for U.S. consumers in both online and offline transactions. In most other countries of the world, consumers overwhelmingly prefer to use cash. Because cash does not work well for online transactions, electronic cash fills an important need for consumers in those countries as they conduct B2C electronic commerce. This type of need does not exist in the United States because U.S. consumers already use payment cards for traditional commerce, and these payment cards work well for electronic commerce.

KDD Communications (KCOM) is the Internet subsidiary of Kokusai Denshin Denwa, which is Japan's largest global phone company. KCOM has its own NetCoin electronic cash system and offers electronic cash through its NetCoin Center. Shoppers can go to the Net-Coin Center and obtain electronic cash that can be stored on their computers. Then, they can shop online for recipes or travel directories (10 cents each), or download MP3 music for less than a dollar per song. Other content providers, such as Japanese newspapers, provide access to their newspaper archives and charge a small fee to retrieve articles. Japan even has a donation site where visitors can donate electronic coins to charitable organizations.

Specific reasons for past failures of electronic cash systems in the United States are not completely clear. Some industry observers blame the failure on the way that many electronic cash systems were implemented. Most of these systems required the user to download and install complicated client-side software that ran in conjunction with the browser. Also, there were a number of competing technologies; therefore, no standards were ever developed for the entire electronic cash system. The absence of electronic cash standards means that consumers are faced with choosing from an array of proprietary electronic cash alternatives—none of which are interoperable. **Interoperable software** runs transparently on a variety of hardware configurations and on different software systems.

Despite their rough start, not all electronic cash ventures have failed. Next, you will learn about some of the Internet companies that currently offer electronic cash services and bill presentment and payment systems.

CheckFree

CheckFree, the largest online bill processor in the world, provides online payment processing services to both large corporations and individual Internet users. CheckFree provides infrastructure and software that permits users to pay all their bills with online electronic checks. CheckFree provides part of the technology that the Web portal Yahoo! uses to provide its **Yahoo! Bill Pay** service (see Figure 11-5).

505

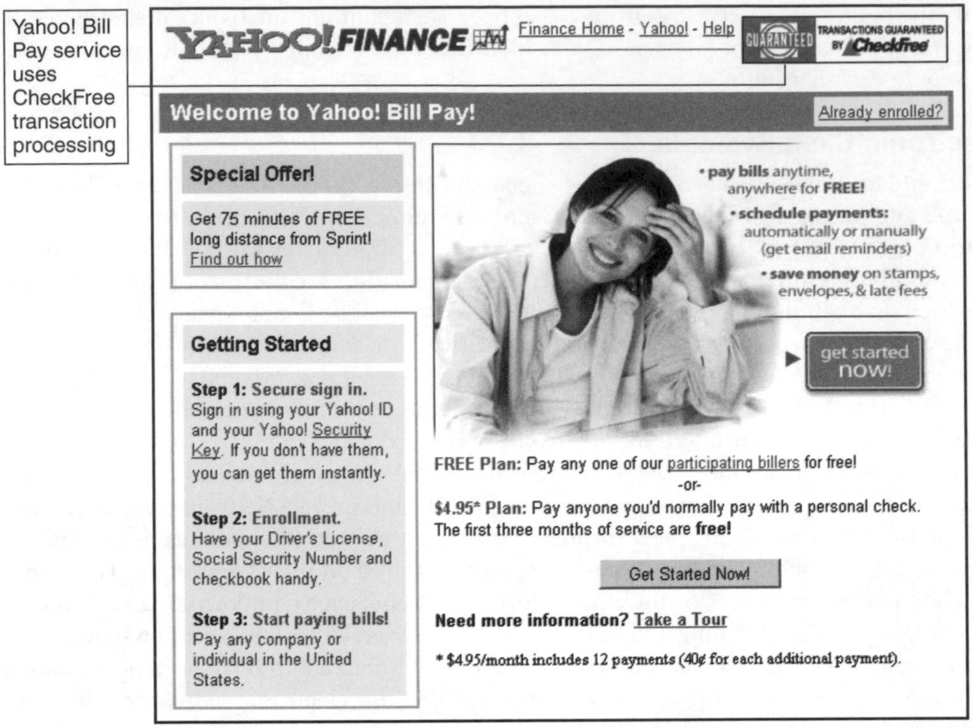

Yahoo! Bill Pay service uses CheckFree transaction processing

FIGURE 11-5   Yahoo! Bill Pay service

Clickshare

Clickshare is an electronic cash system aimed at magazine and newspaper publishers. Clickshare's technology has occasionally been called a micropayment-only system; however, the ability to make micropayments is only one of Clickshare's features. Users with an ISP that supports Clickshare are registered automatically with Clickshare. When users click links leading to other sites that are registered with Clickshare, they can make purchases on those sites without having to register again. Clickshare keeps track of transactions and bills the user's ISP. The ISP, which already has an account relationship with the user, then bills the user for his or her purchases.

Another feature of Clickshare is that it tracks where a user travels on the Internet. This feature has significant value to advertisers and marketers that want to measure audience preferences; however, it does defeat anonymity, and anonymity is one reason that consumers might want to use Clickshare. The micropayment capability is, according to the company, a by-product of the core functionality of tracking identified users. Clickshare tracks users with the standard HTTP Web protocol and does not require cookies or software wallets. Clickshare claims to be the only company that can do this. (Click the **Clickshare Example Scenario** link in the Online Companion for a diagram and explanation of how users are billed for the hyperlinks that they click.)

PayPal

PayPal is the electronic cash payment system that you read about in the opening case of this chapter. PayPal was founded in 1999, and in 2000 it merged with another payment processing service, X.com. PayPal provides payment processing services to businesses and to individuals. PayPal earns a profit on the **float**, which is money that is deposited in PayPal accounts and not used immediately. After two years in business, PayPal began charging a transaction fee to businesses that use the service to collect payments. Individuals who use PayPal to send money to other individuals do not pay a transaction fee. The free payment clearing service that PayPal provides to individuals is called a **peer-to-peer (P2P) payment system** because the payments are from one type of entity to another of the same type.

PayPal eliminates the need to pay for online purchases by writing and mailing checks or using payment cards. PayPal allows consumers to send money instantly and securely to anyone with an e-mail address, including an online merchant. PayPal is a convenient way for auction bidders to pay for their purchases, and sellers like it because it eliminates the risks posed by other types of online payments. PayPal transactions clear instantly so that the sender's account is reduced and the receiver's account is credited when the transaction occurs. Anyone with a PayPal account—online merchants or eBay auction participants alike—can withdraw cash from their PayPal accounts at any time by requesting that PayPal send them a check or make a direct deposit to their checking accounts. Figure 11-6 shows PayPal's home page.

To use PayPal, merchants and consumers first must register for a PayPal account. There is no minimum amount that a PayPal account must contain, and customers add money to their PayPal accounts by authorizing a transfer from their checking accounts or by using a credit card. Once members' payments are approved and deposited into their PayPal accounts, they can use their PayPal money to pay for purchases.

Merchants must have PayPal accounts to accept PayPal payments. Using PayPal to pay for auction purchases is very popular. A consumer can use PayPal to pay a seller for purchases even if the seller does not have a PayPal account. PayPal sends the seller an e-mail message indicating that a payment is waiting at the PayPal Web site. To collect PayPal cash, the seller or merchant that received the e-mail message must register and provide PayPal with payment instructions. PayPal then either sends the merchant a check or deposits funds directly into the merchant's checking account.

PayPal grew rapidly by serving the needs of buyers and sellers on auction sites such as eBay, Yahoo! Auctions, and Amazon Auctions. This success and its potential for profits did not go unnoticed by the management team at eBay. In May 1999, eBay purchased a small electronic payments company and, one year later, sold a 35 percent stake in that company to Wells Fargo bank. This company, Billpoint, was operated as a joint venture between eBay and Wells Fargo. Billpoint grew rapidly, but PayPal maintained its first-mover advantage and remained the most widely used payment processing system on eBay. After unsuccessfully battling PayPal with its Billpoint service for three years, eBay finally gave up and decided to buy PayPal, as you learned in this chapter's opening case.

Other companies have entered the peer-to-peer payments business as well. First Data Corporation, which owns Western Union, offers what it calls electronic money orders that customers can use to settle auction transactions through its **BidPay** site. Traditional banks have also created Internet payment sites, such as Citibank's c2it payments service, but

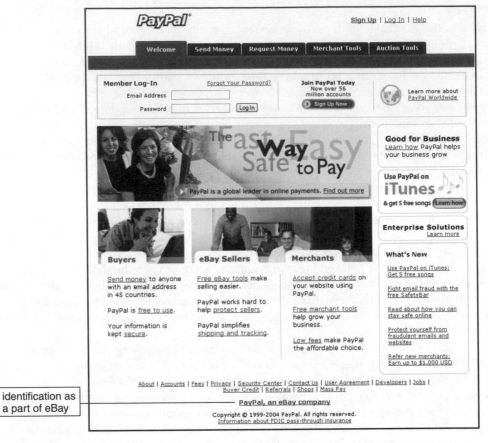

identification as a part of eBay

FIGURE 11-6   PayPal home page

these have been less successful than nonbank entries into the business. In 2003, Citibank closed its c2it operation. Citibank does allow its customers to make peer-to-peer transfers from their checking accounts; however, that service requires the customer initiating the transfer to have a checking account with the bank.

ELECTRONIC WALLETS

As consumers are becoming more enthusiastic about online shopping, they have begun to tire of repeatedly entering detailed shipping and payment information each time they make online purchases. Filling out forms ranks high on online customers' lists of gripes about online shopping. To address these concerns, many electronic commerce sites include a feature that allows a customer to store name, address, and credit card information on the site. However, consumers must enter their information at each site with which they want to do business. An **electronic wallet** (sometimes called an **e-wallet**), serving a function similar to a physical wallet, holds credit card numbers, electronic cash, owner identification, and owner contact information and provides that information at an electronic commerce site's checkout counter. Electronic wallets give consumers the benefit of entering their information just once, instead of having to enter their information at every site with which they want to do business.

Electronic wallets make shopping more efficient. When consumers select items to purchase, they can then click their electronic wallet to order the items quickly. In the future, wallets could serve their owners by tracking purchases and maintaining receipts for those purchases. Maintaining records of a consumer's purchasing habits is something that online giants such as Amazon.com have mastered, but an enhanced digital wallet could reverse that process and use a Web robot to suggest where the consumer might find a lower price on an item that he or she purchases regularly.

Electronic wallets fall into two categories based on where they are stored. A **server-side electronic wallet** stores a customer's information on a remote server belonging to a particular merchant or wallet publisher. The main weakness of server-side electronic wallets is that a security breach could reveal thousands of users' personal information—including credit card numbers—to unauthorized parties. Typically, server-side electronic wallets employ strong security measures that minimize the possibility of unauthorized disclosure.

A **client-side electronic wallet** stores a consumer's information on his or her own computer. Many of the early electronic wallets were client-side wallets that required users to download the wallet software. This need to download software onto every computer used to make purchases is a chief disadvantage of client-side wallets. Server-side wallets, on the other hand, remain on a server and thus require no download time or installation on a user's computer. Before a consumer can use a server-side wallet on a particular merchant's site, the merchant must enable that specific wallet. Each wallet vendor must convince a large number of merchants to enable its wallet before it will be accepted by consumers. Thus, only a few server-side wallet vendors will be able to succeed in the market.

A disadvantage of client-side wallets is that they are not portable. For example, a client-side wallet is not available when a purchase is made from a computer other than the computer on which the wallet resides.

In a client-side electronic wallet, the sensitive information (such as credit card numbers) is stored on the user's computer instead of the wallet provider's central server. This removes the risk that an attack on a client-side electronic wallet vendor's server could reveal the sensitive information. However, an attack on the user's computer could yield that information. Most security analysts agree that storing sensitive information on client computers is safer than storing that information on the vendor server because it requires attackers to launch many attacks on user computers, which are more difficult to identify (even though the user computers are less likely than a vendor server to have strong security features installed). It also prevents the easily identified servers of the wallet vendors from being attractive targets for such attacks.

For a wallet to be useful at many online sites, it should be able to populate the data fields in any merchant's forms at any site that the consumer visits. This accessibility means that the electronic wallet manufacturer and merchants from many sites must coordinate their efforts so that a wallet can recognize what consumer information goes into each field of a given merchant's forms.

Electronic wallets store shipping and billing information, including a consumer's first and last names, street address, city, state, country, and postal code. Most electronic wallets also can hold many credit card names and numbers, affording the consumer a choice of credit cards at the online checkout. Some electronic wallets also hold electronic cash from various providers.

A number of companies entered the electronic wallet business, including major firms such as MasterCard. Most of these companies have abandoned their efforts because current versions of all major browsers now include a feature that remembers names, addresses, and other commonly requested information and provides a one-click completion of fields on Web forms that request that information. Two survivors in the e-wallet arena are Microsoft .NET Passport and Yahoo! Wallet.

Microsoft .NET Passport

Microsoft .NET Passport (often referred to as Passport or Microsoft Passport) is a server-side electronic wallet operated by Microsoft. Anyone who obtains a Hotmail account, which is Microsoft's free e-mail service, is signed up automatically for a Passport account. People who use Microsoft MSN Internet access service also must sign up for a Passport account. Passport functions in the same way as most other electronic wallets—by completing order forms automatically. All of the personal data entered into a Passport wallet is encrypted and password protected.

Passport consists of four integrated services: Passport single sign-in service (SSI), Passport Wallet service, Kids Passport service, and public profiles. The sign-in service allows a user to sign in at a participating Web site using his or her username and password. The Passport Wallet service provides standard electronic wallet functions, such as secure storage and form completion of credit card and address information. When requested by a participating merchant, a consumer's secure information is released to the merchant so that the consumer does not need to enter data into a form. The Kids Passport service helps parents protect and control their children's online privacy, and the public profiles service allows consumers to create a public page of information about themselves. Figure 11-7 shows the Passport home page.

FIGURE 11-7   Microsoft .NET Passport home page

Yahoo! Wallet

Yahoo! Wallet is a server-side electronic wallet offered by the Web portal site Yahoo! The Yahoo! Wallet functions in the same way as most other electronic wallets—by completing order forms automatically with identifying information and credit card payment information. Yahoo! Wallet lets users store information about several major credit and charge cards, along with Visa and MasterCard debit cards.

Yahoo! Wallet is accepted at more than 10,000 Yahoo! Store merchants (these are merchants on the Yahoo! Shopping section of the portal), and also can be used to pay for airplane tickets and hotel reservations booked through the Yahoo! Travel section of the portal. Yahoo! Wallet also works when users pay for premium services at Yahoo!, such as extra mail storage or Web hosting fees on the Yahoo! GeoCities Plus or Website Services portions of the site. Sellers on Yahoo! Auctions can pay their auction fees using the Yahoo! Wallet, too.

Yahoo! has the advantage of hosting a number of services and shops that it can be certain accommodate its own wallet; thus, it is certain to have a large number of merchants (including itself) that accept its wallet. Yahoo! also offers its PayDirect service, which you learned about earlier in this chapter, as a part of Yahoo! Wallet.

Many industry observers and privacy rights activist groups are concerned about electronic wallets because they give the company that issues the electronic wallet access to a great deal of information about the individual using the wallet. Several groups have attempted to enact standards intended to address wallet privacy concerns.

W3C Micropayment Standards Development Activity

Wallet information includes identification of the users and a complete record of their online purchasing activity. An alternative to having individual companies offer electronic wallet services is to have standards for electronic wallets built into the structure of the Web itself. With open standards, many different companies could offer electronic wallet services that would work on many different Web sites. This approach would distribute the information gathering and storage among a number of companies and thus reduce the risk of having one company in control of so much private information.

The World Wide Web Consortium (W3C) conducted an active standards development activity for micropayments in electronic commerce for several years. Although the activity has now been closed, the **W3C Electronic Commerce Interest Group (ECIG)** developed a set of standards called the **Common Markup for Micropayment Per-Fee-Links** before it ended its activities. The proposed ECIG micropayment links standard is a set of guidelines that provides an extensible and interoperable way to embed micropayment information in a Web page. An **extensible system** is one that developers can add to (or extend) without voiding any earlier work on the system.

Merchants that want the widest Internet consumer audience must be willing to offer support for several different payment systems. To do so, the merchants' Web pages must embed, within each Web page, payment information that is specific to each payment system adopted by different consumers. This redundancy has motivated the W3C to develop a common Web page markup system that all micropayment systems can support. The ECIG draft standard proposes an architecture that includes the following steps: The client (the consumer's Web browser) initiates the micropayment activity by contacting the merchant server. The client browser consists of the browser itself, a module called Per-Fee-Link Handler (PFLH), and one or more electronic payment wallets. On the merchant side of the client-server architecture is an HTTP server. The W3C document proposes new HTML tags to identify the embedded micropayment information. Figure 11-8 shows a list of these proposed micropayment tags. It is now up to the individual electronic wallet and electronic cash vendors to develop and revise their software to conform to the new standard.

Field name	Short description	Format	Requirements
merchanturl	Identifies the merchant site.	URL	MUST be provided
merchantname	Specifies a merchant designation.	character string	MAY be provided
buyurl	Identifies what the client is buying.	relative URL	MUST be provided
textlink	Describes textually what the client is buying. The text source of the fee link.	character string	MUST be provided
imagelink	Describes graphically what the client is buying. The graphic source of the fee link.	URL	MAY be provided
price	Specifies amount and currency.	character string	MUST be provided
duration	Indicates the time after purchase any URL can be retrieved with payment.	integer number	SHOULD be provided
longdesc	Describes in detail the content of the client's purchase.	character string	SHOULD be provided
requesturl	Identifies what the client is actually requesting.	relative URL	MAY be provided
expiration	Indicates a date until which the offer from the merchant is valid.	character string: YYYY-MM-DDThh:mm:ssTZD	MAY be provided
specific field	Provides information unique to each payment system.	URL and character string	MAY be provided

FIGURE 11-8   W3C proposed micropayment HTML tags

The ECML Standard

The W3C initiative is not the only attempt to develop standards for the operation of electronic wallets. A consortium of several high-tech companies and credit card companies proposed an alternative standard that would replace the competing electronic wallet standards with a single standard. The consortium of companies, which includes America Online, Compaq, Dell, IBM, Microsoft, Visa U.S.A., and MasterCard, has agreed on a technology called **ECML**, or **Electronic Commerce Modeling Language**. Users can enter their credit card and address information once into an ECML-capable electronic wallet. Users control access to their ECML electronic wallets, and the wallets will be accepted at all commerce sites if the consortium is successful in generating widespread acceptance. So far, several electronic wallet makers have accepted and implemented the standard. They include IBM, Microsoft, and Trintech.

The ECML standard will expedite online processing for customers by simplifying the form-filling procedure. Currently, the field names for various forms vary from one online merchant to another. For example, one merchant might refer to the telephone number field as "phone," while another merchant might call it "PhoneNumber." The ECML format works by providing uniform field names upon which all merchants can build their input forms. These uniform names are implemented as XML tags, which you learned about in Chapter 2. An ECML-compliant Web site allows a customer to enter billing, shipping, and payment information once. If the customer visits another ECML-compliant site, the same customer information is already filled out when the customer proceeds to the checkout counter.

Assuming that an acceptable standard will evolve, the ultimate success of electronic wallets will depend on the confidence that Internet users have in the technology. As the NetBank story (see the Learning from Failures feature) illustrates, customer confidence is an important part of the success of any Internet technology, especially when that technology controls a person's financial welfare.

LEARNING FROM FAILURES

NetBank

CompuBank and NetBank were two of the first Internet banks to open in the United States. They were both pure Internet banks; that is, neither was founded by an existing bank with a physical presence. After four years of operation, CompuBank had about 50,000 accounts and $64 million of deposits and was losing more than $20 million per year. NetBank had done considerably better, with 160,000 accounts and $1 billion of deposits and 10 consecutive quarters of profitability.

In early 2001, CompuBank decided to close its operations and found NetBank to be a willing purchaser of its accounts. When a bank buys accounts from another bank, it performs a series of procedures called due diligence. These **due diligence** procedures include checking the new customers' credit histories and banking records. Due diligence is usually performed before the transaction is completed and before the closing bank's customers look to the buying bank as the institution that will handle their accounts.

For a number of reasons, not all of which are clear, the due diligence process was still under way on the date that the transfer of accounts was to take place. NetBank placed holds on many accounts and sent letters to many account holders explaining that they were not acceptable customers by NetBank standards. For any bank, this would have been a difficult situation, but the nature of the two banks as Internet-only operations made things considerably worse for everyone.

Press accounts of the fiasco included stories of the problems that between 4000 and 8000 CompuBank depositors experienced. Some of the problems were small—online bill payments did not occur, debit and credit cards were rejected at stores and restaurants, and ATMs would not yield cash—while others were much larger. One couple who had kept the money to cover closing costs on a house purchase in a CompuBank account found that NetBank had placed a hold on the money.

Because they could not pay the closing costs, they were forced to find another mortgage lender. In the suit they filed against NetBank, the couple asserted that the increased rate on the mortgage loan would cost them tens of thousands of dollars. Other CompuBank

continued

customers were irritated that they lost access to their money for weeks. Some customers could not determine whether the bills they had set up to be paid automatically had, in fact, been paid.

NetBank admitted failures in customer service related to the incident. Many customers who called to complain or ask for explanations experienced 45-minute waits on hold and then were transferred to the bank's Security Department, where a recording answered and asked callers to leave their Social Security numbers and wait to be called back. None of the customers reported being called back. The timing of NetBank's notification was problematic, too. Many customers reported receiving a letter from NetBank indicating that there were problems with their accounts. The letter, dated April 30, was received by the customers on or after May 14. The letter included a telephone number to call for assistance, but that number had been disconnected on May 12. Many of the unhappy customers found each other on Internet discussion boards and compared notes.

NetBank has not disclosed the number of customers it lost by its handling of this transition; indeed, it may not know. CompuBank's customers were largely experienced Internet users who chose to be part of the leading edge in handling their financial affairs. Many of them, after this experience, have sworn that they will never again do business with a bank that does not have a physical presence. The lesson from NetBank's experience is that customer service and the ability to communicate with customers become extremely important for companies that process electronic payments or are responsible for their customers' finances.

STORED-VALUE CARDS

Today, most people carry a number of plastic cards—credit cards, debit cards, charge cards, driver's license, health insurance card, employee or student identification card, and others. One solution that could reduce all those cards to a single plastic card is called a stored-value card.

A **stored-value card** can be an elaborate smart card with a microchip or a plastic card with a magnetic strip that records the currency balance. The main difference is that a smart card can store larger amounts of information and includes a processor chip on the card. The card readers needed for smart cards are different, too. Common stored-value cards include prepaid phone, copy, subway, and bus cards. Many people use the terms "stored-value card" and "smart card" interchangeably.

Magnetic Strip Cards

Most magnetic strip cards hold value that can be recharged by inserting them into the appropriate machines, inserting currency into the machine, and withdrawing the card; the card's strip stores the increased cash value. Magnetic strip cards are passive; that is, they cannot send or receive information, nor can they increment or decrement the value of cash stored on the card. The processing must be done on a device into which the card is inserted. Although both magnetic strip cards and smart cards can store electronic cash, a smart card is better suited for Internet payment transactions because it has some processing capability.

Smart Cards

A **smart card** is a stored-value card that is a plastic card with an embedded microchip that can store information. Credit, debit, and charge cards currently store limited information on a magnetic strip. A smart card can store about 100 times the amount of information that a magnetic strip plastic card can store. A smart card can hold private user data, such as financial facts, encryption keys, account information, credit card numbers, health insurance information, medical records, and so on.

Smart cards are safer than conventional credit cards because the information stored on a smart card is encrypted. For example, conventional credit cards show your account number on the face of the card and your signature on the back. The card number and a forged signature are all that a thief needs to purchase items and charge them against your card. With a smart card, credit theft is much more difficult because the key to unlock the encrypted information is a PIN; there is no visible number on the card that a thief can identify, nor is there a physical signature on the card that a thief can see and use as an example for a forgery.

Smart cards have been in use for more than a decade. Popular in Europe and parts of Asia, smart cards so far have not been as successful in the United States. In Europe and Japan, smart cards are being used for telephone calls at public phones and for television programs delivered by cable to people's homes. The cards are also very popular in Hong Kong, where many retail counters and restaurant cash registers have smart card readers. The city's transportation companies—subways, buses, railways, trams, and ferries—joined together and created a smart card called the Octopus that lets commuters use one card for all of their public transportation needs. The Octopus can be reloaded at any transportation location or at 7-Eleven stores throughout Hong Kong. The **Hong Kong Citybus** Web page with information about the Octopus Card appears in Figure 11-9.

Smart cards are beginning to appear in the United States. In San Francisco, the Bay Area **Metropolitan Transportation Commission** created a smart card system patterned after the Octopus Card. This system, TransLink, is the first integrated ticketing system for public transportation in the United States. The transportation smart card, implemented in late 2001, allows commuters to ride most modes of public transit available in the city, including trains, buses, cabs, and ferries, by simply waving a single card near a reader device in transit vehicles or in stations. TransLink users can reload their smart cards at several retail outlets or directly from their bank accounts.

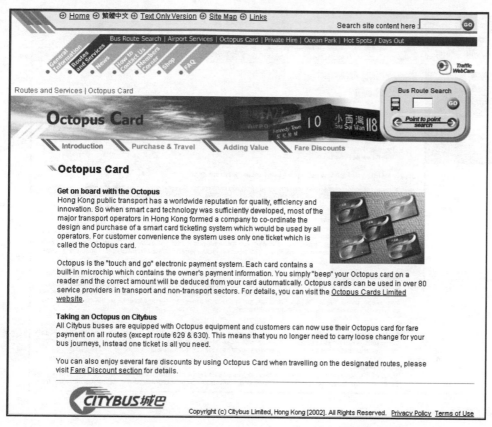

FIGURE 11-9 Octopus smart card information on the Hong Kong Citybus site

Visa recently introduced its smart card, the **smart Visa card**. One of the first promotions of the new Visa card occurred in late 2002 when retailer Target introduced its Target Visa smart card for use in Target stores and on the Target.com Web site. The Target Visa includes electronic wallet and automated login information for the Target.com Web site, but it also functions as a normal Visa card at other merchants. American Express has also released a smart card called **Blue**. Figure 11-10 shows a Web page leading to information about the American Express Blue smart card.

In the United States, the **Smart Card Alliance** promotes the benefits of smart cards. The organization promotes the widespread acceptance of multiple-application smart card technology. Its members include companies in banking, financial services, computer technology, healthcare, telecommunications, and a number of government agencies. The Alliance focuses on information exchange and member interaction. Every member of the Alliance recognizes that smart cards can succeed in the United States only if a critical mass of smart cards supports applications—both physical and Internet-based—of interest to consumers. The Alliance promotes compatibility among smart cards, card reader devices, and applications.

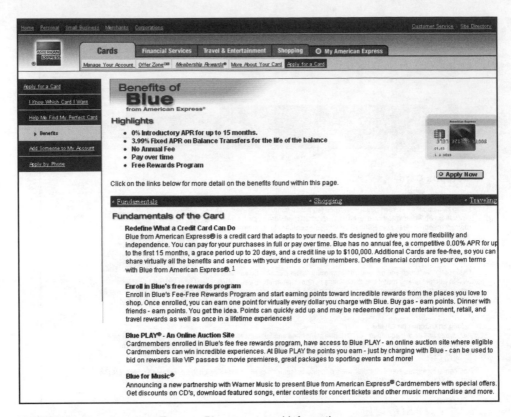

FIGURE 11-10 American Express Blue smart card information page

INTERNET TECHNOLOGIES AND THE BANKING INDUSTRY

As you learned earlier in this chapter, the largest dollar volume of payments today are still made using paper checks. These paper checks are processed through the world's banking system. The other major payment forms in use today also involve banks in one way or another. This section outlines how Internet technologies are providing new tools and creating new threats for the banking industry.

Check Processing

In the past, checks were processed physically by banks and clearinghouses. When a person wrote a check to pay for an item at a retail store, the retailer would deposit the check in its bank account. The retailer's bank would then send the paper check to a clearinghouse, which would manage the transfer of funds from the consumer's bank to the retailer's account. The paper check would then be transported to the consumer's bank, which might then send the cancelled check to the consumer. In recent years, many banks have stopped sending cancelled checks to their consumer account holders to save postage. Despite these savings, the cost of transporting tons of paper checks around the country has grown each year.

In addition to the transportation costs, another disadvantage of using paper checks is the delay that occurs between the time that a person writes a check and the time that check clears the person's bank. This delay (which is similar to the delay you learned about earlier in PayPal accounts, and which is also called float) makes it possible to write checks a few days before money is in the account to cover those checks. In effect, the bank's customer obtains the free use of funds for a few days and the bank loses the use of those funds for the same time period. Although the delay normally lasts only a few days, there are times when it can become significantly longer. Railroad and airline strikes, for example, have caused the float to be extended. The most recent incidents that caused a significant increase in the float were the terrorist attacks of September 11, 2001.

Banks have been working for years to develop technologies that will help them reduce the float. In 2004, a U.S. law went into effect that many bankers believe will eventually eliminate the float. This law, called the **Check Clearing for the 21st Century Act** (or, more simply, **Check 21**), permits banks to eliminate the movement of physical checks entirely. In a Check 21-compliant world, the retailer can scan the customer's check. The scanned image is transmitted instantly through a clearing system and posts almost immediately to both accounts (that is, the withdrawal from the customer's account and the deposit to the retailer's account occur instantly), eliminating any float on the transaction.

You can learn more about the Check 21 law and its implementation by using the links in the Online Companion to the **BAI Check 21 Resource Center**, the **Federal Reserve Bank Check 21 Services** pages, or the **American Bankers Association Check 21 Resource Center**.

Phishing Attacks

In Chapter 10, you learned about the phishing expedition, which is a technique for committing fraud against the customers of online businesses. Although phishing expeditions can be launched against all types of online businesses, they are of particular concern to financial institutions because their customers expect a high degree of security to be maintained over the personal information and resources that they entrust to their online financial institutions.

The basic structure of a phishing attack is fairly simple. The attacker sends e-mail messages (such as the one shown in Figure 11-11) to a large number of recipients who might have an account at the targeted Web site (PayPal is the targeted site in the example shown in the figure). The e-mail message tells the recipient that his or her account has been compromised and it is necessary for the recipient to log in to the account to correct the matter. The e-mail message includes a link that appears to be a link to the login page of the Web site. However, the link actually leads the recipient to the phishing attack perpetrator's Web site, which is disguised to look like the targeted Web site. The unsuspecting recipient enters his or her login name and password, which the perpetrator captures and then uses to access the recipient's account. Once inside the victim's account, the perpetrator can access personal information, make purchases, or withdraw funds at will.

Date: [Date removed] 08:05:42 +0600
From: "Services PayPal" <services@paypal.com>
Subject: PayPal Account sensitive features are access limited!
To: [E-mail addresses removed]

Dear valued **PayPal** member:

PayPal is committed to maintaining a safe environment for its community of
buyers and sellers. To protect the security of your account, PayPal employs
some of the most advanced security systems in the world and our anti-fraud
teams regularly screen the PayPal system for unusual activity.

Recently, our Account Review Team identified some unusual activity in your
account. In accordance with PayPal's User Agreement and to ensure that your
account has not been compromised, access to your account was limited. Your
account access will remain limited until this issue has been resolved. This
is a fraud prevention measure meant to ensure that your account is not compromised.

In order to secure your account and quickly restore full access, we may
require some specific information from you for the following reason:

We would like to ensure that your account was not accessed by an
unauthorized third party. Because protecting the security of your account
is our primary concern, we have limited access to sensitive PayPal account
features. We understand that this may be an inconvenience but please
understand that this temporary limitation is for your protection.

Case ID Number: PP-040-187-541

We encourage you to log in and restore full access as soon as possible.
Should access to your account remain limited for an extended period of
time, it may result in further limitations on the use of your account.

However, failure to restore your records will result in account suspension.
Please update your records within 48 hours. Once you have updated your account
records, your **PayPal** session will not be interrupted and will continue as normal.

To update your **Paypal** records click on the following link:
https://www.paypal.com/cgi-bin/webscr?cmd=_login-run

Thank you for your prompt attention to this matter. Please understand that
this is a security measure meant to help protect you and your account. We
apologize for any inconvenience.

Sincerely,
PayPal Account Review Department

PayPal Email ID PP522

Accounts Management As outlined in our User Agreement, **PayPal** will
periodically send you information about site changes and enhancements.

Visit our Privacy Policy and User Agreement if you have any questions.
http://www.paypal.com/cgi-bin/webscr?cmd=p/gen/ua/policy_privacy-outside

FIGURE 11-11   Phishing e-mail message

The links in phishing e-mails are usually disguised. One common way to disguise the real URL is to use the "@" sign, which causes the Web server to ignore all characters that precede the "@" and only use the characters that follow it. For example, a link that displays:

https://paypal.com@218.36.41.188/fl/login.html

looks like it is an address at PayPal. However, the "@" sign causes the Web server to ignore the "paypal.com" and instead takes the victim to a Web page at the IP address "218.36.41.188."

In the e-mail shown in the figure, the link appears in the victim's e-mail client software as:

https://paypal.com/cgi-bin/webscr?cmd=_login-run

but when the victim clicks the link, the browser opens a completely different URL:

http://leasurelandscapes.com/snow/webscr.dll

Instead of the URL it shows in the e-mail client, the link in the phishing e-mail actually includes following JavaScript code:

```
<A onmouseover="window.status='https://www.paypal.com/cgi-bin/webscr?
cmd=_login-run'; return true" onmouseout="window.status='https://www.
paypal.com/cgi-bin/webscr?cmd=_login-run'"href="http://
leasurelandscapes.com/snow/webscr.dll">https://www.paypal.com/cgi-bin/
webscr?cmd=_login-run</A>
```

This code is invisible in most e-mail clients, so the victim might never know that the Web browser has opened a phony site. Phishing attack perpetrators use a variety of other tricks to hide the URLs, including code that pops up windows that look exactly like a browser address bar. The window is coded to pop up over the browser's address bar. You can learn more about the details of phishing techniques by visiting the Web sites of the **First Conference on Email and Anti-Spam**. and the **Anti-Phishing Working Group**.

Phishing Attack Countermeasures

In Chapter 8, you learned that several groups are working on ways to improve the Internet's mail transport protocols so that spam senders can be identified. Since spam is a key element of phishing attacks, any protocol change that improves e-mail recipients' ability to identify the source of an e-mail message will also help to reduce the threat of phishing attacks.

The most important step that companies can take today, however, is to educate their Web site users. Most online banking sites continually warn their customers that the site never sends e-mail that asks for account information or that asks the recipient to log in to their Web site and make changes to his or her account information. PayPal occasionally interrupts its own log-in screen sequence to insert a page that provides information about phishing attacks.

Many companies, especially those that operate financial Web sites, have contracted with consulting firms that specialize in anti-phishing work. These consultants monitor the

Web for new Web sites that use the company's name or logo and move quickly to shut down those sites. Most phishing perpetrators set up their entrapping Web sites a few days before they launch their e-mail campaign, so this technique can be effective. Another anti-phishing technique is to monitor online chat rooms that are used by criminals. By watching for offers of stolen credit card information and other phishing exploits, consultants can identify phishing schemes that are under way.

The most recent estimates of annual losses caused by phishing expeditions exceed $1.2 billion. The incidence of phishing attacks has grown rapidly over the past two years and most industry analysts expect that phishing will be a problem that will plague online businesses for the near future. Phishing can be a very profitable criminal activity and as more companies increase their defenses, analysts expect phishing perpetrators to become even better at working around those defenses.

522

Summary

Online stores can accept a variety of forms of payment. Credit, debit, and charge cards (payment cards) are the most popular forms of payment on the Internet. They are ubiquitous, convenient, and easy to use.

Electronic cash, one form of online payment, has been slow to catch on in the United States. A number of companies have faltered in recent years as they attempted to introduce electronic cash to the online world. Electronic cash is especially useful for making micropayments because the cost of processing payment cards for small transactions is greater than the profit on such transactions. Electronic cash shares several benefits with real cash: it is portable, anonymous, and usable for international transactions. Electronic cash can be stored online or offline. A third party, such as a bank, stores online electronic cash. The consumer holds offline cash in specially designed wallets.

Electronic wallets provide convenience to online shoppers because they hold payment card information, electronic cash, and personal consumer identification. Electronic wallets eliminate the need for consumers to reenter payment card and shipping information at a site's electronic checkout counter. Instead, the electronic wallet automatically fills in form information at sites that recognize the particular wallet software's technology. One persistent problem with electronic wallets is the lack of an internationally accepted standard. Both the W3C and the ECML standards group have created standards; however, neither has seen wide adoption by merchants, consumers, or wallet providers. With a single wallet standard, merchants would be more willing to install electronic, wallet-friendly software on their commerce sites.

Stored-value cards, including smart cards and magnetic strip cards, are physical devices that hold information, including cash value, for the cardholder. Magnetic strip cards have limited capacity. Smart cards can store greater amounts of data on a microchip embedded in the card and are intended to replace the collection of plastic cards people now carry, including payment cards, driver's licenses, and insurance cards. Trials of smart cards in a few U.S. cities have proved disappointing; however, smart cards are popular in other parts of the world. Visa and American Express have introduced smart cards. Unlike electronic cash or payment cards, smart cards require merchants to install new hardware that can read the smart cards.

Banks still process most monetary transactions, and a large part of the dollar volume of those transactions is still done by writing checks. Increasingly, banks are using Internet technologies to process those checks. Phishing expeditions create a significant threat to online financial institutions and their customers. If not controlled, this threat could reduce the general level of confidence that consumers have in online business and hurt the growth of electronic commerce.

Key Terms

Acquiring bank	Check 21
Anonymous electronic cash	Client-side electronic wallet
Automated Clearing House (ACH)	Closed loop system
Card not present transactions	Credit card
Charge card	Credit card association
Chargeback	Customer issuing bank

Debit card

Double-spending

Due diligence

Electronic cash

Electronic Commerce Modeling
Language (ECML)

Electronic wallet (e-wallet)

EMV standard

Extensible system

Float

Interoperable software

Merchant account

Merchant bank

Micropayments

Money laundering

Open loop system

Payment card

Payment processing service provider

Peer-to-peer (P2P) payment system

Scrip

Server-side electronic wallet

Single-use card

Small payments

Smart card

Stored-value card

Review Questions

RQ1. Write two paragraphs in which you define "scrip" and outline the advantages and disadvantages of scrip for consumers.

RQ2. In about 100 words, describe the difficulties that can arise for merchants that want to process "card not present" credit card transactions.

RQ3. In about 200 words, outline the reasons why a consumer who owns a credit card would want to use an electronic payment system, such as PayPal, for an Internet transaction. In an additional 200 words, outline the reasons that a small merchant might want to use an electronic payment system in addition to, or instead of, accepting credit cards.

RQ4. In one paragraph, outline the problems that a company might encounter if it were to conduct international transactions using electronic cash.

RQ5. In about 100 words, explain what electronic wallets are and how they can be useful to consumers.

RQ6. In about 200 words, outline the advantages and disadvantages of smart cards for online merchants.

Exercises

E1. Matt Remes has formed a small business and has just completed building an electronic commerce Web site that sells subscriptions to special-interest newsletters. The titles range from *Apple Growers Digest* to *Wilderness Backpacking Newsletter*. Many organizations and individuals produce the newsletters, and Matt's role is to raise the visibility of these somewhat obscure publications. The newsletters are published and available either biweekly or monthly. Unlike traditional subscription services, Matt's business has an agreement from all newsletter publishers that he can sell subscriptions for single issues or for periods of up to three years. He does not want to allow subscribers to use their payment cards to purchase a subscription that is less than two years in duration. But he finds that nearly 60 percent of the first-time customers on his site prefer to order a sample issue before committing to a subscription of a year or more. Discuss this case and present possible solutions to the problem. In about 200 words, describe existing systems that Matt could use to provide his subscribers with a system that does not depend on payment cards.

E2.   Bonnie Carson has owned and managed her gift and card shop in the Central Shopping Mall for three years. Business has been good, but Bonnie wanted to expand her business. One year ago, she hired a Web designer and built a Web site hosted by a national Internet service provider. Part of the monthly ISP fee for her merchant site includes the software needed to process credit card purchases. She has obtained a merchant account with a national credit card processing company. Bonnie's Web-based business is beginning to pick up. She wants to provide more payment options to her customers. Write a report in which you advise Bonnie on the use of payment processing services such as **PayPal**. Identify at least three reasons that Bonnie should use such a service and at least three reasons why she should not.

E3.   Evan Moskowitz has formed an Internet training company called Teach-U-Comp to market and sell computer courses online. The first courses the company will offer online are introductions to computer programming languages, including Visual Basic .NET, Java, and C++. Students can sign up for as many courses as they want, and each course takes four weeks to complete. Each course costs $95, and students receive continuing education units (CEUs) based on the duration of the course and its level of difficulty. Evan is busy creating the online content and installing the course delivery software, and he hired you to investigate the feasibility of implementing an electronic wallet payment system in addition to the site's existing credit card payment system. Investigate available electronic wallet software, such as **Microsoft Passport** and **Gator**. You should also review the current status of the Electronic Commerce Modeling Language **(ECML)**. Write a 400-word report of your findings for Evan. Conclude your report with specific recommendations.

E4.   During the Internet business expansion of the late 1990s, several major banks launched peer-to-peer payment systems. None of these systems was successful in competing with the PayPal system you learned about in this chapter. Two of the bank systems were Citibank's c2it and Bank One's eMoneyMail. In about 300 words, outline the reasons why you believe these two banks were unable to overcome PayPal's first-mover advantage. You can use your library, links in the Online Companion, and your favorite search engines to conduct your research.

Cases

C1. First Internet Bank of Indiana

During the first wave of electronic commerce, many established banks opened online branches and a considerable number of new, completely online, banks were formed. Many of these online banking initiatives were closed, sold, or merged into other operations after the first wave of electronic commerce had subsided. By 2001, many notable names that had dominated the first wave were gone. For example, Bank One had closed its online subsidiary Wingspan Bank and merged its operations into its existing retail banking department. Royal Bank of Canada had done the same thing with its Security First Network Bank (generally believed to have been the first online bank). CompuBank and G&L Internet Bank were both sold to other banks and USABancshares.com was closed in a flurry of fraud accusations and regulatory concerns.

Many of these early online banks faced similar challenges. They often bought loans instead of originating them. Purchased loans yield lower interest income because the originating bank always charges a fee or discount. They also tended to pay higher rates on customer deposits to attract new customers. These routes to rapid growth can significantly reduce profitability. Physical banks with many branches gain customers and market share because people walk or drive by a branch office and see the bank's name. Online banks must buy advertising that establishes them as viable brands in a highly competitive market. The need to purchase advertising also reduces profits. Small businesses were reluctant to deal with online banks in the early years of their existence. Small businesses generate considerable profits for banks because they tend to borrow money at relatively high interest rates and also tend to keep large balances in their checking accounts. Thus, there were a number of challenges that made survival difficult for online banks.

In 2004, the U.S. Federal Deposit Insurance Corporation (FDIC) issued a report on "limited-purpose banks" (which included Internet banks) in its *Future of Banking Study* series. The FDIC report concluded that the economics of operating an online bank were not attractive and that very few such banks could ever expect to be successful in the long term. The FDIC maintains an informal record of banks that operate primarily as Internet banks. That list recently included a meager 15 bank names. Of those 15, only three operate with no physical branch offices. One of those three is the **First Internet Bank of Indiana** (often called First IB).

First IB was launched in early 1999. By 2001, the bank had become profitable and had more than $200 million in assets. Compared to the large international banks that dominate the industry, $200 million is a relatively small amount (for example, the Bank of America has more than $500 billion in assets), but First IB was able to operate efficiently and with low costs because it had no physical branch offices and very few employees compared to traditional banks.

First IB invested its resources in building the best Web site it could design and then followed a process of continually adjusting the site's design and the services offered to respond to customer comments and requests. For example, First IB created a frequently asked questions (FAQ) feature that reduced customer inquiries dramatically. It was also one of the first banks to offer statements and check images online. In 2004, the bank began to make check images available online the day after the check cleared (the industry average delay at that time was four to seven days). The bank has consistently received excellent reviews of its services by online business rating agencies and in the press.

Required:

1. Create a list of 10 specific concerns that a consumer might have when considering an online bank. Write a paragraph for each concern that describes how First IB addresses or fails to address it.

2. Evaluate how well the design of the First IB Web site meets the needs of a potential small business customer. In about 300 words, discuss the elements of the site that work particularly well in meeting the needs of this type of site visitor. In about 300 words, outline specific changes you would make to the site to better meet the needs of a potential small business customer.

3. Assume you are a security consultant hired by First IB. The president of the bank has become concerned about the potential damage that a phishing expedition directed at First IB customers could do to the bank's reputation. In about 500 words, analyze the phishing threat that faces First IB and outline steps that First IB should take to counter the threat.

Note: Your instructor might assign you to a group to complete this case, and might ask you to prepare a formal presentation of your results to your class.

C2. The Moose Hut

Rod and Martha Nelson started The Moose Hut (TMH), a gift shop in Calgary, Alberta, more than 15 years ago. The Nelsons have capitalized on the tourist trade drawn by the Calgary Stampede, which is one of the largest rodeos in the world. The shop sells a wide range of Canadian-themed items to rodeo fans and other tourists who visit central Alberta throughout the year. TMH's offerings range from inexpensive food items, such as pure Canadian maple syrup and smoked salmon, to much more expensive handcrafted gifts, including Inuit and First Nations artwork. The company's trademark product, the Moose Mug, is one of its biggest-selling items.

Many of TMH's customers return to the store whenever they visit Calgary. TMH's line of Canada Day Party Favours is especially popular with homesick Canadians who have moved to other countries, and TMH has been selling those products by mail order for the past several years. After reviewing the sales numbers for these mail order items, Martha has decided that it might be a good idea to expand the mail order operation and begin accepting orders through a Web site. Many of the store's items have a high value-to-weight ratio and would be easy to ship to customers around the world.

TMH currently accepts only checks denominated in Canadian or U.S. currency in its mail order operation; however, taking orders on a Web site will probably require the company to be more flexible in accepting multiple payment methods. Rod and Martha asked you to help them examine payment processing alternatives for TMH's new Web business.

To be acceptable, a payment processing method needs to handle all major credit cards, perform currency conversions, and be available to a Canadian merchant. Most important is that the payment processing method must be reasonably priced. The margins on most gift items at TMH are between 10 percent and 30 percent of the selling price, but the extra costs of shipping and handling items sold through the Web site will reduce those margins. TMH would like to keep the payment processing costs below 4 percent of the selling price, if possible.

Required:

1. Using the links in the Online Companion for this case, identify at least three payment processing options that might be suitable for TMH. Write a report of about three double-spaced pages in which you describe each of the three payment processing options. Include specific advantages and disadvantages for each option.

2. Prepare a one-page memorandum in which you make a specific recommendation to Rod and Martha. Include an explanation of the reasons for your recommendation.

Note: Your instructor might assign you to a group to complete this case, and might ask you to prepare a formal presentation of your results to your class.

For Further Study and Research

American Banker. 2002. "First Internet of Indiana Turns a Profit Again," 167(95), May 17, 13.

Bach, D. 2001. "Web Stand-alone Model Gets a Lift," *American Banker*, 166(213), November 6, 1–2.

Barlas, P. 2003. "PayPal Pushes for Business Use," *Investor's Business Daily*, October 24, A4.

Bills, S. 2001. "Microsoft Says Aggregation on Site Doesn't Make It a Foe," *American Banker*, 166(167), August 29, 1–2.

Boss, S., D. McGranahan, and A. Mehta. 2000. "Will the Banks Control Online Banking?" *The McKinsey Quarterly*, June, 70–77.

Card News. 2003. "TouchCredit Founder Speaks Out On Biometrics And Online Payment Processing," 18(14), July 9, 1–2.

Chakravorti, S. and T. McHugh. 2002. "Why Do We Use So Many Checks?" *Federal Reserve Bank of Chicago Economic Perspectives*, Third Quarter, 44–59.

Credit Card Management, 2003. "A Dubious Honor for Online Payments," 15(13), March, 14.

DeCarmo, L. 2000. "Security Protocols and Performance," *Dr. Dobb's Journal*, 25(11), November, 40–48.

Dragoon, A. 2004. "Fighting Phish, Fakes, and Frauds," *CIO Magazine*, 17(22), September 1, 33–38.

Drake, C., J. Oliver, and E. Koontz. 2004. "Anatomy of a Phishing Email," *Proceedings of the First Conference on Email and Anti-spam*. Mountain View, CA, July 30.

Dreazen, Y. 2002. "Money Transfers: Too User-Friendly? Legislation Aimed at Stopping Terrorism Could Have a Devastating Impact on an Innocent Bystander: PayPal," *The Wall Street Journal*, October 21, R9.

Electronic Gaming Business. 2003. "Micropayments Promise New Game Revenue Models," 1(8), July 30, 1–2.

Financial Services Distribution. 2004. "Specialist U.S. Banks: Internet Fails But Cards Shine," August 27, 11.

Galbraith, J. 1995. *Money: Whence it Came, Where it Went*. London: Penguin Books.

Gilbert, J. 2001. "Target's Use of Technology Boosts Its Brand Image," *Business 2.0*, June. (http://www.business2.com/marketing/2001/06/brand_technology.htm)

Glasner, J. 2002. "Who'll Pay, Pal, for This IPO?" *Wired News*, February 5. (http://www.wired.com/news/ ebiz/0,1272,50220,00.html)

Grant, D. 2001. "Internet Banking Nightmare: Couple Sue After Access to Their Funds Was Cut Off for 10 Crucial Days," *EastSideJournal.com*, June 10. (http://www.eastsidejournal.com/sited/story/html/56486)

Grimes, B. 2001. "Forgot Your Site Password? Just Yodlee," *PC World*, 19(2), December, 59.

Hammersley, B. 2003. "Online: Making the Web Pay," *The Guardian*, August 7, 24.

Hisey, P. 2001. "Credit Card Fraud Hurts E-Tailers," *Retail Merchandiser*, 41(9), September, 33–34.

Kingston, J. 2001. "The Tech Scene: Don't Spend Your Last Flooz on Web Money," *American Banker*, 166(157), August 15, 1–2.

Kingston, J. 2003. "E-Pay Overtaking Paper; Clients Want More Integration," *American Banker*, 168(81), April 29, 21.

Kuykendall, L. 2003. "Citi to Pull the Plug on c2it Next Month," *American Banker*, October 1, 7.

Lewis, H. 2001. "NetBank, CompuBank Merge, Customers Get Squashed," *Bankrate.com*, May 22. (http://www.bankrate.com/bzrt/news/ob/20010521a.asp)

528

Magnusson, P. 2001. "Yes, They Certainly Will," *Business Week*, November 5, 90–91.

Mantel, B. and T. McHugh. 2002. *Changing E-Payment Payment Networks in the U.S.: The Strategic, Competitive & Innovative Implications.* Chicago: Federal Reserve Bank of Chicago.

Markoff, J. 2002. "Vulnerability Is Discovered in Security for Smart Cards," *The New York Times*, May 13. (http://www.nytimes.com/2002/05/13/technology/13SMAR.html)

Marlin, S. 2003. "Who Needs Cash?" *Information Week*, December 29, 20–22.

McHugh, T. 2002. "The Growth Of Person-To-Person Electronic Payments," *Chicago Fed Letter*, August, Number 180. (http://www.chicagofed.org/publications/fedletter/2002/)

Mearian, L. 2005. "Wells Fargo Buys into Check Image Sharing," *Computerworld*, January 14. (http://www.computerworld.com/databasetopics/data/story/0, 10801,98966,00.html)

Miles, S. 2002. "What's a Check? After Years of False Starts, Online Banking Is Finally Catching On," *The Wall Street Journal*, October 21, R5.

Mulligan, P. and S. Gordon. 2002. "The Impact of Information Technology on Customer and Supplier Relationships in the Financial Services," *International Journal of Service Industry Management*, 13(1), 29–46.

Nahnybida, S. 2003. "Expectations Unfulfilled on E-Billing, E-Payments," *Bank Technology News*, 16(10), October, 62–63.

Nerurkar, U. 2000. "Security Analysis & Design," *Dr. Dobb's Journal*, 25(11), November, 50–56.

Orr, B. 2002. "EPN Wants To Be the Payments Backbone of E-Commerce," *ABA Banking Journal*, 94(12), December, 52–54.

Ptacek, M. 2001. "CompuBank's Demise May Signal a New Era," *American Banker*, 166(63), April 2, 16.

Quain, J. 2003. "Can You Spare Some Change?" *PC Magazine*, 22(23), December 30, 25.

Ramsaran, C. 2004. "Catch of the Day: Banks Face New Phishing Scams," *Bank Systems & Technology*, December 1, 13.

Ramstad, E. 2004. "Hong Kong's Money Card Is a Hit," *The Wall Street Journal*, February 19, B3.

Rist, C. 2003. "Making Bank on Small Change," *Business 2.0*, 4(10), November, 56–57.

Rob, M. and E. Opara. 2003. "Online Credit Card Processing Models: Critical Issues to Consider by Small Merchants," *Human Systems Management*, 22(3), 133–142.

Roberts-Witt, S. 2001. "Show Me the Money," *PC Magazine*, 20(6), March 20, 13–15.

Robinson, B. 2001. "Is It Too Late for Smart Cards?" *Information Week*, March 19, 81–83.

Rosato, D. 2004. "Why Are You Still Writing Checks?" *Money*, 33(1), January, 94–97.

Roth, A. 2001. "CompuBank Merge Nettles NetBank," *American Banker*, 166(119), June 21, 1–2.

Rush, L. 2003 "Get Paid: P2P Payments Gaining Consumer Trust," *E-Commerce Guide*, November 18. (http://ecommerce.internet.com/how/paid/article/0,10364_3110831,00.html)

Sapsford, J. 2000. "American Express Credit Cards to Offer Disposable Numbers for Web Shopping," *The Wall Street Journal*, September 8, B6.

Schwartz, E. 2001. "Digital Cash Payoff," *Technology Review*, 104(10), December, 62–68.

Scucka, D. 2001. "Charging Into Japan: eCharge Thinks Japan's Consumers Will Take to Its Net-Based System," *J@pan Inc*, 3(4), April, 70–72.

Smith, G. 2002. "Account Aggregation Falls Apart," *Business Week*, July 2. (http://www.businessweek.com/ technology/content/jul2002/tc2002072_3404.htm)

Stoneman, B. 2003. "FAQs Lighten Service Load at First Internet Bank of Indiana," *American Banker*, 168(2), January 13, 12.

Sturgeon, J. 2003. "Electronic Payments," *CFO Magazine*, 19(15), Winter, 52–53.

529

Payment Systems For Electronic Commerce

Tedeschi, B. 2001. "Online Billing, An Idea That Looked Like a Winner, Has Held Its Own," *The New York Times*, June 4, C8.

United States Census Bureau. 2004. *2004-2005 Statistical Abstract of the United States*. Washington, D.C.: U.S. Census Bureau.

Walker, L. 2001. "The Slow Evolution of Online Bill Paying," *The Washington Post*, September 6, E1.

Wingfield, N. and J. Sapsford. 2002. "eBay to Buy PayPal for $1.4 Billion," *The Wall Street Journal*, July 9, A6.

Yom, C. 2004. "Limited-purpose Banks: Their Specialties, Performance, and Prospects," *FDIC Future of Banking Study Series*, June, 1–45. Washington, D.C.: Federal Deposit Insurance Corporation (FDIC).

PART 4

INTEGRATION

CHAPTER **12**

PLANNING FOR ELECTRONIC COMMERCE

LEARNING OBJECTIVES

In this chapter, you will learn about:

- Planning electronic commerce initiatives
- Strategies for developing electronic commerce Web sites
- Managing electronic commerce implementations

INTRODUCTION

AlliedSignal (now **Honeywell**) is a diversified manufacturing and technology business selling products in the aerospace, automotive, chemicals, fibers, and plastics industries. In 1999, the company had more than 70,000 employees and annual sales exceeding $15 billion. Although some of AlliedSignal's products used new technologies, or helped other firms create new technologies, many of the products were commodity items that were manufactured and sold just as they had been for decades. In early 1999, AlliedSignal's CEO, Larry Bossidy, called together the heads of the company's business units for a one-day conference. He invited Michael Dell, chairman and CEO of **Dell Computers**, and John Chambers, CEO of **Cisco Systems**, to speak about their companies' electronic commerce implementation successes.

At the end of the day, Bossidy gave the business unit heads their marching orders. They were to take what they had learned and create a strategy for implementing electronic commerce in their business units—in two months. Bossidy told the room full of rather stunned managers that, although most of their business units were at or near the top of their industries, the Internet would change everything. He believed that the kinds of electronic commerce strategies that had worked so well for Dell and Cisco in the computer industry could also work in many of AlliedSignal's businesses. He wanted to make sure that AlliedSignal was the first to exploit those strategies and any other strategies that the business managers could devise. In two months, each manager reported back with a strategy that included multiple electronic commerce projects, such as Web sites for selling products, providing customer service, improving corporate infrastructure, managing supply chains, coordinating logistics, holding auctions, and creating virtual communities. These plans were evaluated in the company's annual strategic planning process, and the best ones were chosen for funding and immediate implementation. In a matter of months, one of the largest industrial enterprises in the world had drastically altered its course, setting sail for the uncharted waters of electronic commerce.

PLANNING ELECTRONIC COMMERCE INITIATIVES

The ability of companies to plan, design, and implement cohesive electronic commerce strategies will make the difference between success and failure for the majority of them. The tremendous leverage that firms can gain by being the first to do business a new way on the Web has caught the attention of top executives in many industries. The keys to successful implementation of any information technology project are planning and execution. This chapter provides some useful guidelines for those readers who will manage the planning, implementation, and continuing operations of electronic commerce initiatives. A successful business plan for an electronic commerce initiative should include activities that identify the initiative's specific objectives and link those objectives to business strategies (strategies that you learned about in Chapters 3, 4, 5, and 6).

In setting the objectives for an electronic commerce initiative, managers should consider the strategic role of the project, its intended scope, and the resources available for executing it. In this section, you will learn how to identify objectives and link those business objectives to business strategies. In later sections of this chapter, you will learn about Web site development strategies and how to manage the implementation of an electronic commerce initiative.

Identifying Objectives

Businesses undertake electronic commerce initiatives for a wide variety of reasons. Objectives that businesses typically strive to accomplish through electronic commerce include: increasing sales in existing markets, opening new markets, serving existing customers better, identifying new vendors, coordinating more efficiently with existing vendors, or recruiting employees more effectively.

The types of objectives vary with the size of the organization. For example, small companies might want a Web site that encourages site visitors to do business using existing channels rather than through the Web site itself to reduce the cost of the site. A site that offers only product or service information is much less expensive to design, build, and maintain than a site that offers transaction handling, bidding, communications, or other capabilities. Decisions regarding resource allocations for electronic commerce initiatives should consider the expected benefits and costs of meeting the objectives. These decisions should also consider the risks inherent in the electronic commerce initiative and compare them to the risks of inaction—a failure to act could concede a strategic advantage to competitors.

Linking Objectives to Business Strategies

Businesses can use tactics called **downstream strategies** to improve the value that the business provides to its customers. Alternatively, businesses can pursue **upstream strategies** that focus on reducing costs or generating value by working with suppliers or inbound shipping and freight service providers.

In earlier chapters of this book, you learned about many of the things that companies are doing on the Web. The Web is a tremendously attractive sales channel for many firms; however, companies can use electronic commerce to do much more than selling. They can use the Web to complement their business strategies and improve their competitive positions. Electronic commerce opportunities can inspire businesses to undertake activities such as:

- Building brands
- Enhancing existing marketing programs
- Selling products and services
- Selling advertising
- Developing a better understanding of customer needs
- Improving after-sale service and support
- Purchasing products and services
- Managing supply chains
- Operating auctions
- Building virtual communities and Web portals

The success of these activities can be difficult to measure. In the first wave of electronic commerce, many companies engaged in these activities on the Web without setting specific, measurable goals. In the mid-1990s companies that had good ideas could find plenty of investors and start a business activity on the Web. These early activities usually did not face much competition. Successes and failures were measured in broad strokes. A company would either become the eBay of its industry or it would disappear by slipping into bankruptcy or being acquired by another company.

In the second wave of electronic commerce, more companies have begun taking a closer look at the benefits and costs of their electronic commerce projects. Measuring both benefits and costs is becoming more important. A good implementation plan should set specific objectives for benefits to be achieved and costs to be incurred. In many cases, a company will create a pilot Web site to test an online business idea and then release a production version of the site when it works well. Companies must specify clear goals for their pilot tests so that they know when the site is ready to go into full operation.

Measuring Benefits

Some benefits of electronic commerce initiatives are tangible and easy to measure. These include such things as increased sales or reduced costs. Other benefits are intangible and can be much more difficult to measure, such as increased customer satisfaction. When identifying benefit objectives, managers should try to set objectives that are measurable, even when those objectives are for intangible benefits. For example, success in achieving a goal of increased customer satisfaction might be measured by counting the number of first-time customers who return to the site and buy.

Many companies create Web sites to build brands or enhance their existing marketing programs. These companies can set goals in terms of increased brand awareness, as measured by market research surveys and opinion polls. Companies that sell goods or services online can measure sales volume in units or dollars. A complication that occurs in measuring either brand awareness or sales is that the increases can be caused by other things that the company is doing at the same time or by a general improvement in the economy. A good marketing staff or outside consulting firm can help a company sort out the effects of marketing and sales programs. Firms may need these groups to help set and evaluate these kinds of goals for electronic commerce initiatives.

Companies that want to use Web sites to improve customer service or after-sale support might set goals of increased customer satisfaction or reduced costs of providing customer service or support. For example, **Philips Lighting** wanted to use the Web to provide an ordering system for its smaller customers that did not use EDI. The primary goal for this initiative was to reduce the cost of processing smaller orders. Philips had identified that responding to inventory availability and order status inquiries accounted for over half the cost of processing smaller orders. Customers who placed small orders often called or sent faxes asking for this information.

Philips built a pilot Web site and invited a number of its smaller customers to try it. The company found that customer service phone calls from the test group of customers dropped by 80 percent. Based on that measurable increase in efficiency, Philips decided to invest in additional hardware and personnel to staff a version of the Web site that could handle virtually all of its smaller customers. The reduction in the cost of handling small orders justified the additional investment.

Companies can use a variety of similar measurements to assess the benefits of other electronic commerce initiatives. Supply chain managers can measure supply cost reductions, quality improvements, or faster deliveries of ordered goods. Auction sites can set goals for the number of auctions, the number of bidders and sellers, the dollar volume of items sold, the number of items sold, or the number of registered participants. The ability to track such numbers is usually built into auction site software. Virtual communities

535

and Web portals measure the number of visitors and try to measure the quality of their visitors' experiences.

Some sites use online surveys to gather this data; however, most settle for estimates based on length of time that each visitor remains on the site and how often visitors return. A summary of benefits and measurements that companies can make to assess the value of those benefits (these measurements are often called **metrics**) appears in Figure 12-1.

Electronic commerce initiatives	Common measurements of benefits provided
Build brands	Surveys or opinion polls that measure brand awareness
Enhance existing marketing programs	Change in per-unit sales volume
Improve customer service	Customer satisfaction surveys, quantity of customer complaints
Reduce cost of after-sale support	Quantity and type (telephone, fax, e-mail) of support activities
Improve supply chain operation	Cost, quality, and on-time delivery of materials or services purchased
Hold auctions	Quantity of auctions, bidders, sellers, items sold, registered participants; dollar volume of items sold
Provide portals and virtual communities	Number of visitors, number of return visits per visitor, and duration of average visit

FIGURE 12-1   Measuring the benefits of electronic commerce initiatives

No matter how a company measures the benefits provided by its Web site, it usually tries to convert the raw activity measurements to dollars. Having the benefits measured in dollars lets the company compare benefits to costs and compare the net benefit (benefits minus costs) of a particular initiative to the net benefits provided by other projects. Although each activity provides some value to the company, it is often difficult to measure that value in dollars. Usually, even the best attempts to convert benefits to dollars yield only rough approximations.

Managing Costs

At first glance, the task of identifying and estimating costs may seem much easier than the task of setting benefits objectives. However, many managers have found that information technology project costs can be as difficult to estimate and control as the benefits of those projects. Since Web development uses hardware and software technologies that change even more rapidly than those used in other information technology projects, managers often find that their experience does not help much when they are making estimates. Most changes in the cost of hardware are downward, but the increasing sophistication of software provides an ever increasing demand for more of the newer, cheaper hardware. This often yields a net increase in overall hardware costs. The more sophisticated software, of course, usually costs more than the amount originally budgeted, too. Even though electronic commerce initiatives tend to be completed within a shorter time frame than many other information technology projects, the rapid changes in Web technology can quickly destroy a manager's best-laid plans.

Total Cost of Ownership

In addition to hardware and software costs, the project budget must include the costs of hiring, training, and paying the personnel who will design the Web site, write or customize the software, create the content, and operate and maintain the site. Many organizations now track costs by activity and calculate a total cost for each activity. These cost numbers, called **total cost of ownership (TCO)**, include a wide variety of costs related to the activity. The TCO of an electronic commerce implementation includes the costs of hardware (server computers, routers, firewalls, and load balancing devices), software (licenses for operating systems, Web server software, database software, and application software), design work outsourced, salaries and benefits for employees involved in the project, and the costs of maintaining the site once it is operational. A good TCO number would, for example, include assumptions about how often the site would need to be redesigned in the future. You can learn more about TCO by visiting the **Computerworld Total Cost of Ownership Quickstudy** Web page.

537

Change Management

Any information system project involves change, and change can be upsetting to people. As employees of an organization become accustomed to their specific duties, many of them draw comfort from their knowledge and develop a sense of security because they know their jobs well and are good at doing them. When changes are introduced into a workplace, employees become concerned about their abilities to cope with the changes and with their ability to continue to do good work. They often become worried that they might lose their jobs. These concerns can lead to increased stress that can be damaging to morale and work performance. Management researchers have developed strategies for **change management**, which is the process of helping employees cope with these changes. Change management techniques include communicating the need for change to employees, including employees in the decision processes leading up to the change, allowing employees to participate in the planning for the change, and other tactics designed to help employees feel that they are a part of the change. This helps employees overcome the feelings of powerlessness that can lead to stress and reduced work performance.

Opportunity Costs

For many companies, one of the largest and most significant costs associated with electronic commerce initiatives is the cost of not undertaking such an initiative. The foregone benefits that a company could have obtained from an electronic commerce initiative that they chose not to pursue are costs. Managers and accountants use the term **opportunity cost** to describe such lost benefits from an action not taken.

Web Site Costs

Based on data collected in separate recent surveys, International Data Corporation and Gartner, Inc. both estimated that the cost for a large company to build and implement an adequate entry-level electronic commerce site was about $1 million. About 79 percent of this cost was labor related; 10 percent was the cost of software and 11 percent was the cost of hardware. A Gartner study concluded that it would take between $2 million and $5 million to build a site that would compare favorably to leading sites. International Data Corporation noted that 10 of the top 100 electronic commerce sites had spent over $10 million for development and implementation.

Although a small company can put a Web site online for under $4000, the TCO for an electronic commerce implementation with full transaction and payment processing capabilities is difficult to keep under $8000 per year. In fact, recent surveys of smaller companies showed that their expenditures on construction of new electronic commerce Web sites averages $110,000. Industry analysts have pegged the minimum dollar amount needed to open a complete electronic commerce Web site at $100,000.

Gartner estimates that establishing a basic electronic commerce operation on the Web today will cost a company between $100,000 and $1 million, and creating a site that is noticeably ahead of most competitors' sites will cost a minimum of $15 million. Figure 12-2 summarizes industry estimates for the cost of creating a Web business at three different levels: a basic entry level, a level comparable to most existing Web competitors, and a level that makes the Web site stand out as noticeably different from competitors' sites (in Gartner's terms, a "true differentiator").

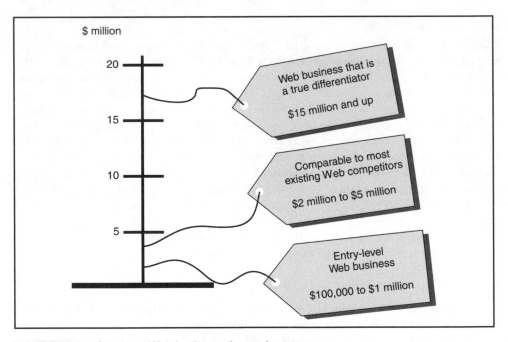

FIGURE 12-2   Starting a Web business: three price tags

Using the initial cost of building an electronic commerce site to make decisions about launching online business activities can be misleading. Web technology continues to evolve at a rapid pace and most businesses want to take advantage of what that technology offers to remain competitive. Most experts agree that the annual cost to maintain and improve a site once it is up and running—whether it is a small site or a large site—will be between 50 percent and 200 percent of its initial cost. Thus, ongoing maintenance costs can be a more significant factor in making implementation decisions than the initial cost of building a site.

In a 2001 article, members of the management consulting firm McKinsey & Company reported a study that estimated start-up and ongoing costs for magazine publishers' Web sites (see the Barsh, et al. reference in the For Further Study and Research section at the end of this chapter for a reference to the full report). The McKinsey study estimated costs for two types of magazine sites: a full portal site that served as a destination in itself, and a more limited magazine companion site that complemented a printed magazine. The full portal site cost estimate was $2.4 million to build and $4.3 million per year to maintain, with a staff of 35 people. The companion site cost estimate was $150,000 to build and $270,000 per year to maintain, with a staff of two people. Both of these estimates exclude the cost of developing content for the site and assume that the magazine publisher already has an existing IT infrastructure for a print publishing business serving a subscriber base of 300,000. Figure 12-3 shows the approximate breakdown of these costs. Although these estimates are now a few years old, they still provide a rough idea of the range of costs that can be incurred for different types of online business operations.

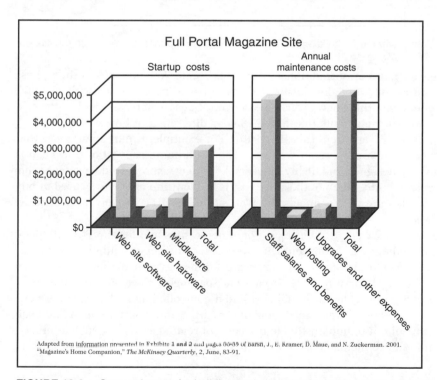

FIGURE 12-3   Cost estimates for building and operating magazine publisher Web sites

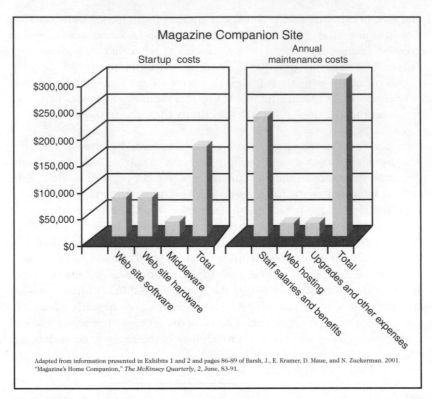

Adapted from information presented in Exhibits 1 and 2 and pages 86-89 of Barsh, J., E. Kramer, D. Maue, and N. Zuckerman. 2001. "Magazine's Home Companion," *The McKinsey Quarterly*, 2, June, 83-91.

FIGURE 12-3   Cost estimates for building and operating magazine publisher Web sites (continued)

As an increasing number of traditional businesses create Web versions of their physical stores, the cost to build an online business that is a true differentiator—with a site that stands out and offers something new to customers—continues to increase. Much of the cost for such a Web site is for elements that make a major difference in how well the site works, but are not readily apparent to a site visitor. For example, **Kmart** spent more than $140 million to create its online retail Web site. The site's home page, shown in Figure 12-4, is certainly well designed and highly functional, but few visitors would ever guess how much this site cost to build. Much of the site's cost is hidden; the money was used to buy and customize middleware that connects the Web site to Kmart's vast inventory and logistics databases.

The high price tags for creating electronic commerce sites and for the TCO of operating and maintaining them can be discouraging to smaller businesses and organizations. However, as you learned in Chapter 9, smaller organizations can control their costs by using a combination of a third-party hosting service and packaged electronic commerce software. These options provide low initial cost and a controlled annual TCO. However, organizations that use these lower cost options for creating and maintaining online businesses must be careful not to underestimate the costs of related activities, such as creating and maintaining a product catalog or Web site content.

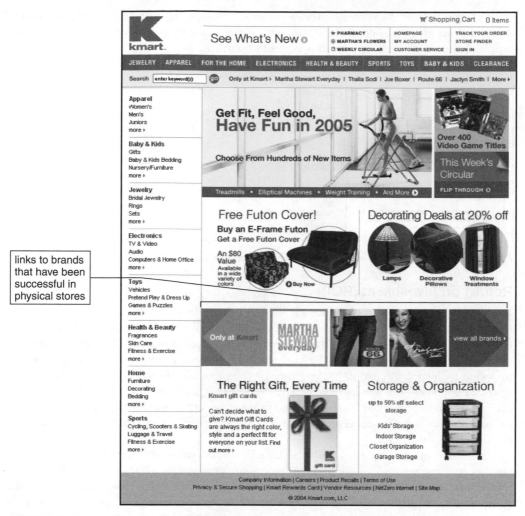

links to brands that have been successful in physical stores

541

FIGURE 12-4 Kmart's online store home page

Comparing Benefits to Costs

Most companies have procedures that call for an evaluation of any major expenditure of funds. These major investments in equipment, personnel, and other assets are called **capital projects** or **capital investments**. The techniques that companies use to evaluate proposed capital projects range from very simple calculations to complex computer simulation models. However, no matter how complex the technique, it always reduces to a comparison of benefits and costs. If the benefits exceed the costs of a project by a comfortable margin, the company invests in the project.

A key part of creating a business plan for electronic commerce initiatives is the process of identifying potential benefits (including intangibles such as employee satisfaction and company reputation), identifying the costs required to generate those benefits, and evaluating whether the benefits exceed the costs. Companies should evaluate each element of their electronic commerce strategies using this cost/benefit approach. A representation of the cost/benefit approach appears in Figure 12-5.

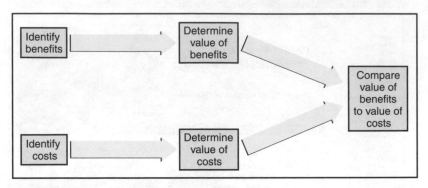

FIGURE 12-5   Cost/benefit evaluation of electronic commerce strategy elements

Return on Investment (ROI)

You might have learned techniques for capital project evaluation, such as the payback method, the net present value method, or the internal rate of return method, in your accounting or finance courses. These evaluation approaches are called **return on investment (ROI)** techniques because they measure the amount of income (return) that will be provided by a specific current expenditure (investment). ROI techniques provide a quantitative expression of a comfortable benefit-to-cost margin for a specific company. They can also mathematically adjust for the reduced value of benefits that the investment will return in future years (benefits received in future years are worth less than those received in the current year).

Although most companies evaluate the anticipated value of electronic commerce initiatives in some way before approving them, many companies see these projects as absolutely necessary investments. Thus, businesses might not subject these initiatives to the same close examination and rigid requirements as other capital projects. These companies fear being left behind as competitors stake their claims in the online marketspace. The value of early positioning in a new market is so great that many companies are willing to invest large amounts of money with few near-term profit prospects.

Newspaper Web sites are a good example of this desire to establish a foothold in the online marketspace. In the first wave of electronic commerce, there were only a few profitable newspaper sites (such as Gannet's **USA Today** and *The Wall Street Journal's* **WSJ.com** sites). Most newspaper sites took several years to become profitable. As you learned earlier in this book, an increasing number of newspaper sites now charge subscriptions for access to parts of their sites, and other newspaper sites charge for access to archived articles. Despite their early losses, most newspaper companies believed that they could not afford to ignore the long-term potential of the Web. These companies calculated their opportunity costs of not being present on the Web (for example, the loss of future profits to be earned from the Web site or the risk of losing market share to competitors) to be greater than the losses they experienced when they started their sites.

In the second wave of electronic commerce, more companies are taking a hard look at any expenditure related to the Web. Many companies have turned to ROI as the measurement tool for evaluating new electronic commerce projects because that is what they used for other IT projects in the past. ROI is a simple-to-understand tool that is easily applied; however, managers should be careful when using it to evaluate online business initiatives. ROI has some built-in biases that can lead managers to make poor decisions.

First, ROI requires that all costs and benefits be stated in dollars. Because it is usually easier to quantify costs than benefits, ROI measurements can be biased in a way that gives undue weight to costs. Second, ROI focuses on benefits that can be predicted. Many electronic commerce initiatives have returned benefits that were not foreseen by their planners. The benefits developed after the initiatives were in place. For example, Cisco Systems created online customer forums to allow customers to discuss product issues with each other. The main benefits from this initiative were to reduce customer service costs and increase customer satisfaction regarding the availability of product information. In addition, the forums turned out to be a great way for Cisco engineers to get feedback from customers on new products that they were developing. This second use was not foreseen by the project's planners and has become the most important and beneficial outcome of the customer forums. An ROI analysis would have missed this benefit completely.

Yet another weakness of ROI is that it tends to emphasize short-run benefits over long-run benefits. The mathematics of ROI calculations do account for both correctly, but short-term benefits are easier to foresee, so they tend to get included in the ROI calculations. Long-term benefits are harder to imagine and harder to quantify, so they tend to be included less often and less accurately in the ROI calculation. This biases ROI calculations to weigh short-term costs and benefits more heavily than long-term costs and benefits. This can lead managers who rely on ROI measures to make incorrect decisions. You can learn more about this topic at the *CIO* **E-business Research Center on ROI** and the ***Computerworld* ROI Knowledge Center**.

STRATEGIES FOR DEVELOPING ELECTRONIC COMMERCE WEB SITES

When companies began establishing their presences on the Web, the typical Web site was a static brochure that was not updated frequently with new information and seldom had any capabilities for helping the company's customers or vendors transact business. As Web sites have become the home not only of transaction processing but also of automated business processes of all kinds, these Web sites have become important parts of companies' information systems infrastructures. The evolution of Web site functions—from the static brochures of the early days of electronic commerce, to transaction-processing tools, to today's automated homes for business processes of all kinds—appears in Figure 12-6.

This transformation occurred rapidly—taking only a year or two in most companies. Because the change in the focus of Web sites happened so fast, very few businesses were able to change the way they developed and managed their Web sites to meet the demands of this new focus. Although the purposes and scope of business Web sites have increased greatly, few businesses manage their Web sites as the dynamic business applications they have become. Fortunately, large companies have over the years developed tools that they use to manage their software development projects. As companies begin to see their Web sites

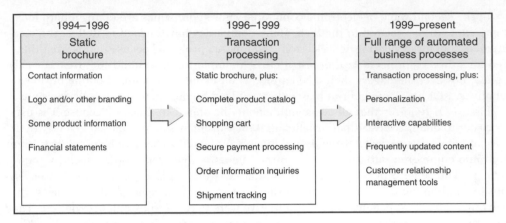

1994–1996	1996–1999	1999–present
Static brochure	**Transaction processing**	**Full range of automated business processes**
Contact information	Static brochure, plus:	Transaction processing, plus:
Logo and/or other branding	Complete product catalog	Personalization
Some product information	Shopping cart	Interactive capabilities
Financial statements	Secure payment processing	Frequently updated content
	Order information inquiries	Customer relationship management tools
	Shipment tracking	

FIGURE 12-6   Increasing complexity of Web site functions

as collections of software applications, they are beginning to use these tools to manage the development and maintenance of their Web sites.

Many large and midsize companies have found it extremely difficult to develop new information systems and Web sites that work with such systems to create new markets or reconfigure their supply chains. In the past, companies that have had success in exploring new ways of working with their customers and suppliers by reconfiguring supply chains have had the luxury of time—in many cases, years—to complete those reconfigurations. However, the speed at which the Internet has changed markets and marketing channels throughout entire industry value chains precludes lengthy reconfigurations. Now, companies that want to successfully adapt to the changed business environment of the information age must explore alternatives to traditional systems development methods. These alternatives include the incubator and fast venturing approaches that you will learn about later in this chapter.

Internal Development vs. Outsourcing

Although many companies would like to think that they can avoid electronic commerce site development problems by outsourcing the entire project, savvy leaders realize that they cannot. No matter what kind of electronic commerce initiative a company is contemplating, the initiative's success depends on how well it is integrated into and supports the activities in which the business is already engaged. Using internal people to lead all projects helps to ensure that the company's specific needs are addressed and that the initiative is congruent with the goals and the culture of the organization. Outside consultants are seldom able to learn enough about an organization's culture to accomplish these objectives. However, few companies are large enough or have sufficient in-house expertise to launch an electronic commerce project without some external help. Even Wal-Mart, with annual sales of more than $150 billion, did not undertake its 2000 Web site relaunch alone. The key to success is finding the right balance between outside and inside support for the project. Hiring another company to provide the outside support for all or part of the project is called **outsourcing**.

The Internal Team

The first step in determining which parts of an electronic commerce project to outsource is to create an internal team that is responsible for the project. This team should include people with enough knowledge about the Internet and its technologies to know what kinds of things are possible. Team members should be creative thinkers who are interested in taking the company beyond its current boundaries, and they should be people who have distinguished themselves in some way by doing something very well for the company. If they are not already recognized by their peers as successful individuals, the project may suffer from lack of credibility.

Some companies make the mistake of appointing as electronic commerce project leader a technical wizard who does not know much about the business and is not well-known throughout the company. Such a choice can greatly increase the likelihood of failure. Business knowledge, creativity, and the respect of the firm's operating function managers are all much more important than technical expertise in establishing successful electronic commerce. Project leaders need a good sense of the company's goals and culture to manage an implementation effectively.

Measuring the achievements of this internal team is very important. The measurements do not have to be monetary. Achievement can be expressed in whatever terms are appropriate to the objectives of the initiative. Customer satisfaction, number of sales leads generated, and reductions in order-processing time are examples of metrics that can provide a sense of the team's level of accomplishment. The measurements should show how the project is affecting the company's ability to provide value to the consumer. Many consultants advise companies to set aside between 5 percent and 10 percent of a project's budget for quantifying the project's value and measuring the achievement of that value.

Increasingly, companies are recognizing the value of the intellectual capital built up in the form of employees' knowledge about the business and its processes. In the past, many companies ignored the value of their human assets because such resources did not appear in the accounting records or financial statements.

Leif Edvinsson has pioneered the use of human capital measures at Skandia Group, a large financial services company in Sweden. In addition to acknowledging employees' competencies, Edvinsson's measures include the value of customer loyalty and business partnerships as part of a company's intellectual capital. This networking approach to evaluating intellectual capital shows promise as a tool for assessing and tracking the value of internal teams and their connections to external consultants. These measurements are now being adapted for use in measuring systems development efforts. You can learn more about the use of human capital measurements by reading the books by Edvinsson and Max Boisot, another proponent of human capital measurement, which are included in the For Further Study and Research section at the end of this chapter.

The internal team should hold ultimate and complete responsibility for the electronic commerce initiative, from the setting of objectives to the final implementation and operation of the site. The internal team decides which parts of the project to outsource, to whom those parts are outsourced, and what consultants or partners the company needs to hire for the project. Consultants, outsourcing providers, and partners can be extremely important early in the project because they often develop skills and expertise in new technologies before most information systems professionals.

Early Outsourcing

In many electronic commerce projects, the company outsources the initial site design and development to launch the project quickly. The outsourcing team then trains the company's information systems professionals in the new technology before handing the operation of the site over to them. This approach is called **early outsourcing**. Since operating an electronic commerce site can rapidly become a source of competitive advantage for a company, it is best to have the company's own information systems people working closely with the outsourcing team and developing ideas for improvements as early as possible in the life of the project.

Late Outsourcing

In the more traditional approach to information systems outsourcing, the company's information systems professionals do the initial design and development work, implement the system, and operate the system until it becomes a stable part of the business operation. Once the company has gained all the competitive advantage provided by the system, the maintenance of the electronic commerce system can be outsourced so that the company's information systems professionals can turn their attention and talents to developing new technologies that will provide further competitive advantage. This approach is called **late outsourcing**. Although for years late outsourcing has been the standard for allocating scarce information systems talent to projects, electronic commerce initiatives lend themselves more to the early outsourcing approach.

Partial Outsourcing

In both the early outsourcing and late outsourcing approaches, a single group is responsible for the entire design, development, and operation of a project —either inside or outside the company. This typical outsourcing pattern works well for many information systems projects. However, electronic commerce initiatives can benefit from a partial outsourcing approach, too. In **partial outsourcing**, which is also called **component outsourcing**, the company identifies specific portions of the project that can be completely designed, developed, implemented, and operated by another firm that specializes in a particular function.

Many smaller Web sites outsource their e-mail handling and response functions. Customers expect rapid and accurate responses to any e-mail inquiry they make of a Web site with which they are doing business. Many companies send the customer an automatic order confirmation by e-mail as soon as the order or credit card payment is accepted. A number of companies provide e-mail autoresponse functions on an outsourcing basis.

Another common example of partial outsourcing is an electronic payment system. Many vendors are willing to provide complete customer payment processing. These vendors provide a site that takes over when customers are ready to pay and returns the customers to the original site after processing the payment transaction.

One of the most common elements of electronic commerce initiatives that companies outsource using this approach is the Web hosting activity that you learned about in Chapter 9. Providers of Internet connectivity, applications, and business services (including ISPs, CSPs, MSPs, and ASPs) offer Web hosting services to companies that want to operate electronic commerce sites, but that do not want to invest in the hardware and staff needed to create their own Web servers. These service providers are usually willing to accommodate requests for a variety of service levels. Small businesses can rent space on an existing server at the ISP's location. Larger companies can purchase the server hardware and have the service provider install and maintain it at the service provider's location. The service provider has the continuous staffing and expertise needed to keep an electronic commerce site up and running 24 hours a day, seven days a week (this kind of service is often called **24/7 operation**). Most service providers offer a wide range of services, including personal Web access for individuals. Some service providers specialize in services to business. These larger service providers cater to companies that want to operate electronic commerce sites. They typically offer wider bandwidth connections to the Internet than smaller service providers and also offer more reliable continuous service.

A number of service providers offer services beyond basic Internet connectivity to companies that want to do business on the Web. Many of these services were described earlier as candidates for partial outsourcing strategies and include automated e-mail response, transaction processing, payment processing, security, customer service and support, order fulfillment, and product distribution.

LEARNING FROM FAILURES

Nordisk Aviation

Nordisk Aviation is a subsidiary of the Norwegian Norsk Hydro Group. It designs, manufactures, and repairs air cargo containers for both freight and passenger baggage for major airlines throughout the world and for freight carriers such as FedEx and UPS. It also designs and sells handling systems and pallets that work with the containers. The company has annual sales of more than $200 million and employs more than 500 people at its 22 locations around the world.

Nordisk was a strong believer in using the outsourcing approach for its IT projects—its IT Department included only two people. These two IT staff members worked as the overseers of every IT design and implementation project for the company. They also managed the ongoing IT services provided to Nordisk by other companies.

In late 2000, Manfred Gollent, the president of Nordisk, decided it was time to upgrade the company's Web site—which had been operating as an information site for several years—to include portal features that would allow Nordisk customers to check order status and learn about current developments in container and container-handling systems design. The logical approach for Nordisk was to find a company to which it could outsource the project.

The two members of Nordisk's IT staff went to work finding suitable Web developers. The previous Web developer had disappeared; they were unable to find any trace of the

continued

Planning for Electronic Commerce

person who had created the existing Web site. The developer had created the Web site so that it used a number of programs to deliver dynamic pages. Unfortunately, the developer had given Nordisk only the executable code and not the actual programs. He also did not provide Nordisk with any documentation of the programs.

When the Web site was initially created, it was not an important strategic project for Nordisk. The IT staff members, who were busy with other important projects, did not ensure that the application code and documentation were received. Nordisk had to hire a company to rebuild the site completely to obtain the additional portal functions it wanted to add to the site. The lesson from the Nordisk case is that even when a company is outsourcing virtually all of its Web development, it must have procedures in place to ensure that the project is internally managed and documented.

Selecting a Hosting Service

The internal team should be responsible for selecting the ISP that will provide the site's hosting service. For smaller electronic commerce projects, teams can consult an ISP directory such as **The List**, which you learned about in Chapter 9. These sites provide a search engine that helps visitors choose an ISP, Web hosting service, or ASP that meets their needs from the sites' thousands of listings.

For larger Web site implementations, the team should obtain the advice of consultants or other firms that rate service providers (ISPs, ASPs, and CSPs), such as **HostCompare.com** and **Keynote Systems**. The most important factors to evaluate when selecting a hosting service include:

- Functionality
- Reliability
- Bandwidth and server scalability
- Security
- Backup and disaster recovery
- Cost

Because the company's information on customers, products, pricing, and other data will be placed in the hands of the service provider, the vendor's security policies and practices are very important, as you learned in Chapter 10. No matter what security guarantees the service provider offers, the company should monitor the security of the electronic commerce operation through its own personnel or by hiring a security consulting firm.

New Methods for Implementing Partial Outsourcing

In the past five years, new ways of implementing the partial outsourcing strategy have evolved specifically for Web businesses. The next two sections describe two of the more popular of these methods; incubators and fast venturing.

Incubators

An **incubator** is a company that offers start-up companies a physical location with offices, accounting and legal assistance, computers, and Internet connections at a very low monthly cost. Sometimes, the incubator offers seed money, management advice, and marketing assistance as well. In exchange, the incubator receives an ownership interest in the company, typically between 10 percent and 50 percent.

When the company grows to the point that it can obtain venture capital financing or launch a public offering of its stock, the incubator sells all or part of its interest and reinvests the money in a new incubator candidate. One of the first Internet incubators was **Idealab**, which helped companies such as **CarsDirect.com**, **Overture**, and **Tickets.com** get their starts.

Some companies have created internal incubators. A number of companies used internal incubators in the past to develop technologies that the companies planned to use in their main business operations. Most of these programs, such as the Kodak internal venturing program of the 1980s, were unsuccessful and, ultimately, were shut down. Employees in internal incubators found it difficult to maintain an entrepreneurial spirit when they knew that the technology they were developing would ultimately be taken away and controlled by the parent company.

More recently, companies such as Matsushita Electric's U.S. Panasonic division have started internal incubators to help launch new companies that will grow to become important strategic partners. The companies launched in the incubator will retain their individual management teams and the assets they develop. The prospects for these strategic partner incubators appear to be much brighter than those of the old-style technology development incubators.

Fast Venturing

Often, large companies struggle to emulate the entrepreneurial spirit of smaller companies as they launch their Internet business initiatives. Many of these companies are trying to expand the internal incubator model and create an effective support system for new business and technology ideas, such as electronic commerce initiatives. One approach that is becoming popular is called fast venturing.

In **fast venturing**, an existing company that wants to launch an electronic commerce initiative joins external equity partners and operational partners that can offer the experience and skills needed to develop and scale up the project very rapidly. Equity partners are usually banks or venture capitalists that sometimes offer money, but are more likely to offer experience gained from guiding other start-ups that they have funded. Operational partners are firms, such as systems integrators, consultants, and Web portals, that have experience in moving projects along and scaling up prototypes. The roles of each participant in fast venturing are described in Figure 12-7.

The venture sponsor is the existing company that wants to launch the electronic commerce initiative. The equity partners are entities that have provided start-up money to new ventures in the past and have developed knowledge about operating new ventures. The equity partners provide advice based on this knowledge to the venture sponsor, which typically has little experience in developing new ventures. The operational partners are people and companies that previously have built Web business sites. Thus, they can provide

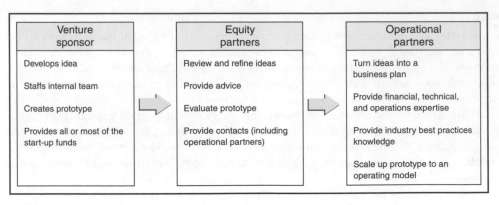

Venture sponsor	Equity partners	Operational partners
Develops idea	Review and refine ideas	Turn ideas into a business plan
Staffs internal team	Provide advice	Provide financial, technical, and operations expertise
Creates prototype	Evaluate prototype	Provide industry best practices knowledge
Provides all or most of the start-up funds	Provide contacts (including operational partners)	Scale up prototype to an operating model

FIGURE 12-7   Elements of fast venturing

expertise in the technologies and business practices needed to create a successful operating electronic commerce site.

MANAGING ELECTRONIC COMMERCE IMPLEMENTATIONS

The best way to manage any complex electronic commerce implementation is to use formal management techniques. Project management, project portfolio management, specific staffing, and postimplementation audits are methods businesses use to efficiently administer their electronic commerce projects.

Project Management

Project management is a collection of formal techniques for planning and controlling the activities undertaken to achieve a specific goal. Project management was developed by the U.S. military and the defense contractors that worked with the military in the 1950s and the 1960s to develop weapons and other large systems. Not only was defense spending increasing in those years, but individual projects were becoming so large that it became impossible for managers to maintain control over them without some kind of assistance.

The project plan includes criteria for cost, schedule, and performance—it helps project managers make intelligent trade-off decisions regarding these three criteria. For example, if it becomes necessary for a project to be completed early, the project manager can compress the schedule by either increasing the project's cost or decreasing its performance.

Today, project managers use specific application software called **project management software** to help them manage projects. Project management software products, such as **Microsoft Project** and **Primavera Project Planner**, give managers an array of built-in tools for managing resources and schedules. The software can generate charts and tables that show, for example, which parts of the project are critical to its timely completion, which parts can be rescheduled or delayed without changing the project completion date, and where additional resources might be most effective in speeding up the project. Figure 12-8 shows an activity tracking screen from Primavera Project Planner.

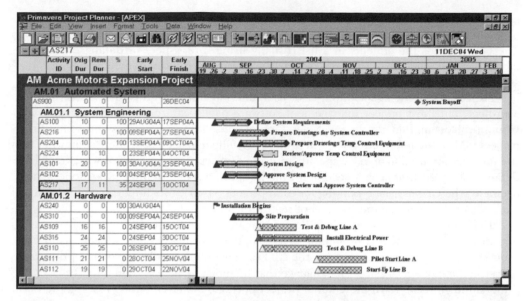

FIGURE 12-8   Tracking activities in Primavera Project Planner

In addition to managing the people and tasks of the internal team, project management software can help the team manage the tasks assigned to consultants, technology partners, and outsourced service providers. By examining the costs and completion times of tasks as they are completed, project managers can learn how the project is progressing and continually revise the estimated costs and completion times of future tasks.

Information systems development projects have a well-deserved reputation for running out of control and ultimately failing. They are much more likely to fail than other types of projects, such as building construction projects. The main causes for information systems project failures are rapidly changing technologies, long development times, and changing customer expectations. Because of this vulnerability, many teams rely on project management software to help them achieve project goals.

Although electronic commerce certainly uses rapidly changing technologies, the development times for most electronic commerce projects are relatively short—often they are accomplished in under six months. This gives both the technologies and the expectations of users less time to change. Thus, electronic commerce initiatives are, in general, more successful than other types of information systems implementations.

You can learn more about project management by reading the references listed in the For Further Study and Research section at the end of this chapter, or by clicking the Online Companion link for the **Project Management Institute**, a not-for-profit organization devoted to the promotion of professional project management practices.

Project Portfolio Management

Larger organizations often have many IT implementation projects going on simultaneously—a number of which could be electronic commerce implementations or updates. Some chief information officers (CIOs) of larger companies now use a portfolio approach to managing these multiple projects. **Project portfolio management** is a technique in which each project is monitored as if it were an investment in a financial portfolio. The CIO records the projects in a list (usually using spreadsheet or database management software) and updates the list regularly with current information about each project's status.

Project management software performs a function similar to this for the tasks within a project, but most project management software packages are designed to handle individual projects and do not do a very good job of consolidating activities across multiple projects. Also, the information used in project portfolio management differs somewhat from the information used to manage specific projects. Project management software tracks the details of how each project is accomplishing its specific goals. In project portfolio management, the CIO assigns a ranking for each project based on its importance to the strategic goals of the business and its level of risk (probability of failure).

To develop these rankings, the CIO can use any of the methods that financial managers use to evaluate the risk of making investments in business assets. Indeed, using the tools of financial management helps the CIO to explain electronic commerce projects as investments in assets—using the language that financial managers (and often the CEO) understand. You can learn more about project portfolio management by reading the Berinato article cited in the For Further Study and Research section at the end of this chapter.

Staffing for Electronic Commerce

Regardless of whether the internal team decides to outsource parts of the design and implementation activity, it must determine the staffing needs of the electronic commerce initiative. The general areas of staffing that are most important to the success of an electronic commerce initiative include:

- Business managers
- Project managers
- Account managers
- Applications specialists
- Web programmers
- Web graphics designers
- Content creators
- Content managers or editors
- Customer service reps
- Systems administrators
- Network operators
- Database administrators

The business management function should include internal staff. The **business manager** should be a member of the internal team that sets the objectives for the project. The business manager is responsible for implementing the elements of the business plan and reaching the objectives set by the internal team. If revisions to the plan are necessary as the project proceeds, the business manager develops specific proposals for plan modifications and additional funding and presents them to the internal team and top management for approval.

The business manager should have experience and knowledge related to the business activity that is being implemented on the electronic commerce site. For example, if business managers are assigned to a retail consumer site, they should have experience managing a retail sales operation.

In addition to including the business manager, the business management function in large electronic commerce initiatives may include other individuals who carry out specialized functions, such as project management or account management, that the business manager does not have time to handle personally. A **project manager** is a person with specific training or skills in tracking costs and the accomplishment of specific objectives in a project. Many project managers are certified by organizations such as the Project Management Institute (which you learned about earlier in this chapter) and have skills in the use of project management software.

An **account manager** keeps track of multiple Web sites in use by a project or keeps track of the projects that will combine to create a larger Web site. Most larger projects will have a test version, a demonstration version, and a production version of the Web site located on different servers. The test version is the "under construction" version of a Web site. Because most sites are frequently updated with new features and content, the test version gives the company a place to make sure that each new feature works before exposing it to customers. The demonstration version has features that have passed testing and must be demonstrated to an internal audience (for example, the Marketing Department) for approval. The production version is the full operating version of the site that is available to customers and other visitors. The account manager supervises the location of specific Web pages and related software installations as they are moved from test to demonstration to production. In smaller projects, the business manager handles the project and account management functions.

As more vendors provide packaged software solutions for electronic commerce, such as those you learned about in Chapter 9, companies need information systems staff that can install and maintain the software. Most large businesses have **applications specialists** who maintain accounting, human resources, and logistics software. Similarly, electronic commerce sites that buy software to handle catalogs, payment processing, and other features need applications specialists to maintain the software. Although the installation of these software packages can be outsourced, most companies prefer to train internal staff to serve in this function when the site becomes operational.

Web sites have evolved from static HTML to more complex designs built with dynamic Web page generation technologies and XML data integration. As Web sites have become more complicated, the need for **Web programmers**, who design and write the underlying code for dynamic database-driven Web pages has increased. Good Web programmers understand several different dynamic Web page generation technologies and are highly skilled in at least one of them. Many Web programmers also have database manipulation and query skills, such as the ability to write SQL or PHP code.

Because the Web is a visual medium, the role of graphic elements on individual Web pages is important. A company must either retain the services of a graphics design firm, a Web design firm that includes graphics designers, or must hire employees with graphics design skills. A **Web graphics designer** is a person trained in art, layout, and composition and who also understands how Web pages are constructed. The Web graphics designer, or design team for larger sites, must ensure that the Web pages on the site are visually appealing, easy to use, and make consistent use of graphics elements from page to page.

Most larger sites and many smaller sites include content created specifically for the Web site. Other sites adapt content from existing sources within the company for use on the Web site, or purchase content to use on the site. These activities require that the company hire **content creators** to write original content and **content managers** or **content editors** to purchase existing material and adapt it for use on the site.

The Web offers businesses a unique opportunity to reach out to their customers. Thus, business-to-consumer and business-to-business sites that want to capitalize on that opportunity must include a customer relationship management function. **Customer service** personnel help design and implement customer relationship management activities in the electronic commerce operation. They can, for example, issue and administer passwords, design customer interface features, handle customer e-mail and telephone requests for service or follow-up action, and conduct telemarketing for the site. Companies strive to provide the best possible service to satisfy the demands of their customers. The increasing power of customers to organize and express their expectations on the Web is a natural extension of the increase in consumerism that has occurred over the past two decades.

Some companies outsource parts of their customer relationship management operation to independent call centers. A **call center** is a company that handles incoming customer telephone calls and e-mails for other companies. Using a call center often makes sense for smaller companies that do not have the volume of customer inquiries to justify creating an internal call center operation. Some call centers work with a variety of businesses; others focus on one specialty area. For example, a specialized call center might contract with software manufacturers to provide installation help for their software products. Call center employees who are skilled in helping customers install one software package are often able to learn how to support other software packages very quickly.

A systems administrator who understands the server hardware and operating system is an essential part of a successful electronic commerce implementation. The **systems administrator** is responsible for the system's reliable and secure operation. If the site operation is outsourced to a service provider, the service provider supplies this function. If the site is hosted by the company, it needs to devote at least one person to this job. In addition, the internal system administrator needs sufficient staff to maintain full 24/7 operation and site security. These **network operations** staff functions include load estimation and load monitoring, resolving network problems as they arise, designing and implementing fault-resistant technologies, and managing any network operations that are outsourced to service providers or telephone companies.

Most electronic commerce sites require some kind of **database administration** function to support activities such as transaction processing, order entry, inquiry management, or shipment logistics. These activities require either an existing database into which the site is being integrated, or a separate database established for the electronic commerce initiative. It is important to have a database administrator who can effectively manage the design and implementation of this function.

554

Postimplementation Audits

After an electronic commerce site is successfully launched, most of the project's resources are devoted to maintaining and improving the site's operations. However, an increasing number of businesses are realizing the value of a postimplementation audit. A **postimplementation audit** (also called a **postaudit review**) is a formal review of a project after it is up and running.

The postimplementation audit gives managers a chance to examine the objectives, performance specifications, cost estimates, and scheduled delivery dates that were established for the project in its planning stage and compare them to what actually happened. In the past, most project reviews focused on identifying individuals to blame for cost overruns or missed delivery dates. Because many external forces in technology projects can overwhelm the best efforts of managers, this blame identification approach was generally unproductive, as well as uncomfortable, for the managers on the project.

A postimplementation audit allows the internal team, the business manager, and the project manager to raise questions about the project's objectives and provide their "in-the-trenches" feedback on strategies that were set in the project's initial design. By agreeing beforehand not to lay blame, the company obtains valuable information that it can use in planning future projects and gives the participants a meaningful learning experience.

The audit should result in a comprehensive report that analyzes the project's overall performance, how well the project was administered, whether the organizational structure was appropriate for the project, and the specific performance of the project team(s). Each section of the report should compare actual results to the project's objectives. Many companies modify their project management organization structure after completing each project based on the contents of postaudit review reports. Many companies also include a confidential section in the report that evaluates each team member's performance on the project. Summaries of member performance can help managers decide which employees should be included in future team projects.

555

Summary

This chapter provided an overview of key elements that are typically included in business plans for electronic commerce implementations. The first step is setting objectives. Specific objectives derive from the initiative's overall goals and include planned benefits and planned costs. The benefit and cost objectives should be stated in measurable terms, such as dollars or quantities. Before undertaking an electronic commerce project, most companies will evaluate its estimated costs and benefits.

Businesses use a number of evaluation techniques; however, most businesses calculate projects' return on investment to gauge their value. In the early days of electronic commerce, many companies undertook electronic commerce projects without evaluating their costs and benefits in detail because they feared being left out of the Internet boom. In the second wave of electronic commerce, fewer companies are undertaking electronic commerce initiatives without subjecting them to the same quantitative analysis they use for other IT projects. However, the benefits of electronic commerce projects can be harder to define and quantify than the benefits expected from most other IT projects, so managers should be careful when using these quantitative measures to evaluate electronic commerce projects.

Companies must decide how much, if any, of an electronic commerce project to outsource. The first step in determining an outsourcing strategy is to form an internal team that includes knowledgeable individuals from within the company. The internal team develops the specific project objectives and is responsible for meeting those objectives. The internal team designs an outsourcing strategy, selects a hosting service (or decides to have the company host its own Web server), and supervises the staffing of the project.

Project management is a formal way to plan and control specific tasks and resources used in a project. It provides project managers with a tool they can use to make informed trade-offs among the project elements of schedule, cost, and performance. Large organizations are beginning to use project portfolio management techniques to track and make trade-offs among multiple ongoing projects. Electronic commerce initiatives are usually completed within a short time frame and thus are less likely to run out of control than other information systems development projects.

The company must staff the electronic commerce initiative regardless of whether portions of the project are outsourced. Critical staffing areas include business management, application specialists, customer service staff, systems administration, network operations staff, and database administration. A good way for all participants to learn from project experiences is to conduct a postimplementation audit that compares project objectives to the actual results.

Key Terms

24/7 operation	Capital project
Account manager	Change management
Applications specialist	Component outsourcing
Business manager	Content creator
Call center	Content editor
Capital investment	Content manager

556

Customer service
Database administration
Downstream strategies
Early outsourcing
Fast venturing
Incubator
Late outsourcing
Metrics
Network operations
Opportunity cost
Outsourcing
Partial outsourcing

Postimplementation audit (postaudit review)
Project management
Project management software
Project manager
Project portfolio management
Return on investment (ROI)
Systems administrator
Total cost of ownership (TCO)
Upstream strategies
Web graphics designer
Web programmer

Review Questions

RQ1. Name three benefit objectives that a business might decide to measure in an electronic commerce business plan.

RQ2. In two paragraphs, explain why some firms approved online business initiatives without taking a close look at the return on investment numbers of those projects during the first wave of electronic commerce.

RQ3. In one paragraph, explain why late outsourcing is seldom used in electronic commerce projects.

RQ4. In about 200 words, name and briefly describe four factors that a company should evaluate when selecting an ISP, ASP, or CSP to provide Web hosting services.

RQ5. In about 250 words, explain why the head of the business management function of an electronic commerce initiative should be an employee of the company implementing the project even if most of the work is outsourced.

Exercises

E1. The Grover Cams Company manufactures cams and other components for diesel engines. As Web site manager for Grover, you created an attractive Web site that includes information about the company's history, its financial statements, and digitized depictions of the company's main products. You have been talking with your manager, chief information officer Tom Buckles, for several months about adding electronic commerce features to the Web site that will allow your smaller customers to order directly from Grover instead of through their local distributors. Tom finally created a capital budget proposal for the Web site expansion and submitted it to Grover's board of directors. The board always calculates and evaluates a capital project's return on investment before approving it. The board told Tom that the project did not provide a high enough financial return to approve it. However, the board realized that electronic commerce initiatives could be important to Grover's future strategic position in the business; thus, it is willing to consider nonmonetary factors as a basis for approving the project. Tom would like to take the project back to the board next month, but he does not have a good sense of what nonmonetary factors might persuade the board to approve the project. He wants you to write a memo that outlines some of those factors and explains why they are important to Grover's future strategic position. In addition to considering the discussion in this chapter, you may want to use the Online Companion and draw on resources at *Business Week*'s **e.biz**, *CIO*'s **E-Business Research Center**, Internet.com's **Electronic Commerce Guide**, or ZDNet's **eBusiness Update** as you prepare your memo.

E2. You are working for International Delicacies, which has become successful selling unusual food and other gift items through its mail order catalog. Most customers call the toll-free telephone number on the catalog, but some still send in orders by mail. Your manager, Jagdish Singh, wants to add an online store that will complement the company's existing mail order and telephone sales channels. He wants you to lead the internal team for the project. Write a memo to Jagdish of about 500 words in which you outline the steps you will take to staff the internal team, make decisions about internal development vs. outsourcing, and choose a hosting service. Be sure to include your thoughts on whether an incubator or a fast venturing strategy might make sense in this case.

E3. As manager of networks and computing operations for Fashion Land, a retailer of women's clothing and accessories, you have seen the business grow from seven stores in Kansas City to over 100 stores located throughout the Midwest. Fashion Land's marketing research team has found that many members of its target customer group—females between the ages of 15 and 35—are becoming regular users of the Web. The researchers have asked you for help in developing an electronic commerce initiative for Fashion Land. Alone, or in a team assigned by your instructor, do the following:

a. Outline a business strategy for Fashion Land's electronic commerce initiative. The outline should include a list of specific objectives and the costs and benefits of accomplishing each objective. The outline should also include recommendations regarding what to outsource, what Web hosting services are needed, and what staff should be hired.

b. Prepare a memo that lists and briefly describes the major hardware, software, security, payment processing, advertising, international, legal, and ethics issues that might arise in the development of this electronic commerce site.

Cases

C1. Idealab

Bill Gross started his first company (a solar-powered device manufacturer) when he was 15 years old. After graduating from Caltech, he started a software company, GNP, that he later sold to Lotus. Gross had made a considerable amount of money and was interested in exploring better ways of getting ideas converted into profitable businesses. He became fascinated by the idea of business incubators about the same time he became fascinated with the business potential of the Internet. In 1996, he pooled some of his wealth with contributions from several partners to create Idealab.

Idealab was one of the first companies to provide an incubator that was open to individual entrepreneurs. Idealab provided venture capital and gave entrepreneurs a place to work and develop their business ideas alongside other entrepreneurs. In the first wave of electronic commerce, Idealab was very successful. Although many of its incubated companies eventually failed, enough of them succeeded that Idealab was able to fund several generations of new businesses through its operations. In its first year, it supported 10 new businesses, including the very successful CitySearch Web site. In its second year, Idealab helped create another 10 businesses, including the successful sites Shopping.com, Tickets.com, and WeddingChannel.com. In subsequent years, Idealab incubated companies such as NetZero, Cooking.com, CarsDirect.com, Picasa, and GoTo.com (which later became Overture and was eventually acquired by Yahoo!). Not all of Idealab's companies were successful, however. One of the most dramatic failures of the first wave of electronic commerce, eToys, had been an Idealab company. Idealab had more winners than losers, though; by early 2000, the company had more than $4 billion in assets.

In 2000, Gross devised a new strategy that would go beyond Idealab's original purpose as an incubator. He developed a plan to compete with Amazon.com using existing Idealab companies. His plan was to combine about 10 of the companies in the incubator (including specialty retailer Eve.com and online jewelry store Ice.com) and promote them (using large amounts of money that would be raised from outside investors) as a single marketplace under the name Big.com. However, just as Gross began raising money to support the launch of this new marketplace, the pool of dot-com investment funds dried up. The new combined company quickly failed. Eve.com and Big.com no longer exist. The founders of Ice.com bought their company back from Idealab and moved it to their home in Montreal (where the company is now operating profitably). Within a few months, the failure of Big.com and the lower stock market valuations of Idealab's holdings reduced the value of the company's assets from $4 billion to $200 million.

Idealab's investors were upset by Gross' change in strategy and by the drop in their company's value. In January 2002, 44 of them sued Gross and other Idealab managers for $750 million. The suit alleged mismanagement of the funds invested and further alleged that Gross had used Idealab funds to pay personal expenses. Eighteen months later, a court held that the allegations were without merit and the suit was dismissed. Gross was once again able to devote his time to operating Idealab as an incubator.

Gross laid off more than two-thirds of Idealab's employees and stopped accepting outside venture capital. Idealab no longer provides incubator space for entrepreneurs who have developed ideas on their own. The company only funds ideas generated by the Idealab management team. Idealab's asset value has rebounded somewhat and is now around $700 million.

Required:

1. In its first three years of operation, Idealab recruited entrepreneurs to its incubator who had business experience, but who did not know much about the Internet. In about 300 words, explain what benefits Idealab was able to provide to these entrepreneurs and why the incubator environment was beneficial to them.

2. In about 200 words, analyze Idealab's 2000 decision to change its focus from being an incubator to merging its companies in an attempt to compete with Amazon.com. In your analysis, discuss whether the decision was a strategic error or just a case of bad timing.

3. In about 200 words, explain why Gross decided to devote Idealab's resources to the development of internally generated ideas in 2003.

Note: Your instructor might assign you to a group to complete this case, and might ask you to prepare a formal presentation of your results to your class.

C2. Davis Humanics

Davis Humanics (DH) is a company founded in 1982 that provides human resources services to about 7000 companies with a total of nearly 100,000 employees. These services include payroll processing, tax filing, health insurance and claims management, and retirement plan management. DH has annual sales of $2 billion and about 1000 employees. DH has grown rapidly and has clients of all sizes, ranging from smaller companies with fewer than 50 employees to Fortune 500 companies.

As DH grows, it is having trouble maintaining a consistent quality of service. Account managers each must handle more clients, and it is becoming difficult for those account managers to maintain a high degree of personal contact with the human resources executives who control DH's contracts. In the past, account managers worked with a small set of client contact people, but now account managers must work with more people, many of whom they have never met. In addition to account managers, client personnel have regular contacts with DH operations staff (who handle input tasks), DH systems staff (who help customize the interfaces between DH systems and client systems), and DH professional staff (lawyers, actuaries, and human resources professionals who consult with DH clients and their legal counsel regarding the operation of their retirement and benefits plans).

Because DH's clients are so different in size and how they operate, DH has to be flexible in handling input data. For example, DH's payroll-processing service allows clients many different ways to send in time card data. The largest clients arrange for customized computer-to-computer transfer of information. Some large clients use EDI transfers. Most medium and smaller-sized clients e-mail or fax the time card information, but a significant number of them mail paper lists that DH must scan into its systems. The health insurance claims-handling operation is even more troublesome. In addition to having clients send information in various formats, the insurance companies demand that information be submitted in specific formats, each of which is different.

The complexity of DH's operations is growing as rapidly as the company adds new clients. Sandi Higbee, DH's director of Operations, asks for your help in outlining a Web-based customer relationship management (CRM) system that will help manage the account managers' ever increasing levels of customer contact. Sandi reviewed the products offered by several leading CRM vendors and believes that one might work as a base product, but no matter which product is chosen, she believes that substantial customization will be necessary because DH's operations are so complex and different from most companies that sell products or simple services to customers. A good CRM system for DH would need to monitor all types of customer interactions with DH account managers, operations staff, systems staff, and professional staff. In addition, the system's Web interface should allow DH clients to access parts of the CRM system so they can track DH's follow-up on their work requests and pending inquiries.

DH evaluates all capital projects, including IT projects, using ROI. Sandi is worried about this because she believes that many of the benefits of this CRM project will be hard to quantify. On the other hand, the costs of the CRM project (software and hardware purchase and cost of consultants who will customize the CRM software to meet DH's specific needs) will be very easy to quantify and will be large. Sandi expects the vendor-consultant teams to submit bids of between $1 million and $2 million for this project.

Required:

1. Prepare an outline of the benefits that DH might expect to obtain from this CRM project. Use categories to organize your list of benefits; for example, you might identify benefits that will accrue to DH's account managers, operations staff, IT staff, and professional staff. Because DH's clients will also benefit, you might be able to identify benefits that will accrue to DH's Marketing and Sales departments or to DH's New Product Development department. Be sure to include any long-term benefits that you think might occur after the CRM system has been in place for several years.

2. Estimate the dollar value of each benefit you identified in the first part of your answer.

3. Prepare a one-page memorandum to the DH board of directors in which you argue against using ROI as the primary method for evaluating this project. Keep in mind that these directors have little time to review your arguments and are very much inclined to use ROI for all project evaluations.

Note: Your instructor might assign you to a group to complete this case, and might ask you to prepare a formal presentation of your results to your class.

For Further Study and Research

Aragon, L. 2004. "Idealab: Bubble Fund Finds Itself Back at Square One," *Venture Capital Journal*, 44(6), June, 20.

Abdel-Hamid, T. and S. Madnick. 1991. *Software Project Dynamics: An Integrated Approach*. Englewood Cliffs, NJ: Prentice Hall.

Abdel-Hamid, T., K. Sengopta, and C. Sweet. 1999. "The Impact of Goals on Software Project Management: An Experimental Investigation," *MIS Quarterly*, 23(4), December, 531–555.

Bannan, K. 2004. "Entrepreneur Learns Why It's Best to Optimize Site Before It Launches," *B to B*, 89(15), December 13, 19.

Barias, D. 2002. "Gevity HR," *Line56: The E-Business Executive Daily*, July 26. (http://www.line56.com/articles/default.asp?articleid-3879)

Barsh, J., E. Kramer, D. Maue, and N. Zuckerman. 2001. "Magazine's Home Companion," *The McKinsey Quarterly*, April, 83–91.

Beach, G. 2002. "ROI Is DOA," *CIO*, 15(12), April 1, 98.

Berinato, S. 2001. "Do the Math," *CIO*, 15(1), October 1, 53–60.

Bernard, A. 2004. "Majority of IT and Business Plans Still Not Linked," *Internet News*, March 5. (http:// www.internetnews.com/stats/article.php/3305971)

Berry, J. 2001. "Sometimes It's OK to Skip ROI Model," *InternetWeek*, October 22, 41.

Berry, J. 2003. "Assume Nothing. Audit Instead," *Computerworld*, 37(14), April 7, 43.

Blazier, A. 2003. "Far from Dead, Idealab Continues to Build for Future," *San Gabriel Valley Tribune*, July 12, C1.

Boisot, M. 1999. *Knowledge Assets: Securing Competitive Advantage in the Information Economy*. New York: Oxford University Press.

Borck, J. 2001. "A Balancing Act to ROI," *InfoWorld*, 23(30), July 23, 54.

Brooks, F. 1995. *The Mythical Man-Month: Essays on Software Engineering, Anniversary Edition*. Reading, MA. Addison-Wesley.

Buderi, B. 2005. "Conquering the Digital Haystack: New Start-ups Are Changing the Way People Search the Web," *Inc.*, January, 34–35.

Canadian Business. 2003. "Dot-com Wonder Boys," 76(7), April 14, 30–36.

Canadian Business. 2003. "It Seemed Like a Good Idea," 76(7), April 14, 34.

Copeland, R. 2001. "ROI: The IT Department's Moving Target," *Information Week*, August 6, 45–47.

Edvinsson, L. and M. Malone. 1997. *Intellectual Capital: Realising Your Company's True Value by Finding its Hidden Brainpower.* New York: HarperCollins.

Fleming, Q. and J. Koppelman. 2003. "What's Your Project's Real Price Tag?" *Harvard Business Review*, 81(9), September, 20–21.

Gido, J. and J. Clements. 1999. *Successful Project Management.* Cincinnati: South-Western College Publishing.

Glass, R. 1997. *Software Runaways: Lessons Learned from Massive Software Project Failures.* Upper Saddle River, NJ: PTR Prentice Hall.

Goldratt, E. 1997. *Critical Chain.* Great Barrington, MA: North River Press.

Grimes, A. 2004. "Court Deals Blow to Investors' Suit Against Idealab," *The Wall Street Journal*, June 30, B6.

Hellweg, E. and S. Donahue. 2000. "The Smart Way to Start an Internet Company," *Business 2.0*, March 1, 64–66.

Heun, C. 2000. "No Web Bargains for Kmart," *InformationWeek*, August 21, 18.

Hsu, M. 2003. "How to Prepare for an Information Technology Audit," *Community Banker*, 12(9), September, 60–62.

Kambil, A., E. Eselius, and K. Monteiro. 2000. "Fast Venturing: The Quick Way to Start a Web Business," *Sloan Management Review*, 41(4), Summer, 55–67.

Kara, D. 1999. "Sourcing Solutions for Wired World Emerging," *Software Magazine*, 19(1), June, 60–71.

Karpinski, R. 2001. "Vanished into Thin Air," *InternetWeek.com*, June 13. (http://www.internetweek.com/transtoday01/ttoday061301.htm)

Karpinski, R. 2001. "Vanishing Vendors Are Common Concern," *InternetWeek*, June 25, 15.

Keefe, P. 2003. "Backing Up ROI," *Computerworld*, 37(12), March 24, 22.

Keen, P. 2000. "Six Months—or Else," *Computerworld*, 34(15), April 10, 48.

Keil, M. and D. Robey, 1999. "Turning Around Troubled Software Projects: An Exploratory Study of the De-Escalation of Commitment to Failing Courses of Action," *Journal of Management Information Systems*, 15(4), 63–87.

Keil, M., P. Cule, K. Lyytinen, and R. Schmidt. 1998. "A Framework for Identifying Software Project Risks," *Communications of the ACM*, 41(11), November, 76–83.

Kerzner, H. 2000. *Advanced Project Management: Best Practices.* New York: John Wiley & Sons.

Leung, L. 2003. "Managing Offshore Outsourcing," *Network World*, 20(49), December 8, 59.

McConnell, S. 1996. *Rapid Development: Taming Wild Software Schedules.* Redmond, WA: Microsoft Press.

Melymuka, K. 2000. "Born to Lead Projects," *Computerworld*, 34(13), March 27, 62–63.

Mollison, C. 2002. "To Outsource or Not to Outsource: That Is the Question," *Internet World*, January 1, 23–42.

Murthi, S. 2002. "Managing the Strategic IT Project," *Intelligent Enterprise*, 5(18), November 15, 49–52.

Neuwirth, R. 1998. "Race into Cyberspace Gushes $80M Red Ink," *Editor & Publisher*, 131(51), December 19, 12–13.

Nocera, J. and E. Florian. 2001. "Bill Gross Blew Through $800 Million in Eight Months (and He's Got Nothing to Show for It): Why Is he Still Smiling?" *Fortune*, 143(5), March 5, 70–77.

Ramsey, C. 2000. "Managing Web Sites as Dynamic Business Applications," *Intranet Design Magazine*, June. (http://idm.internet.com/articles/200006/wm_index.html)

Randall, L. 1999. "Average E-Commerce Web Site Costs US $1 Million," *Computing Canada*, 25(24), June 18, 11.

Rogers, A. 1999. "Up-Front Web Costs Are Half the Story," *Computer Reseller News*, June 7, 3.

Sacks, D. 2005, "The Accidental Guru," *Fast Company*, January, 64–71.

Sawhney , M. 2002. "Damn the ROI, Full Speed Ahead: 'Show Me the Money' May Not Be the Right Demand for E-Business Projects," *CIO*, 15(19), July 15, 36–38

Schindler, M. and M. Eppler. 2003. "Harvesting Project Knowledge: A Review of Project Learning Methods and Success Factors," *International Journal of Project Management*, 21(3), April, 219–228.

Schwalbe, K. 2002. *Information Technology Project Management*. Second Edition. Boston, MA: Course Technology.

Siebel, T. and P. House. 1999. *Cyber Rules: Strategies for Excelling at E-Business*. New York: Currency-Doubleday.

Southgate, D. 2002. "Keeping ROI in Sight Fosters Strong E-business Results," *TechRepublic*, July 29. (http://www.techrepublic.com/article.jhtml?id=r00520020724dcs01. htm&FROM=w057)

Stewart, T. 1999. "Larry Bossidy's New Role Model: Michael Dell," *Fortune*, 139(7), April 12, 166–167.

Stoiber, J. 1999. "Maximizing IT Investments," *CIO Enterprise Magazine*, July 15. (http://www.cio. com/archive/enterprise/071599_checks.html)

Tan, B., N. Tang, and P. Forrester. 2004. "Application of Quality Function Deployment for e-Business Planning," *Production Planning & Control*, 15(8), December, 802–815.

The Wall Street Journal. 1999. "Spending Campaign Is Set for Newspaper's Web Site," June 22, B16.

United States Department of Justice Inspector General. 2002. *Audit Report No. 03-09: Federal Bureau of Investigation's Management Of Information Technology Investments*. Washington, D.C.: U.S. Department of Justice.

United States General Accounting Office. 2002. *Desktop Outsourcing: Positive Results Reported, But Analyses Could Be Strengthened*. Washington, D.C.: U.S. General Accounting Office.

Varon, E. 2002. "How to Take Control of Your Web Site," *CIO*, 15(6), January 1, 90–92.

Violino, B. 2000. "Payback Time for E-Business—Net Projects No Longer Too 'Strategic' for ROI," *InternetWeek*, May 1, 1.

Webster, J. 2002. "Calculating Web Site Payoff," *Computerworld*, 36(6), February 4, 34.

Wexler, J. 2000. "Lands of Opportunity," *Computerworld*, 34(26), June 26, 72–73.

Wilder, C. 1999. "ROI: E-Business Strategic Investment," *InformationWeek*, May 24, 48–56.

Wysocki, B. 2000. "U.S. Incubators Help Japan Hatch Ideas," *The Wall Street Journal*, June 12, A1.

Yourdon, E. 2000. "Success in E-Projects," *Computerworld*, 34(34), August 21, 36.

Yourdon, E. and P. Becker. 1997. *Death March: The Complete Software Developer's Guide to Surviving "Mission Impossible" Projects*. Upper Saddle River, NJ: Prentice Hall.

24/7 operation   The operation of a site or service 24 hours a day, seven days a week.

802.11a, 802.11b, 802.11g, 802.11n   An improved version of Wi-Fi; introduced in 2002; it is capable of transmitting data at speeds up to 54 Mbps.

A

Acceptance   An expression of willingness to take an offer, including all of its stated terms.

Access control list (ACL)   A list of resources and the user names of people who are permitted access to those resources within a computer system.

Account aggregation   A feature of online banks that allows a customer to obtain bank, investment, loan, and other financial account information from multiple Web sites and display it all in one location at the bank's Web site.

Account manager   A person who keeps track of multiple Web sites in use by a project or keeps track of the projects that combine to create a larger Web site.

Accredited Standards Committee X12 (ASC X12)   A committee that develops and maintains uniform EDI standards in the United States.

Acquiring bank   Synonymous with merchant bank, which is a bank that does business with merchants who want to accept credit cards.

Acquisition cost   The total amount of money that a site spends, on average, to draw one visitor to the site.

Active ad   A Web ad that generates graphical activity that "floats" over the Web page itself instead of opening in a separate window.

Active content   Programs that are embedded transparently in Web pages that cause action to occur.

Active Server Pages (ASP)   Applications that generate dynamic content within Web pages using either Jscript code or Visual Basic.

Active wiretapping   An integrity threat that exists when an unauthorized party can alter a message stream of information.

ActiveX   An object, or control, that contains programs and properties that are put in Web pages to perform particular tasks.

Activity   A task performed by a worker in the course of doing his or her job.

Ad view   A Web site visitor page request that contains an advertisement.

Ad-blocking software   A program that prevents banner ads and pop-up ads from loading.

Addressable media   Advertising efforts sent to a known addressee; these include direct mail, telephone calls, and e-mail.

Advanced Encryption Standard (AES)   The new encryption standard designed to keep government infor-mation secure using the Rijndael algorithm. Introduced in February 2001 by the National Institute of Standards and Technology (NIST).

Advertising-subscription mixed revenue model   A revenue model in which subscribers pay a fee and accept some level of advertising.

Advertising-supported revenue model A revenue model in which Web sites provide free content along with advertising or messages provided by other companies that pay the Web site operator for delivering the advertising or messages.

Affiliate marketing An advertising technique in which one Web site (called an "affiliate") includes descriptions, reviews, ratings, or other information about products that are sold on another Web site. The affiliate site includes links to the selling site, which pays the affiliate site a commission on sales made to visitors that arrived from a link on the affiliate site.

Affiliate program broker A company that serves as a clearinghouse or marketplace for sites that run affiliate programs and sites that want to become affiliates.

American National Standards Institute (ANSI) The coordinating body for electrical, mechanical, and other technical standards in the United States.

Anchor tag The HTML tag used to specify hyperlinks.

Animated GIF Animated Web ad graphics that are attention-grabbing.

Anonymous electronic cash Electronic cash that cannot be traced back to the person who spent it.

Anonymous FTP A protocol that allows users to access limited parts of a remote computer using FTP without having an account on the remote computer.

Antivirus software Software that detects viruses and worms and either deletes them or isolates them on the client computer so they cannot run.

Applet A program that executes within another program; it cannot execute directly on a computer.

Application (application program, application software) A program that performs a specific function, such as creating invoices, calculating payroll, or processing payments received from customers.

Application integration The coordination of all of a company's existing systems to each other and to the company's Web site.

Application server A middle-tier software and hardware combination that lies between the Internet and a corporate backend server.

Application service provider (ASP) A Web-based site that provides management of applications such as spreadsheets, human resources management, or e-mail to companies for a fee.

Application software Synonymous with application.

Applications specialist The member of an electronic commerce team who is responsible for maintenance of software that performs a specific function, such as catalog, payment processing, accounting, human resources, and logistics software.

Ascending-price auction A type of auction in which bidders publicly announce their successively higher bids until no higher bid is forthcoming; also called an English auction.

Asymmetric connection An Internet connection that provides different bandwidths for each direction.

Asymmetric digital subscriber Line (ADSL) Internet connections using the DSL protocol with bandwidths from 16 to 640 Kbps upstream and 1.5 to 9 Mbps downstream.

Asymmetric encryption Synonymous with public-key encryption, which is the encoding of messages using two mathematically related but distinct numeric keys.

Asynchronous transfer mode (ATM) Internet connections with bandwidths of up to 622 Gbps.

Attachment A data file (document, spreadsheet, or other) that is appended to an e-mail message.

Auctioneer   The person who manages an auction.

Auction consignment services   Companies that take an item and create an online auction for that item, handle the transaction, and remit the balance of the proceeds after deducting a fee. These services are performed on behalf of people and small businesses who want to use an online auction, but do not have the skills or the time to become a seller.

Authority to bind   The ability or authority of an individual to commit his or her company to a contract.

Automated clearing house (ACH)   One of several systems set up by banks or government agencies, such as the U.S. Federal Reserve Board, that process high volumes of low dollar amount electronic fund transfers.

B

Backbone routers   Computers that handle packet traffic along the Internet's main connecting points and that can each handle more than 50 million packets per second.

Backdoor   An electronic hole in electronic commerce software left open by accident or intentionally.

Bandwidth   The amount of data that can be transmitted in a fixed amount of time. Also, the number of simultaneous site visitors that a Web site can accommodate without degrading service.

Banner ad   A small rectangular object on a Web page that displays a stationary or moving graphic and includes a hyperlink to the advertiser's Web site.

Banner advertising network   An organization that acts as a broker between advertisers and Web sites that carry ads.

Banner exchange network   An organization that coordinates ad-sharing so that other sites run your ad and your site runs other exchange members' ads.

Base 2 (binary)   A number system in which each digit is either a 0 or a 1, corresponding to a condition of either "off" or "on." Also known as a binary system.

Bayesian revision   A statistical technique in which additional knowledge is used to revise earlier estimates of probabilities.

Behavioral segmentation   The creation of a separate experience for customers based on their behavior.

Benchmarking   Testing that compares hardware and software performances.

Bid   An offer of a certain price made on an item that is up for auction.

Bidder   A potential buyer at an auction; one who places bids.

Bill presentment   A Web site feature that allows customers to view and pay bills online.

Biometric security device   A security device that uses an element of a person's biological makeup to confirm identification. These devices include writing pads that detect the form and pressure of a person writing a signature, eye scanners that read the pattern of blood vessels in a person's retina, and palm scanners that read the palm of a person's hand (rather than just one fingerprint).

Black hat hackers   Hackers who use their skills for ill purposes.

Black list spam filter   Software that looks for From addresses in incoming messages that are known to be spammers. The software can delete the message or put it into a separate mailbox for review.

Blade server   A server configuration in which small server computers are each installed on a single computer board and then many of those boards are installed into a rack-mounted frame.

Bluetooth   A wireless standard that is used for short distances and lower bandwidth connections.

Border router   The computers located at the border between the organization and the Internet that decide how best to forward each packet of information as it travels on the Internet to its destination. Synonymous with router, router computer, and gateway computer.

Brand   Customers' perceptions of a product.

Brand leveraging   A strategy in which a well-established Web site extends its dominant positions to other products and services.

Broadband   Connections that operate at speeds of greater than about 200 Kbps.

Buffer   An area of a computer's memory that is set aside to hold data read from a file or database.

Buffer overrun (buffer overflow)   An error that occurs when programs filling buffers malfunction and overfill the buffer, spilling the excess data outside the designated buffer memory area.

Business logic   Rules of a particular business.

Business manager   The member of an electronic commerce team who is responsible for implementing the elements of the business plan and reaching the objectives set by the internal team. The business manager should have experience in and knowledge of the business activity being implemented in the site.

Business model   A set of processes that combine to yield a profit.

Business process patent   A patent that protects a specific set of procedures for conducting a particular business activity.

Business processes   The activities in which businesses engage as they conduct commerce.

Business-to-business (B2B)   Transactions conducted between businesses on the Web.

Business-to-consumer (B2C)   Transactions conducted between shoppers and businesses on the Web.

Business-to-government (B2G)   A category of electronic commerce that includes business transactions with government agencies, such as paying taxes and filing required reports.

Business unit   A unit within a company that is organized around a specific combination of product, distribu-tion channel, and customer type. Synonymous with strategic business unit.

Byte   An 8-bit number (in most computer applications).

C

Call center   A company that handles customer telephone calls and emails for other companies.

Cannibalization   The loss of traditional sales of a product to its electronic counterpart.

Capital investment   A major outlay of funds made by a company to purchase fixed assets such as property, a factory, or equipment.

Capital project   Synonymous with capital investment.

Card not present   A credit card transaction in which the merchant's location and the purchaser's location are different.

Cascading Style Sheets (CSS)   An HTML feature that allows designers to apply many predefined page display styles to Web pages.

Catalog   On electronic commerce sites, a listing of goods or services that may include photographs and descriptions, often stored in a database.

Catalog model A revenue model in which the seller establishes a brand image, then uses the strength of that image to sell through printed catalogs mailed to prospective buyers. Buyers place orders by mail or by calling the seller's toll-free telephone number.

Category manager A company that handles responsibility for a particular product line within a retail store. Synonymous with channel distribution manager or fulfillment manager.

Cause marketing An affiliate marketing program that benefits a charitable organization.

Centralized architecture A server structure that uses few very large and fast computers.

Certification authority (CA) A company that issues digital certificates to organizations or individuals.

Challenge-response A content-filtering security technique that requires an unknown sender to reply to a challenge presented in an e-mail. These challenges are designed so that a human can respond easily, but a computer would have difficulty formulating the response.

Change management The process of helping employees cope with changes in the workplace.

Channel conflict The problem that arises when a company's sales in one related set of sales outlets interferes with its sales in another related set of sales outlets; for example, when sales through the company's Web site interfere with sales in that company's retail store.

Channel cooperation A strategy that coordinates sales and credit among various sales outlets, including online, catalog, and brick-and-mortar sales.

Channel distribution manager A company that handles responsibility for a particular product line within a retail store. Synonymous with category manager or fulfillment manager.

Charge card A card with no preset spending limit. The entire amount charged to the card must be paid in full each month.

Chargeback The process in which a merchant bank retrieves the money it placed in a merchant account as a result of a cardholder successfully contesting a charge.

Check 21 A U.S. law that permits banks to replace the physical movement of checks with transmission of scanned images.

Cipher text Text that comprises a seemingly random assemblage of bits. Cipher text is what messages become after they are encrypted.

Circuit A specific route between source and destination along which data travels.

Circuit switching A way of connecting computers or other devices that uses a centrally controlled single connection. In this method, which is used by telephone companies to provide voice telephone service, the connection is made, data is transferred, and the connection is terminated.

Click Synonymous with click-through.

Clickstream Data about site visitors.

Click-through The loading of an advertiser's Web page that results from a visitor clicking on a banner advertisement on another Web page.

Client-level filtering An e-mail content filtering technique in which the filtering software is placed on the individual user's computer.

Client/server architecture A combination of client computers running Web client software and server computers running Web server software.

Client-side electronic wallet An electronic wallet that stores a consumer's information on the consumer's own computer.

Client-side scripting The embedding of script languages in HTML documents.

Closed architecture The use of proprietary communication protocols by computer manufacturers in the early days of computing, preventing computers made by different manufacturers from being connected to each other. Also called proprietary architecture.

Closed loop system   A payment card arrangement involving a consumer, a merchant, and a payment card company (such as American Express or Discover) that processes transactions between the consumer and merchant without involving banks.

Closing tag   The second half of a two-sided HTML tag; it is identified by a slash (/) that precedes the tag's name.

Collision   The occurrence of two messages resulting in the same hash value; the probability of this happening is extremely small.

Co-location (collocation, colocation)   An Internet service arrangement in which the service provider rents a physical space to the client to install its own server hardware.

Colon hexadecimal (colon hex)   The shorthand notation system used for expressing IPv6 addresses that uses eight groups of 16 bits ($8 \times 16 = 128$). Each group is expressed as four hexadecimal digits and the groups are separated by colons.

Commerce service provider (CSP)   A Web host service that also provides commerce hosting services on its computer.

Commodity item   A product or service that has become so standardized and well-known that buyers cannot detect a difference in the offerings of various sellers; buyers usually base their purchase decisions for such products and services solely on price.

Common law   The part of English and U.S. law that is established by the history of law.

Communication modes   Ways of identifying and reaching customers.

Company   A business engaged in commerce; synonymous with firm.

Component outsourcing   Synonymous with partial outsourcing; the outsourcing of the design, development, implementation, or operation of specific portions of an electronic commerce system .

Component-based application system   A business logic approach that separates presentation logic from business logic.

Computer forensics   The field responsible for the collection, preservation, and analysis of computer-related evidence.

Computer forensics expert   An individual hired to access client computers to locate information that can be used in legal proceedings.

Computer network   Any technology that allows people to connect computers to each other.

Computer security   The protection of computer resources from various types of threats.

Computer virus   Synonymous with virus, which is software that attaches itself to another program and can cause damage when the host program is activated.

Configuration table   Information about connections that lead to particular groups of routers, specifications on which connections to use first, and rules for handling instances of heavy packet traffic and network congestion.

Consideration   The bargained-for exchange of something valuable, such as money, property, or future services.

Constructive notice   The idea that citizens should know that when they leave one area and enter another, they become subject to the laws of the new area.

Consumer-to-business   An industry term for electronic commerce that occurs in general consumer auctions; bidders at a general consumer auction might be businesses.

Consumer-to-consumer (C2C)   A category of electronic commerce that includes individuals who buy and sell items among themselves.

Content creator   A person who writes original content for a Web site.

Content editor   A person who purchases and adapts existing material for use on a Web site.

Content management software Software used by companies to control the large amounts of text, graphics, and media files used in business.

Content manager Synonymous with content editor.

Contextual advertising An advertising technique in which ads are placed in proximity to related content.

Contract An agreement between two or more legal entities that provides for an exchange of value between or among them.

Contract purchasing Direct materials purchasing in which the company negotiates long-term contracts for most of the materials that it will need. Also called replenishment purchasing.

Conversion The transition of a first-time visitor to a customer.

Conversion cost The total amount of money that a site spends, on average, to induce one visitor to make a purchase, sign up for a subscription, or (on an advertising-supported site) register.

Conversion rate Used in advertising to calculate the percentage of recipients that respond to an ad or promotion.

Cookie Bits of information about Web site visitors created by Web sites and stored on client computers.

Cookie blocker A third-party program that prevents cookie storage selectively.

Copy control An electronic mechanism for providing a fixed upper limit to the number of copies that one can make of a digital work.

Copyright A legal protection of intellectual property.

Cost per thousand (CPM) An advertising pricing metric that equals the dollar amount paid to reach 1000 people in an estimated audience.

Countermeasure A physical or logical procedure that recognizes, reduces, or eliminates a threat.

Cracker A technologically skilled person who uses their skills to obtain unauthorized entry into computers or network systems, usually with the intent of stealing information or damaging the information, the system's software, or the system's hardware.

Crawler Synonymous with spider, which is the first part of a search engine.

Credit card A payment card that has a spending limit based on the card holder's credit limit. A minimum monthly payment must be made against the balance on the card, and interest is charged on the unpaid balance.

Credit card associations Member-run organizations that issue credit cards to individual consumers. Also called customer issuing banks.

Cryptography The science that studies encryption, which is the hiding of messages so that only the sender and receiver can read them.

Culture The combination of language and customs that are unique to a particular population.

Customer issuing banks Member-run organizations that issue credit cards to individual consumers. Also called credit card associations.

Customer life cycle The five stages of customer loyalty.

Customer portal A corporate Web site designed to meet the needs of customers by offering additional services such as private stores, part number cross-referencing, product use guidelines, and safety information.

Customer relationship management Synonymous with technology-enabled relationship management, it is the obtaining and use of detailed customer information.

Customer relationship management (CRM) software   Software that collects data on customer activities; this data is then used by managers to conduct analytical activities.

Customer service   The people within an electronic commerce team who are responsible for managing customer relationships in the electronic commerce operation.

Customer value   The cost that a customer pays for a product, minus the benefits the customer gains from the product.

Customer-centric   The Web site development approach of putting the customer at the center of all site designs.

Cybervandalism   The electronic defacing of an existing Web site page.

Cybersquatting   The practice of registering a domain name that is the trademark of another person or company with the hope that the trademark owner will pay huge amounts of money for the domain rights.

D

Data Encryption Standard (DES)   An encryption standard adopted by the U.S. government for encrypting sensitive information.

Data mining   Looking for hidden patterns in data.

Database   The storage element of a search engine.

Database administration   The function within an electronic commerce team that is responsible for defining the data elements in the database design and the operation of the database management software.

Database manager   Software that stores information in a highly structured way.

Database server   The server computer on which database management software runs.

Data-grade line   The quality of telephone wiring in most urban and suburban areas; made more carefully of higher grade copper than voice-grade lines with better capability for carrying data.

Dead link   A Web link that when clicked displays an error message instead of a Web page.

Debit card   A payment card that removes the amount of the charge from the cardholder's bank account and transfers it to the seller's bank account.

Decentralized architecture   A server structure that uses a large number of less-powerful computers and divides the workload among them.

Decrypted   Information that has been decoded. The opposite of encrypted.

Decryption program   A procedure to reverse the encryption process, resulting in the decoding of an encrypted message.

Dedicated hosting   A Web hosting option in which the hosting company provides exclusive use of a specific server computer that is owned and administered by the hosting company.

Defamatory statement   A statement that is false and injures the reputation of a person or company.

Demographic information   Characteristics that marketers use to group visitors, including address, age, gender, income level, type of job held, hobbies, and religion.

Demographic segmentation   The grouping of customers by characteristics such as age, gender, family size, income, education, religion, or ethnicity.

Descending-price auction   Synonymous with Dutch auction, which is an open auction in which bidding starts at a high price and drops until a bidder accepts the price.

Dictionary attack program   A program that cycles through an electronic dictionary, trying every word in the book as a password.

Digital certificate (digital ID)   An attachment to an e-mail message or data embedded in a Web page that verifies the identity of a sender or Web site.

Digital signature   An encryption message digest.

Digital Subscriber Line (DSL)   Telephone-line ISP connectivity that is a higher grade than standard 56K connectivity.

Digital watermark   A digital code or stream embedded undetectably in a digital image or audio file.

Direct connection EDI   The form of EDI in which EDI translator computers at each company are linked directly to each other through modems and dial-up telephone lines or leased lines.

Direct materials   Materials that become part of the finished product in a manufacturing process.

Disintermediation   The removal of an intermediary from a value chain.

Distributed architecture   Synonymous with decentralized architecture, which is a server structure that uses a large number of less-powerful computers and divides the workload among them.

Distributed database system   A database within a large information system that stores the same data in many different physical locations.

Distributed information system   A large information system that stores the same data in many different physical locations.

Domain name   The address of a Web page, it can contain two or more word groups separated by periods. Components of domain names become more general from right to left.

Domain name hosting   A service that permits the purchaser of a domain name to maintain a simple Web site so that the domain name remains in use.

Domain name ownership change   The changing of owner information maintained by a public domain registrar in the registrar's database to reflect the new owner's name and business address.

Domain name parking   Synonymous with domain name hosting.

Domain name server (DNS)   A computer on the Internet that maintains directories that link domain names to IP addresses.

Dotted decimal   The IP address notation in which addresses appear as four separate numbers separated by periods.

Double auction   A type of auction in which buyers and sellers each submit combined price-quantity bids to an auctioneer. The auctioneer matches the sellers' offers (starting with the lowest price, then going up) to the buyers' offers (starting with the highest price, then going down) until all of the quantities are sold.

Double-spending   The spending of the same unit of electronic cash twice by submitting the same electronic currency to two different vendors.

Download   To receive a file from another computer.

Downstream bandwidth (downlink bandwidth)   The connection that occurs when information travels to your computer from your ISP.

Downstream strategies   Tactics that improve the value that a business provides to its customers.

Due diligence   Background research procedures.

Dutch auction   A form of open auction in which bidding starts at a high price and drops until a bidder accepts the price.

Dynamic catalog   An area of a Web site that stores information about products in a database.

Dynamic content   Nonstatic information constructed in response to a Web client's request.

Dynamic page   A Web page whose content is shaped by a program in response to a user request.

E

Early outsourcing   The hiring of an external company to do initial electronic commerce site design and development. The external team then trains the original company's information systems professionals in the new technology, eventually handing over complete responsibility of the site to the internal team.

Eavesdropper   A person or device who is able to listen in on and copy Internet transmissions.

EDI for Administration, Commerce, and Transport (EDIFACT)   The 1987 publication that summarizes the United Nations' standard transaction sets for international EDI.

EDI-capable banks   Banks that are able to exchange payment and remittance through value-added networks.

EDI compatible   Firms that are able to exchange data in specific standard electronic formats with other firms.

Effect   The impact of an action.

E-government   The use of electronic commerce by governments and government agencies to perform business-like activities.

Electronic business (e-business)   Another term for electronic commerce; sometimes used as a broader term for electronic commerce that includes all business processes, as distinguished from a narrow definition of electronic commerce that includes sales and purchase transactions only.

Electronic cash   A form of electronic payment that is anonymous and can be spent only once.

Electronic commerce (e-commerce)   Business activities conducted using electronic data transmission over the Internet and the World Wide Web.

Electronic Commerce Modeling Language (ECML)   A proposed standard for electronic wallets that provides universal standard field names.

Electronic customer relationship management (eCRM)   Synonymous with technology-enabled relationship management, it is the obtaining and use of detailed customer information.

Electronic data interchange (EDI)   Exchange between businesses of computerreadable data in a standard format.

Electronic funds transfer (EFT)   Electronic transfer of account exchange information over secure private communications networks.

Electronic mail (e-mail)   Messages that are sent from one user to another (or multiple recipients) using particular mail programs and protocols.

Electronic wallet (e-wallet)   A software utility that holds electronic cash, credit card information, owner identification and address information and provides this data automatically at electronic commerce sites.

E-mail client software   Programs used to read and send e-mail.

E-mail server   A computer that is devoted to handling e-mail.

EMV standard   A single standard for the handling of payment card transactions developed cooperatively by Visa, MasterCard, and MasterCard Europe.

Encapsulation   The process that occurs when VPN software encrypts packet contents, then places the encrypted packets inside an IP wrapper in another packet.

Encryption The coding of information using a mathematical-based program and secret key, it makes a message illegible to casual observers or those without the decoding key.

Encryption algorithm The logic that implements an encryption program.

Encryption program A program that transforms plain text into cipher text.

English auction A type of auction in which bidders publicly announce their successively higher bids until no higher bid is forthcoming.

Enterprise application integration The coordination of all of a company's existing systems to each other and to the company's Web site.

Enterprise-class software Commerce software used by large-scale electronic commerce businesses.

Enterprise resource planning (ERP) Business software that integrates all facets of a business, including planning, manufacturing, sales, and marketing.

Entity body The part of a message from a client that contains the HTML page requested by the client and passes bulk information to the server.

E-procurement The use of Internet technologies in a company's purchasing and supply management functions.

E-procurement software Software that allows a company to manage its purchasing function through a Web interface.

Escrow service An independent third party who holds an auction buyer's payment until the buyer receives the purchased item and is satisfied that it is what the seller represented it to be.

E-sourcing The use of Internet technologies in the activities a company undertakes to identify vendors that offer materials, supplies, and services that the company needs.

Ethical hacker A computer security specialist hired to probe PCs and locate information that can be used in legal proceedings.

Extensible Hypertext Markup Language (XHTML) A new markup language proposed by the WC3 that is a reformulation of HTML version 4.0 as an XML application.

Extensible Markup Language (XML) A language that describes the semantics of a page's contents and defines data records on a page.

Extensible Stylesheet Language (XSL) A language that formats XML code for viewing in a Web browser.

Extensible system Any system that can be easily enhanced without voiding earlier work done on the system.

Extranet A network system that extends a company's intranet and allows it to connect with the networks of business partners or other designated associates.

E-zine An electronic magazine.

F

Fair use The approved limited use of copyright material when certain conditions are met.

False positive An e-mail message that is incorrectly rejected by an e-mail filter as being spam when it is actually valid e-mail.

Fast venturing The joining of an existing company that wants to launch an electronic commerce initiative with external equity partners and operational partners who provide the experience and skills needed to develop and scale up the project very rapidly.

Fee-for-service revenue model A revenue model in which payment is based on the value of the service provided.

Fee-for-transaction revenue model A revenue model in which businesses charge a fee for services based on the number or size of the transactions they process.

File Transfer Protocol (FTP)   A protocol that enables users to transfer files over the Internet.

Financial EDI (FEDI)   The EDI transaction sets that provide instructions to a trading partner's bank.

Financial VANS (FVANS)   Value-added networks that are not banks but can translate financial transaction sets into ACH formats and transmit them to banks that are not EDI capable.

Finger   An Internet utility program that runs on UNIX computers and allows a user to obtain limited information about other network users.

Firewall   A computer that provides a defense between one network (inside the firewall) and another network (outside the firewall, such as the Internet) that could pose a threat to the inside network. All traffic to and from the network must pass through the firewall. Only authorized traffic, as defined by the local security policy, is allowed to pass through the firewall.

Firm   A business engaged in commerce.

First-party cookie   A cookie that is placed on the client computer by the Web server site.

First-price sealed-bid auction   A type of auction in which bidders submit their bids independently and privately, with the highest bidder winning the auction.

Fixed-point wireless   A data transmittal service that uses a system of repeaters to forward a radio signal from an ISP to customers.

Flat-rate access   A telephone usage system in which the consumer or business pays one monthly fee for unlimited telephone line usage.

Float   Money deposited in a customer's account that earns interest for the merchant.

Forum selection clause   A statement within a contract that dictates that the contract will be enforced according to the laws of a particular state; signing a contract with a forum selection clause constitutes voluntary submission to the jurisdiction named in the forum selection clause.

Four Ps of marketing   The essential issues of marketing: product, price, promotion, and place.

Fractional T1   High bandwidth telephone company connections that operate at speeds between 128 Kbps and 1.5 Mbps in 128-Kbps increments

Frame relay   A routing technology.

Fulfillment manager   A company that handles responsibility for a particular product line within a retail store. Synonymous with category manager or channel distribution manager.

Full-privilege FTP   A protocol that allows users to upload files to and download files from a remote computer using FTP.

G

Gateway computers   Synonymous with routers, which are computers that determine the best way for data packets to move forward.

Gateway server   A firewall that filters traffic based on applications requested by clients on the trusted network.

Generalized Markup Language (GML)   An early markup language result-ing from efforts to create standard formatting styles for electronic documents.

Geographic segmentation   The grouping of customers by location of home or workplace.

Graphical user interface (GUI)   Computer program control functions that are displayed using pictures, icons, and other easy-to-use graphical elements.

Group purchasing site   A type of auction Web site that negotiates with a seller to obtain lower prices on an item as individual buyers enter bids on that item.

H

Hacker   A dedicated programmer who writes complex code that tests the limits of technology; usually meant in a positive way.

Hash algorithm   A security utility that mathematically combines every character in a message to create a fixed-length number (usually 128 bits in length) that is a condensation, or fingerprint, of the original message.

Hash coding   The process used to calculate a number from a message.

Hash value   The number that results when a message is hash coded.

Hexadecimal (base 16)   A number system that uses 16 digits.

Hierarchical business organization   Firms that include a number of levels with cumulative responsibility. These organizations are typically headed by a top-level president or officer. A number of vice presidents report to the president. A larger number of middle managers report to the vice presidents.

Hierarchical hyperlink structure   A hyperlink structure in which the user starts from a home page and follows links to other pages in whatever order they wish.

High-speed DSL (HDSL)   An Internet connection service that provides 768 Kbps of symmetric bandwidth.

Home page   In a hierarchical Web page structure, the introductory page of a Web site. Synonymous with start page.

Hot spot   A wireless access point (WAP) that is open to the public.

HTML extensions   Developer-created Web page features that only work in certain browsers.

Hyperlink   A type of tag that points to another location in the same or another HTML document. Also called a hypertext link.

Hypertext   A system of navigating between HTML pages using links.

Hypertext elements   HTML text elements that are related to each other within one document or among several documents.

Hypertext link (hyperlink)   A pointer in an HTML document to another location within the same document or to a different HTML document.

Hypertext Markup Language (HTML)   The language of the Internet; it contains codes attached to text that describe text elements and their relation to one another.

Hypertext server   Synonymous with Web server, which is a computer that is connected to the Internet and that stores files written in HTML that are publicly available through an Internet connection.

Hypertext Transfer Protocol (HTTP)   The Internet protocol responsible for transferring and displaying Web pages.

I

Implied contract   The agreement between two parties stating that a contract exists, even if no contract has been written and signed.

Impression   The loading of a banner ad on a Web page.

Income tax   Taxes that are levied by national, state, and local governments on the net income generated by business activities.

Incubator   A company that offers start-up businesses a physical location with offices, accounting and legal assistance, computers, and Internet connections at a very low monthly cost.

Independent exchange   A vertical portal that is not controlled by a company that was an established buyer or seller in the industry.

Independent industry marketplace   A vertical portal that is focused on a specific industry.

Index   A list containing every Web page found by a spider, crawler, or bot.

Indirect connection EDI   The form of EDI in which each company transmits and receives EDI messages through a value-added network.

Indirect materials   Materials and supplies that are purchased by a company in support of the manufacturing of an item, but not directly used in the production of the product.

Industry   Multiple firms selling similar products to similar customers.

Industry consortia-sponsored marketplace   A marketplace formed by several large buyers in a particular industry.

Industry marketplace   A vertical portal that is focused on a single industry.

Industry value chain   The larger stream of activities in which a particular business unit's value chain is embedded.

Integrated Services Digital Network (ISDN)   High-grade telephone service that uses the DSL protocol and offers bandwidths of up to 128 Kbps.

Integrity   The category of computer security that addresses the validity of data; confirmation that data has not been modified.

Integrity violation   A security violation that occurs whenever a message is altered while in transit between sender and receiver.

Intellectual property   A general term that includes all products of the human mind, including tangible and intangible products.

Intelligent software agent (software robot or bot)   A program that performs information gathering, information filtering, and/or mediation on behalf of a person or entity.

Interactive Mail Access Protocol (IMAP)   A newer e-mail protocol with improvements over POP.

Interactive marketing unit (IMU) ad format   The standard banner sizes that most Web sites have voluntarily agreed to use.

Internet   A global system of interconnected computer networks.

Internet access provider (IAP)   Synonymous with Internet service provider.

Internet backbone   Routers that handle packet traffic along the Internet's main connecting points.

Internet EDI   EDI on the Internet.

Internet host   A computer that is directly connected to the Internet.

Internet Protocol   See TCP/IP.

Internet Protocol version 4 (IPv4)   The version of IP that has been in use for the past 20 years on the Internet; it uses a 32-bit number to identify the computers connected to the Internet.

Internet Protocol version 6 (IPv6)   The protocol that will replace IPv4.

Internet service provider (ISP)   A company that sells Internet access rights directly to Internet users.

Internet2   A successor to the Internet used for conducting research, it offers bandwidths in excess of 1 Gbps.

Interoperability   The coordination of a company's information systems so that they all work together.

Interoperable software   Software that will run transparently on a variety of hardware and software configurations.

Interstitial ad   An intrusive Web ad that opens in its own browser window, instead of the page that the user intended to load.

Intranet   An interconnected network of computers operated within a single company or organization.

Intrusion detection system   A part of a firewall that monitors attempts to login to servers and analyzes those attempts for patterns that might indicate a cracker's attack is under way.

IP address   The 32-bit number that represents the address of a particular location (computer) on the Internet.

IP tunneling   The creation of a private passageway through the public Internet that provides secure transmission from one extranet partner to another.

IP wrapper   The outer packet in the encapsulation process.

J

Java sandbox   A Web browser security feature that limits the actions that can be preformed by a Java applet that has been downloaded from the Web.

JavaServer pages (JSP)   A server-side scripting program developed by Sun Microsystems.

Java servlet   An application that runs on a Web server and generates dynamic content.

JavaScript   A scripting language developed by Netscape to enable Web page designers to build active content.

Judicial comity   An agreement between courts from different nations to voluntarily enforce the other nation's laws or judgments out of a sense of friendly civility.

Jurisdiction   A government's ability to exert control over a person or corporation.

K

Key   A number used to encode or decode messages.

Knowledge management   The intentional collection, classification, and dissemination of information about a company, its products, and its processes.

Knowledge management (KM) software
Software that helps companies collect and organize information, share the information among users, enhance the ability of users to collaborate, and preserve the knowledge gained for future use.

L

Late outsourcing   The hiring of an external company to maintain an electronic commerce site that has been designed and developed by an internal information systems team.

Law of diminishing returns   The characteristic of most activities to yield less value as the amount of consumption increases.

Leased line   A permanent telephone connection between two points; it is always active.

Legitimacy   The idea that those subject to laws should have some role in formulating them.

Life-cycle segmentation   The use of customer life cycle stages to create groups of customers that are in each stage.

Linear hyperlink structure   A hyperlink structure that resembles conventional paper documents in which the user reads pages in serial order.

Link checker   A site management tool that examines each page on the site and reports on any URLs that are broken, that seem to be broken, or that are in some way incorrect.

Liquidation broker   An agent that finds buyers for unusable and excess inventory.

Load-balancing switch   A piece of network hardware that monitors the workloads of servers attached to it and assigns incoming Web traffic to the server that has the most available capacity at that instant in time.

Local area network (LAN)   A network that connects workstations and PCs within a single physical location.

Localization A type of language translation that considers multiple elements of the local environment, such as business and cultural practices, in addition to local dialect variations in the language.

Localized advertising Online advertising in which ads are generated in response to a search for products or services in a specific geographic area.

Lock-in effect The inherent greater value to customers of existing companies than new sites.

Log file A collection of data that shows information about Web site visitors' access habits.

Logical security The protection of assets using nonphysical means.

Long-arm statute A state law that creates personal jurisdiction for courts.

M

Machine translation Language translation that is done by software; such translation can reach speeds of 400,000 words per hour.

Macro virus A virus that is transmitted or contained inside a downloaded file attachment; it can cause damage to a computer and reveal otherwise confidential information.

Mail bomb A security attack wherein many people (hundreds or thousands) each send a message to a particular address, exceeding the recipient's allowable mail limit and causing mail systems to malfunction.

Mail order model Synonymous with catalog model.

Mailing list An e-mail address that forwards messages to certain users who are subscribers.

Maintenance, repair, and operating (MRO) Commodity supplies, including general industrial mer-chandise and standard machine tools that are used in a variety of industries.

Managed service provider (MSP) A Web site hosting service firm; synonymous with ASP and CSP.

Man-in-the-middle exploit A message integrity violation in which the contents of the e-mail are changed in a way that negates the message's original meaning.

Many-to-many communications A model of communications in which a number of entities communicate with a number of other entities.

Many-to-one communications model A model of communications in which a number of entities communicate with a single other entity.

Market A real or virtual space in which potential buyers and sellers come into contact with each other and agree on a medium of exchange (such as currency or barter).

Market segmentation The identification by advertisers of specific subsets of their markets that have common characteristics.

Marketing mix The combination of elements that companies use to achieve their goals for selling and promoting their products and services.

Marketing strategy A particular marketing mix that is used to promote a company or product.

Marketspace A market that occurs in the virtual world instead of in the physical world.

Markup tags (tags) Web page code that provides formatting instructions that Web client software can understand.

Masquerading (spoofing) Pretending to be someone you are not (for example, by sending an e-mail that shows someone else as the sender) or representing a Web site as an original when it is an imposter.

Mass media The method of contacting potential customers through the distribution of broadcast, printed, billboard, or mailed advertising materials.

Merchandising The combination of store design, layout, and product display intended to create an environment that encourages customers to buy.

Merchant account An account that a merchant must hold with a bank that allows the merchant to process payment card transactions.

Merchant bank A bank that does business with merchants who want to accept credit cards.

Mesh routing A version of fixed-point wireless that directly transmits Wi-Fi packets through hundreds of short-range transceivers that are located close to each other.

Message digest The number that results from the application of an encryption algorithm to plain text information.

Meta language A language that comprises a set of language elements and can be used to define other languages.

Metrics Measurements that companies use to assess the value of site visitor activity.

Micromarketing The practice of targeting very small and well-defined market segments.

Micropayments Internet payments for items costing very little—usually $1 or less.

Middleware Software that handles connections between electronic commerce software and accounting systems.

Minimum bid In an English auction, the price for an item at which the auctioning begins.

Minimum bid increment The amount by which one bid must exceed the previous bid.

Mobile commerce (m-commerce) Resources accessed using devices that have wireless connections, such as stock quotes, directions, weather forecasts, and airline flight schedules.

Monetizing The conversion of existing regular site visitors seeking free information or services into fee-paying subscribers or purchasers of services.

Money laundering A technique used by criminals to convert money that they have obtained illegally into cash that they can spend without having it identified as the proceeds of an illegal activity.

Multipurpose Internet Mail Extension (MIME) An e-mail protocol that allows users to attach binary files to e-mail messages.

Multivector virus A virus that can enter a computer system in several different ways.

N

Naive Bayesian filter E-mail filtering software that classifies messages based on learned patterns indicated by the e-mail user's categorization of incoming mail. The filter eventually learns to recognize spam and filter it out.

Name changing A problem that occurs when someone registers purposely misspelled variations of well-known domain names. These variants sometimes lure consumers who make typographical errors in entering a URL.

Name stealing Theft of a Web site's name that occurs when someone, posing as a site's administrator, changes the ownership of the domain name assigned to the site to another site and owner.

National Center for Supercomputing Applications (NCSA) Housed at the University of Illinois, Urbana-Champaign, the NCSA is one of the five original centers in the National Science Foundation's Supercomputer Centers Program. Mosaic, the first Internet browser program and predecessor to the Netscape browser, was invented at NCSA.

Necessity The category of computer security that addresses data delay or data denial threats.

Necessity threat   The disruption of normal computer processing or denial of processing. Also called delay, denial, or denial-of-service threat (DoS).

Net bandwidth   The actual speed information travels, taking into account traffic on the communication channel at any given time.

Network access points (NAPs)   The four primary connection points for access to the Internet backbone in the United States.

Network access providers   The few large companies that are the primary providers of Internet access; they, in turn, sell Internet access to smaller Internet service providers.

Network Address Translation (NAT) device   A computer that converts private IP addresses into normal IP addresses when they forward packets to the Internet.

Network Control Protocol (NCP)   Used by ARPANET in the early 1970s to route messages in its experimental wide area network.

Network economic structure   A business structure wherein firms coordinate their strategies, resources, and skill sets by forming a long-term, stable relationship based on a shared purpose.

Network effect   An increase in the value of a network to its participants, which occurs as more people or organizations participate in the network.

Network operations   Web site staff whose responsibilities include load estimation and monitoring, resolving network problems as they arise, designing and implementing fault-resistance technologies, and managing any network operations that are outsourced to ISPs, CSPs, or telephone companies.

Network specification   The set of rules that equipment connected to a network must follow.

Newsgroup   A topic area in Usenet where people read and post articles.

Nexus   The association between a taxpaying entity and a governmental taxing authority.

Nonrepudiation   Verification that a particular transaction actually occurred; this prevents parties from denying a transaction's validity or its existence.

Notice   The expression of a change in rules (usually, legal or cultural rules) typically represented by a physical boundary.

N-tier architecture   Higher-order client-server architectures that have more than three tiers.

O

Occasion segmentation   Behavioral segmentation that is based on things that happen at a specific time or occasion.

Octet   An 8-bit number.

Offer   A declaration of willingness to buy or sell a product or service; it includes sufficient details to be firm, precise, and unambiguous.

One-to-many communication model   A model of communications in which one entity communicates with a number of other entities.

One-to-one communication model   A model of communications in which one entity communicates with one other entity.

One-to-one marketing   A highly customized approach to offering products and services that match the needs of a particular customer.

One-way function   An algorithm that cannot be converted back to its original value.

Online community   Synonymous with virtual community, which is an electronic gathering place for people with common interests.

Ontology   A set of standards that defines, in detail, the relationships among RDF standards and specific XML tags within a particular knowledge domain.

Open architecture   The philosophy behind the Internet that dictates that independent networks should not require any internal changes to be connected to the network, packets that do not arrive at their destinations must be retransmitted from their source network, routers do not retain information about the packets they handle, and no global control exists over the network.

Open auction (open-outcry auction)   An auction in which bids are publicly announced (such as an English auction).

Open EDI   EDI conducted on the Internet instead of over private leased lines.

Open loop system   A payment card arrangement involving a consumer and his or her bank, a merchant and its bank, and a third party (such as Visa or MasterCard) that processes transactions between the consumer and merchant.

Open-outcry double auction   A double auction in which buy and sell offers are announced publicly. Typically conducted in exchange floor or trading pit environments for items of known quality, such as securities or graded agricultural products, that are regularly traded in large quantities.

Open session   A continuous connection that is maintained between a client and server on the Internet.

Open source   Freely available source code for software.

Open-source software   Software that is developed by a community of programmers who make the software available for download and use at no cost.

Opening tag   An HTML tag that precedes the text that a tag affects.

Opportunity cost   Lost benefits from an action not taken.

Optical fiber   An Internet connectivity material with bandwidths up to 10 Gbps.

Opt-in   A personal information collection policy in which the company collecting the information does not use the information for any other purpose (or sell or rent the information) unless the customer specifically chooses to allow that use.

Opt-in e-mail   The practice of sending e-mail messages to people who have requested information on a particular topic or about a specific product.

Opt-out   A personal information collection policy in which the company collecting the information assumes that the customer does not object to the company's use of the information unless the customer specifically chooses to deny permission.

Orphan file   A file on the Web site that is not linked to any page.

Outsourcing   The hiring of another company to perform design, implementation, or operational tasks for an information systems project.

P

Packet-filter firewall   A firewall that examines all data flowing back and forth between a trusted network and the Internet.

Packets   The small pieces of files and e-mail messages that travel over the Internet.

Packet-switched   A network in which packets are labeled electronically with their origin, sequence, and destination addresses. Packets travel from computer to computer along the interconnected networks until they reach their destination. Each packet can take a different path through the interconnected networks and the packets may arrive out of order. The destination computer collects the packets and reassembles the original file or e-mail message from the pieces in each packet.

Page view   A page request made by a Web site visitor.

Page-based application system   Application server software that returns pages generated by scripts that include the rules for presenting data on the Web page with the business logic.

Paid placement (sponsorship)   The purchasing of a top listing in results listings for a particular set of search terms.

Partial outsourcing   The outsourcing of the design, development, implementation, or operation of specific portions of an electronic commerce system.

Patent   An exclusive right to make, use, and sell an invention granted by a government to the inventor.

Payment card   A general term for plastic cards used instead of cash to make purchases, including credit cards, debit cards, and charge cards.

Payment processing service provider   A third-party company that handles payment card processing for online businesses.

Pay-per-click model   A revenue model in which an affiliate earns payment each time a site visitor clicks a link to load the seller's page.

Pay-per-conversion model   A revenue model in which an affiliate earns payment each time a site visitor is converted from a visitor into either a qualified prospect or a customer.

Peer-to-peer (P2P) payment system   Payments from one type of entity to another of the same type.

Per se defamation   A legal cause of action in which a court deems some types of statements to be so negative that injury is assumed.

Perimeter expansion   The increase in firewall limits beyond traditional borders caused by telecommuting.

Permission marketing   A marketing strategy that only sends specific information to people who have indicated an interest in receiving information about the product or service being promoted.

Persistent cookie   A cookie that exists indefinitely.

Personal area network (PAN)   A small, low-bandwidth Bluetooth network of up to 10 networks of eight devices each. It used for tasks such as wireless synchronization of laptop computers with desktop computers and wireless printing from laptops, PDAs, or mobile phones. Synonymous with piconet.

Personal contact   A method of identifying and reaching customers that involves searching for, qualifying, and contacting potential customers.

Personal firewall   A software-only firewall that is installed on an individual client computer.

Personal jurisdiction   A court's authority to hear a case based on the residency of the defendant; a court has personal jurisdiction over a case if the defendant is a resident of the state in which the court is located.

Personal shopper   An intelligent agent program that learns a customer's preferences and makes suggestions.

Phishing expedition   A masquerading attack that combines spam with spoofing. The perpetrator sends millions of spam e-mails that appear to be from a respectable company. The e-mails contain a link to a Web page that is designed to look exactly like the company's site. The victim is encouraged to enter his or her username, password, and sometimes credit card information.

PHP: Hypertext Preprocessor (PHP)   A Web programming language that can be used to write server-side scripts that generate dynamic Web pages.

Physical security   Tangible protection devices such as alarms, guards, fireproof doors, fences, and vaults.

Piconet A small, low-bandwidth Bluetooth network of up to 10 networks of eight devices each. It used for tasks such as wireless synchronization of laptop computers with desktop computers and wireless printing from laptops, PDAs, or mobile phones. Synonymous with personal area network.

Ping (Packet Internet Groper) A program that tests the connectivity between two computers connected to the Internet.

Place (distribution) The need to have products or services available in many different locations.

Plain old telephone service (POTS) The network connecting telephones; it provides a reliable data transmission bandwidth of about 56 Kbps.

Plain text Normal, unencrypted text.

Plug-in An application that helps a browser to display information (such as video or animation) but that is not part of the browser.

Pop-behind ad A pop-up ad that is followed very quickly by a command that returns the focus to the original browser window, resulting in an ad that is parked behind the user's browser waiting to appear when the browser is closed.

Pop-up ad An ad that appears in its own window when the user opens or closes a Web page.

Portal (Web portal) A Web site that serves as a customizable home base from which users do their searching, navigating, and other Web-based activity.

Post Office Protocol (POP) The protocol responsible for retrieving e-mail from a mail server.

Postimplementation audit (postaudit review) A formal review of a project after it is up and running.

Power A form of control over physical space (such as a state) and the people and objects that reside in that space.

Presence The public image conveyed by an organization to its stakeholders.

Pretty Good Privacy (PGP) A popular technology used to implement public-key encryption to protect the privacy of e-mail messages.

Price The amount a customer pays for a product.

Primary activities Activities that are required to do business: design, production, promotion, marketing, delivery, and support of products or services.

Privacy The protection of individual rights to nondisclosure of information.

Private company marketplace A marketplace that provides auctions, requests for quotes postings, and other features to companies that want to operate their own marketplace.

Private IP addresses A series of IP numbers that have been set aside for subnet use and are not permitted on packets that travel on the Internet.

Private key A single key that is used to encrypt and decrypt messages. Synonymous with symmetric key.

Private network A private, leased-line connection between two companies that physically links their individual computers or intranets.

Private store A password-protected area of a Web site that offers individual customers negotiated price reductions on a limited selection of products and other customized features.

Private valuation The amount a bidder is willing to pay for an item that is up for auction.

Private-key encryption The encoding of a message using a single numeric key to encode and decode data, it requires both the sender and receiver of the message to know the key, which must be guarded from public disclosure.

Procurement   The business activity that includes all purchasing activities plus the monitoring of all elements of purchase transactions.

Product   The physical item or service that a company is selling.

Product disparagement   A statement that is false and injures the reputation of a product or service.

Project management   Formal techniques for planning and controlling activities undertaken to achieve a specific goal.

Project management software   Application software that provides built-in tools for managing people, resources, and schedules.

Project manager   A person with specific training or skills in tracking costs and the accomplishment of specific objectives in a project.

Project portfolio management   A technique in which each project is monitored as if it were an investment in a financial portfolio.

Promotion   Any means of spreading the word about a product.

Property tax   Taxes levied by states and local governments on the personal property and real estate used in a business.

Proprietary architecture   The use of vendor-specific communication protocols by computer manufacturers in the early days of computing, preventing computers made by different manufacturers from being connected to each other. Also called closed architecture.

Prospecting   The part of personal contact selling in which the salesperson identifies potential customers.

Protocol   A collection of rules for formatting, ordering, and error-checking data sent across a network.

Proxy bid   In an electronic auction, a predetermined maximum bid submitted by a bidder.

Proxy server firewall   A firewall that communicates with the Internet on behalf of the trusted network.

Psychographic segmentation   The grouping of customers by variables such as social class, personality, or their approach to life.

Public key   One of a pair of mathematically related numeric keys, it is used to encrypt messages and is freely distributed to the public.

Public marketplace   A vertical portal that is open to new buyers and sellers just entering an industry.

Public network   An extranet that allows the public to access its intranet or when two or more companies link their intranets.

Public-key encryption   The encoding of messages using two mathematically related but distinct numeric keys.

Purchasing card (p-card)   Payment cards that give individual managers the ability to make multiple small purchases at their discretion while providing cost tracking information to the procurement office.

R

Radio frequency identification device (RFID)   Small chips that include radio transponders; they can be used to track inventory as it moves through an industry value chain.

Rational branding   An advertising strategy that substitutes an offer to help Web users in some way in exchange for their viewing an ad.

Reintermediation   The introduction of a new intermediary into a value chain.

Remote server administration   Control of a Web site by an administrator from any Internet-connected computer.

Repeat visits   Subsequent visits a Web site visitor makes to a particular page.

Repeater   A transmitter-receiver device used in a fixed-point wireless network to forward a radio signal from the ISP to customers. Synonymous with transceiver.

Replenishment purchasing   Direct materials purchasing in which the company negotiates long-term contracts for most of the materials that it will need. Also called contract purchasing.

Request header   The part of an HTTP message from a client to a server that contains additional information about the client and more information about the request.

Request line   The part of an HTTP message from a client to a server that contains a command, the name of the target resource (without the protocol or domain name), and the protocol name and version.

Request message   The HTTP message that a Web client sends to request a file or files from a Web server.

Reserve price (reserve)   The minimum price a seller will accept for an item sold at auction.

Resource description framework   A set of standards for XML syntax.

Response header field   In a client/-server transmission, it follows the response header line and returns information describing the server's attributes.

Response header line   The part of a message from a server to a client that indicates the HTTP version used by the server, status of the response, and an explanation of the status information.

Response message   The reply that a Web server sends in response to a client request.

Response time   The amount of time a server requires to process one request.

Retained customer   A customer who returns to a site one or more times after making his or her first purchase.

Retention costs   The costs of inducing customers to return to a Web site and buy again.

Return on investment (ROI)   A method for evaluating the potential costs and benefits of a proposed capital investment.

Revenue model   The combination of strategies and techniques that a company uses to generate cash flow into the business from customers.

Reverse auction (seller-bid auction)   A type of auction in which sellers bid prices for which they are willing to sell items or services.

Reverse bid   The process in which an auction customer seeks products by describing an item or service in which he or she is interested, and then entertains responses from merchants who offer to supply the item at a particular price.

Reverse link checker   A Web site management program that checks on sites with which a company has entered a link exchange program and ensures that link exchange partners are fulfilling their obligation to include a link back to the company's Web site.

Rich media ad   Web ads that generate graphical activity that "floats" over the Web page itself instead of opening in a separate window. Also called active ads.

Rich media objects   Programming components of attention-grabbing Web banner ads.

Roaming   The shifting of Wi-Fi devices from one WAP to another without requiring intervention by the user.

Robot (bot, spider, crawler)   A program that automatically searches the Web to find Web pages that might be interesting to people.

Router   A computer that determines the best way for data packets to move forward to their destination.

Router computers (routing computers) The computers that decide how best to forward each packet of information as it travels on the Internet to its destination. Synonymous with gateway computers and routers.

Routing algorithm The program used by a router to determine the best path for data packets to travel.

Routing table Synonymous with configuration table, which is information about connections that lead to particular groups of routers, specifications on which connections to use first, and rules for handling instances of heavy packet traffic and network congestion.

S

Scalable A system's ability to be adapted to meet changing requirements.

Scaling problem The exponential increase in cost that results from the expansion of a private network.

Scrip A limited-use digital or paper value store issued by a private company rather than a government. It generally must be exchanged for goods or services with the company that issued it and usually cannot be exchanged for cash.

Sealed-bid auction An auction in which bidders submit their bids independently and are usually prohibited from sharing information with each other.

Search engine Web software that finds other pages based on key word matching.

Search engine optimization (search engine positioning, search engine placement) The combined art and science of having a particular URL listed near the top of search engine results.

Search engine placement broker A company that aggregates inclusion and placement rights on multiple search engines and then sells those combination packages to advertisers.

Search engine ranking The weighting of the factors that search engines use to decide which URLs will appear first on searches for a particular search term.

Search term sponsorship The option of purchasing a top listing on results pages for a particular set of search terms. Also called paid placement or sponsorship.

Search utility The part of a search engine that finds matching Web pages for search terms.

Second-price sealed-bid auction A type of auction in which bidders submit their bids independently and privately; the highest bidder wins the auction but pays only the amount bid by the second-highest bidder.

Secrecy The category of computer security that addresses the protection of data from unauthorized disclosure and confirmation of data source authenticity.

Secure envelope A security utility that encapsulates a message and provides secrecy, integrity, and client/server authentication.

Secure Sockets Layer (SSL) A protocol for transmitting private information securely over the Internet.

Security policy A written statement describing assets to be protected, the reasons for protecting the assets, the parties responsible for protection, and acceptable and unacceptable behaviors.

Segment Also called a market segment; a subset of a company's potential customer pool that has common demographic characteristics.

Self-hosting A system of Web hosting in which the online business owns and maintains the server and all its software.

Semantic Web A project developed by Tim Berners-Lee intended to blend technologies and information to create a next-generation Web that would result in words on Web pages being tagged (using XML) with their meanings.

Server   A powerful computer dedicated to managing disk drives, printers, or network traffic.

Server architecture   The different ways that servers can be connected to each other and to related hardware such as routers and switches.

Server farm   A large collection of electronic commerce Web site servers.

Server-level filtering   An e-mail content filtering technique in which the filtering software resides on the mail server.

Server software   The software that a server computer uses to make files and programs available to other computers on the same network.

Server-side electronic wallet   An electronic wallet that stores a customer's information on a remote server that belongs to a particular merchant or to the wallet's publisher.

Server-side scripting (server-side includes or server-side technologies)   A Web page response approach in which programs running on the Web server create Web pages before sending them back to the requesting Web clients as parts of response messages.

Service mark   A distinctive mark, device, motto, or implement used to identify services provided by a company.

Session cookie   A cookie that exists only until you shut down your browser.

Session key   A key used by an encryption algorithm to create cipher text from plain text during a single secure session.

Shared hosting   A Web hosting arrangement in which the hosting company provides Web space on a server computer that also hosts other Web sites.

Shill bidder   An individual employed by a seller or auctioneer who makes bids on behalf of the seller, sometimes artificially inflating an item's price. Shill bidders may be prohibited by the rules of a particular auction.

Shipping profile   The collection of attributes that affect how easily a product can be packaged and delivered.

Shopping cart   An electronic commerce utility that keeps track of selected items for purchase and automates the purchasing process.

Short message service (SMS)   A protocol used to transmit short text messages to cell phones and other wireless devices.

Signature   Any symbol executed or adopted for the purpose of authenticating a writing.

Signed (message or code)   The status of a message or Web page when it contains an attached digital certificate.

Simple Mail Transfer Protocol (SMTP)   A standardized protocol used by a mail server to format and administer e-mail.

Simple Object Access Protocol (SOAP)   A message-passing protocol that defines how to send marked up data from one software application to another across a network.

Single-use card   A payment card with disposable numbers, which gives consumers a unique card number that is valid for one transaction only.

Site map   On a hierarchically structured Web site, a page that contains a map or listing of the Web pages in their hierarchical order.

Site sponsorship   The opportunity for an advertiser to sponsor part or all of a Web site to promote its products, services, or brands. Site sponsorships are more subtle than banner or pop-up ads.

Skyscraper ad   A large banner ad on the side of a Web page that remains visible as the user scrolls down through the page.

Small payment   Any payment of less than $10.

Smart card   A plastic card with an embedded microchip that contains information about the card owner.

Sniffer program   A program that taps into the Internet and records information that passes through a router from the data's source to its destination.

Snipe   The act of placing a winning bid in an online auction at the last possible moment.

Sniping software   Auction software that observes auction progress until the last second or two of the auction clock, then places a bid high enough to win the auction.

Software agent   A program that performs information gathering, information filtering, and/or mediation on behalf of a person or entity. Synonymous with intelligent software agent.

Sourcing   The part of procurement devoted to identifying suppliers and determining the qualifications of those suppliers.

Spam (unsolicited commercial e-mail or bulk mail)   Electronic junk mail.

Spend   The total dollar amount of the goods and services that a company buys during a year.

Spider   The first part of a search engine, it automatically and frequently searches the Web to find pages and updates its database of information about old Web sites.

Spot market   A loosely organized market within a specific industry.

Spot purchasing   Direct materials purchasing that occurs within a spot market.

Stakeholders   The various entities involved in a business; these include customers, suppliers, employees, stockholders, neighbors, and the general public.

Standard Generalized Markup Language (SGML)   An old, complex text markup language used to cre-ate frequently revised documents that need to be printed in various formats.

Start page   In a hierarchical Web page structure, the introductory page of a Web site. Synonymous with home page.

Stateless connection   A connection between a client and server over the Internet in which each transmission of information is independent; no continuous connection is maintained.

Static catalog   A simple list of products written in HTML and displayed on a Web page or a series of Web pages.

Static page   A Web page that displays unchanging information retrieved from disk.

Statute of Frauds   State laws that specify that contracts for the sale of goods worth over $500 and contracts that require actions that cannot be completed within one year must be created by a signed writing.

Statutory law   That part of British and U.S. law that comprises laws passed by elected legislative bodies.

Steganography   The hiding of information (such as commands) within another piece of information.

Stickiness   The ability of a Web site to keep visitors at its site and to attract repeat visitors.

Sticky   The condition of having stickiness.

Stored value card   Either an elaborate smart card or a simple plastic card with a magnetic strip that records currency balance, such as a prepaid phone, copy, subway, or bus card.

Strategic alliance   The coordination of strategies, resources, and skill sets by companies into long-term, stable relationships with other companies and individuals based on shared purposes.

Strategic business unit (business unit)   A unit within a company that is organized around a specific combination of product, distribution channel, and customer type.

Strategic partners   The entities taking part in a strategic alliance.

Strategic partnership   Synonymous with strategic alliance.

Style sheet   A set of instructions used for Web page formatting. It is stored in a separate file and lets designers apply specific formatting styles to a page.

Subject-matter jurisdiction   A court's authority to decide a dispute between entities based on the issue of dispute.

Subnetting   The use of reserved private IP addresses within LANs and WANs to provide additional address space.

Supply alliances   Long-term relationships among participants in the supply chain.

Supply chain   The part of an industry value chain that precedes a particular strategic business unit. It includes the network of suppliers, transportation firms, and brokers that combine to provide a material or service to the strategic business unit.

Supply chain management (SCM)   The process of taking an active role in working with suppliers and other participants in the supply chain to improve products and processes.

Supply chain management (SCM) software   Software used by companies to coordinate planning and operations with their partners in the industry supply chains of which they are members.

Supply management   Synonymous with procurement, which is the business activity that includes all purchasing activities plus the monitoring of all elements of purchase transactions.

Supply web   An industry value chain that includes many participants that are interconnected in a web or network configuration.

Supporting activities   Secondary activities that back up primary business activities. These include human resource management, purchasing, and technology development.

SWOT analysis   Evaluation of the strengths and weaknesses of a business unit, and identification of the opportunities presented by the markets of the business unit and threats posed by competitors of the business unit.

Symmetric connection   An Internet connection that provides the same bandwidth in both directions.

Symmetric encryption   The encryption of a message using a single numeric key to encode and decode data. Synonymous with private-key encryption.

Systems administrator   A member of an electronic commerce team who understands the server hardware and software and is responsible for the system's reliable and secure operation.

T

T1   High-bandwidth telephone company connections that operate at 1.544 Mbps.

T3   High-bandwidth telephone company connections that operate at 44.736 Mbps.

TCP/IP   The set of protocols that provide the basis for the operation of the Internet. The TCP protocol includes rules that computers on a network use to establish and break connections. The IP protocol determines routing of data packets.

Technology-enabled customer relationship management   Synonymous with technology-enabled relationship management.

Technology-enabled relationship management   The business practice of obtaining detailed information about a customer's behavior, preferences, needs, and buying patterns and using that information to set prices, negotiate terms, tailor promotions, add product features, and provide other customized interactions.

Telecommuting An employment arrangement in which the employee logs in to the company computer from an offsite location through the Internet instead of traveling to an office.

Telework Synonymous with telecommuting.

Telnet A program that allows users to log on to a computer and access its contents from a remote location.

Telnet protocol The set of rules used by Telnet programs.

Terms of service (ToS) Rules and regulations intended to limit the Web site owner's liability for what a visitor might do with information obtained from the site.

Text markup language A language that specifies a set of tags that are inserted into the text.

Third-generation (3G) cell phone A cell phone that incorporates the latest transmission technologies to achieve data speeds of up to 2 Mbps and also uses the SMS protocol to send and receive text messages.

Third-party assurance provider An independent organization that assures privacy policies of Web sites.

Third-party cookie A cookie that originates on a Web site other than the site being visited.

Third-party logistics (3PL) provider A transportation or freight company that operates all or most of a customer's material movement activities.

Threat An act or object that poses a danger to assets.

Three-tier architecture A client/server architecture that builds on the two-tier architecture by adding applications and their associated databases that supply non-HTML information to the Web server on request.

Throughput The number of HTTP requests that a particular hardware and software combination can process in a unit of time.

Tier-one suppliers The capable suppliers that work directly with and have long-term relationships with businesses.

Tier-three suppliers Suppliers that provide components and raw materials to tier-two suppliers.

Tier-two suppliers Suppliers that provide components and raw materials to tierone suppliers.

Top-level domain (TLD) The last part of a domain name; the most general identifier in the name.

Tort An action taken by a legal entity that causes harm to another legal entity.

Total cost of ownership (TCO) Business activity costs including the costs of hiring, training, and paying the personnel who will design the Web site, write or customize the software, create the content, and operate and maintain the site. TCO also includes hardware and software costs.

Touchpoint Online and offline customer contact points.

Touchpoint consistency The provision of similar levels and quality of service in all of a company's interactions with its customers, whether those interactions occur in person, on the telephone, or online.

Tracert A route-tracing program that sends data packets to every computer on the path (Internet) between one computer and another computer and clocks the packets' round-trip times, providing an indication of the time it takes a message to travel from one computer to another and back, pinpointing any data traffic congestion, and ensuring that the remote computer is online.

Trade name The name (or a part of that name) that a business uses to identify itself.

Trademark A distinctive mark, device, motto, or implement that a company affixes to the goods it produces for identification purposes.

Trademark dilution   The reduction of the distinctive quality of a trademark by alternative uses.

Trading partners   Businesses that engage in EDI with one another.

Transaction   An exchange of value.

Transaction costs   The total of all costs incurred by a buyer and seller as they gather information and negotiate a transaction.

Transaction processing   Processes that occur as part of completing a sale; these include calculation of any discounts, taxes, or shipping costs and transmission of payment data (such as a credit card number).

Transaction sets   Formats for specific business data interchanges using EDI.

Transaction taxes   Sales taxes, use taxes, excise taxes, and customs duties that are levied on the products or services that a company sells or uses.

Transceiver   A transmitter-receiver device used in a fixed-point wireless network to forward a radio signal from the ISP to customers. Synonymous with repeater.

Transmission Control Protocol   See TCP/IP.

Trial visit   The first visit a Web site visitor makes to a particular page.

Trigger word   Keywords used to jog the memory of visitors and remind them of something they want to buy on the site.

Triple Data Encryption Standard (3DES)   A robust version of the Data Encryption Standard used by the U.S. government that cannot be cracked even with today's supercomputers.

Trojan horse   A program hidden inside another program or Web page that masks its true purpose (usually destructive).

Trusted (network)   A network that is within a firewall.

Two-tier client/server architecture   A client/server architecture in which only a client and server are involved in the requests and responses that flow between them over the Internet.

U

Ultimate consumer orientation   A focus on the needs of the consumer who is at the end of an industry value chain.

Uniform resource locator (URL)   Names and abbreviations representing the IP address of a particular Web page. Contains the protocol used to access the page and the page's location. Used in place of dotted quad notations.

Universal Description, Discovery and Integration (UDDI) specification   The set of protocols that identify locations of Web services and their associated WSDL descriptions.

Untrusted (network)   A network that is outside a firewall.

Untrusted Java applet   A Java applet that is not known to be secure.

Upload bandwidth   Synonymous with upstream bandwidth.

Upstream bandwidth   The connection that occurs when you send information from your connection to your ISP.

Upstream strategies   Tactics that focus on reducing costs or generating value by working with suppliers or inbound logistics.

URL broker   A business that sells or auctions domain names that it believes others will find valuable.

Usability testing   The testing and evaluation of a company's Web site for ease of use by visitors.

Usage-based market segmentation   Customizing visitor experiences to match the site usage behavior patterns of each visitor or type of visitor.

Use tax   A tax levied by a state on property used in that state that was not purchased in that state.

Usenet (User's News Network)   One of the first mailing lists; it allows subscribers to read and post articles within topic areas.

V

Value chain   A way of organizing the activities that each strategic business unit undertakes to design, produce, promote, market, deliver, and support the products or services it sells.

Value system   Synonymous with industry value chain.

Value-added bank (VAB)   A bank that offers value-added network services for nonfinancial transactions.

Value-added network (VAN)   An independent company that provides connection and EDI transaction forwarding services to businesses engaged in EDI.

Vertical integration   The practice of an existing firm replacing one of its suppliers with its own strategic business unit that creates the supplied product.

Vertical portal (vortal)   A vertically integrated Web information hub focusing on an individual industry.

Vicarious copyright infringement   The violation of an organization's rights that occurs when a company capable of supervising the infringing activity fails to do so and obtains a financial benefit from the infringing activity.

Vickrey auction   Synonymous with second-price sealed-bid auction. Named for William Vickrey, who won the 1996 Nobel Prize in Economics for his studies of the properties of this auction type.

Viral marketing   Tactics that rely on existing customers to tell other persons—the company's prospective customers—about the products or services they have enjoyed using.

Virtual community   An electronic gathering place for people with common interests.

Virtual company   A strategic alliance occurring among companies that operate on the Internet.

Virtual host   Multiple servers that exist on a single computer.

Virtual learning community   A virtual community used for distance learning.

Virtual model   A graphic image built from customer measurements and physical traits on which customers can try clothes. Typically found on sites selling clothing and accessories.

Virtual private network (VPN)   A network that uses public networks and their protocols to transmit sensitive data using a system called "tunneling" or "encapsulation."

Virtual server   Synonymous with virtual host.

Virus   Software that attaches itself to another program and can cause damage when the host program is activated.

Visit   The request of a Web site visitor for a page from a Web site.

Voice-grade line   Telephone wiring that costs less than lines designed to carry data, is made of lower-grade copper, and was never intended to carry data. These lines can only carry limited bandwidth—usually less than 14 Kbps.

W

Warchalking   The practice of placing a chalk mark on a building that has an easily entered wireless network.

Wardrivers   Network attackers who drive around in cars using their wireless-equipped laptop computers to search for accessible networks.

Warranty disclaimer A statement indicating that the seller will not honor some or all implied warranties.

Web See World Wide Web.

Web browser (Web browser software) Software that lets users read HTML documents and move from one HTML document to another using hyperlinks.

Web bug A tiny, invisible Web page graphic that provides a way for a Web site to place cookies.

Web catalog revenue model A revenue model of selling goods and services on the Web wherein the seller establishes a brand image that conveys quality and uses the strength of that image to sell through catalogs mailed to prospective buyers. Buyers place orders by mail or by calling the seller's toll-free telephone number.

Web client computer A computer that is connected to the Internet and used to download Web pages.

Web client software Software that sends requests for Web page files to other computers.

Web community Synonymous with virtual community.

Web directory A listing of hyperlinks to Web pages that is organized into hierarchical categories.

Web EDI EDI on the Internet.

Web graphics designer A person trained in art, layout, and composition who also understands how Web pages are constructed and who ensures that the Web pages are visually appealing, easy to use, and make consistent use of graphics elements from page to page.

Web programmer A programmer who designs and writes the underlying code for dynamic database-driven Web pages.

Web server A computer that receives requests from many different Web clients and responds by sending HTML files back to those Web client computers.

Web server software Software that makes files available to other computers on the Internet.

Web services A combination of software tools that let application software in one organization communicate with other applications over a network using the SOAP, UDDI, and WSDL protocols.

Web Services Description Language (WSDL) A language that describes the characteristics of the logic units that make up specific Web services.

White hat hackers Hackers who use their skills for positive purposes.

White list spam filter Software that looks for From addresses in incoming messages that are known to be good addresses.

Wide area network (WAN) A network of computers that are connected over large distances.

Wi-Fi (wireless ethernet, 802.11b, 802.11a, 802.11g, 802.11n) The most common wireless connection technology for use on LANs ; it can communicate through a wireless access point connected to a LAN to become a part of that LAN.

Winner's curse A psychological phenomenon that causes bidders to become caught up in the excitement of competitive bidding and bid more than their private valuation.

Wire transfer Synonymous with electronic funds transfer, which is the electronic transfer of account exchange information over secure private communications networks.

Wireless access point (WAP) A device that transmits network packets between Wi-Fi equipped computers and other devices that are within its range.

Wireless Application Protocol (WAP) A protocol that allows Web pages formatted in HTML to be displayed on devices with small screens, such as PDAs and mobile phones.

World Wide Web (Web) The subset of Internet computers that connects computers and their contents in a specific way, and that allows for easy sharing of data using a standard interface.

World Wide Web Consortium (W3C) A not-for-profit group that maintains standards for the Web.

Worm A virus that replicates itself on other machines.

Writing A tangible representation of the terms of a contract.

X

XML parser A program that can format an XML file so it can appear on the screen of a computer, a wireless PDA, a mobile phone, or other device.

XML vocabulary A set of XML tag definitions.

Y

Yankee auction A type of English auction that offers multiple units of an item for sale and that allows bidders to specify the quantity of items they want to buy.

Z

Zombie A program that secretly takes over another computer for the purpose of launching attacks on other computers. Zombie attacks can be difficult to trace to their creators.

Bold page numbers indicate where a key term is defined in the text.

I

P

S